Worse than the disease
Pitfalls of medical progress

Worse than the disease
Pitfalls of medical progress

DIANA B. DUTTON

with contributions by
THOMAS A. PRESTON
NANCY E. PFUND

CAMBRIDGE
UNIVERSITY PRESS

362.1
D98lw
1992

Published by the Press Syndicate of the University of Cambridge
The Pitt Building, Trumpington Street, Cambridge CB2 1RP
40 West 20th Street, New York, NY 10011-4211, USA
10 Stamford Road, Oakleigh, Victoria 3166, Australia

First published 1988
First paperback edition 1992
Reprinted 1992

Printed in the United States of America

Library of Congress Cataloging-in-Publication Data
Dutton, Diana Barbara.
Worse than the disease.
Bibliography: p.
Includes index.
1. Medical innovations – Social aspects – Case aspects.
2. Medical innovations – Social aspects.
3. Medical innovations – Moral and ethical aspects – Case studies.
4. Medical innovations – Moral and ethical aspects.
5. Medical innovations – Economic aspects – Case studies.
6. Medical innovations – Economic aspects.
I. Preston, Thomas A., 1933–   . II. Pfund, Nancy E.
III. Title.   IV. Title: Pitfalls of medical progress.
[DNLM: 1. Ethics, Medical.   2. Health Policy – trends – United States.
3. Human Experimentation.   4. Medicine – trends – United States.
5. Technology Assessment, Biomedical.   W 50 D98lW]
RA418.5.M4D88   1988   362.1   87-34203
ISBN 0-521-34023-3

British Library Cataloguing-in-Publication Data
Dutton, Diana B.
Worse than the disease : pitfalls of medical progress.
1. Medicine.   Innovations.   Social aspects
I. Title   II. Preston, Thomas A.   III. Pfund, Nancy E.
306'.46

ISBN 0-521-34023-3 hardback
ISBN 0-521-39557-7 paperback

This material is based upon work supported by the National Science Foundation under Grant No. OSS-7726502. Any opinions, findings, conclusions or recommendations expressed in this publication are those of the authors and do not necessarily reflect the views of the Foundation.

For Hal and Geoffrey, with gratitude and love

Hal, my husband, has been my major source of inspiration for every phase of this book, from the project's initial conception, through the challenges of research, teaching, public outreach, and writing, to the final grueling stages of editing and revision. He gave generously and unstintingly, whether the need was for critical editorial counsel, cooking and shopping so I could write, or all that fell in between. My son Geoffrey's help was less direct but also vital. He allowed me nine uneventful months to work on the book, timing his arrival perfectly a few days after the manuscript was sent off. I am deeply grateful to them both.

Contents

Contributors

DIANA B. DUTTON, Ph.D., is Senior Research Associate at the Stanford University School of Medicine, where she is Associate Director of the Robert Wood Johnson Clinical Scholar Program, a postgraduate training program in the social sciences for physicians. She holds a Ph.D. in social policy analysis from the Massachusetts Institute of Technology, and has taught health policy and medical sociology at the undergraduate and graduate levels at Stanford and at the University of California at Berkeley School of Public Health. She has published numerous articles on the relation between social class and health, the nature of medical care in different practice settings, and the theory and practice of public participation in science policymaking.

THOMAS A. PRESTON, M.D., is Chief of Cardiology, Pacific Medical Center and Professor of Medicine, University of Washington, Seattle, Washington. He has written widely about the artificial heart, contributing articles to the *Hastings Center Report*, major newspapers, and magazines. In 1983, he won the National Association of Science Writers' Science in Society Journalism Award for his work on the artificial heart. He is also author of *The Clay Pedestal* (Charles Scribners and Sons, 1986).

NANCY E. PFUND, M.A., M.P.P.A., received a masters degree in anthropology from Stanford University, where she studied the culture of modern science, and training in management at Yale University. In the late 1970s, she served as a lobbyist for the Sierra Club on Capitol Hill during the height of the debate over recombinant DNA research regulation. Since then she has written a position paper on genetic engineering for the California State Office of Appropriate Technology and has participated in other state policy activities, and has contributed articles to *Biotechnology Newswatch*, a trade journal. She is presently a Senior Analyst for Hambrecht and Quist in San Francisco, where she follows the environmental service industry.

Preface

This book explores the social, ethical, and economic dilemmas society faces as a result of medical innovation, and the role of the general public in resolving them. It is the product of a multidisciplinary research project based at the Stanford University School of Medicine and funded by the National Science Foundation's Program in Ethics and Values in Science and Technology, with additional support from the Rockefeller Foundation.

Many aspects of the book and the project that produced it are, by conventional academic standards, rather unusual. First, the book is written for a general audience rather than for specialists, although we hope that specialists will also find in it new ideas and information. Its primary purpose is to communicate the central issues raised by medical and scientific innovation in a manner that is widely accessible, yet also accurate. We have therefore tried to keep specialized terminology to a minimum and to avoid the dry and often obscure style characteristic of much academic writing, without sacrificing scholarly standards of analytic rigor and documentation. We have also tried to bridge the sometimes considerable chasm between abstract intellectual theories and the practical concerns of policymakers and citizens.

Second, the book is broadly interdisciplinary in outlook. At one point or another, the project included faculty, staff, and students from medicine, history, political science, sociology, economics, philosophy, anthropology, education, regional planning, and law. In analyzing the history and implications of the four cases, we have used the varied perspectives of these different disciplines to enrich our understanding of both the problems of medical innovation and their potential solutions. The issues involved are complex and intertwined. By looking at them through multiple "lenses," we have tried to delineate their principal dimensions.

To have restricted our investigation to the vantage point of a single discipline, or to the "testing" of a particular theoretical model or hypothesis, would, we felt, have imposed unnecessary and arbitrary constraints on our inquiry. For much as one might wish otherwise, the present level of knowledge about medicine's dilemmas and what to do about them is woefully inadequate. Such questions do not respect the traditional boundaries of academic disciplines, nor does any one discipline possess a theoretical framework sufficiently powerful to provide comprehensive hypotheses about expected relationships. The best that can be done under these circumstances is to approach the investigation with an open mind – unfettered by intellectual blinders and cognizant of as many of the potentially relevant theories and facts as possible – and then, putting it all together, try to interpret what has been learned. The merits of the insights gained will be determined not, as in the hard sciences, through replication by independent investigators, but through the test of time and actual experience.

The traditional norm in academe is depth, not breadth. Expertise, as one wag put it, involves knowing more and more about less and less, until one finally knows everything about nothing. Obviously, specialization has allowed enormous efficiencies in the pursuit of knowledge. But, in the process, the "big picture" is often lost. Experts are prepared to expound on one small piece of the puzzle, but not on how the various pieces fit together. In this book we reverse that emphasis. Focusing on innovations in four different areas, we try to show how the present structure and process of biomedical decisionmaking relates to such broad questions as the justice of resource allocation, the nature of judgments about risk, legal and social responsibility for medical injuries, and the governance of science and medicine. We do not pretend to present an exhaustive analysis of any of these issues; readers who wish more extensive or detailed discussions will find a huge volume of literature on each topic. Instead, we have tried to illuminate the central policy questions raised by the cases in light of the major theoretical arguments in the literature, distilling the theory and evidence in each area and leaving to interested readers the task of filling in the details. Such a synthesis, we believe, is essential in attempting to understand the problems confronting medicine and the choices that lie ahead. It is, after all, the broad contours and interrelationships of policy questions rather than their isolated details that must ultimately inform our choices.

Implicit in policy questions, as well as answers, are social values. The values reflected in this book are, generally speaking, egalitarian and humanitarian; they are concerned with respect for individual autonomy and with social justice. Whether one agrees that the arguments and recommendations presented would in fact promote these ideals depends not only on values but also on one's assumptions about social, economic, and political dynamics. Some of the dynamics that influence medical innovation are described in the chapters that follow; others are merely hinted at. The important point here is that our analysis is not, and could not be, "value free." Social values, whether acknowledged or not, form the core of all historical interpretation and policy analysis; the decision to report one event and not another, for instance, or to analyze particular policy consequences and not others, involves value choices. Furthermore, value choices are inescapable in interpreting the human meaning of the events described.

Several unusual aspects of the project that produced this book are also worth mentioning. The project was, in a sense, itself an experiment in participatory, collaborative research, melding investigation with teaching and community involvement. Our goal was to gain a better understanding of the advantages and disadvantages of public participation in biomedical policy decisions through direct observation of the role of nonexperts in our own research, and to involve students in our learning process as well. There were two main forms of public participation. Throughout the early stages of the project, we met regularly with a volunteer "Public Advisory Board," consisting of members of local labor, professional, civic, religious, public interest, educational, and environmental groups. In addition, we undertook an active program of outreach to the community, holding discussions with civic, medical, public health, religious, labor and environmental groups as well as with individual government officials, scientists, academics working on related topics, and groups of lay citizens. The Public Advisory Board gave us ongoing feedback from people familiar with the project as a whole, while the outreach discussions offered a fresh perspective on people's reactions to particular cases or issues.

The responses of these predominantly lay audiences were often remarkably perceptive. It was generally understood, for example, that the health care budget cannot be evaluated separately from overall public expenditures, and that many nonmedical programs may yield greater health benefits than some medical innovations. Referring to the artificial heart program, one person commented: "It seems to me that there are many possible projects which could prevent more deaths – deaths of potentially much healthier populations; for example, highway safety, air traffic control." Another, stressing the finite volume of health resources, commented that "the government cannot continue to finance every 'good idea' that comes along..." One person urged that since the artificial heart program "was done in secret and the cost is astronomical, anything we can do to stop it we should do." Another asked bluntly, "Why are we just sitting here?"

These experiences gave us a firsthand view of public reactions to the case studies and related policy questions, and certainly influenced our own perspective on policy choices. They confirmed our expectation that broader involvement in the research would enrich and deepen our understanding of the issues we were studying. We were impressed, and heartened, by the extent of interest among lay citizens in the problems posed by medical innovation and the desire of many to help with the search for possible solutions.

Predictably, perhaps, this multidisciplinary, participatory experiment in research and teaching also ran into a number of problems. A subject as large and amorphous as that of this book requires a breadth of knowledge and background that neither I, as project director, nor any of the other investigators had at the outset, and we had trouble providing adequate guidance for students and lay volunteers. While many students felt they had received an unusual and much-appreciated educational experience, the mixture of teaching and investigation was clearly not the most efficient way to conduct research. Even more time-consuming was our experimentation with public participation. Although contacts with diverse community groups yielded valuable insights for the project while also providing useful public education, they inevitably slowed the pace of the research and writing.

We expected that most students and lay volunteers, given their inexperience with research, would need considerable direction in order to take an active part in the investigation. They did. We were less prepared for the converse problems of overspecialization among many of the experts we consulted. Although highly knowledgeable about a given field, many experts had difficulty relating the field's major theories to the empirical realities of the four cases, and in expressing their ideas in terms comprehensible to nonexperts (including specialists in other areas). There were also some attempts at "disciplinary imperialism," in which particular experts, convinced of the unique merits of their discipline, sought to extend its sway over the rest of the project. These and other difficulties may be part of the price one must pay for the breadth of vision and cross-fertilization of ideas that can come from interdisciplinary communication and cooperation.

Looking back, many things could perhaps have been done differently. Certainly this was not the most efficient way to write a book. But the book was only one of the project's goals; equally important were our community outreach, teaching, and experimentation with public participation in the study itself. The book is informed by all of these experiences, although in many ways that at this point would be difficult to trace.

Acknowledgments

Because the book emerges from a collaborative research process, it represents the fruits of many people's labor. Students who took affiliated courses, community volunteers, project staff, and expert consultants all helped to develop and interpret much of the necessary information and to formulate some of the basic arguments and ideas. The book could not have been written without their contributions, even though their names may not appear on the final chapters. I am deeply indebted to them, as are the book's other contributors.

A number of individuals deserve special mention. John Bunker, M.D., Professor of Family, Community and Preventive Medicine, helped conceive of and design the original study, served as my co-principal investigator on the National Science Foundation grant, and participated in our many research meetings and teaching activities. He worked on early drafts of the chapters on risks and compensation, and offered valuable help and counsel on many aspects of the book and its completion. Equally instrumental in the original conception of the project was Halsted Holman, M.D.,

Professor of Medicine, whose commitment to human development and political empowerment has been a continuing source of inspiration. He, too, participated in all aspects of the project, and his ideas helped shape many of the arguments contained in the book. Without his encouragement, guidance, and support, neither the project nor I would have weathered the various storms encountered. Thomas Preston, M.D., although not one of the original project members, willingly contributed his own considerable background on the artificial heart to that case study, going well beyond the material obtained by the project. Nancy Pfund, M.A., M.P.P.A., my coauthor on the genetic engineering case study, had worked on Capitol Hill during the debate over the regulation of recombinant DNA research, and brought to the study the unique insights of a participant observer. Committed to keeping the project's research in touch with the real world, she was a key figure in most of our public outreach activities. Finally, Nancy Adess, former president of the national consumer group DES Action, has deepened and enriched our portrayal of the recent history of DES with her own firsthand knowledge of both the triumphs and travails of a grass-roots movement.

Many other people played a major role in the work leading to the final written products. Barton Bernstein, Ph.D., helped with the design and development of the project, attending research meetings, critiquing chapter drafts and conducting extensive research on the artificial heart and swine flu immunization programs. Ralph Silber, M.P.H., served as administrator for the project in its early years, maintaining order among the chaos of its disparate participants and organizing and documenting our public outreach program. Along with Malcolm Goggin, Ph.D., he performed much of the original investigation relating to public participation. Malcolm Goggin, in turn, was actively involved in project teaching, and also conducted the original research on the swine flu program. Richard Gillam, Ph.D., unearthed valuable information on the history of DES and prepared drafts for the project. Lawrence Molton, J.D., furnished the project with his impressive knowledge about legal aspects of the cases, making important contributions to our research on compensation, risks, and DES. Other project participants whose work helped make this book possible include Robin Baker, M.P.H., Randy Bean, Dennis Florig, Ph.D., Seth Foldy, M.D., Jinnet Fowles, Ph.D., Kenneth Freedberg, M.D., Susan Friedland, M.D., Jane Grant, Ph.D., Thomas Grey, J.D., Deborah Lubeck, Ph.D., Carla Lupi, M.D., Becky O'Malley, Linda Schilling, J.D., David Schnell, M.D., Clara Sumpf, and Carol White, Ph.D. We are grateful to Natalie

Fisher, Kathy McFadden Chewey, Bonnie Obrig, and Becca Pronchik for administrative and secretarial services, and to Patti Osborn for feats of bibliographic research that make Sherlock Holmes look like an amateur. Also enormously helpful were the reactions and insights of students too numerous to name who took courses based on the project or undertook related research.

Many professional colleagues gave generously of their time and ideas in critiquing chapter drafts, although they are of course in no way responsible for any shortcomings in the final products. I am especially indebted to Robert Alford and Theodore Marmor for their comprehensive reviews of the entire manuscript, and to Frank Smith, our editor at Cambridge University Press, for his advice, support, and understanding. I also want to thank Susan Bell and Judy Turiel for extensive and valuable suggestions on the DES chapter. We also received helpful comments and other assistance from Spyros Andreopoulos, Sherrie Arnstein, Jeffrey Axelrad, David Banta, Bernard Barber, Lane Bauer, Dan Beauchamps, Philip Bereano, Eli Bernzweig, Henrik Blum, Gert Brieger, B. William Brown, E. Richard Brown, Michael Carella, Daryl Chubin, Pat Cody, Norman Daniels, Richard de Neufville, Marc Franklin, Alan Freeland, Mark Friedlander, Count Gibson, Thomas Gieryn, Amnon Goldworth, Bradford Gray, Neil A. Holtzman, Albert Jonsen, Jay Katz, Dan Kent, Sheldon Krimsky, Marc Lappé, William Lowrance, Barbara MacNeil, Susan Moffitt, Frederic Mosteller, Dorothy Nelkin, Richard Neustadt, Charles Perrow, James Peterson, Joel Primack, Karen Reeds, Stanley Reiser, Arie Rip, Eugene Sandberg, Lawrence Schonberger, Richard Scott, Sybil Shainwald, Milton Silverman, Paul Slovic, Paul Stolley, Arthur Viseltear, Wesley Wagnon, Charles Weiner, Peter Weiner, Phillis Wetherill, Alice Whittemore, Susan Wright, Michael Yesley, and Burke Zimmerman.

The project could not have been undertaken without the support of several funding agencies, most notably the National Science Foundation's Program in Ethics and Values in Science and Technology (EVIST). William Blanpied, director of the EVIST Program when the project was first funded, shepherded it through multiple layers of review and continued to offer constructive criticism and counsel about the project's design and implementation. We are also grateful to the Rockefeller Foundation, which, through Kenneth Warren, provided additional funding, allowing us to complete critical portions of work on the book. Finally, we thank the Spencer Foundation and the California Council for the Humanities for supporting selected public outreach and teaching activities.

Abbreviations

ACIP	Advisory Committee on Immunization Practices of the Public Health Service (in practice, of CDC)
AIA	American Insurance Association, a trade organization representing casualty insurers
AMA	American Medical Association, a national professional medical organization
CDC	Centers for Disease Control, Atlanta, Georgia, an agency of the Public Health Service
E. coli	*Escherichia coli*, a common type of bacteria that inhabits the human gut
FDA	Food and Drug Administration, an agency of HHS
GAO	U.S. General Accounting Office, an agency of Congress
HEW	U.S. Department of Health, Education and Welfare (now called HHS)
HHS	U.S. Department of Health and Human Services (formerly called HEW)
HSAs	Health Systems Agencies, regional agencies established by Congress in the mid-1970s to plan local health care services
JAMA	*Journal of the American Medical Association*
NAS	National Academy of Sciences, an honorary society that advises the government on scientific issues
NIH	National Institutes of Health, an agency of HHS
Ob-Gyn	Obstetrics and Gynecology, as in Ob-Gyn Advisory Committee, an advisory committee to the FDA
OCAW	Oil, Chemical, and Atomic Workers Union
OMB	Office of Management and Budget, a fiscal agency in the Executive Office of the President
OTA	Office of Technology Assessment, an advisory agency to Congress
RAC	Recombinant DNA Molecule Program Advisory Committee, an advisory committee to the NIH
USDA	U.S. Department of Agriculture
WHO	World Health Organization, Geneva, Switzerland

PART I

Overview

1

Introduction

Good afternoon ladies and gentlemen. This is your pilot speaking. We are flying at an altitude of 35,000 feet and a speed of 700 miles an hour. I have two pieces of news to report, one good and one bad. The bad news is that we are lost. The good news is that we are making excellent time.   – Anon.

Not so long ago, it would have seemed obvious where medical science was headed. Medicine's historic contributions to human health are legendary – vaccines, insulin, anesthesia, electronic heart pacemakers, the heart–lung machine, and many others. Entire diseases, such as smallpox and polio, have been eradicated, or nearly so. Such impressive achievements gave no reason to question clinical progress; indeed, they fueled expectations for still greater accomplishments in the future.

In many ways, medicine has met those expectations, even surpassed them. Modern medicine's technical capabilities are truly extraordinary. Many major organs can now be transplanted. Mechanical devices can partially or substantially replace a person's failing joints, heart, lungs, and kidneys, and there are synthetic substitutes for blood, veins, and skin. Machinery can sustain bodily functions after vital signs have ceased. Researchers are working on developing fully implantable artificial hearts, lungs, eyes, and bladders, and are experimenting with human brain transplants. Through gene splicing, scientists can modify the genetic makeup of living organisms, literally creating new forms of life with desired traits in the laboratory. We are poised on the threshold of applying these techniques to humans – human genetic engineering.

But amid these spectacular accomplishments are new worries and concerns, troubling signs that all is not well in the house of medicine. One of the most visible is the issue of costs. The United States now spends over $1 billion per day on medical care. Total medical expenditures constitute nearly 11 percent of the nation's gross national product, double the fraction spent on health in 1960.

Despite sustained governmental and private efforts to control costs, the price of medical care in 1986 rose at *seven* times the rate of inflation.¹ This relentless cost escalation has led some observers to look more critically at the resources being devoted to biomedical research and development, which supply the future armamentarium of medical practice. There is growing agreement that we simply cannot afford an endless parade of fabulously expensive new lifesaving or lifemaking technologies. Somewhere we have to draw the line.

Soaring medical costs have also prompted new scrutiny of what we are getting for our health dollars. The answers are disturbing. Too often, enthusiasm for the latest scientific breakthroughs has led to exaggerated expectations and uncritical acceptance. Medical history is full of examples of promising new techniques that later proved disappointing, if not dangerous.² With improved methods of statistical assessment, it has been possible to evaluate medical innovation more rigorously, and many have been found wanting. According to a federally commissioned study, less than half of the drugs sold between 1938 and 1962 were effective for their claimed therapeutic purposes.³ Similarly, careful evaluation of a broad range of modern medical and surgical innovations has shown that only half offered improvements over standard practice, even without considering costs.⁴ Ineffective drugs and therapies still find their way into general use because adequate assessment is lacking.

Problems have arisen from seemingly safe, even conventional medical practices. Many "miracle drugs," potent weapons in the war against disease, have proven to be double-edged. Indiscriminate use of antibiotics, for example, has led to the development of new strains of antibiotic-resistant bacteria, which are more difficult and costly to control. The tragic consequences of diethylstilbestrol (DES) and thalidomide, drugs given to pregnant mothers that led to death or disfigurement of their children, serve as a poignant reminder that no proof of risk is not the same as proof of *no* risk – an elementary principle of statistics too often ignored amid the hopes and hoopla accompanying the latest therapeutic advance.

Furthermore, we now know that medicine's contribution to the health of the population as a whole is really rather small in comparison to the role of social and environmental conditions. Analyzing trends in morbidity and mortality over the past three centuries, Thomas McKeown has shown that for most diseases, the introduction of effective medical procedures had little if any detectable effect on death rates, whose downward course seemed to be governed primarily by improvements in nutrition, living standards, and personal behaviors such as reproduction.[5] McKeown contends, therefore, that Western medicine's preoccupation with technological and therapeutic intervention is misguided, and that more attention should be given to the effects of social and economic circumstances, which he predicts will be the dominant determinants of health in the future as well. Marc Lalonde, the Canadian Minister of National Health and Welfare, offered a similar recommendation in a widely circulated book proposing a redirection of Canadian health resources based on current patterns of mortality and morbidity. "There is little doubt," Lalonde concluded, "that future improvements in the level of health of Canadians lie mainly in improving the environment, moderating self-imposed risks and adding to our knowledge of human biology."[6]

Although modern medicine no longer seems the panacea it once did, there is little agreement about what to do about it. Some would redouble our commitment to basic research, arguing that important breakthroughs are just around the corner. Others call for a more comprehensive and rigorous system of evaluation for new medical technologies to prevent dissemination of those that are ineffective or harmful. The present federal administration has instituted sweeping changes in health care financing to try to curb medical inflation. None of these strategies, whatever their merits as partial solutions, gets to the heart of what is ailing medical science. New issues continue to emerge, while old problems remain unsolved. Meanwhile, public confidence in medicine, once the object of almost boundless hopes, has plummeted.

What do we want for our medical future? The question is rarely asked. Yet decisions about medical innovation and care involve not only scientific and economic considerations, but also social and moral choices: They influence who will live, and who will die. They are, ultimately, society's alone to make. A forest products company advertises that gene splicing could produce "miracles" in the future, including "the flowering of a 'better' human being." "But who," asks the advertisement, "is going to decide what makes a 'better' human being?"[7]

Of all the dilemmas facing modern medicine, this is the greatest: finding suitable ways to ensure that medical innovation responds to and reflects the interests of all sectors of society. When a reporter asked Dr. Robert Jarvik, one of the leading developers of the artificial heart, whether the news media had overlooked any issues in the publicity surrounding the program, he thought for a moment and then replied that there was one: "What does the public really want? The assumption is that the lab is working in the public interest . . . It's always *assumed* that [the artificial heart] is needed and wanted . . . but no one has really asked."[8]

Many of the questions raised by modern medical technology are new, and require new answers. We can no longer rely on old assumptions and approaches to solve the current problems; indeed, many of these problems flow directly from the failures of past policies. This book attempts to draw lessons from four innovations that illustrate these new problems and uncertainties. We have chosen examples from very different areas: a drug – DES; a medical device – the artificial heart; public health – the 1976 swine flu immunization program; and basic science – genetic engineering. These innovations have all been, to varying degrees and at varying times, controversial. DES and genetic engineering have evoked concerns about actual or potential physical harms, while swine flu and the artificial heart raise questions of cost, equity, need, and effectiveness. These controversies – their origin, trajectory, and ultimate impact on policy – illuminate the nature of public concerns about medical innovation and the most effective forms of public influence and action. We focus especially on the implications of the attempted interventions for medicine's present dilemmas and potentially fruitful new directions.

The cases

DES is a synthetic estrogen that was prescribed to millions of pregnant women in the 1940s through early 1970s in the belief that it helped prevent miscarriages. Although studies published in the early 1950s indicated that DES did nothing of the kind (it may even have increased the likelihood of complications), its use declined only slowly. In 1971, physicians discovered a previously rare form of cancer – clear-cell vaginal adenocarcinoma – in daughters of women who had taken DES during pregnancy. Other reproductive disorders were subsequently discovered in both DES sons and daughters, as well as higher rates of breast cancer and precancerous changes in the women themselves. Despite the discovery in 1971 of the association with clear-cell cancer, two other uses of DES continued throughout the 1970s: It was prescribed as a "morning-after pill" contraceptive in many hospital and college clinics, and was also widely used by livestock producers as an animal growth stimulant. Spanning the entire post–World War II era, the DES story illustrates how differing social climates, government policies, and attitudes about professional behavior can shape drug development and regulation. It raises important questions about the role of the pharmaceutical industry in federal decisionmaking, the ethics of human experimentation, and compensation for victims injured in the course of medical and scientific innovation.

The second case study describes the quest to develop an artificial heart – a totally implantable mechanical device capable of replacing a failing human heart. Government-funded research began in the mid-1960s amid high expectations. Initial plans called for mass production of mechanical hearts by 1970, at a total development cost of $40 million. More than twenty years and $200 million later, the program had not yet produced a satisfactory totally implantable device, and the implants that were performed were marred by persisting technical problems and the occurrence of multiple, crippling strokes among most of the recipients. With its history of exaggerated promises and unforeseen scientific difficulties, the story of the artificial heart offers useful insights into the process of technology development and the ethical questions posed by clinical experimentation on desperately ill patients. It highlights the societal dilemma of allocating resources to high-cost new devices of unproven benefit in an era of cost-control efforts and cutbacks in both public and private medical care.

Third is the tale of a public health effort of unprecedented scope, the 1976 swine flu immunization program. Fearing a "killer" pandemic of swine influenza, the government launched a massive campaign to produce and distribute enough vaccine to inoculate the entire U.S. population, all within a matter of months. From the beginning, the campaign floundered on technical and policy problems; yet, despite growing doubts among experts and increasing public criticism, the mass immunization campaign went ahead. The swine flu epidemic never occurred. However, the vaccine itself proved to have serious side effects, including the paralyzing Guillain–Barré syndrome and some fatalities. All in all, the swine flu program was a fiasco. The case study explores how and why things went wrong, examining the interaction between politics and policy, the snowballing of perceived public health risks, and the dogged determination of program officials, once committed, to carry on in the face of mounting evidence that victory would spell defeat.

The final case study is of genetic engineering – a pathbreaking scientific technique developed in the early 1970s that allows scientists to modify the genes of living organisms. This technique and related methods are providing exciting new understanding of basic genetic function, and at the same time are transforming production processes in such diverse industries as energy, agriculture, and pharmaceuticals. From the outset, however, genetic engineering has been embroiled in controversy over the enormous new powers it conveys and their unknown implications. Scientists sounded the first alarm about possible dangers shortly after the technique's discovery. The debate soon escalated as local communities, and the federal government, sought a voice in determining proper safeguards. Although many of these early fears later proved groundless, uncertainties remain, and controversy has flared periodically throughout the technology's rapid development. In the early 1980s, the mushrooming of commercial biotechnology led a number of communities to pass legislation requiring private companies to follow federal guidelines. More recently, field tests involving the deliberate release of genetically engineered organisms into the environment have touched off a new round of public protest and local opposition. This case study illustrates the changing forces affecting expert and lay perceptions of risk and benefit, and conflicting notions of the accountability of science to society.

A one-time wonder drug that caused unforeseen harms, a sophisticated new medical technology whose financial and ethical costs could outweigh its medical benefits, a nationwide public health effort against a phantom epidemic, and a revolutionary laboratory technique that alters life itself – the four case studies paint a troubling picture of brilliant medical and scientific accomplishments intermingled with unexpected problems, unrealistic expectations, and deepening moral and social dilemmas. Viewed separately, each case provides intriguing clues about what inspires and directs the search for medical and scientific knowledge, and about how and why that search can go astray. Taken together, the cases have much to say about the dilemmas of modern medicine.

Although the cases involve innovations in diverse areas, they share many common patterns of decisionmaking that characterize contemporary biomedical science in the United States and elsewhere. A particularly striking feature is the dominant role of technical and scientific experts in decisionmaking, even on issues with important social or ethical components. The case studies reveal some of the weaknesses of that approach. Also apparent in the cases is the increasingly blurred distinction between "science" and "technology"; for genetic engineering and some aspects of the artificial heart, they are often one and the same. In all four cases, there were aspects of scientific activity, and of uncertainty, that were critical to the technology being developed or applied.

All of the innovations ran into difficulties or setbacks of one sort or another, ranging from the health problems associated with DES and swine flu vaccine, to the technical miscalculations and delays of the artificial heart program, to the public opposition and controversy provoked by genetic engineering. Obviously, these problems vary greatly in magnitude. Despite its occasionally stormy history, for example, genetic engineering is widely viewed as a major scientific advance with far-reaching implications. Rarely is the path of any medical innovation entirely smooth or trouble-free.

Careful scrutiny of the problems encountered in the cases can shed new light on how and why the policy process works by clarifying how and where it breaks down. The study of dysfunction is a time-honored tradition in many fields, including biomedical science itself. As a noted biologist put it in 1912, many important discoveries have emerged from "probing the actual causes of bodily disturbances and the actual removal of such causes."[9] In psychological research, the study of visual illusions and judgment biases has improved understanding of the process of perception, and studies of forgetting have helped explain the way that memory operates. So, too, at the level of organizations and societies, a better understanding of failures in the biomedical policy process can reveal important dynamics in the structure and process of decisionmaking and can suggest possible remedies.

Our investigation of the four cases focuses on the chain of events, including both decisions and "nondecisions," that led ultimately to difficulties or disappointments. When there was uncertainty, did it result in skepticism and caution or in optimism and risk-taking? What were the consequences? Most important, were the problems encountered inevitable, given the evidence and understanding of the time, or could they have been avoided?

Knowledge of many of the outcomes of key policy decisions suggests where to look for errors in decisionmaking, but not, of course, whether they will be found or what they will be. Such "errors" must be judged in light of the information and options available at the time, not by today's standards or with the clarity of hindsight. The task of historical analysis, according to one respected historian of science, is "to see the past in the same terms as those who lived in it, yet at the same time stand apart from those perceptions and evaluate their implications for the functioning of a social system or the initiation of change in that system."[10] Merely because certain adverse outcomes occurred does not mean, a priori, that they were inevitable. Indeed, one of the dangers of hindsight is, in the words of historian R. H. Tawney, that it "give[s] an appearance of inevitability to an existing order by dragging into prominence the forces which have triumphed and thrusting into the background those which they have swallowed up."[11] Each of the case studies therefore gives special attention to attempts to oppose the prevailing course of decisions, including those that were sooner or later "swallowed up" by more powerful forces. Subsequent events reveal the merits of some of these attempted interventions.

In drawing lessons from these cases for the future, it is useful to distinguish between errors in the *way decisions were made* and those caused by the *state of available information*. Did problems occur, in other words, because certain facts were simply beyond the reach of existing knowledge or, rather, because known information was disregarded or knowable information not obtained? The circumstances that prevented effective use of available information are crucial to understand if future policymaking is to be improved.

The case studies examine both formal policy processes and also the more informal side of policymaking – the lobbying and liaisons that took place behind the scenes. In general, we emphasize bureaucratic and political roles rather than personalities or other idiosyncratic factors. This "structural" interpretation of events is usually more easily reconstructed from historical documents and other sources, and is likely to be more helpful in defining remedial organizational and procedural policies.

On the other hand, there are certain individuals in the cases whose own personal stamp on history must be recognized. For example, the efforts of a single DES mother, Pat Cody, launched a nationwide network of "DES Action" groups which have provided critical services and assistance to DES victims. The courageous decision of a few scientists to alert the general public to the possible risks of genetic engineering opened an unprecedented chapter in American science policy. These examples remind us that individual people *altered* the course of past events in these cases, and that such individual influence is possible in the future as well. Structural and political constraints are essential in understanding the forces that shape behavior, but some individuals will always stand out and apart.

It is worth emphasizing that this investigation does not, and cannot, try to assess how common the problems and uncertainties in the four cases are, or will be, in other areas of medical innovation. Unfortunately, few studies exist that trace in detail the evolution of biomedical decisionmaking, including options not pursued, and the responses of different groups as problems emerged.[12] It seems safe to say that general questions concerning costs, risks, efficacy, and equity have wide application, whereas some of the issues raised by the creation of new life forms, the appearance of second-generation harms from DES, and the development of a mechanical heart pose questions that are new in degree if not in kind. By examining how both familiar and novel questions have been handled in four different cases, this book seeks to clarify the range of dilemmas we face as a society and the choices we cannot avoid.

Organization of the book

To set the stage for the case studies, the book begins, after this introduction, by describing the enormous changes that have taken place in both medicine and the larger society over the past few decades and the policy dilemmas they present. This discussion serves as a brief synopsis of the issues raised by the cases and an overview of the main lines of argument running through the book. The following four chapters trace the histories of the four innovations, from their inception up through roughly the mid-1980s. These four case studies present the "facts" of each case as accurately as possible, describing the events and decisions relevant to what we consider the critical policy questions.

The remaining chapters then look in greater detail at some of these policy questions. Chapter 7 discusses the way risks are – and are not – identified, communicated, and controlled in individual medical encounters and for society as a whole. It shows how the concerns of patients and citizens are systematically undervalued relative to the interests of professionals and private industry, and suggests some steps to correct this imbalance. Chapter 8 examines what happens to individuals when risks become reality, and victims of medical or technological injury turn to the courts for financial compensation. For most such injured individuals, the American legal system offers small hope of relief, as the plight of many DES and swine flu vaccine victims attests. The chapter outlines some recent legal trends and alternative systems of compensation that address these problems. Chapter 9 takes up the important question of distributive justice. Using the egalitarian standard of allocating medical resources according to health needs, it considers the ethical implications of recent procedures for selecting candidates for experimental procedures, the growing private influence in publicly funded research, and the widening gap between biomedical priorities and the nation's health problems.

Having framed the policy questions, we turn to possible answers. Central to all of the dilemmas discussed is the notion of the public interest. There are, of course, many competing ideas of what the public interest actually is, but surprisingly little empirical evidence on the subject. Chapter 10 describes the way different public groups and their allies viewed their interests in the four cases, as reflected in the concerns they introduced, or tried to introduce, into the policy arena. The final chapter attempts to sketch, in very general terms, the directions in which policy might move to address these various medical and social challenges. The single most important lesson is the need for new and more effective forms of public accountability, if medical and scientific advances are to reflect the needs and priorities of society as a whole.

2

Where are we and how did we get there?

To understand medicine's present predicaments, it is useful to begin with a brief overview of their historical context. The character of medical science has changed dramatically since the turn of the century as a result of increasing public and private investment and expanding scientific capabilities. These changes, in turn, have fundamentally transformed medicine's role in American social and political life. They have created new moral rights on the part of society, and new responsibilities on the part of biomedical science.

There are also countervailing trends. The more esoteric medical advances have become, the more difficult it is for lay citizens and their elected representatives to exercise meaningful influence over policy choices. At the federal level, biomedical and regulatory priorities are increasingly permeated with the government's economic agenda for promoting private technological innovation and development. Opportunities for direct public participation in governmental decisionmaking have been curtailed, and public confidence in traditional forms of representative government has declined sharply in recent decades. Together, these developments lie at the heart of medicine's present dilemmas and represent the central challenge to their resolution.

The nature of modern medical science

Medicine's evolution during this century has altered not only its internal character but also its relation to the larger society. Both types of change have left modern medicine more beholden than ever to the society that supports it and that is affected by its risks and benefits.

I. Overview

Increased scale and costs

Prior to the second World War, medical care and research were largely private and individual matters, financed by personal resources. Today, they are typically massive enterprises involving high-cost capital equipment and skilled personnel, financed predominantly by public funds and private health insurance. Along with federal support of research in physics and engineering, government funding for biomedical research, largely through the National Institutes of Health (NIH), rose sharply in the postwar years, far exceeding both industrial and philanthropic funding. Public financing of medical care services also mushroomed, especially in the mid-1960s after the enactment of Medicaid and Medicare. In 1986, the government paid over 40 percent of the total national expenditures on health of $458 billion, and funded nearly 60 percent of all health-related research ($8.1 billion of the nation's $14.3 billion health research budget).[1] Medical innovation and care now clearly constitute a major public financial investment.

Compared to these figures, the $200-odd million of federal research funds spent to date on the artificial heart program seems miniscule. However, successful technologies add geometrically to overall expenditures on health care, since they create new products and services that reinforce the inflationary spiral of medical costs. It is estimated that health technologies account for anywhere from 33 to 75 percent of the increase in the cost of hospital care over the past decade.[2] Yet, although policymakers search desperately for ways to curtail aggregate health expenditures, the public ranks health care at the top of a list of areas in which *more* government spending is desired.[3]

Increased public funding for medicine and biomedical research and the use of mass marketing and production have also expanded the scale of modern medicine. The swine flu program's goal was, in President Ford's words, to immunize "every man, woman, and child in America." Growing numbers of large private conglomerates are getting involved in both medical care and biomedical research. In both the public and private sectors, decisionmaking about medical policy issues has gradually shifted away from the involved individuals and the local community and has become increasingly centralized and bureaucratized. Health is becoming a national or even multinational enterprise.

Enhanced power of medicine and science

The extraordinary technological achievements of modern medical science have expanded medicine's capacity for doing good, but also for doing harm. A million people are hospitalized every year because of adverse reactions to drugs; an estimated 130,000 people die from these reactions.[4] It is simply not possible to discover all of the adverse reactions to drugs and vaccines before marketing, as the swine flu program demonstrated. Drugs such as tranquilizers are overused, leading to widespread dependency. Many potent medical therapies entail the risk of serious, even fatal, side effects – witness the debilitating consequences of radiation and chemotherapy for cancer. Even medicine's ability to cure many acute illnesses has proven a mixed blessing: Keeping people alive longer has led to higher rates of disability in the population, rising medical care costs, and an increasing prevalence of chronic conditions and hereditary genetic abnormalities. Here again, ironically, medicine's current problems reveal the "failures of success."[5]

The products of genetic engineering – new life-forms artificially bred in the laboratory – pose a special risk. Unlike many other hazards of modern technology, genetically engineered organisms, once out in the environment, might no longer be controllable. "You can stop splitting the atom," a critic of the research said in 1976, "stop using aerosols, [and] even decide not to kill entire populations by the use of a few bombs. But you cannot recall a new form of life..."[6] A decade later, open-air field testing of gene-altered organisms evoked anew these concerns. At the same time, laboratory-based genetic engineering has been extended to increasingly dangerous organisms. Scientists have attempted to produce deadly diphtheria toxin, a poison so powerful that a single molecule can kill a healthy cell, in the hope that its fatal mechanisms will prove useful in the treatment of some skin cancers.[7] Nevertheless, despite the sobering lessons of past technologies whose dangers were discovered only after the fact, regulatory controls for genetic engineering have been steadily relaxed. There is now little federal oversight of laboratory or production procedures, and environmental release is proceeding even though the potential ecological effects remain in dispute.

In nearly every area, in short, it appears that the enormous power of modern science and technology has outstripped our ability, or willingness, to control its consequences. We have the proverbial tiger by the tail.

Complex social policy issues

Medicine's increasing scale, costs, and capabilities have created qualitatively new policy dilemmas. Many of these arise from the social and human consequences of technical achievements, which are often ignored until it is too late to do anything about them. The development of the artificial heart proceeded for almost a decade with scarcely any consideration of the quality of life of recipients, or of the enormous cost and distribution problems that a successful device would entail. In 1973, an outside review panel did call attention to these issues, but with no apparent impact. The research went ahead, while problems of cost and equitable access remained unsolved. The more sophisticated technological accomplishments become, the more we are confronted with these vexing policy questions – the so-called soft or human consequences of medical innovation, which often prove the hardest to accommodate.

Another area in which technical achievements challenge our ability to cope with them is in pharmaceutical innovation. Mass marketing of drugs creates large-scale exposures to risk that strain traditional legal doctrines in personal injury suits. The huge volume of drugs on the market also taxes the capacity of individual physicians as well as the Food and Drug Administration (FDA) to ensure therapeutic safety and efficacy. Many drugs are prescribed for indications never reviewed by the FDA, a practice that is ardently defended by some medical practitioners.[8] DES was widely prescribed as a postcoital contraceptive in the 1970s, even though no drug company ever received FDA approval to market it for this purpose. While a certain amount of clinical discretion is obviously essential, allowing a new use to become widespread defeats the purpose of federal regulation of drug safety and efficacy.

The growing gap between technical achievements and society's capacity to deal with them has been recognized for some time. In 1968, Congress asked a presidential commission to study "whether social institutions, national resources and national policies would be able to keep pace with medical advances," giving special attention to organ transplantation, genetic engineering, financing of research, experimentation on humans, and the problems of applying knowledge gained from research.[9] In 1983, several presidential commissions later, another set of recommendations dealt with many of these same subjects.[10] These persisting concerns bear witness to the complex demands that technological developments place on society and the inability of present institutional mechanisms to meet them.

Unprecedented governmental and industrial control

Modern science and medicine are increasingly dependent on both government and industry to support the massive biomedical research establishment created by federal postwar investment. Indeed, many federal officials now openly view science as an "instrument of the state."[11] And, as it is said, he who pays the piper calls the tune.

The scientific community welcomed federal support, but strongly resisted related efforts to expand public influence over research policy. When Congress was considering regulating genetic engineering research in the mid-1970s, for example, scientists argued, successfully, that externally imposed controls were unnecessary. Scientific and medical organizations maintain powerful and well-funded lobbyists in Washington and in state capitals to protect their interests in government policy decisions. Increasingly, scientists make personal appearances on Capitol Hill to argue their case. In 1980, when the Reagan administration proposed to slow the rate of increase of federal research spending, a group of American Nobel Prize winners visited Congress to plead for expanded funding.[12] Faced with the political exigencies of federal sponsorship, the ivory tower added a Washington annex.

As the postwar boom in government research funding began to taper off, the biomedical sciences also forged new links with private industry. Such ties, although long commonplace in engineering and other applied fields, had been essentially unknown in the life sciences prior to genetic engineering. Biology had always seemed too "pure" to be of commercial interest. Today, it is a rare university biologist of major stature who does not have some type of corporate connection, either an equity position or consulting arrangement. Through these connections, many academic scientists have been able to turn the fruits of decades of public funding into personal millions. Universities, too, are looking for profit from what, in the world of corporate-style academe, is officially called "intellectual property." A major shift in practices and priorities is also occurring within medicine itself, under the impetus of federal efforts to expand corporate involvement in both health care delivery and medical research. The proportion of for-profit hospitals has been growing rapidly, and is expected to double or triple by the 1990s.

These developments raise troubling questions. Will for-profit medicine meet the often highly unprofitable needs of the poor? How will the lure of commerce and the competitive ethos of business affect medical research priorities and educational values? How should profits be divided among researchers, industrial sponsors, universities, and the taxpaying public? More broadly, who does – or should – own the knowledge being sold? The long-range consequences of the "corporatization" of science and medicine are still unknown, but it is clear that private industry, as well as the federal government, now plays a major role in determining the course of medical innovation.

New ethical questions

With the increasing power of modern medicine comes the responsibility to use this power wisely and justly, and to assure that resources for medical research and care are allocated fairly across all parts of society. During the 1980s, the combination of drastic cuts in most federally funded health and social programs, together with continuing medical and technological innovation, resulted in a significant shift in medical resources away from the medically needy toward the more privileged, and away from proven, low-cost preventive and primary care services toward expensive technologies. Reacting in part to this trend, a 1985 government advisory panel reviewing the artificial heart program recommended against government coverage of artificial heart implants, noting that the projected costs – $2.5–$5 billion per year – might divert resources away from more important, and more cost-effective, uses.[13]

In an era of restrictions in the availability of basic preventive and primary care services, we can no longer afford to ignore considerations of cost-effectiveness. There is little doubt that the money required to pay for artificial heart implants would yield greater net health benefits for society if it were spent in other ways. On the other hand, throughout the 1970s, government advisory panels, including several reviewing the artificial heart program, concluded that when therapies were developed with public funding, the government had a moral obligation to guarantee access to all those in need, regardless of ability to pay.[14] Without government coverage, artificial hearts will be out of reach of the poor and uninsured – the very groups with the highest rates of serious heart disease.

To deny a potentially lifesaving technology to people who cannot afford it, particularly one developed at taxpayer expense, is not something we or other societies have comfortably accepted, at least in the past. Most people believe in some form of a "right" to health care. The only way to avoid these "tragic choices," as legal scholars Guido Calabresi and Philip Bobbitt called them, is to decide, from the start, what advances we want to develop – and are willing to pay for.[15] If we do not make such decisions, we are in effect asking society as a whole to fund medical options that are then available only to the affluent.

Questions of justice also arise in the reparation of injuries resulting from medical and technological innovation. By its nature, innovation involves risk, quite apart from intentional wrongdoing or negligence. Traditional tort law remedies were not designed to cover these inevitable, albeit unpredictable, injuries. Over the past several decades, the courts have been gradually recasting traditional doctrine to expand compensation for injured victims, regardless of fault or responsibility. In response to rising liability insurance costs, there has been a growing movement to limit physicians' and manufacturers' liability through state and federal legislation. Wherever these trends lead, the issue of just compensation for the victims of medical and technological injury is not likely to be resolved quickly or easily.

Genetic engineering raises a different, and in some ways unique, set of ethical questions. Researchers have used genetic engineering to produce "supermice" almost twice the normal adult size (they used mice rather than rats for, as one senior scientist delicately put it, "public relations reasons"). Rat growth hormone genes have been introduced into fish to produce giant "transgenic" fish, and research is underway to produce oversized cattle and other livestock.[16] While humans have always used animals for their own purposes, such developments represent a qualitatively new level of domination over nature: human control of the very design of life itself. Scientists intend to program all manner of desirable traits into farm animals, and crops, and. . . . Where will it end? No one can say.

Genetic engineering has now been attempted on humans, although so far without success.[17] The prospect of human genetic engineering has inspired radically different reactions. The general secretaries of Protestant, Jewish, and Catholic organizations have warned that the temptation to "correct" mental or social traits by genetic means "poses a potential threat to all of humanity."[18] By contrast, some biologists and theologians see the hope of using this technology to cure genetic diseases as a way to "return parts of the creation to their original state," restoring perfection to a world flawed by sin.[19] To skeptics, applying this new technology to humans represents "man playing God," the ultimate act of hubris; others view it as a laudable example of humans using their intelligence for their own betterment.

Whatever its moral ramifications, human genetic engineering epitomizes medicine's supreme confidence in the eventual conquest of nature through science. Implicit in this confidence is a highly secular attitude toward life and human values. Expressing a view held by many of today's medical pioneers, artificial heart researcher Robert Jarvik described life as "an aggregate of a bunch of mechanical and chemical things. It's very ultimately definable."[20] This mechanistic outlook permeates medicine, from the research laboratory to bedside care. It is modern medicine's triumph, and also the root of many of its most perplexing problems.

The changing context: public attitudes and the governance of science

While the character and social role of medical science have been evolving, the larger society has itself undergone changes that are equally profound. The shifting popular mood provides a dramatic index of these changes. Public confidence in almost every major social institution has fallen precipitously over the past several decades. Between 1966 and 1982, the proportion of the population expressing "a great deal of confidence" in Congress dropped from 42 to 13 percent, while confidence in the heads of large corporations fell from 55 to 18 percent. In 1978, six out of every ten Americans indicated "distrust" of government in general.[21]

The causes of growing public alienation and distrust are not fully understood, but seem to reflect a pervasive sense of powerlessness and hopelessness on the one hand, and of dashed expectations on the other. Contributing to this sense have been the accumulating failures of technology and the evident inability, or unwillingness, of leaders and "experts" to prevent them: Chernobyl and the near-disaster at Three Mile Island, the discovery of acid rain and lethal toxic chemical dumps, the intractable problem of radioactive waste disposal, and the apparently deliberate suppression by companies of risk data on prescription drugs, pesticides, and other chemicals. Such failures threaten deeply held beliefs that technology and progress are indistinguishable and that both are unquestionably good. To some observers, they confirm the worst fears of governmental and corporate incompetence or malfeasance.

Public attitudes toward medicine and science also seem to reflect growing doubts. Between 1966 and 1982, the proportion of Americans expressing "a great deal of confidence" in medicine fell from 73 to 32 percent, although medicine remains the most esteemed of all professions.[22] Confidence in science has also declined substantially, and public opinion studies indicate a growing wariness toward technology.[23] In 1985, only 58 percent of the population thought that science and technology did more good than harm, compared with 88 percent in 1957.[24] Faith in medical research remains somewhat higher, yet almost half of the population thinks scientists consider their own research interests more than public benefits in deciding on scientific pursuits.[25]

These trends are part of the increasing cynicism about traditional social institutions. At the same time, public *interest* in science and medicine has never been greater. The 1970s and 1980s have seen a virtual explosion of popular journals and feature sections in newspapers devoted to scientific and medical subjects. National television networks showed President Ford getting his swine flu shot. Newspapers and TV were full of stories on the status of Barney Clark and other early artificial heart recipients. Also adding to the heightened visibility of science are many well-publicized disputes among experts over technical and scientific controversies. Genetic engineering has received an avalanche of publicity, ranging from warnings of potentially irreversible harms to tantalizing claims that the new techniques might eventually be able to "challenge the aging process and even death."[26] In light of such mixed messages and the mounting evidence of technological failures, increasing public interest, and also skepticism, are understandable.

The disillusionment with government and other institutions that arose during the 1960s led to sweeping demands for more broadly based decisionmaking in all areas of government and also direct challenges to the rule of experts: the participatory mandate of the War on Poverty, the egalitarian efforts of the civil rights and women's movements, grassroots protests against the war in Vietnam, popular opposition to nuclear power, the environmental and consumer movements, the nuclear freeze initiative, medical self-help groups, and so forth. Congress passed laws mandating greater public access to government records, broader representation on federal advisory committees, and provisions for direct public intervention in administrative proceedings. In the mid-1970s, local health planning bodies (Health Systems Agencies) were established, with a majority of consumers required on their governing boards.[27] Congress established technical advisory groups such as the Congressional Office of Technology Assessment (OTA) to provide legislators with greater competence in dealing with matters of science and technology.

Many of these federal reforms were abandoned in the 1980s under the Reagan administration, whose free-market philosophy was incompatible with a vigorous, community-oriented public sector. Public access to information has been restricted, consumer participation in administrative decisionmaking has been reduced, and the growing role of special interests in Washington and state capitals has diminished the accountability of elected officials and federal administrators. A more subtle change has also occurred that has had an even more profound effect on biomedicine's allegiance and mission: Federal health policy has become increasingly suffused with the government's agenda for private economic development.

Research for economic growth or the public's health?

The driving force for the nation's economy, according to federal officials and business leaders, is its ability to develop new and better technologies. Technological innovation is vital to economic growth and international competitiveness; and scientific research – particularly the "biologic revolution" heralded by genetic engineering – provides the intellectual base for continuing innovation and development.[28] To promote technological innovation, the federal government, beginning in the 1970s and accelerating in the 1980s, has reduced regulation across the board, encouraged closer ties between academic science and private industry, and enacted a variety of policies designed to "reindustrialize" America and foster greater competitiveness. Health research and regulation have been directly affected by these changes.

One target was federal drug regulation. In response to long-standing complaints from the drug industry that unnecessarily stringent government requirements for drug testing and approval had slowed pharmaceutical innovation in the United States, resulting in a "drug lag," federal regulations were modified to shorten the review process and reduce reporting requirements.[29] Donald Kennedy, former FDA head, questioned the so-called drug lag, suggesting that it was largely a reflection of a worldwide slowdown in important pharmaceutical discoveries, which he attributed more to limitations in scientific knowledge than to the regulatory climate.[30] Even if there is a lag, of course, this may be what society chooses to pay for protection against ineffective – or dangerous – drugs. It was public shock over thalidomide that led, in 1962, to a major strengthening of federal food and drug laws.

Some observers see few if any conflicts between promoting technological innovation to bolster the economy on the one hand, and health and social goals on the other. "Richer is safer," says political scientist Aaron Wildavsky; more economic and scientific development offers the surest way to improve the level of social well-being.[31] He dismisses concerns about the risks of new drugs and technologies as the utopian pursuit of a risk-free society by overcautious government bureaucrats, and laments the loss of the freewheeling, risk-taking spirit of the frontier. Others disagree. In a book entitled *A Nation of Guinea Pigs*, legal scholar Marshall Shapo argues that with the mass marketing of drugs, pesticides, and other chemical products, we are already the unwitting victims of mass chemical experimentation.[32] According to Shapo and others, without greater public accountability, more "progress" along current lines, emphasizing chemical and technological innovation, will only create greater risks and more unexpected problems.

The growing alliance between science and industry has been promoted on the grounds that it will accelerate academic research and enhance its utility for commercial enterprise. Medical services have been singled out as having one of the highest "growth potentials" in the private sector.[33] When the Reagan administration threatened to cut biomedical research funding, scientists in academe as well as industry argued forcefully that continued federal support was critical to national security and a strong economy. As evidence, NIH produced a report demonstrating "that biomedical research and knowledge generation have provided important industrial applications and, as a result, generated substantial income for private-sector industry."[34] The report estimated that ten biomedical discoveries alone had contributed more to the gross national product annually than the total combined appropriations for NIH since its inception in 1937.

Not everyone is sanguine about the benefits of medical science's expanding ties with industry, however, at least following the economic policies of recent federal administrations. Many health analysts contend that our society is already too biased toward medical intervention – whether chemical, biological, or surgical – instead of prevention; they fear that closer ties with industry will further distort health priorities, leading to still greater emphasis on the development of new and ever more expensive technologies. Too often such after-the-fact technological solutions leave untouched the economic and environmental conditions that breed disease, while deflecting attention away from needed social, behavioral, or political changes.

There is also concern about what close ties with business mean for the direction of academic research, the types of personnel trained, and the norms of scientific practice. It is no secret that the quest for profits sometimes encourages corporate behavior that is incompatible with the public's health needs. Drug companies continued to promote DES for prenatal use for more than a decade after its lack of efficacy had been demonstrated. The cattle industry managed to resist federal efforts to ban DES in animal feed for nearly a decade after the link with clear-cell cancer had been discovered. More recently, scientists have complained that the involvement of private companies in genetic engineering has led to a new air of secrecy and competition in the research community which is at odds with the collaborative needs of scientific investigation.[35]

Economic policies to promote technological development clearly conflict with the goal of controlling health care costs. In 1979, in an attempt to curb spiraling health costs, the Department of Health, Education and Welfare decided to "wage war on runaway medical technology" by limiting the availability of patents for new technologies resulting from government-funded research.[36] At the same time, both Congress and federal officials, worried about the lagging economy, were searching for ways to *facilitate* the availability of patents in order to stimulate technological innovation. The cure devised for the economy ran headlong into HEW's efforts at cost control. The larger economic interests prevailed, and HEW quickly bowed to pressures from Congress to increase patent availability.

When health and economic goals are merged, these various conflicts are inevitably obscured. More importantly, the fundamental purpose of medical and scientific innovation becomes confused. Economic considerations begin to shape health priorities, even though the types of innovations needed to stimulate the economy may *not*, in many cases, be those that would yield the greatest benefits to health. Indeed, our most important health needs today require not more "technology," in the conventional sense – we cannot even seem to manage what we have – but rather such things as more creative and efficient forms of medical care, new kinds of workplace arrangements, innovative approaches to risk detection and management, more effective forms of patient education and self-care, novel social service and community support programs, new channels of communication between science and the public, and revitalized forms of government. Although promising initiatives along these lines were developed in the late 1960s and early 1970s, they have never been seriously implemented on a large-scale basis. Today, few discussions of the "innovation lag" even mention such alternative measures. So deeply imbedded is the role of technology in our culture that the term "innovation" is often used as if it were synonymous with *technological* innovation. A new term was even coined, "technovation," making the synthesis explicit.[37]

This is not to suggest that economic goals can or should be ignored. Obviously, we must strike a reasonable balance between health and the economy. But it is important not to confound health priorities with economic ones if we are to understand the choices we face and the trade-offs involved. Otherwise, we end up with a muddled set of priorities which, in trying to pursue two potentially incompatible objectives, may in the end serve neither very well.

Who shall decide?

Decisions about our medical future must reconcile competing social, political, and economic goals as well as contradictory technical advice. While medical information is essential to such decisions, it is far from sufficient; choices must necessarily reflect the social *values* we place on various risks and benefits. Values determine not only the priorities given to different options but, equally important, the range and types of options considered. Thus, we come back again to the underlying question: Who will make these decisions about risks and priorities, and what values – substantive and procedural – should they reflect?

Despite the demands for greater public accountability that swept the nation in the 1960s and 1970s, most citizens today still have no direct voice in biomedical policy, and little meaningful influence through elected officials. In the federal government, most health policy matters are decided by administrative agencies, whose appointed staffs are insulated from the electoral process. The main agency is the Department of Health and Human Services, an umbrella organization which includes specialized research institutes such as the National Institutes of Health, as well as other agencies concerned primarily with regulation and health protection, such as the Food and Drug Administration and the Centers for Disease Control. These agencies rely heavily on the scientific experts on their staffs, as well as outside scientists in universities and private industry. As the case studies will show, powerful coalitions often develop among agency officials, scientific and industry interest groups, and sympathetic members of Congress – a "cozy triumvirate" of issue networks which presents further obstacles to public influence.[38]

Although Congress legally controls the budgets for biomedical research, most legislators have neither the time nor the technical information necessary to use the budgetary process effectively, and they are hampered by the increasingly complex technical issues. In the early 1980s, Congress tried to tighten the reins on NIH by requiring periodic budgetary review, customary for most federal agencies, but was thwarted by intense opposition from the biomedical research community.[39] Congress also exerts some influence over administrative agencies through oversight committees, public hearings, advisory committees, and special investigations, but these mechanisms are usually confined to particular problems or issues.[40]

Public influence is still more sharply circumscribed in private industry and academic institutions, in whose laboratories many critical decisions about scientific and technological innovation are made. The public has no voice in corporate decisions other than through analysts' projections of the likely market for new products and services. Developments not seen as commercially viable are generally not pursued, no matter how socially worthwhile they may be. Some academic scientists explicitly disavow responsibility for the practical implications of their work. And in both industry and government, the technical complexity of many biomedical issues has given specialists and experts, or those who can afford to hire them, increasing control over science and technology policy.

We face a paradox. Society has a growing stake in the conduct and consequences of medical science and foots most of the bill, yet has a shrinking voice in the choices required. Ultimately, this problem strikes at the very heart of democracy. Central to democratic government is the principle that important social decisions should be made by bodies representative of, and accountable to, the citizenry. One may argue that the large share of tax dollars devoted to biomedical research earns the public this right. The moral argument for public accountability is even stronger. It is society as a whole that bears the risks of medical innovation as well as stands to gain from advances. Medical and scientific research will shape the way that we as well as future generations experience illness, death, and some aspects of life itself.

But does the general public have anything useful to say about policy questions as complex and esoteric as those raised by medical innovation? One view, held by many scientists, is that the issues are too difficult for nonexperts to make competent judgments; the public mistrust that greeted electricity, the power loom, and the automobile is sometimes cited as evidence of ignorance and irresponsibility. Some scientists, recalling stories of Lysenko, fear that greater public influence in biomedical policy could lead to political repression.[41] Other observers argue that to the contrary, the consideration of multiple perspectives leads to greater scrutiny, richer dialogue, and better – or at least more representative – decisions. Harold Green, a law professor and chair of the 1973 Artificial Heart Assessment Panel writes:

> The beginning of wisdom is the recognition that no one – no matter how wise and objective – has the capacity to state reliably and in a universally acceptable manner what benefits the public truly wants and what costs (and risks) it is willing to bear for the sake of having those benefits. The best that we can do is to ventilate the benefits and risks openly and candidly so that public sentiment can somehow, albeit imprecisely, be reflected in the electoral, legislative and political processes to instruct, or at least signal, public officials.[42]

As we shall see in the following pages, many citizens, public representatives, and dissident scientists clearly wanted more of a voice in policymaking in the four cases and were capable of making useful and responsible contributions. Where broader participation

occurred, it generally widened the range of policy options considered, increased understanding of possible risks and practical
problems, suggested new kinds of scientific experiments, led to
more careful consideration of existing evidence, and focused attention on the human or ethical consequences of policy choices.
As it turned out, the more cautious attitude toward risks of many
lay groups, and their more humanistic and practical orientation,
might have provided a beneficial antidote for the optimistic and
predominantly scientific-technical bent of decisionmakers. Had the
policy process been more open to such perspectives, some of the
lurking problems might have been uncovered and avoided. In
general, however, these concerns were minimized or repressed
rather than exploited as a source of potentially valuable ideas and
insights.

Intolerance for the views and values of nonexperts reflects a
conception of science as an autonomous, largely private activity
without immediate, practical social consequences. That conception, as we have seen, is long outdated. Modern medical science
is inextricably intertwined with society at virtually every level,
from its funding and priority-setting to the social and economic
impacts of its products. Nowhere is the social character of modern
science more evident than in the alliance between science and industry, forged in the interests of private economic development.

Health policies and priorities, and medical innovation itself, cannot avoid favoring some social goals over others. The question is
which goals these will be and how they will be chosen. Here,
biomedical policy converges directly with social policy and political decisionmaking; modern medical science should be subject to
the rules and constraints that properly apply to other social activities in contemporary democracies. The four case studies that follow suggest that science and medicine may, in the long run, have
as much to gain from such a relationship as society.

PART II

Four case studies

3

DES and the elusive goal of
drug safety

Really? Yes . . . desPLEX [a brand of DES] to prevent abortion, mis-
carriage and premature labor.
Recommended for routine prophylaxis in ALL pregnancies . . . big-
ger and stronger babies too.
— Grant Chemical Company advertisement,
American Journal of Obstetrics and Gynecology, June 1957

In February 1938, a brief letter appeared in the British scientific
journal *Nature* announcing the synthesis of a nonsteroidal compound with the properties of natural estrogen.[1] The report was
by a British scientist-physician, Sir Edward Charles Dodds of the
Courtald Institute of Biochemistry in London, and colleagues at
Oxford. The researchers named their new compound "stilboestrol"; it later came to be called diethylstilbestrol, or simply DES.

The announcement ran scarcely over a page, yet it sent ripples
of excitement throughout the medical community. Here was a
substance that could be used instead of the scarce and expensive
natural estrogens in treating a wide range of problems thought to
be related to imbalances in sex hormones. Clinical investigators
around the world lost no time in launching experiments on patients, and most raved about the results. One researcher, speaking
at a national medical meeting in the United States in 1939, called
DES "the most valuable addition to our therapy in recent years."[2]
A noted gynecologist declared that it had "tremendous clinical possibilities."[3] By 1940, the response to DES and other synthetic estrogens had been so favorable that Dodds observed that
"their immediate practical value is now established." But he

Special appreciation to Nancy Adess, former President of DES Action National, for important contributions to the latter half of this chapter. The discussion of the early history
of DES (pages 33–47) draws heavily on Susan E. Bell's excellent analysis, *The Synthetic
Compound Diethylstilbestrol 1938–1941: The Social Construction of a Medical Treatment*, Ph.D.
dissertation, Brandeis University, May 1980.

also had grander visions. "It is perhaps not too much to hope," he mused in 1940, "that just as aspirin marked the beginning of a new era in medicine, so these compounds may be the first steps along a road leading to great advances in the treatment of disease."[4]

The road, we now know, led elsewhere. In 1971, doctors reported a previously rare form of cancer – clear-cell vaginal adenocarcinoma – in daughters of women who had received DES during pregnancy. Since then, other health problems have surfaced among DES daughters, including structural abnormalities of the reproductive tract and difficulty conceiving and carrying their own pregnancies. Studies of DES sons indicate that they are more likely than nonexposed men to have genital tract abnormalities, and studies of DES mothers show an increased risk of breast cancer.

DES did mark the beginning of a new era in medicine, but not the one Dodds envisioned. It represented the first known human occurrence of transplancental carcinogenesis – the development of cancer in offspring due to exposure in utero to a substance that crossed the mother's placenta. This outcome was all the more poignant because it resulted from medical intervention intended to help patients, not to harm. It was a grim reminder of the hazards of modern drug therapy.

Dodds was right about one thing, however. DES would indeed find many uses, justified or not. Medically, it has been used to treat problems of menopause, acne, gonorrhea in children, and certain types of cancer, to suppress lactation after childbirth, to stunt growth in adolescent girls, and to prevent miscarriages. Despite the 1971 discovery of the link with vaginal cancer, DES was widely prescribed in many hospital and college clinics as a "morning-after pill" contraceptive during the 1970s and early 1980s. And, because of its growth-stimulating properties, it was used by most U.S. livestock producers to fatten cattle and other animals from the 1950s through the late 1970s, until this practice was finally banned. Women who took DES during pregnancy and their children are therefore not the only ones who have been exposed to DES. Also potentially at risk are unknown numbers of patients treated with DES for other medical problems, workers in chemical plants and feed mills where the drug and feed additives are produced, and – to one degree or another – everyone who has eaten U.S. meat. DES is still on the market today for a few medical conditions.

The transformation of DES from one-time wonder drug to

modern-day disaster offers a fascinating glimpse into the forces propelling medical and technological innovation onward and the limits of both regulatory mechanisms and medical science in protecting the public against the risks of unsafe and ineffective drugs. Rarely were the different medical and agricultural uses of DES linked, even though some of the hazards they revealed were directly relevant to the other applications. Many analyses of the history of DES suffer from similar compartmentalization, treating each use of DES as if it were a separate episode. Here we view the DES story as a whole. We trace the development of each use from its inception up through the present, focusing on the warning signs that emerged in each arena and the way in which they were discounted or dismissed. The uncritical embrace of DES by scientific, medical, and governmental authorities is sobering. The following account attempts to suggest some of the reasons for this seeming myopia and the role of the different participants in the disaster that was taking shape.

To unravel the various threads of that disaster, we begin our tale a half-century ago, with a brief but momentous announcement of Sir Edward Charles Dodds.[5]

The discovery of DES

The synthesis of DES in 1938 represented the culmination of more than a decade of research by Dodds and other scientists aimed at creating a synthetic estrogen. Dodds had a dual goal: to understand and simplify the basic chemical properties of estrogenic substances, and to create a clinically useful substitute for natural estrogens. His research was part of a broader resurgence of interest in the field of sex endocrinology. By the 1930s, various scientific developments had enabled scientists to isolate, purify, and synthesize a number of hormones, including estrogens, leading to previously unknown possibilities for clinical investigation.

Much of the work in sex endocrinology focused on what researchers and clinicians viewed as "female hormonal disturbances," including almost all problems associated with menstruation and menopause. Although estrogen administration appeared to be useful clinically, particularly for menopausal symptoms such as hot flashes and depression, natural estrogens were scarce and costly; moreover, they had to be injected, which involved the added expense and inconvenience of doctor visits, often over an extended time period.

DES, however, could be produced in much greater quantities than previously possible, was comparatively cheap ($2/gram versus $300/gram for natural estrogens), was effective when taken by mouth, and was two to three times as potent as natural estrogen.[6] Furthermore, because it was not patented, it was available to any interested scientist or manufacturer. (The British Government's Medical Research Council, which had sponsored Dodds's work, prohibited patenting any discoveries it had funded in order to ensure that they would be freely available to all.) The only notable drawback of DES was that, like natural estrogens, it appeared to be carcinogenic under certain conditions. Numerous studies in the 1930s had shown that animals exposed to large doses of natural estrogen developed cancer or other abnormal growths. Nevertheless, despite the greater potency of DES, most U.S. medical authorities considered it to be as safe for humans as natural estrogen if prescribed appropriately, and dangerous only if used "unscientifically."[7]

Experimentation with DES in both animals and humans began simultaneously, almost immediately after its synthesis in 1938. Early animal research foreshadowed some of the most important future uses of DES. For example, one group of scientists investigated the effects of DES on cows and other animals because of "the probability that diethylstilboestrol will find important applications in clinical medicine and also in veterinary practice and animal husbandry."[8] Another line of research was a direct forerunner of the morning-after pill. Studies begun in the 1920s had shown that natural estrogens given to animals after mating interfered with pregnancy.[9] Following up on these findings, Dodds and two colleagues reported in 1938 that DES, too, prevented implantation of fertilized ova and also ended established pregnancies in rabbits and rats. They pointed out that, in view of similarities between the menstrual cycle in primates and lower animals, their conclusions should "be applicable to women," but hinted that such human applications were better not pursued.[10]

In general, the idea of interfering medically with pregnancy was anathema to physicians in those days. Up until 1937, the American Medical Association had opposed doctors giving any kind of contraceptive advice to patients.[11] Dodds himself apparently wanted no part of any effort to develop a DES contraceptive pill, and single-handedly kept one British pharmaceutical company from marketing a birth control pill because, according to the former president, "he was convinced that later in life women would get breast cancer."[12] Not until the socially liberated 1960s, when birth control had achieved broad acceptance, would work on the contraceptive properties of DES resume. In the interim, researchers seemed largely to ignore the worrisome implications of these early animal findings for the use of DES in pregnant women.

DES tests new drug regulations

The synthesis of DES immediately elicited great interest among clinicians in such fields as endocrinology, gynecology, obstetrics, and oncology. The first two British human clinical trials sought to determine whether DES could replace natural estrogens in treating postmenopausal disturbances, amenorrhea, and dysmenorrhea; they reported generally favorable results.[13] By 1941, 257 papers had been published demonstrating the value of DES not only for the treatment of menopausal symptoms but for a wide range of other clinical problems as well. These papers showed, a literature review concluded, that DES was "as effective if not more so than the natural estrogens in any of the clinical conditions in which it has been tried."[14]

The prospective market for DES looked extremely promising, especially in the United States. Pharmaceutical companies began actively trying to promote this potentially lucrative new drug (it was, in fact, a drug company scientist who wrote the above-mentioned review of DES documenting its numerous clinical applications). The main tactic companies used was to distribute free "research samples" of DES liberally to physicians and researchers in order to acquaint them with the drug and, in the process, obtain favorable clinical testimony. Before manufacturers could actually sell DES in the United States, however, it had to be approved by the federal Food and Drug Administration (FDA).

The drug industry and its regulation in the early 1900s were quite different from what they are today. In those days the bulk of the drug market consisted of proprietary or "patent" medicines, which were advertised directly to the public and, except for narcotics, could be purchased without prescription. Many of these nostrums, touted as "miracle cures," were at best worthless, at worst decidedly dangerous.[15] Public anger over the exaggerated and misleading claims for many patent medicines had helped lead, in 1906, to the passage of the Pure Food Act, which prohibited the "misbranding" of foods and drugs. Although this Act was a

major achievement, it did not significantly diminish the popularity
of patent medicines, nor did it stop the deceptive promotion of
unsafe drugs. Indeed, drug sales soared in the early part of the
century.[16]

By the early 1930s, pressure had developed for more stringent
drug controls, much of it emanating from top FDA officials, with
strong support from national women's organizations.[17] For five
years, successive drug reform bills were introduced in Congress
but failed to pass because of strenuous opposition from patent
medicine manufacturers and other vested interests. The turning
point came in 1937, when over 100 people died from a liquid sulfa
drug, elixir of sulfanilamide. It turned out that the manufacturer
had never tested the safety of the solvent used in the drug, since
the 1906 Act did not require proof of safety. Provisions were
quickly added to the proposed reform bill making it illegal to sell
a drug before its safety had been documented. Galvanized by popular outrage over the sulfanilamide catastrophe, Congress finally
passed the Federal Food, Drug, and Cosmetic Act of 1938, giving
the FDA jurisdiction over the safety of new drugs.[18]

Under the 1938 Act, manufacturers were required to submit
a New Drug Application (NDA) to the FDA consisting of studies testing the product's safety, samples of the product and its labeling, and a description of manufacturing standards. The FDA
then had 180 days to review the application; if no action was taken
within that time, the application automatically became effective. (This passive form of approval–through–inaction was
designed to avoid the appearance that the FDA had actively "approved" a new drug – an indication of the agency's skittishness
about its new responsibilities.) If the FDA intended to reject an
NDA, it would announce a hearing to which the company was
invited. Such hearings were usually "very perfunctory," according
to one informed observer, and the outcome was invariably denial
of the application.[19]

If the procedures for getting new drugs approved under the
1938 Act were straightforward, the judgments upon which they
were based were not. What exactly did proof of "safety" mean?
The 1938 Act did not specify the types of investigations required,
how many, or who was qualified to perform them. Nor did it
indicate how the quality of different investigations was to be
judged, or when an application could be considered complete.
Answers to these and other questions remained to be worked out
in the early years of the law's implementation.

DES would provide the first major test case of just such questions.

How should the FDA define "safety"?

Amid the generally favorable clinical evaluations of DES, a good deal of evidence of its potential risks was beginning to accumulate. By 1939 over forty articles documenting the carcinogenic effects of both synthetic and natural estrogens in animals had appeared in the medical literature, including several specifically of DES.[20] A number of other studies reported that animal fetuses exposed in utero to DES and other estrogens developed various types of physical abnormalities, mainly abnormalities of the reproductive tract.[21] New findings corroborated the earlier work showing that DES was harmful to the fertilized ova of mice and other animals, and DES was also found to have a deleterious effect upon the corpus luteum of early pregnancy (which produces progesterone essential to pregnancy maintenance).[22] Although these findings were all based on studies of animals rather than humans, a number of the papers expressly stated that the findings might also be applicable to humans.

Nor were studies of DES in humans entirely reassuring. Many, though not all, investigators reported that DES caused unpleasant side effects in patients, including nausea, vomiting, abdominal distress, loss of appetite, lassitude, skin rashes, and even mental disorders.[23] The reasons for such symptoms were unclear, and some researchers speculated that they might indicate damage to internal organs such as the liver or kidneys (in at least one animal study, some DES-exposed mice had died of liver damage).[24] In addition, there were reports that some male workers in British companies producing DES had developed gynecomastia (growth of breasts and other female traits) and sexual impotence.[25]

The increasing experimental use of DES together with the mounting evidence of its hazards led the American Medical Association's prestigious Council on Pharmacy and Chemistry to issue, in December 1939, an official "preliminary report" on DES. The Council noted that the "remarkably effective . . . therapeutic results and the convenience of administration are conducive to widespread interest," but warned that "stilbestrol may be carcinogenic under certain conditions" and that "toxic reactions are associated with the use of this substance." DES should not be adopted for routine use, the Council concluded, until further research had been conducted:

> Because the product is so potent and because the possibility of harm must be recognized, the Council is of the opinion that it should not be recognized for general use . . . and that its use by the general medical profession should not be undertaken until further studies have led to a better understanding of the proper functions of such drugs.[26]

An editorial in the same December 1939 issue of the *Journal of the American Medical Association* (*JAMA*) underscored the Council's concerns. It advised "against long continued and indiscriminate therapeutic use of estrogens," warning that "the possibility of carcinoma induced by estrogens cannot be ignored." The editorial noted that carcinogenic effects might not be confined to the sex organs "when the doses are excessive and over too long a period." "This point should be firmly established," it concluded,

> since it appears likely that in the future the medical profession may be importuned to prescribe to patients large doses of high potency estrogens, such as stilbestrol, because of the ease of administration. . . . [27]

Four months later, another *JAMA* editorial warned that even small doses of estrogens might not be safe if administered over a sustained period.[28]

Risks notwithstanding, many drug companies were eager to market DES in view of its growing list of clinical uses, modest price, and lack of patent (which meant firms did not have to pay costly licensing fees). Numerous manufacturers filed initial applications with the FDA, but many withdrew or abandoned their applications when it became clear that the FDA's requirements for proving the "safety" of DES might not be easy to meet. By the end of 1940, only a dozen or so NDAs, all from larger companies, were still active.[29]

The FDA took the DES applications quite seriously for a number of reasons. Officials in the agency were understandably worried about the adverse reactions associated with the drug, especially given the AMA Council's official disapproval of DES and *JAMA*'s repeated editorial warnings. Such troubling evidence and opinions, one informed source suggested at the time, led Dr. James J. Durrett, then chief of the New Drug Section of the FDA, to become "very cautious" about DES and to demand "a substantial amount of additional study" before any applications would be approved.[30]

On the other hand, the enthusiastic reports from physicians and researchers about the clinical prospects of DES, and the receipt of ten or more separate NDAs within a year, could not fail to impress FDA officials.[31] The extraordinary level of interest generated by DES clearly placed it in a category of drugs classified as "distinctly new and possibly highly valuable." The 1938 Act instructed the FDA to give such drugs "very serious consideration" and to reconcile two potentially conflicting objectives:

> no valuable drug should be unnecessarily withheld from public use; the release of drugs without sufficient testing to establish their safety must be avoided.[32]

For DES, the conflict between these two objectives was virtually unavoidable, since there was ample evidence documenting both the clinical utility of DES in estrogen replacement therapy as well as its risks. Moreover, the volume of data being submitted in connection with ten separate NDAs threatened to overwhelm the limited capabilities of the FDA's New Drug Section, which at the time included only two medical officers.[33] The FDA's first response was to seek more time. "We asked [the manufacturers] to withdraw their New Drug Applications," recalled Dr. Theodore G. Klumpp, then head of the Drug Division, in which the New Drug Section was located. "It was very clear to us that we were going to run into the end of the 180-day period and in those days we took the legislative mandate very seriously and did not want to go on beyond that time."[34]

Then they came up with a better idea. Even with more time, considering the ten NDAs separately still posed, in Klumpp's words, "a very serious administrative problem."[35] Why not require the manufacturers to combine their NDAs into one master application to expedite the review process? The FDA had used such a procedure several times in the past and found it quite satisfactory.[36] More important, such a joint application would allow the manufacturers to present the strongest possible proof of safety, which Klumpp at least saw as an important goal.

FDA: public protector or corporate coach?

To sound out this idea, FDA officials consulted a trusted spokesman for the drug industry, Carson Frailey, the Washington representative of the American Drug Manufacturers' Association.

Through the years, Frailey had earned the confidence of both the FDA and pharmaceutical manufacturers as a reliable conduit of information. He became indispensable in the negotiations over DES. "This arrangement saved us a great deal of time and effort," Klumpp later explained. "We could find out what we wanted through Mr. Frailey. Instead of making . . . twelve telephone calls to twelve different companies, we just had to call Mr. Frailey, and he made the twelve telephone calls."[37] Frailey's services were equally appreciated by the drug industry. "Mr. Frailey was in constant touch with the Food and Drug Administration and was able to tell us what was going on there," recalled an official with Eli Lilly, one of the larger pharmaceutical companies. "It saved us having people running back and forth finding out for ourselves."[38]

Frailey informed the FDA that the companies "didn't like [the idea of collaborating] one bit" and that some might refuse to go along with it.[39] But the FDA remained adamant. A hearing was scheduled, a sure sign that the NDAs would be rejected. On December 30, 1940, Durrett, chief of the New Drug Section, and Commissioner Walter Campbell, head of the entire FDA, met with Frailey and interested DES manufacturers and told them that the existing NDAs contained insufficient proof of safety (Klumpp was not present). Durrett made it clear that the only hope the companies had of getting their applications approved was to withdraw them ("without prejudice") and then resubmit them jointly, augmented and reorganized, in a single "Master File." That same day, FDA Commissioner Campbell received telegrams from all of the prospective DES manufacturers withdrawing their pending NDAs.[40]

Representatives of the interested DES manufacturers met the following month and eight decided, somewhat reluctantly, to go along with the FDA's plan (four more joined the group later on). They agreed to obtain detailed clinical reports from various medical experts whom the FDA had identified as "key men" on DES, and to ask their opinions about whether it should be released. Each company was to contact the experts whom it had been supplying with DES. A committee of four was formed to coordinate the effort, with Dr. D. C. Hines, of Eli Lilly, as head. It became known as the "Small Committee."[41]

Both the Small Committee and the FDA then proceeded with their respective tasks. However, it was the drug company representatives who shaped the data on which the FDA was to determine the safety of DES. The Small Committee devised a questionnaire focusing on clinical evaluations to be sent to the identified experts. Unobtrusively but decisively, this questionnaire set the most favorable possible terms of discourse for DES, highlighting its strengths – clinical benefits to patients suffering from menopausal symptoms and other hormonal problems – while ignoring its most obvious weakness – that it caused cancer, physical abnormalities, and termination of pregnancy in test animals. This shift in focus was crucial in defusing the previous criticisms and warnings regarding DES, and was to be pivotal in its ultimate approval.[42]

To gather information and bolster support for DES, Lilly's Hines paid personal visits to many of the leading clinical investigators around the country. Both Klumpp and Durrett, of the FDA, did the same, in an effort to form their own opinions. Everyone heard more or less the same thing: All of the experts were enthusiastic about DES, with the exception of four doctors in New York – Drs. Shorr, Salmon, Geist, and Kurzrock. These four, who came to be called the "New York group," had observed a disturbingly high rate of nausea and vomiting among their patients (as many as 60–80 percent had such reactions), which they feared might be a sign of liver damage or other internal problems. Until the mechanisms of these toxic effects were better understood, the four New York doctors opposed the approval of DES.[43]

By contrast, the rest of the clinical reports coming in to the Small Committee looked very encouraging – even with regard to toxicity. By March 1941, Hines judged the Master File strong enough to warrant another approach to the FDA, but he was still worried about what he called "the apparent antagonism of the New York group."[44] On March 5, 1941, he wrote to one drug company official:

> Because all the opinions which we have received to date are extremely favorable, it seems to me that we are justified in pushing Stilbestrol before the Food and Drug Administration, but, of course, it will be much easier if the New York group is willing to soften its opinions.

And he added, "We know that one of the group is at least willing to be open-minded. . . ."[45]

For help in counteracting the New York group's objections, the Small Committee turned to Dr. Elmer Sevringhaus, an eminent academic endocrinologist and known DES enthusiast. (Sevringhaus, like many other principals in the DES story, later found employment with a pharmaceutical company.)[46] Severinghaus came up with a seemingly simple explanation for why the New York doctors were finding higher rates of toxic reactions than he and others had: Their dosage levels were too high. With lower doses of DES, Severinghaus maintained, most of the toxic effects would disappear. Shorr gave the FDA a different explanation for the divergence in findings, suggesting that perhaps other investigators had "not observed their patients with sufficient diligence."[47] But it was Severinghaus's version that ultimately prevailed at the FDA.

Neutralization of the New York group's objections proved to be critical to the quest for approval of DES. The Small Committee's next meeting with FDA officials, held on March 25, was, as Hines described it, "marked by an entirely different attitude on the part of Dr. Klumpp and his assistants than was shown by Dr. Durrett and Dr. Campbell at the previous meeting on December 30, 1940."[48] Part of this change undoubtedly resulted simply from Klumpp's presence, as Klumpp had for some time been more supportive of DES than either Durrett or Campbell. But it also reflected a genuine shift of sentiment within the agency itself, brought about by the voluminous clinical data on DES accumulating in the Master File along with some strategic lobbying by the medical community.[49] Klumpp informed the Small Committee that the FDA was "anxious to approve Stilbestrol" as soon as it had sufficient evidence of safety, and that it "expects to grant permits as soon as the data warrant."[50]

This turned out to be a mere month and a half later. On May 12, 1941, Frailey wrote Hines that the "time now seems propitious" for submission of the Master File, adding, somewhat mysteriously, that although he could not guarantee approval, his suggestion had "official background."[51] As usual, Frailey's advice was sound. There had been a "gradual convergence of thinking" among FDA officials regarding DES, Klumpp later reported, and by mid-June they had reached the decision to approve it (although no announcement was made until September). Even the skeptical Durrett had come to favor approval because, according to Klumpp, "the data was so overwhelming."[52]

The Master File had indeed reached impressive proportions, including fifty-four completed questionnaires from the selected "experts," plus over 5,300 clinical reports.[53] This was substantially more data than had ever been submitted for any previous drug application. Moreover, the quality of the data, in Klumpp's view, was also higher.[54]

Yet the data were hardly without flaws. Many of the case reports amounted to little more than the traditional physician "testimonials" – favorable case reports from clinicians who had "tried" the drug on their patients – rather than scientifically controlled studies. Even the studies by specialists were not necessarily free of bias since all of these investigators were to some extent indebted to the drug companies that had given them free DES research samples. Physicians' case reports to manufacturers not uncommonly closed with thinly veiled pleas for funding or supplies. One physician ended his letter to Sharpe and Dohme: "If the Lord will send me a guardian angel to pay for the pathological studies, which I cannot personally do, I will continue the investigation."[55]

The most serious flaw in the Master File, however, was not what it contained but what it omitted – namely, evidence from any animal studies. The drug manufacturers had conducted little if any independent research on either the carcinogenic or teratogenic effects of DES, even though the medical and scientific literature contained studies documenting both; nor did they include questions about such effects on the questionnaire they distributed to the experts. The entire file focused exclusively on clinical results in humans. The evidence that had previously evoked such concern regarding whether DES would cause cancer in the women taking it or what effects it might have on the offspring was simply ignored.[56]

The FDA's willingness to accept this evidence, carefully constructed by the drug manufacturers, as proof of DES's safety stems from a combination of factors. Certainly attitudes toward chemical risks in those days were more tolerant than they are in many quarters today. Confidence in the emerging powers of "scientific" new drugs, of which DES was only one example, was pervasive. Furthermore, DES was assumed to be no more dangerous than natural estrogens already on the market, despite its greater potency, and it was a good deal cheaper and easier to use. (Of course, the natural estrogens had been approved before the passage of the 1938 Act requiring proof of safety, but that fact was usually overlooked.) In addition, proponents of DES had succeeded in bringing considerable pressure to bear on the FDA from the medical profession as well as the drug industry. For these or other reasons, FDA officials not only accepted the Master File as sufficient but felt they had subjected DES to the most comprehensive and demanding review to date.

And so they may have. Unfortunately, they were willing to permit the exclusion of the very evidence that signaled the tragedy now in the making.

FDA approval and the failure of regulatory safeguards

The Small Committee forwarded the Master File to Carson Frailey on May 28, 1941, for submission to the FDA, and the applications of the twelve cooperating firms became effective on September 12, 1941.[57] The Small Committee's work had paid off. "I think we can congratulate ourselves," Hines wrote to one of the participating companies in dissolving the Committee, "that [the approval of Stilbestrol] came when it did."[58] Klumpp left the FDA in June 1941, shortly after the agreement on DES had been reached, and spent six months with the AMA's Council on Pharmacy and Chemistry. He then became president of Winthrop Chemical Company, one of the firms involved in the DES Master File. In June 1942, DES received the AMA Council's stamp of approval, although Klumpp professed to have had nothing to do with this.[59]

The FDA's approval of DES included various restrictions. First, it was approved for the treatment of four – and only four – indications: menopausal symptoms (the most common use), gonorrheal vaginitis, senile vaginitis, and suppression of lactation after pregnancy. Other "experimental" uses could not even be mentioned in product literature. The FDA was fully aware that many specialists were using DES for a wide range of unapproved indications, but did not want to encourage experimentation by general practitioners, who were generally considered less qualified than specialists to evaluate new drug therapies.[60] Second, tablet strengths were limited to 5 milligrams (mg) or less, since higher dosages were still viewed as dangerous. Finally, warnings were required on DES packaging stating: "CAUTION: To be used only by or on the prescription of a physician." Product literature describing treatment regimens and contraindications was provided to physicians, but not to patients – in fact, the law strictly prohibited companies from giving patients any information at all on the uses or risks of prescription drugs lest they be tempted to engage in the then–highly disapproved of practice of "self-medication."[61] The assumption was that only physicians could be trusted to exercise proper medical judgment and restraint, and that they would help ensure the safe use of DES.

These restrictions were intended to limit the risks DES might pose for patients. To many officials, they represented a "conservative" approach to this admittedly potent drug.[62] Nevertheless, an important precedent was established: DES became the first potentially dangerous drug approved under the 1938 Act which did not pretend to be lifesaving – as the sulfa drugs were, for example – but merely life-enhancing.[63] "Safety," as it was defined for DES, depended not so much on the actual hazards involved – which were known to exist – as on the adequacy of intended safeguards. Yet these safeguards, as we shall see, would soon crumble under the pressure of expanding clinical usage.

The approval of DES also revealed the subtle interplay between the FDA's judgments about drug safety and efficacy. Although formally charged under the 1938 Act with reviewing only drug safety, in fact the FDA routinely considered efficacy as well in its assessments. "We do not tolerate reactions of the same character and extent where a drug has some insignificant therapeutic effect," Klumpp told a meeting of the American Pharmaceutical Manufacturers Association in 1941. He went on to illustrate:

> the drug arsphenamine [an arsenic compound] may be considered a safe drug in the treatment of syphilis, even though it kills occasionally and causes many instances of damage. On the other hand, would we consider the intravenous use of arsphenamine a safe treatment for dandruff? I think not.[64]

Klumpp later confirmed that in approving DES, the FDA had deemed it *both* "safe and effective" for the four indications listed.[65] Since approval was limited to those four indications, the FDA presumably did not, at that time, consider DES safe and effective for other uses. But once it was on the market, physicians were free to use it for whatever purpose they chose; the 1938 Act (and indeed FDA regulations today) explicitly avoided any interference with "the practice of medicine."[66] As the case of DES would quickly demonstrate, the problem of the unapproved use of approved drugs was, and remains, one of the major loopholes of federal drug regulation.

Predictably perhaps, the FDA's restrictions on DES proved no match for the growing interest in this potent and inexpensive new compound. The public appetite had been whetted by tantalizing reports from medical meetings hailing the discovery of new "age-reversing" hormones, which, according to one *New York Times* story, were capable of "pushing back the calendar an equivalent of twenty-five years."[67] Hopes and expectations for DES ran high. An article in the AMA-sponsored journal *Hygeia* entitled "Help for Women Over 40" announced the release of DES and told of the "extraordinary clinical possibilities" offered by this "sensational drug" which could "mitigate suffering for millions of women."[68]

Neither before nor after its approval by the FDA did DES usage remain confined to the four specified indications. By 1941, studies had appeared in the medical literature describing the use of DES in the treatment of over eleven different conditions, including, in addition to the approved four, amenorrhea, dysmenorrhea, hypoestrinism, uterine bleeding, pregnancy complications, mental problems, and a group of "miscellaneous" cases.[69] With the release of DES for general use, such experimentation became even more widespread, FDA restrictions notwithstanding. In 1943, Dodds, the creator of DES, announced that he had used it successfully to cure prostate cancer in men (one of the few uses of DES still approved today). Clinicians were anxious to explore the possible benefits of this new "wonder drug" for everything from female hormonal problems to chronic arthritis (improvements were noted in both). One enterprising physician even used it to stunt growth in adolescent girls who seemed to be getting too tall![70]

Not surprisingly, pharmaceutical companies actively encouraged such experimentation in the hope of expanding the drug's prospective market. Although unapproved uses could not be listed on drug labeling, the manufacturer's "detail men" – sales representatives who visited physicians and hospitals personally to promote company products – could and did suggest new applications to physicians. To keep their detail men apprised of new developments, some companies compiled internal summaries of the latest research findings. By late 1943, for example, the "McNeil-O-Gram," an inhouse sales publication of McNeil Laboratories, had described the use of DES for such ailments as prostate cancer, feminine "psychosis," nausea and toxemia of pregnancy, miscarriages, extrauterine retained placenta, infantilism, amenorrhea or hypoplasia of the genitalia or breasts, and even herpes.[71]

The burgeoning use of DES and other hormones in the early 1940s, for almost every conceivable female disorder, did lead to renewed warnings. "A tidal wave of endocrine therapy is sweeping the country," one endocrinologist observed with dismay in 1945. "I am expecting daily to hear a radio announcer say, 'Get estrogens at your neighborhood drug store, they are great for your hot flashes or any other ills peculiar to the female sex.' "[72] An AMA Council member complained about "the flood of advertising material constantly washing up on [the physician's] desk" promoting one or another estrogen.[73] Articles in the medical literature charged physicians with "overzealousness" in their use of the new preparations, "indiscriminate and excessive dosage with diethylstilbestrol," and "ill advised and injudicious endocrine therapy." If such indiscriminate use continued, one expert warned, "it is reasonable to expect that sometime a growth, perhaps a malignant growth, may result."[74]

For the most part, however, these warnings did little to dampen the medical community's enthusiasm for DES and related hormones. Clinical investigators had found a powerful new therapeutic key to the female reproductive system and were anxious to find out just what secrets it might finally unlock. They were drawn, ineluctably, to the mystery of birth . . .

The new antimiscarriage drug

With the synthesis of DES in 1938, researchers were no longer hampered by the scarcity of natural estrogens, and experimentation in treating problems of pregnancy had begun almost immediately. The American pioneer of this use was a Houston gynecologist named Karl J. Karnaky.[75]

The idea of using DES in pregnancy apparently came not from Karnaky but from scientists at E. R. Squibb and Sons, one of the original DES manufacturers. Karnaky's research had led him to believe that premature labor and miscarriages were due to inadequate estrogen levels and thus might be prevented by estrogen replacement. Squibb representatives did not have to do much arm-twisting to convince him to use free samples of DES instead of natural estrogens. Two Squibb researchers "came to Houston, fed me and dined me . . . and I started using it," Karnaky recalled.[76]

The initial results were not encouraging. Karnaky began giving DES to dogs, but soon found, as he put it, that "the dang dogs were dying like flies."[77] Nevertheless, when he told Squibb officials he was going to abandon these experiments, they urged him not to give up and to try DES on women instead, and sent him experimental doses of up to 25 mg directly from Dodds in England. And indeed, DES did work better on women than dogs; not only did it not kill them, but injections of the drug directly into the cervix appeared to stop many cases of unexplained uterine bleeding. So "a rule was made" at his hospital, Karnaky reported, "that all women coming to the Emergency Room, Gynecological or Menstrual Disorder Clinic were to receive an injection."[78]

Observing that DES seemed to inhibit uterine contractions, Karnaky decided to see if it might also help prevent threatened abortion, although he realized that this ran contrary to the findings of earlier research that estrogens were toxic to fertilized eggs. "Much to our surprise," he wrote, "labor pains stopped almost immediately or within 30 to 60 seconds" after the DES injections. "The stopping of labor pains by an estrogen is just opposite to what we have been taught to expect," he acknowledged, but offered no direct explanation for this apparent contradiction.[79]

Karnaky presented a "preliminary report" of his findings at a 1941 medical meeting, consisting of a series of case descriptions accompanied by largely unsubstantiated assertions that DES had improved pregnancy maintenance without harm to the fetus. Although dosages varied, most were quite high; some women received up to 6,000 mg of DES. "We can give too little stilbestrol," Karnaky marveled, "but we cannot give too much."[80] Physicians asked to comment on his paper were more guarded, but only one was openly skeptical. Here was a drug that looked like it just might help save babies. The worthiness of the goal may have helped mute criticisms of the less than rigorous evidence Karnaky had presented.

Karnaky's research on DES was soon overshadowed by work from a more prestigious source of biomedical investigation: Harvard Medical School. Well-regarded scientists affiliated with Harvard, George Smith, M.D., a gynecologist, and his wife Olive Watkins Smith, Ph.D., an endocrinologist, along with a colleague, Priscilla White, M.D., had also been exploring the role of endocrine therapy in the management of pregnancy complications. The Smiths published a series of papers beginning in the mid-1930s developing a seemingly plausible theory of how two critical hormones – estrogen and progesterone – interact during pregnancy.[81] According to this theory, elevated estrogen levels during pregnancy stimulate the secretion of progesterone, which is essential in preparing the uterus to receive and sustain the fertilized egg.

Deficient levels of either estrogen or progesterone would lead to complications or failure of the pregnancy. Although additional progesterone could be supplied directly, it was generally more feasible (and much less expensive) to stimulate the body's own progesterone production by giving the woman additonal estrogen. For both practical and theoretical reasons, the Smiths came to consider DES the most suitable progesterone-stimulating agent.[82]

Priscilla White applied the Smith's theory clinically to diabetic women, among whom problems of prematurity, miscarriage, and fetal death were relatively common. For women with an "abnormal hormonal balance," as she defined it, White used a combination of estrogen therapy, insulin, special diet, and early induced labor. She reported impressive results: fetal survival rates of over 90 percent for the treated women compared with 40–60 percent for untreated women with abnormal hormonal balances.[83] By 1940, White claimed that estrogen therapy for pregnant diabetic women was "no longer a physiologic curiosity but one well within the field of practical therapeutics."[84]

Encouraged by White's clinical results, the Smiths decided to test estrogen therapy on a much broader group of pregnant women whom they deemed, using a very liberal definition, to be at risk for complications. This included all women who had failed to conceive during two years of trying, who had any history of prior miscarriage or premature delivery, any bleeding during pregnancy, or fibroids, hypertension, diabetes, or previous reproductive surgery.[85] In a 1946 paper, the Smiths reported that the use of DES to stimulate progesterone production in such women looked "promising," and recommended "preventive" therapy with DES for accidents of both early and late pregnancy.[86] Based on data from a single patient, they recommended specific dosage regimens for prenatal therapy which quickly became the accepted standard: 30 mg of DES daily in the sixteenth week of pregnancy with an increase of 5 mg each week, reaching 125 mg daily in the thirty-fifth week. These were massive dosages of DES, hundreds of times the amount used in treating menopausal symptoms, but the Smiths reported "no evidence of harmful effects." They noted that their proposed regimen was "being tried by a number of individuals throughout the country and the results reported to us," adding, "We would appreciate case reports from any others stimulated by this publication to try it."[87]

There is no doubt that many physicians were indeed "stimulated" by the Smiths' article to try prenatal DES therapy. By the late 1940s, the use of large doses of DES to prevent "accidents of pregnancy" had gained widespread acceptance among clinicians, despite the absence of FDA approval of either this use of the drug or the large dosages required.

The FDA approves DES for use in pregnancy

Physicians were not the only ones who responded to the Smiths' call for experimentation with DES in preventing problems of pregnancy. With the postwar baby boom now well underway, drug companies were eager to supply what promised to be a major new market for DES.

At first manufacturers furnished 25-mg tablets of DES as free "research" samples, but they soon decided that this new application was likely to be important – and lucrative – enough to justify regular marketing. Official FDA approval of this new use was required, however, since the original approval of DES had covered only four indications, none of them prenatal. Accordingly, beginning in the spring of 1947, many of the major DES manufacturers submitted supplemental applications to the FDA requesting permission to produce and market 25-mg tablets for prevention of complications of pregnancy. As evidence of safety, they cited primarily the work of Karnaky, White, and the Smiths.[88]

This work clearly represented the leading research on the prenatal use of DES, yet it was by no means universally accepted by all experts in the field. In discussions at medical meetings, commentators had raised questions about the work of all three investigators concerning possible biases in the selection of the patients treated, the possibility that the benefits observed resulted not from the hormone therapy itself but from "the excellent antepartum care" that the patients received, the fact that untreated patients had not been given a placebo treatment (an identical-looking sugar pill given to a group of subjects in an experiment to reduce spurious psychological effects), and other methodological criticisms that might account for what one speaker termed the "rather startling statistics" reported. Several commentators recalled the lethal effects of estrogens on animal fetuses. One expressed "a degree of bewilderment" that DES was recommended for pregnant diabetic women since it seemed likely only to exacerbate the problems they already had in delivering overly large babies. To some experts, the whole concept of hormonal management of threatened miscarriage was at best still controversial. "We speak of threatened abortion as if it actually were an endocrine disease," one physician noted. "It may be, but I do not think there is sufficient evidence yet to draw that conclusion."[89]

There were also significant inconsistencies among the Karnaky, Smith, and White studies that might have raised additional questions within the FDA about this proposed new use. Karnaky and the Smiths used DES alone; White used DES in combination with progesterone. Moreover, Karnaky claimed that massive injections of DES into the cervix would stop miscarriages already underway, whereas the Smiths recommended graduated doses of DES taken orally, early in pregnancy, to *prevent* later miscarriages and other complications. Indeed the Smiths themselves had pointed out the inconsistency between their results and Karnaky's in one of their papers cited in manufacturers' supplemental applications to the FDA. "Although our findings indicate that diethylstilbestrol stimulates progesterone secretion," the Smiths had written, "it hardly seems possible that this effect would be an immediate result of local injection. We wonder if Karnaky's observations may not be due to some reflex nervous reaction, and if oil alone, so administered, might not elicit the same response."[90] The FDA let such inconsistencies pass without comment.

Again, however, the strongest indictment of this proposed new use of DES, as in the initial 1941 application, lay in what was not submitted to the FDA rather than in what was. By 1947 more than 300 studies had been published in the medical and scientific literature documenting carcinogenic effects of both natural and synthetic estrogens in animals and, under some conditions, in humans.[91] Yet no mention of the carcinogenic potential of DES appeared in the manufacturers' supplemental applications to the FDA.

Nor was there any hint of the warnings experts had continued to issue about the dangers of DES and other estrogens for the women receiving them. "It is hoped," a 1940 editorial in the *Journal of the American Medical Association* had stated pointedly, "that it will not be necessary for the appearance of numerous reports of estrogen-induced cancer to convince physicians that they should be exceedingly cautious in the administration of estrogens."[92] A 1942 review of the role of estrogens in carcinogenesis by a renowned scientist had concluded that "hormonal treatment at high levels for long periods is followed by appearances of atypical growths, including tumors and cancers in some experimental animals in such high incidences that the endocrine stimulation may play a considerable, rather than an incidental, role."[93] A paper published a few years later argued, moreover, that cancerous and precancerous changes were induced by "prolonged and *not necessarily excessive* administration of oestrogens" (emphasis in original).[94] A 1944 article in the *New England Journal of Medicine* had warned physicians "against indiscriminate and excessive dosage with diethylstilbestrol," concluding: "Since our knowledge of the possible carcinogenic role of the estrogens is still very incomplete, it is a wise policy to avoid unnecessarily large doses and to avoid such treatment altogether or hold it to a minimum for persons who [are particularly susceptible]."[95] (By 1947 Karnaky was giving some women as much as 24,050 mg of DES.)[96] And in 1947, the *Dispensatory of the United States of America*, an encyclopedia of medical and pharmacological information, issued a prophetic warning:

> To date no national catastrophy [sic] has been recognized, but it is perhaps too early for any deleterious effect on the incidence of carcinoma of the female generative tract or breast to appear.[97]

Also not mentioned in the manufacturers' supplemental applications was the possibility that DES could be hazardous not only to the women taking it but to the exposed fetus. By 1947 it was well known, at least among endocrinologists, that hormones and other substances could cross the placenta and affect the developing fetus.[98] Karnaky himself, in one of the studies cited by manufacturers, had observed that all of the DES-exposed babies carried to term "exhibited a darkening of the areolae around their nipples, labia, and linea alba, similar in intensity to that of their mothers, *indicating that this effect of diethylstilbestrol also is shared by the fetus*" (emphasis added).[99] Research on animals had shown that fetuses exposed in utero to exogenous hormones and other substances frequently developed various physical abnormalities. Beginning in the late 1930s, a number of studies on "experimental intersexuality" described what happened to animal offspring when pregnant females were treated with estrogens, including DES. The findings revealed, among other things, that female rat offspring developed enlarged uteri and structural changes in the vagina and ovaries, and that male rat offspring had small and inadequately developed penises and other sexual deformities.[100] Even the possibility of intrauterine exposure leading to carcinogenic changes in animal offspring had been demonstrated, albeit not specifically for DES or other hormones.[101]

The FDA could hardly have been unaware of the continuing controversy surrounding DES. The agency received a number of letters from concerned experts in the spring of 1947 recommending against approval of DES for use in pregnancy.[102] Furthermore, one of the studies cited in most of the later manufacturers' supplemental applications, although generally favorable to DES, stated specifically that there were "many questions which have not as yet been completely answered," including: "Is diethylstilbestrol in such large doses carcinogenic, and as such unsafe to give even to pregnant women?" and "Can diethylstilbestrol in any way affect the glandular balance of the child in utero, particularly the male child?"[103]

Had FDA officials been inclined toward skepticism, such objections and uncertainties might at least have given pause. But they showed no such inclination, accepting the Smith–Karnaky–White studies as sufficient proof of safety and approving most supplemental NDAs within a month or two of their receipt.[104] The approvals, issued from 1947 on, allowed manufacturers to market DES in tablets of up to 25 mg for the treatment of habitual abortion (i.e., miscarriage), threatened abortion, premature labor, and pregnancy complicated by diabetes (100-mg tablets were later approved). The only restriction placed on these new higher-dosage tablets was a required statement on their label to prevent them from being mistaken for the ordinary dosage: "Warning: This dosage is too great for the customary use of diethylstilbestrol."[105]

Implicit in the FDA's actions, once again, was the judgment that the newly approved uses of DES would be not only safe but also effective. Some proposed new uses were in fact rejected as clinically unverified. Officials refused to permit manufacturers to recommend DES for the treatment of preeclampsia, eclampsia, and intrauterine death, for example, explaining: "We are not aware of extensive work which clearly establishes these uses as sound."[106] It was not that such uses were any more dangerous than those that were approved – indeed, they could not have been, since all involved comparable dosages of DES over roughly the same time period; it was merely that the agency was not persuaded of their effectiveness. The FDA apparently had fewer reservations about the prenatal indications for which DES was approved. Yet even these uses, one official acknowledged to a manufacturer in 1947, were "decidedly in an experimental stage."[107]

The FDA's approval of prenatal DES was thus predicated on the assumption that the drug's clinical value, which was admittedly still uncertain, would be confirmed by subsequent evidence and experience. Put bluntly, the agency had authorized what amounted to mass experimentation on pregnant women – the sanctioned use of a drug with known risks whose effectiveness had not yet been fully proven. Yet only a short while later, as we shall see, the FDA would do nothing to rescind that authorization despite compelling evidence that DES was *not* in fact effective for its newly approved prenatal uses.

Prenatal use persists despite evidence of inefficacy

The popularity of DES grew steadily over the next few years as its array of reported achievements continued to expand. The popular press carried glowing stories of its success in treating not only female problems such as menopause and miscarriage but also of "dramatic" and "striking" results for prostate cancer and even mumps in men.[108] Prenatal use of DES underwent similar expansion. At a 1949 medical meeting, the Smiths announced findings that appeared to show that DES benefited *all* first-time pregnancies, even those with no history of previous problems. "The drug stimulated better placental function and hence bigger and healthier babies" in premature deliveries, they reported, while in pregnancies that went to term, DES seemed to "render normal gestation 'more normal,' as it were." In another study, the Smiths observed that "the placentas from stilbestrol treated patients were grossly more healthy looking and the babies unusually rugged. . . . "[109]

Although some medical experts remained skeptical, DES was well on its way to becoming a standard part of prenatal care among the nation's obstetricians. ("Recommended for routine prophylaxis in *ALL* pregnancies," read one advertisement in the mid-1950s.)[110] Sales of 25-mg tablets rose rapidly in the early 1950s as more and more practicing physicians came to view DES as a virtual "panacea" for female ills.[111]

The growing popularity of prenatal DES therapy and the Smiths' escalating claims about its medical benefits against miscarriages prompted a new wave of research. Many of the results were disquieting at best. First the previously accepted theoretical rationale for prenatal DES therapy began to crumble. Several studies failed to confirm the Smiths' observation that DES led to increased levels of pregnanediol in the urine (thought to be an index of progesterone production), implying that DES did not in fact stimulate the body to produce more progesterone as the Smiths had theorized. If the Smiths' theories about DES effects were untrue, one group of researchers concluded, "its use in threatened abortion has no theoretical basis."[12]

Moreover, a number of clinical trials cast serious doubt on the actual effectiveness of prenatal DES therapy in practice. By 1952, four separate studies had concluded that DES offered no advantages in cases of threatened miscarriage over such alternatives as progesterone injections, bed rest and sedation, or even no treatment at all.[13] These studies – unlike others with favorable results also published during the same period – all incorporated methodological improvements that reflected the growing sophistication of medical research. As noted previously, the Smiths had been criticized at medical meetings for failing to compare prenatal DES therapy with an analogous placebo therapy given to a "control group" of randomly selected, equivalent patients. Only such a controlled study can eliminate all of the extraneous variables that might affect the outcome, including the extra attention that an investigator might unwittingly give to patients receiving an experimental treatment. As Dr. James Ferguson, the author of the most carefully controlled of the four negative studies of DES, observed: "It is impossible to give the same obstetric care to women treated with a special medicine as you give to patients who do not receive that medicine. The special group will inevitably receive more careful thought, more time, more sympathy and, unconsciously, decisions on management may be swayed."[14]

Ferguson's methods and negative conclusions were persuasive. Following his presentation at a 1952 medical meeting, one commentator remarked: "Dr. Ferguson has, I believe, driven a very large nail into the coffin that we will use some day to bury some of the extremely outsized claims for the beneficial effects of stilbestrol."[15]

Seven months later, at the annual meeting of the American Gynecological Society at Lake Placid, N.Y., Dr. William Dieckmann presented results that should have nailed the coffin shut. Dieckmann and his colleagues at the University of Chicago had conducted the largest and most methodologically sophisticated trial of DES to date, comparing 840 women who had been given DES with 806 matched controls who had received an identical-looking placebo. To prevent any possibility of subjective bias, the study was "double-blind": Neither the patients nor the investigators knew which patients were getting the placebos and which DES. (Some of the patients later claimed they were told they were getting "vitamins" and were unaware of participating in a study at all, as discussed in Chapters 7 and 8.) The results of this painstaking study seemed unequivocal: DES did not decrease the incidence of miscarriages, prematurity, or toxemia, nor did it increase the size of premature babies or improve perinatal mortality. In sum, the authors concluded, DES had "no therapeutic value in pregnancy."[116] They could have gone further: A later reanalysis showed that the data reported actually indicated that DES had significantly *increased* miscarriages, neonatal deaths, and premature births.[117]

The Smiths and their supporters at the Lake Placid meeting sprang to the defense of DES. The Dieckmann study was not really comparable to theirs, the Smiths argued, because it included some women who were not pregnant for the first time, whereas their final study had been limited to first-time pregnancies. (They ignored the absence of such a limit in their earlier work and the fact that Dieckmann's results for first-time pregnancies were comparable to those for the whole group.) They also objected to the University of Chicago team's inclusion of patients who experienced early miscarriages or who had a history of late miscarriages on the grounds that these women could not be helped by prophylactic treatment with DES. To some in the audience, such arguments had a distinctly hollow ring compared with the solid and meticulous evidence Dieckmann had presented. Indeed, the Dieckmann study would eventually come to be viewed as decisive in discrediting prenatal DES therapy. But the Smiths had many supporters, especially in the Boston area, who needed more convincing. "As a former Bostonian," one physician declared defending the Smiths' conclusions, "I would be entirely lacking in civic loyalty if I had not used stilbestrol in my private practice."[118]

Although other studies soon provided additional evidence that DES was useless in preventing miscarriages, it continued to be a popular prenatal therapy among practicing obstetricians for many years.[119] Sales peaked in 1953, the year the Dieckmann study was published, but it is estimated that doctors in the United States wrote an average of 100,000 prescriptions for DES per year for pregnant women throughout the 1960s. At these rates, somewhere between 20,000 to 100,000 babies (depending on assumptions about DES usage and fertility) were exposed to DES each year for nearly two decades after it had been shown beyond any reasonable scientific doubt to be ineffective.[120]

Why were physicians so slow to abandon this now highly controversial drug? For one thing, the Smiths still ardently defended its use and many clinicians were undoubtedly at a loss to know which of the contending experts to believe.[121] Many others were probably too busy with patient care to keep abreast of the medical literature or follow the debate unfolding at professional meetings. Even then, it was difficult for practicing physicians to keep up with the steadily expanding array of new drugs. In a 1949 editorial entitled "Too Many Drugs?" the AMA's Council on Pharmacy and Chemistry warned that "the gold rush to stake out claims and make a killing" in new fields such as "estrogenic preparations" interferes "with the acquisition of the precise information that is essential to the proper evaluation of the scope and of the dangers of new medications."[122] With a drug like DES, such information was not likely to come from a physician's own clinical experience, as it was too easy to attribute successful outcomes to the DES therapy rather than to any of the numerous other factors that might have been involved – bed rest, sedation, or simply good luck. Many busy practitioners probably continued to prescribe DES because they thought it could do no harm and might possibly help.

That impression was deliberately fostered by DES manufacturers. Drug companies remained largely silent about the recent negative findings on DES. No word of these findings appeared in the *Physicians' Desk Reference* (PDR), a manual distributed free of charge annually to practitioners containing manufacturers' descriptive information about their products. Every year until 1968, DES appeared in the *PDR* index as an indicated treatment for "habitual and threatened abortions," and the product descriptions listed "pregnancy accidents" as one of the indicated uses.[123] Then as now, the *PDR* was a major source of therapeutic information for practicing physicians. A 1960 survey found that over 90 percent of the doctors interviewed kept the *PDR* on or near their desks, and over half of the busiest doctors consulted it at least once a day.[124] The literature distributed to physicians by the pharmaceutical companies was equally lopsided. For example, Lilly's 1953 brochure on prenatal DES devoted several pages to a detailed description of the positive evidence, all based on studies from the 1940s, and only a single sentence to the later, better-controlled studies reporting negative results.[125] Such a presentation was clearly designed to sell DES, not to inform physicians.

Drug company detail men were probably even more influential. Numerous studies have shown that detail men are the physician's most common and valued source of information about drugs.[126] Unfettered by the FDA's constraints on written information, detail men were free to slant their discussions of DES in any way they chose, and it is safe to assume that many continued to advocate prenatal therapy. It was not uncommon for detail men to make what one manager euphemistically termed a "real expanded claim" about products (i.e., gross exaggerations of therapeutic benefit), targeting doctors known to be high prescribers. "Tell 'Em Again and Again and Again – Tell 'Em Until They're Sold and Stay Sold," exhorted one inhouse sales document.[127] According to a medical director for Squibb, the unstated rule was that "anything that helps to sell a drug is valid, even if it is supported by the crudest testimonial, while anything that decreases sales must be suppressed, distorted and rejected because it is not absolutely conclusive proof."[128] A letter from one of Squibb's detail men to his home office suggests just such an approach to DES:

> I got into a discussion today at Jewish Memorial Hospital with a group of outstanding Gyn. & Obs. men. . . . These men are all in agreement that the Smith & Smith approach to the problem of threatened abortion with increasing doses of Stilbetin [Squibb's brand of DES] is of no value at all. Dr. Goodfriend has been in touch with outstanding men over the country and he says they all hold the same opinion. . . . For myself I will continue plugging Stilbetin.[129]

If many physicians failed to appreciate the significance of the studies casting doubt on prenatal DES therapy, the lay public was kept even more in the dark. Popular journals continued to carry optimistic reports about DES long after Dieckmann and other investigators had exposed it as worthless. For example, a 1960 article in *Good Housekeeping* entitled "Why You Won't Lose Your Baby" stated: "today many obstetricians are convinced that hormonal treatments should be an integrated part of any concentrated attack on miscarriage."[130] Reading such things, many pregnant women naturally sought these treatments from their doctors, putting additional pressure on practitioners to continue prescribing prenatal DES. "The public has been so frequently told of the virtue of this drug," wrote two critics, "that it now requires a courageous physician to refuse this medication. The mass of pharmaceutical literature, extolling the wonders of this drug, has also rendered most practitioners amenable to his [sic] patient's demands." This may have been a bit of an overstatement; patients are typically quite subordinate to physicians, especially female patients to male physicians. Nevertheless, it does suggest that patients, too, were caught up in the tide of pro-DES sentiment.[131]

The FDA took no visible part in the growing debate over pre-natal DES therapy. From its past actions, in which drug efficacy had played an integral – and acknowledged – role in judgments about safety, one might reasonably have expected some response from the FDA to the news that the benefits of DES in pregnancy were at best dubious. Certainly the risks of DES had been amply documented in animal studies. Although DES could not be taken off the market as an "imminent hazard" since it had other, valid medical uses, the FDA could have removed accidents of pregnancy from the list of approved indications, or insisted on a more objective account of efficacy from manufacturers, or even forced companies to send out what later became known as "Dear Doctor" letters, correcting previous misinformation given to physicians.[132] Instead, the FDA did nothing. Under the 1938 Food and Drug laws, the FDA was still formally responsible only for ensuring drug safety, not efficacy, and no compelling proof of human harms from DES had yet been presented. The new FDA Commissioner, George P. Larrick, an appointee of President Eisenhower, had a close and cordial relationship with the drug industry and had no desire to pick any unnecessary fights.[133] Besides, the still understaffed FDA had its hands full simply keeping up with the 100–200 new drug applications it received each year.[134] It was in the agency's interest, as Commissioner Larrick defined it, to interpret its legal mandate narrowly, which meant ignoring evidence of inefficacy.

In short, almost all of the parties involved – doctors, drug companies, patients, and the FDA – had a stake in prenatal DES living up to its claimed benefits as well as their own reasons for dismissing the discomforting evidence of its possible dangers. It was this unfortunate convergence of perceived interests that helped perpetuate prenatal DES therapy long beyond the time when, by all rights, it should have ceased.

Plumper chickens, bulkier beef

As human uses of DES multiplied during the 1940s and 1950s, another entirely different role for DES was also gaining increasing popularity. Agricultural researchers had discovered that, just as DES seemed to produce bigger babies, so too it could be used to fatten up chickens and other animals being bred for slaughter. DES had been approved for use in veterinary practice by the FDA in 1942, a year after its release for human use, and in 1947 was approved as a growth promotant for chickens.[135] With the FDA's approval, poultry producers began to implant in the necks of young chickens 15-mg DES pellets, which slowly dissolved and acted as a chemical castrator on male birds, turning them into the tender, plump equivalents of capons, which fetched a premium price. Farmers had long been castrating male birds surgically in order to eliminate the production of male sex hormones associated with leaner, harder bodies; DES did the same thing chemically, but faster and with greater total weight gain.

To prevent people from consuming any remaining DES residues, the heads and necks of these "caponettes," as they were called, were removed before marketing. The heads and necks were then sold, at the Department of Agriculture's suggestion, to mink ranchers as cheap food for their minks. But mink ranchers soon discovered that the DES-fed chicken wastes were making their female minks sterile, and in 1950 brought charges against the Department of Agriculture. There were also reports that male kitchen workers who consumed large quantities of chicken necks were being "caponized" and developing large breasts.[136] The Canadian government banned the sale of DES-fed chickens, but marketing continued in the United States – indeed, agricultural scientists were already exploring the use of DES in other animals. In 1954, in response to studies showing that steers fed 5–10 mg of DES gained weight 35 percent more rapidly than usual and consumed 20 percent less food, the FDA approved the addition of DES to cattle feed.[137] Approval of DES ear implants for cattle and also DES implants and feed additives for sheep and pigs followed shortly thereafter.[138]

Cattle men rushed to take advantage of this moneymaking new discovery. "Only two months ago the first feed-lot cattle started getting 'stilbestrol-added' supplements," the March 1955 issue of *Farm Journal* announced. "Right now *close to two million head* are getting the stuff in their daily feed. Amazing? You bet it is! Nothing has ever hit the meat-animal business with the impact of stilbestrol."[139]

Farmers had joined doctors in the national infatuation with DES, science's latest wonder drug.

In Congress, meanwhile, the growing use of chemicals in food production was generating increasing concern about food safety in general, and cancer in particular. In 1951 Representative James Delaney (D–New York) held a series of hearings on food safety, initiating a decade-long campaign against chemical contamination of foods. Several witnesses attacked the use of DES to fatten poultry, citing possible carcinogenic risks to humans.[140] And there were further unsettling disclosures. In late 1955, a report appeared in the journal *Science* warning that inadvertent contamination of mouse feed with DES in a mill previously used to prepare cattle supplements had led to "serious reproductive disturbances" in the mice.[141] At a 1956 symposium on medicated feeds, several scientists urged against feeding livestock DES, citing twenty years of experiments which showed that continuing administration of minute doses of estrogen was actually *more* effective in inducing cancer than intermittent injections of larger doses.[142] In 1957, as part of his effort to enact legislation controlling chemical additives in foods and cosmetics, Representative Delaney placed a letter in the Congressional Record stating that DES residues in marketed poultry were 342,000 times the level found to cause cancer in mice and constituted a hazard to consumers.[143]

These troubling disclosures had no discernible impact on farmers – or, for that matter, doctors. By 1970, nearly 75 percent of all U.S. cattle were being given DES.[144] And the medical community seemed unperturbed – if it was even aware – that one of its favorite drugs for pregnant women was being attacked as a human carcinogen. In both arenas, trouble was clearly brewing, but for the most part the signposts of danger were rationalized, discounted, or simply ignored.

In Congress, however, pressure was building to strengthen federal food and drug regulations. Between 1956 and 1961, Congressional appropriations for the FDA tripled and the agency doubled in size.[145] In 1958, Congress passed the first of a series of laws to expand the FDA's powers, adopting, at Representative Delaney's insistence, what came to be called the Delaney Clause, which prohibited the use of any food additive "found to induce cancer when ingested by man or animal."[146] Few realized at the time how controversial this prohibition would become as more sensitive tests for detecting residues were developed and the list of cancer-causing substances grew.

One of the first targets of the Delaney Clause was DES-treated chickens. By 1959, improved testing procedures had revealed small amounts of DES residues in the skin and liver of caponettes treated with stilbestrol. Under the Delaney Clause, the FDA had no choice but to ban the use of DES in fattening fowl. (The Department of Agriculture obligingly bought up all of the caponettes already on the market and distributed them, after appropriate processing, to "welfare agencies.") In announcing this ban, the FDA gave public notice that

> the drug diethylstilbestrol is capable of producing and has produced cancer in animals and that this drug may be expected to produce, excite or stimulate the growth of certain cancers in human beings.[147]

The use of DES to fatten cattle and sheep continued, since no residues had yet been detected in these meats. Cattle growers were nevertheless apprehensive, fearing that it was only a matter of time before the ban was extended to beef. By 1962, they had generated enough support in Congress to obtain special legislation modifying the Delaney Clause, designed specifically to allow them to continue using DES. Under this new legislation, known carcinogens could be given to meat animals so long as no residue remained in the meat when the chemical was used according to label directions that were "reasonably certain to be followed in practice."[148] For DES, this meant withdrawing feed additives at least 48 hours before slaughter. Much of the looming battle over the use of DES in livestock would revolve around this important loophole.

The loophole remained unused until 1965, because only then did the FDA or the Department of Agriculture (USDA) begin testing meat regularly for residues. Even then only a tiny fraction of the cattle slaughtered annually were tested (1,000 out of 30 million). Furthermore, the testing methods employed were capable of detecting DES only down to levels of 10 parts per billion (ppb), yet DES had been shown to cause tumors in mice at feeding levels of 6.3 ppb.[149] (Indeed, a proven "no effect" level has still not been discovered.) So even meat deemed free of residues by these methods might still contain potentially hazardous amounts of DES.

Residues did turn up in beef tested every year between 1965 and 1970. However, under the special DES loophole, this did not necessarily mean that the required 48-hour withdrawal period was insufficient; some cattle growers might simply not be obeying it. Yet instead of strengthening enforcement procedures, in 1970 the USDA cut its residue testing program in half, and officials admitted that they could no longer assure consumers that the meat they were eating was hormone-free. At the same time, the FDA doubled the permissible amount of DES that could be given to cattle, from 10 mg to 20 mg per day.[150]

Key members of Congress watched this apparent collapse of regulatory control over the agricultural use of DES with dismay. Representative L. H. Fountain (D–North Carolina), Chairman of the House Intergovernmental Relations Subcommittee of the Committee on Government Operations, was already alarmed at the escalating use of medicated animal feeds, and held hearings in March 1971 on the regulation of feed additives. DES was a major topic. Much of the testimony underscored Fountain's concerns, especially that of a public interest lawyer who informed the committee that the use of hormones for animal growth promotion was banned in twenty-one countries.[151] The new FDA Commissioner under President Nixon, Charles Edwards, admitted that the agency did "not have a good control method for detecting DES residues in meat," and, under questioning, FDA officials promised to increase the residue sampling program using improved methods.[152] There things might have rested but for the fateful discovery that Dr. Arthur Herbst and colleagues would report the following month.

The FDA – reluctant regulator

A similar process of expansion and retreat during the 1960s occurred in the regulation of human drugs. In 1959, Senator Estes Kefauver held a series of Congressional hearings on profiteering and other abuses in the pharmaceutical industry. These hearings attracted wide attention, but Kefauver could get little support for proposed drug reform legislation until – as in 1938 – disaster struck, this time in the form of thalidomide. Shock and horror swept the nation at the news that grossly malformed babies were being born in Europe to mothers who had taken thalidomide, a sedative, during pregnancy. Even though the United States had largely escaped this latest tragedy, there was a groundswell of support for stronger drug regulations. A compromise bill was drafted, known as the Kefauver–Harris Amendments to the 1938 Food and Drug Act. In 1962 President Kennedy signed these amendments into law.[153]

The 1962 Amendments strengthened drug regulations in various ways, the most important being that, for the first time, new drugs had to be proven effective, as well as safe, before they could be marketed. Claims of effectiveness could not be based merely on masses of less-than-scientific physician "testimonials," which had comprised the bulk of earlier new drug applications, but had to be supported by "substantial evidence" from "adequate and well-controlled investigations . . . by experts qualified by scientific training and experience to evaluate the effectiveness of the drug involved."[154] The 1962 Amendments also gave the FDA closer control over the distribution and testing of experimental drugs before they were approved and the power to withdraw approval from marketed drugs for specified reasons.

The new efficacy requirements applied not only to future drug approvals but also to the roughly 4,000 existing drugs, including DES, that had been introduced since 1938. All of these drugs were to be reviewed to determine whether they were effective according to the newly adopted standards; any that were not would be withdrawn from the market. The review was not to start until 1964, in order to give both the FDA and manufacturers time to prepare for the monumental and politically explosive task ahead. Drugs used in pregnancy, FDA Commissioner Larrick assured Congress, would be given "priority attention."[155]

But the burst of regulatory zeal in human drug regulation seemed to run out of steam following the passage of the 1962 Amendments, just as efforts to expand animal drug regulation had in the mid-1960s. In 1963, with the memory of thalidomide still fresh in many minds, the FDA announced that it was contemplating "a series of significant new steps to strengthen the public's protection against potentially birth-damaging drugs," but that it "intend[ed] to move slowly because of the scientific uncertainties surrounding . . . drugs that damage the human embryo."[156] And slowly indeed it moved. By the spring of 1966, only token progress had been made on the drug efficacy reviews mandated by the 1962 Amendments. To speed up the process, the FDA, under the leadership of a new, reform-oriented Commissioner, contracted with the National Academy of Sciences (NAS) to perform the required evaluations.[157]

With over 16,000 different therapeutic claims to consider, this was a formidable undertaking. Thirty panels were assembled, each consisting of six medical or scientific experts. All therapeutic claims were to be rated "effective" or "ineffective," or, where the evidence was ambiguous, "probably" or "possibly" effective. With intermediate ratings, manufacturers had six to twelve months to submit additional evidence once the FDA had published the NAS evaluations in the *Federal Register*.[158] But, harassed by legal challenges from drug companies and other delaying tactics, the FDA released the NAS ratings very slowly, a few at a time. Congressional critics and consumer groups accused the FDA of deliberate foot-dragging to protect drug manufacturers threatened with the loss of profitable drugs, a charge the FDA Commissioner vigorously denied. By June 1970, however, only about 15 percent of the ratings had been published. A coalition of public interest groups brought suit, charging the FDA with violation of the Harris–Kefauver Amendments. The DES rating was one of those not yet published, even though the FDA had received the NAS's evaluation of it in 1967.[159]

The NAS panel reviewing prenatal DES rated it "possibly effective," although the panel acknowledged "that its effectiveness cannot be documented by literature or its own experience." Why the panel did not simply rate it ineffective given the compelling evidence provided by the Dieckmann study (which the panel cited) and other controlled trials remains a mystery.[160] By 1969, even Eli Lilly, the major manufacturer of DES, had stopped listing "accidents of pregnancy" as an indication for DES in the *Physicians' Desk Reference*. Regarding safety, the NAS panel concluded that DES was "not harmful in such conditions as threatened abortion." This, too, was far more sanguine than Lilly's own product literature. From 1969 on, Lilly's descriptions of DES in the *PDR* included the following statement:

> WARNING: Because of possible adverse reaction on the fetus, the risk of estrogen therapy should be weighed against the possible benefits when diethylstilbestrol is considered for use in a known pregnancy.[161]

Although many physicians undoubtedly discounted such risks, the NAS panel's blanket assertion of safety was still surprising in light of the known cancer-causing ability of DES and the animal studies demonstrating its adverse effects on exposed embryos – not to mention the thalidomide disaster, which had so vividly illustrated the damage drugs can cause in utero.

Yet even this most generous of evaluations could have given the FDA an opening to ban the use of DES in pregnancy. It had only to release the NAS rating of DES, wait the required six months for manufacturers to submit additional proof of efficacy, and then withdraw approval. Had the FDA moved quickly when it received the NAS rating in 1967, DES could have been contraindicated for use in pregnant women by 1968. But it did not. By the spring of 1971, the FDA still had not released the DES rating, and prenatal use continued with full FDA approval.

Another use for DES: from pregnancy enhancer to pregnancy terminator

While the FDA struggled with its new responsibilities in the 1960s, DES was finding yet another niche for itself in the American landscape quite separate from its use in livestock and in pregnant women. It became the world's first postcoital contraceptive.

That this development occurred in the sexually liberated 1960s was no coincidence. Attitudes toward birth control had loosened up dramatically from the prewar years. For some, contraception had become a matter of women's rights, for others, the key to survival in a world of rapidly increasing population. Government and private funding flowed into research on contraceptive methods, and researchers began again to explore the long-unexploited ability of estrogens to terminate early pregnancies.

This time Yale, rather than Harvard, led the way. Building on earlier animal studies indicating that estrogens could prevent the implantation of fertilized ova, two researchers at Yale, John Morris and Gertrude van Wagenen, theorized that DES might act as a contraceptive agent in humans if administered in large doses (50 mg/day for five days) shortly after intercourse. Given the now generally accepted clinical role of DES in *promoting* pregnancies, Morris and van Wagenen admitted that they began human testing "with some trepidation," but further experiments on some "courageous volunteers" convinced them that their theory was correct. In 1966 they published their initial findings in the *American Journal of Obstetrics and Gynecology*, and the following year reported further promising results.[162]

Fertility researchers hailed this exciting new form of after-the-fact birth control. "An effective postcoital antifertility agent... quite possibly could be the most significant achievement of the decade in population control," raved one physician. An article in the popular press called it "the biggest revolution yet in birth control," predicting that it might "make 'the pill' look like a minor event."[163]

Once again, the medical community enthusiastically embraced

what appeared to be another valuable new use for DES. Physicians on college campuses and in hospital and family planning clinics around the country began to try out this seemingly promising new remedy for unwanted pregnancies on their patients, using the 25-mg DES tablets already on the market. Technically, the 1962 Drug Amendments required clinical investigators to obtain FDA permission before testing new drugs – or new uses of existing drugs – on humans. With drugs already on the market, however, a certain amount of clinical experimentation was commonplace, and the FDA rarely interfered, especially if the physician's motive was primarily therapeutic. Thus, as with prenatal DES therapy in the early 1940s, in practice there was little to stop doctors from conducting what amounted to informal clinical research on their patients, and many did just that. One morning-after pill enthusiast even suggested that the simplest way to ensure an adequate supply of experimental subjects would be if "campus vending machines could dispense little 'six-packs' of stilbestrol tablets, each packet promising a refund if the user would simply report results promptly to a central registry."[164] By the early 1970s, the morning-after pill had become well enough entrenched, despite its lack of FDA approval, that it would survive largely unscathed the firestorm of controversy that was about to engulf DES.

The cancer link: local initiative, FDA inertia

Clear-cell adenocarcinoma of the vagina was a rare type of cancer before the 1960s, and almost never seen in women under age 50. In fact, in all the world's medical literature, only four cases had been reported in women under 30.[165] So when Dr. Arthur Herbst and his colleagues Howard Ulfelder and Robert Scully saw seven practically identical cases of this disease at Massachusetts General Hospital in Boston between 1966 and 1969, all in young women under 22, they were understandably baffled. They reported this mysterious cluster of cases in the journal *Cancer*, and enlisted the aid of epidemiologist Dr. David Poskanzer to research for possible causes.[166] In the meantime, they learned of an eighth case at a nearby Boston hospital.

The researchers undertook a classic retrospective, case-control epidemiologic study. For each of the eight cancer patients, they selected as "controls" four women who had been born within five days of the patient at the same hospital and on the same type of service (ward or private). A research team interviewed all of the patients, controls, and their mothers about a broad range of factors, including smoking, douching, and use of tampons or birth control pills, to see if there were any systematic differences. They found none. Then, one of the patients' mothers asked if the stilbestrol tablets she had been given during pregnancy might be the culprit. Her question solved the puzzle: Medical records showed that seven of the eight mothers had also taken DES during the first trimester of pregnancy.

This was a groundbreaking discovery. Never before had it been shown in humans that tissue changes caused by prenatal exposure could later become cancerous. It was well known that substances crossed the placenta and affected fetal tissue formation, and that cancer often developed long after exposure to a carcinogenic agent, but these two processes had never before been linked in humans. Furthermore, this discovery had direct implications for medical practice. The dangers of prescribing DES to pregnant women were no longer a matter of speculation.

Herbst and his colleagues immediately prepared a report for publication and submitted it to the *New England Journal of Medicine*. Recognizing the significance of the findings, the journal's editor, Dr. Franz Ingelfinger, decided not to wait for its scheduled publication in April but to alert the FDA immediately so the agency could take appropriate action. In March 1971, Inglefinger sent the FDA prepublication galleys of Herbst's article. A month later the FDA requested copies of the raw data from Herbst, which he promptly sent. That was the last either Herbst or Inglefinger heard from the FDA for some six months.[167]

Herbst's article appeared in the *New England Journal of Medicine* on April 22, 1971. In an accompanying editorial, respected epidemiologist Dr. Alexander Langmuir underscored the study's "great scientific importance and serious social implications." The findings add "a new dimension to the whole matter of what drugs are safe and unsafe to administer to pregnant women," Langmuir wrote, and advised physicians to "think more seriously before administering any drug to a pregnant woman."[168]

The editorial concluded with a call for a central surveillance system to collect information on other DES-related cancer cases. Herbst and his colleagues quickly set up a registry to gather data on all cases of clear-cell adenocarcinoma in the United States and around the world. The FDA did not respond to the editorial's appeal.

Concerned by Herbst's article, the Director of New York's Cancer Control Bureau, Peter Greenwald, checked the New York Cancer Registry for reports of clear-cell adenocarcinoma. To his amazement, he found five cases, all in women under age 30 born between 1951 and 1953. Greenwald immediately began reviewing these women's hospital records and interviewing their mothers, and discovered that all of the mothers had been given DES or another synthetic estrogen during pregnancy. Greenwald's findings were published in the *New England Journal of Medicine* on August 12, 1971, four months after the Herbst report, providing important independent corroboration of Herbst's discovery.[169]

Greenwald's study drew a swift reaction from New York officials. In June, before the paper was published, Dr. Hollis Ingraham, New York State's Commissioner of Health, wrote to all of the practicing physicians in New York warning them of the danger of estrogen administration during pregnancy and asking for their cooperation in "this life-preserving campaign."[170] At the same time, he established a monitoring and surveillance program for the state to collect information on additional cases of adenocarcinoma and other types of cancer that might be related to DES.

Ingraham wrote to FDA Commissioner Edwards in June 1971 notifying him of Greenwald's data and of his letter to New York physicians. "We also recommend most urgently" he wrote, "that the Food and Drug Administration initiate immediate measures to ban the use of synthetic estrogens during pregnancy." Ingraham informed Edwards that New York had established "a program of continuous surveillance and monitoring," and urged that "similar surveillance and monitoring programs be conducted nationally."[171] An FDA representative acknowledged receiving the letter, but no one else in the agency contacted either Greenwald or Ingraham again until early August, and then only to ask for more data. Greenwald immediately mailed complete information on all of New York's DES-related cancer patients, and heard from the FDA no further.[172]

Following the Herbst and Greenwald reports, a number of medical journals carried editorials stressing the significance of this new discovery, calling it a "most dramatic development in cancer research," a "time bomb for child," a "stunning observation."[173] Still the FDA remained silent. An internal FDA staff analysis of the Herbst report, completed in April 1971, had recommended deferral of changes in the labeling for DES use during pregnancy "until such time as the raw data can be obtained and thoroughly analyzed."[174] The raw data had arrived just two days later, but the FDA had apparently conducted no further analyses and took no action until, quite literally, it had no choice.

Congress intervenes

As summer stretched into fall, Representative Fountain and others in Congress grew increasingly impatient with the FDA's lack of response to the alarming data it had received. To make matters worse, in October 1971, further evidence of regulatory failure in the livestock arena came to light. Since the Fountain committee's March hearings, USDA officials had reported finding no DES residues in the meat sampled, and in August the Assistant Secretary of the USDA assured Senator William Proxmire of the same thing. But shortly thereafter, the National Resources Defense Council, an environmental group, learned that this was untrue – that the USDA *had* found DES residues of up to 37 ppb in at least ten animals, but that lower officials had withheld these results pending comfirmation by a second method of testing. Since no other method was available, the results had not been reported. Critics accused the department of lying to protect the cattle industry; the USDA blamed the episode on administrative confusion. Nevertheless, the USDA Assistant Secretary wrote Senator Proxmire apologizing for his department's "inexcusable error" and promising a full investigation of this "gross malpractice."[175]

By now many observers had lost all confidence in regulatory solutions to the DES problem. Senator Proxmire introduced legislation to ban all use of DES in cattle and sheep, and Representative Ogden Reid introduced a similar bill in the House. The National Resources Defense Council and other consumer groups filed suit against the FDA charging the agency with violation of the Delaney Clause and requiring an immediate ban on all livestock uses of DES. Representative Fountain, tired of waiting for the FDA to respond to the Herbst and Greenwald reports linking DES with cancer, announced that his committee would reconvene hearings on November 11 to investigate this "startling development, the implications of which have yet to be unraveled."[176]

On October 25, 1971, two weeks before the scheduled hearings, a study of the postcoital use of DES appeared in the *Journal of the American Medical Association* confirming that this use of DES, too, eluded regulatory controls. The author, Dr. Lucile Kuchera, a physician at the University of Michigan health service, reported that she had tested DES as a postcoital contraceptive on 1,000 Michigan coeds, with 100 percent success in preventing pregnancies and no serious adverse reactions.[177] Kuchera's report received broad and favorable coverage in the popular press, with only passing mention of the link with cancer, and of course no indication that this use of DES had never been approved by the FDA. To many legislators, however, it only underscored the lack of effective regulation of a drug now identified as a transplacental carcinogen, providing further impetus for Congress to intervene.

With legislation and a lawsuit pending, and facing a hostile Congressional inquiry, the FDA clearly had to do something. And it did, on at least one of the DES battlefronts. On November 10, 1971, the day before the Fountain hearings were to begin, the FDA finally published the long-overdue NAS rating for DES in the *Federal Register* and announced that DES was contraindicated in pregnancy.[178] This was *eight* months after the FDA had received the galleys of Herbst's study and five months after Ingraham's letter. While the FDA dallied, it is estimated that doctors wrote more than 60,000 prescriptions of DES for pregnant women.[179]

The tone of the Fountain committee's hearings in November was much sharper than in March. FDA Commissioner Edwards was called to testify and came under fire on all three DES fronts. The committee's first concern was why the FDA had taken so long to act after receiving the DES rating from the NAS, and why the agency had not responded to the Herbst and Greenwald findings – even after Ingraham's direct request for federal action – until the eve of a Congressional inquiry. "How in the world can you justify," a committee staff member asked Edwards in frustration, "in view of the hazard which was certainly delineated very, very sharply by the findings in Massachusetts and in New York – how could you justify not warning physicians immediately and not contraindicating a use which you know is not effective and where there is a safety question?"[180]

Edwards's answer was less than satisfying. He waffled on the question of DES's lack of effectiveness in preventing miscarriage, defending his agency's sluggish response on the grounds that the FDA's judgments had not only national but worldwide ramifications. "I feel we have a very significant responsibility," he testified, "to be certain of what we are acting upon." He continued:

> I must, however, say I do not agree with Dr. Ingraham's action. I think he was premature. I think that there were others he should have consulted prior to his taking the actions that he took – not that his actions are necessarily wrong, but I think he could well have studied it a little bit more than he did.

Before the FDA could act, Edwards explained, "there were lots of analyses that had to be carried out" on the data supplied by Herbst and Greenwald.[181]

Yet, when asked to furnish these analyses to the Fountain committee, the FDA came up empty-handed. An FDA official explained lamely:

> While no written record was maintained, we wish to assure you that data provided by the New York State Department of Health, along with that reported to us by Dr. Arthur L. Herbst, was carefully reviewed by our staff.[182]

The FDA never did produce a satisfactory explanation for its eight-month delay in reacting to the reported association with vaginal cancer. But Edwards's effort to justify this delay was revealing. In arguing that the agency could not act until it had convincing evidence that DES was unsafe or ineffective, he was, in effect, turning the FDA's legally mandated responsibility to ensure drug safety and efficacy on its head.[183] His position, simply stated, seemed to be that drugs were safe and effective until proven otherwise.[184] Even after the association with vaginal cancer was discovered, FDA officials still seemed bent on giving DES every benefit of the doubt until forced – either by incontrovertible evidence or Congressional pressure – to acknowledge its risks.

Following the 1971 Fountain hearings, the FDA sent a *Drug Bulletin* to all physicians calling their attention to the Herbst and Greenwald findings and informing them that DES was contraindicated in pregnancy. Yet in spite of this warning and all of the publicity surrounding the discovery of the cancer link, total sales of DES for Eli Lilly, the major supplier, reportedly *increased* by 4 percent between 1971 and 1972.[185] This increase was particularly ironic inasmuch as 1971 was the year when President Nixon declared an all-out "War on Cancer."

A second major topic of the 1971 Fountain committee hearings was the FDA's continuing failure to eliminate DES residues in beef. Edwards contended that the miniscule traces of DES that remained, which were found only in the livers, were harmless to human health, and that banning DES would raise the price of beef by 3.5 cents per pound. By contrast, most of the medical experts called to testify focused on the possible health risks. Dr. Roy Hertz, a leading cancer specialist and longtime advisor to the FDA, called allowing a proven carcinogen to remain in the nation's food supply "really a foolhardy undertaking," whatever cost savings it might yield.[186] FDA officials said they were introducing new regulations lengthening the required withdrawal period from 48 hours to seven days, and promised to ban DES if residues continued to be found. Critics remained deeply skeptical. "Food and Drug hasn't been able to enforce the 48-hour withdrawal, so why should we think it can enforce a 7-day withdrawal?" a National Resources Defense Council lawyer asked.[187] Hertz called the new regulations "unfeasible and impractical and ill-advised."[188]

Finally, the Fountain committee questioned the extensive and still unapproved use of DES as a postcoital contraceptive. In an apparent attempt to preempt criticism, Commissioner Edwards announced that the FDA would be sending a drug bulletin to physicians stating that "the FDA regards this use as investigational and is currently reviewing data to determine [its] safety and efficacy. . . ."[189] (The promised bulletin never appeared.) This did not satisfy Chairman Fountain, who thought that the newfound association between DES and vaginal cancer created "a special urgency that FDA require that all investigational work on this drug be strictly controlled."[190] Under questioning, Edwards could only agree.

The next day an editorial in the *Washington Post* detailed the various lapses of the FDA and the USDA and concluded:

> The slow and incomplete response of these two federal bodies illustrates pointedly the need for an independent consumer protection agency that would be empowered to act against such unacceptable bureaucratic behavior.[191]

This remark was prophetic. "Unacceptable bureaucratic behavior" was to continue for many years, in all three DES arenas.

Prenatal DES: the failure to inform

Amazingly, even the discovery of the link with cancer did not put an end to the prenatal use of DES. Many doctors continued to prescribe DES to pregnant women, either because, despite the

1971 *FDA Drug Bulletin* and other publicity, they remained unaware of its risks, or doubted the evidence. During 1974, U.S. doctors wrote an estimated 11,000 prescriptions for DES during pregnancy.[192] At Senate hearings held in 1975, a pediatrician from Idaho reported that he had been unable to convince some obstetricians to stop using DES prenatally and that the state agencies he had notified were unwilling to intervene.[193]

The FDA's response was true to form. The agency took no action until it again found itself under attack from Congress and consumer advocates for its passivity in the face of a clearly identified hazard. And all it did then was to include, on the last page of a 1975 *FDA Drug Bulletin*, a two-paragraph warning to physicians about the dangers of prescribing hormones during pregnancy.[194] The FDA seemed to assume, as it had all along, that only doctors, not patients, could prevent inappropriate use of DES.

In the 1971 hearings, Representative Fountain had suggested that perhaps, through radio or other media announcements, the women themselves should be alerted to the risk. Commissioner Edwards had balked. The FDA had to be careful "not to create an emotional crisis on the part of American women," Edwards explained, concluding: "I think this is one of the responsibilities that has to be laid directly at the doorstep of the American doctor."[195] This attitude shaped federal policy for many years until, in 1977, the National Cancer Institute finally published the first government-sponsored informational brochures on DES that were aimed at patients rather than doctors.[196]

The medical profession had as little appetite for informing patients who might be at risk as the government did. For example, a 1971 *JAMA* editorial by an AMA staff member stated that in view of the "small" (but admittedly still unknown) risk of cancer, "an organized effort by the medical profession to inform all women who were given estrogen therapy" would be "of questionable advisability." "Meanwhile," the editorial continued, "the fact that a risk exists should be known to the physician and should guide him to act in a careful, responsible fashion. . . ."[197] In other words, doctors should be aware of the potential risks, but not patients. Such attitudes were reminiscent of the medical profession's vehement opposition throughout the early history of DES to the provision of any information to consumers that might encourage "self-medication." A 1974 *JAMA* editorial declared similarly: "Since the risk of [vaginal and cervical] lesions seems small, it may not be wise to stir a national alarm," but warned that physicians might be held liable if their inactivity caused "irreparable harm." The editorial suggested sending notices to mothers at their last known addresses. "Many mothers, lost to follow-up, will never receive these notices," the author allowed, but concluded: "Further attempts to locate them should not be the responsibility of individual physicians, but rather the task of some governmental agency, if any."[198]

Well might the medical profession wish the job on "some governmental agency." Nationally, DES-exposed women numbered in the millions. Many doctors who had used DES did so routinely, and knew that searching their records would be a monumental task. One conscientious physician proved the point. Dr. Joseph Scott employed a medical secretary for 1,500 hours to comb his records for exposed patients, and identified 70 cases.[199] But he was the exception. Numerous physicians who had prescribed DES not only refused to search their records, but actually managed to "lose" the charts of treated patients. Although individual doctors have almost never been sued in DES cases, fear of possible litigation probably led to many unexplained disappearances of office files. More than one mother reported being told by her doctor that he had never given her DES, even when she had concrete evidence, such as an empty pill bottle, to the contrary.[200] To some disillusioned women, it began to seem that many doctors were more interested in *not* finding medical records than in finding them.

If the government and doctors were wary of informing patients of possible risks, drug companies were even more so. In theory, since companies had promoted DES as a safe drug and profited from its sales, they should properly bear some of the burden of alerting those now potentially endangered by their products and funding educational campaigns or screening clinics. Drug companies did in fact devote substantial sums to DES following the discovery of vaginal cancer, but almost all of this went into efforts to defend themselves against the growing number of lawsuits brought by cancer victims (see Chapter 8).

Nor was media coverage very helpful in identifying and informing the DES-exposed. Most major metropolitan dailies reported on the Herbst study in April 1971 and the FDA announcement contraindicating the use of DES during pregnancy that November. But in general, the coverage was superficial, and did not tell women how to find out whether they were at risk, or what to do if they were. Not until the women's health movement started addressing the problems associated with DES in the mid-1970s did the popular press – notably women's magazines – begin to provide practical information useful to DES victims.

As it became increasingly obvious that neither the government, doctors, drug companies, nor the news media was doing much to meet the needs of DES victims, women themselves began to take action. Consumer advocates, DES mothers and daughters, and public health activists appeared at numerous Congressional hearings during the early 1970s to tell of the continuing problems and press for remedies. Small groups of concerned DES mothers and daughters formed around the country and, with help from sympathetic health professionals, developed their own informational pamphlets alerting women to the possibility of DES exposure and instructing them on how to get proper medical care. By 1978, there were enough of these groups to organize a national network, which they called DES Action. The activities and accomplishments of this remarkable grassroots organization are discussed more fully in Chapter 10.

Lobbying and publicity by DES Action and other consumer groups were instrumental in focusing public attention on the plight of DES victims and in mobilizing further governmental action. In 1978, prodded by Dr. Sidney Wolfe of Ralph Nader's Health Research Group and DES activists, the Department of Health, Education and Welfare appointed a National Task Force on DES to study all aspects of DES exposure and advise the government on appropriate actions. The 1978 Task Force's report stressed the importance of continued research on DES and urged doctors to notify all women who had received DES during pregnancy of their exposure. To circulate this message to physicians, the FDA included it in another *FDA Drug Bulletin*, and the U.S. Surgeon General issued a special Physician Advisory on DES to the nation's doctors – the third such advisory on any medical topic ever issued – discussing the risks of DES, its health effects, and recommending medical follow-up for exposed patients.[201] But even this unusual step seemed to have little impact on physicians. In 1980, by one estimate, less than one-tenth of the women who had taken DES had been located and informed of their risk.[202]

In 1984, substantial data emerged on increased rates of breast cancer among DES mothers and squamous cell dysplasia of the cervix and vagina among daughters.[203] The government, again under pressure from Wolfe, DES Action, and others, convened another DES Task Force in 1985 to consider the new findings and offer recommendations. The group completed its report, but for five months the federal government refused to release it. Frustrated scientists on the 1985 Task Force finally gave a copy to Wolfe, who immediately made it public, explaining that it contained "important information which DES mothers, their children and physicians and other health care professionals need to be made aware of immediately."[204] And it did. The report urged physicians to warn patients treated with DES during pregnancy of their increased risk of breast cancer and recommended expanded efforts by physicians "to determine exposure of their patients." It also spelled out areas of needed research and – in what may explain its attempted suppression – listed six "anticipated early actions by the [government] to aid in the implementation of the recommendations from the Task Force." Two years later, several of the actions listed had still not been taken and others were largely perfunctory.[205]

Unfortunately, the need to inform DES victims was not the only issue that remained unaddressed following the 1971 cancer discovery. In the ensuing struggles to stop the use of DES as a morning-after pill and to remove DES from meat, the federal government played a similarly indecisive and ineffectual role. In the battle over meat, in fact, the FDA seemed at times to be as much a part of the problem as the solution.

Getting DES out of meat: winning the battle and losing the war

FDA Commissioner Edwards made no secret of his view that using DES as a growth promotant for livestock made good economic sense and posed little if any medical risk to the meat-eating public. In a 1972 interview with the *Des Moines Register*, he attacked "self-seeking politicians" and "headline-hunting scientists" for exploiting the DES issue, claiming they were making a mountain out of a molehill:

> Common sense tells us that when we find from one-half to two parts of diethylstilbestrol per billion of beef liver, it probably doesn't mean a heck of a lot.[206]

Most of the FDA's subsequent actions were consistent with this view. In June 1972, in response to evidence that DES residues had quadrupled in the past year despite the new seven-day withdrawal period, Edwards announced that the FDA was considering steps to ban DES in animal feed, the first of which would be a public hearing. But, he stated pointedly: "We have not yet concluded that withdrawal of approval for DES is the appropriate course of action."[207] As critics well knew, such hearings often dragged on for years. On June 20, Representative Fountain addressed the full House in ringing tones, decrying the upcoming hearing as "merely a tactic for delaying the regulatory action which the law requires," and calling on Congress "to take firm corrective steps to protect the American people."[208] Senator Edward Kennedy denounced the FDA's inaction in Senate hearings on DES.[209]

The next month brought more incriminating evidence. The USDA reported finding the highest DES residues yet in beef and lamb, and the Director of the National Cancer Institute stated publicly that it would be "prudent" to eliminate DES entirely from the food supply pending the outcome of the FDA's public hearing.[210] A week later a new scientific study employing more sophisticated methods revealed detectable residues of DES in beef liver after the seven-day withdrawal period (and possibly for as long as thirty days). Since the Delaney Clause prohibited any detectable residues of a known carcinogen, Edwards now had no choice but to act, and on August 4, 1972 he announced a five-month phase-out of DES feed additives. The FDA did not hide its disdain for this action. "Under the law," the announcement stated, "there is no alternative but to withdraw approval of the drug, even though there is no known public health hazard resulting from its use."[211]

The news of a pending ban took much of the political heat off the FDA. Senator Proxmire's bill to ban livestock uses of DES died in House Committee. The Fountain team went off to write its report. The National Resources Defense Council withdrew its suit. But declarations of victory proved premature. In January 1974, on appeal by DES manufacturers, the ban was overturned by the U.S. Court of Appeals – an outcome some critics claimed was just what the FDA had intended in the first place. The court ruled that in the expressly stated absence of a public health hazard, and in view of the "benefits of DES in enhancing meat production," the FDA had no reason to withdraw approval without allowing manufacturers the customary public hearing required by law.[212] So DES was back on the market. Moreover, the court ruling permitted the FDA to weigh the claimed benefits of DES feed additives and implants against their potential risks, a flexibility the agency was denied under the Delaney Clause. Some observers suspected that the FDA's seemingly bungled ban had really been a masterful strategy in disguise.

Capitalizing on the publicity generated by the overturned DES ban, the FDA urged Congress to revise the Delaney Clause so that benefits could be weighed against risks, permitting the agency to justify the continued use of DES.[213] Officials argued that the Delaney Clause's strict prohibition against the addition of carcinogens to the food supply was unrealistically rigid, impractical, and unenforceable. Consumer advocates countered that such a change would expose regulatory agencies to unhealthy pressures from affected industries exaggerating the purported "benefits" of toxic chemicals in food production, and defended Delaney's high standards as a desirable counterbalance to the inevitable failures of compliance and monitoring. Debate over these issues continues to this day.[214]

Meanwhile, the use of DES in livestock continued and Congress again took up the issue. "We are back to zero because of the inadequacies of an FDA procedure," Senator Kennedy declared at Senate hearings in September 1974. "They were thrown out of court, and now they have scheduled a hearing but we do not know the time."[215] In fact, it took more than two years for the FDA to proceed with the public hearings, and more than five years for the agency to reinstate the ban. In June 1979, FDA Commissioner Donald Kennedy announced that he was withdrawing approval of the use of DES in livestock. Kennedy's decision covered many of the points that had been hotly contested over the years: that no "safe" residue levels of DES had been established; that the law did not authorize FDA to consider the economic benefits of drugs such as DES; and that in any event manufacturers had not demonstrated that the ban would have any significant adverse environmental or economic impact.[216]

The war against DES in meat was finally over, at least officially, but some illicit use continued. In 1980 the government disclosed that more than 300,000 cattle had been illegally fattened with DES since the ban had begun and warned that feedlot operators caught violating the ban would be subject to criminal penalties.[217] Illegal use may still be occurring. In the mid-1980s, an unusually high incidence of premature sexual development among young girls and boys in Puerto Rico raised questions about whether these cases were caused by the unlawful use of DES and other hormones by livestock producers.[218]

For the future, the legal use of hormones and other chemicals in agriculture undoubtedly poses a greater danger to human health than any remaining illegal use of DES. When DES was banned, alternative estrogenic and other hormone preparations were ready and waiting to take its place, all with the FDA's full blessing. Today virtually every steer raised in the United States is treated with some type of growth-promoting product – synthetic estrogens and other hormones without the stigma of DES but whose long-term effects on human health are unknown.[219] Away from the glare of publicity over DES, the FDA quietly eliminated the sixty-day withdrawal period originally required for Synovex, the most commonly used estrogenic growth stimulant. And, of course, antibiotics, pesticides, fungicides, and other toxic chemicals are routinely used in food production. Renewed concern over these practices emerged in 1987, when a battery of independent studies provided, to quote the *New York Times*, "more convincing evidence than ever before that modern food production, and the inability of Federal agencies to adequately inspect food and detect toxic substances, may be jeopardizing Americans' health."[220]

The ban on DES in meat was, unfortunately, a Pyrrhic victory.

The rise and fall of the DES morning-after pill

The FDA's efforts to control the expanding use of DES as a morning-after pill following the 1971 discovery of vaginal cancer were as halting and ambivalent as those against DES in livestock. Officials again seemed determined to allow this use of DES to continue and pursued a regulatory strategy that ultimately failed.

Initially, the agency appeared to be moving toward disapproving postcoital use of DES. Both the FDA's Advisory Committee on Obstetrics and Gynecology and a special governmental advisory group on DES appointed in 1972 were critical of this experimental therapy and of the evidence documenting its effectiveness. In the fall of 1972, the Ob-Gyn Advisory Committee recommended that the FDA warn physicians that prescribing DES for postcoital contraception was "a non-indicated use because there is insufficient evidence for its efficacy and safety."[221] The majority of the DES advisory group members favored removing DES from the market altogether, since reasonable substitutes existed for most of the approved indications.[222] Consumer groups such as Nader's Health Research Group and women's health activists around the country spoke out publicly and in Congress against the growing use of the morning-after pill and the frequent lack of proper medical follow-up.[223]

But there were also proponents of the morning-after pill within the FDA, notably J. Richard Crout, head of the Bureau of Drugs. Crout and others joined forces with NIH officials responsible for family planning research and influential investigators to push for a regulatory scheme that would not prohibit postcoital DES but would limit its use to "emergency" situations. The idea was, following the same logic as in the original 1941 approval of DES, that strict provisions regarding the drug's use would prevent abuse. One may only wonder why FDA officials thought this strategy would work this time around when it had failed so dismally before.

By early 1973, Crout and other advocates of DES had made some headway within the FDA. And the Ob-Gyn Advisory Committee itself was now more sympathetically inclined toward postcoital DES, having lost several members who had previously been opposed (most of the members were now, as Crout put it, "in the population control business.")[224] FDA officials decided the time had come to try to implement their scheme, which involved approving postcoital use under strictly specified conditions. There was only one problem: No manufacturer had sought FDA approval for this use. Manufacturers had little reason to go through the cumbersome approval process when DES was already being widely used as a morning-after pill based on word of mouth and favorable medical reports.

But the FDA was not to be deterred. Crout and other officials decided that if manufacturers would not initiate the approval process, they would. The FDA would simply ignore the lack of formal requests from manufacturers and go ahead and approve the postcoital use of DES, hoping thereby to clarify the agency's position on the matter and perhaps encourage companies to apply for approval. The first step in this highly unusual maneuver ("where the cart has been put before the horse," as one contraceptive researcher described it) was to win the endorsement of the Ob-Gyn Advisory Committee.[225]

At its January 26, 1973 meeting, the Ob-Gyn Advisory Committee was asked to consider whether existing studies provided sufficient evidence of the safety and efficacy of postcoital DES to justify approval.[226] The meeting had two parts: a morning session open to the public, as required under the recently enacted 1972

Federal Advisory Committee Act, and a closed "executive" session in the afternoon. Testifying at the morning session were consumer advocates such as Sydney Wolfe and Anita Johnson of the Health Research Group, who argued against approval and challenged the scientific validity of the influential Kuchera study. Most of the invited medical and scientific experts, including Kuchera, spoke in favor of approval, although a few expressed serious reservations. Dr. Mortimer Lipsett, a leading cancer specialist, opened his presentation by saying: "I see Dick Crout left – because I was going to mention if he expected a clear forthright statement about the safety of estrogens, it was not to be forthcoming." Lipsett went on to stress the carcinogenic properties of estrogens, noting that no "threshold dose" had been established and that although the morning-after pill was "unlikely" to be carcinogenic, "given the right circumstances and the right individual, it could be."[227]

The afternoon session, which included only members of the Ob-Gyn Advisory Committee and a few FDA officials, had an entirely different character from the morning's discussion. Many of the warnings and reservations that had been expressed earlier were simply ignored as the group worked its way toward a defensible endorsement of postcoital DES. Other objections – such as the Wolfe–Johnson attack on the Kuchera study – had to be carefully skirted, since this study provided important documentation of the morning-after pill's safety and efficacy. Here Crout smoothed the way by offering the committee his own rather creative interpretation of what "substantial evidence" of safety and efficacy actually meant:

> Substantial evidence is somewhere between – this is not in the law – somewhere between a scintilla and a preponderance. So it doesn't say the evidence has to be absolutely overwhelming and free of controversy.... There is hardly any issue upon which there isn't to some degree conflicting evidence. But there has to be substantial [evidence], more than a smidgeon, but not necessarily completely overwhelming.[228]

Reassured by this "smidgeon–scintilla" definition of substantial evidence, the committee voted unanimously, almost immediately thereafter, that postcoital DES was efficacious. Then followed a discussion of safety, which also concluded in the committee's unanimous agreement that "there is no evidence for a significant risk to the patient, except for some immediate minor side reactions, and that a carcinogenic potential from this dosage has not been proven."[229]

In reality both safety and efficacy were certainly debatable. "No question we could have made those studies go either direction," Crout later admitted.[230] Pregnancies had occurred following postcoital DES therapy in several reported trials (Kuchera herself acknowledged a few months later that there had been six pregnancies among subjects excluded from her study purporting to show 100 percent efficacy).[231] And if the treatment failed, then exposure of the fetus to potentially dangerous doses of DES could not be ruled out, meaning that safety, too, was in doubt. (It was later shown, in fact, that some DES daughters who developed cancer had been exposed to smaller total dosages of DES than those of the usual morning-after pill regimen.)[232] By relying heavily on the Kuchera study and the early work of Morris and van Wagenen, both of which claimed that termination of pregnancy was guaranteed, the committee could comfortably dispose of any lingering doubts about risks to the fetus.

With the issues of safety and efficacy thus resolved – however arbitrarily – the outcome was ensured: The committee voted to recommend approval of the postcoital use of DES under the condition that use be limited to emergency situations and with specified provisions for labeling and marketing.

From a position of initial opposition, the FDA had now moved full circle to approval – an about-face all the more remarkable considering that it occurred with full knowledge of the link between DES and vaginal cancer, and over the protests of consumer groups as well as some prominent experts. Congress was again moved to intervene, and over the next few years, both the Senate and the House held hearings sharply questioning the FDA's failure to curb the widening and still uncontrolled use of the morning-after pill.[233]

FDA officials went doggedly ahead with their plan. In May 1973 the agency sent out a *Drug Bulletin* to all physicians advising them that the use of DES as a postcoital contraceptive had been approved for "emergency" situations.[234] This announcement was in error; approval could not occur prior to the specification of required package labeling and marketing conditions, and these were not even proposed until the following September. Nevertheless, the FDA left this error uncorrected for two years, allowing many physicians to remain misinformed about the morning-after pill's true status and contributing directly to expanded usage. Finally, in 1975, the FDA issued its official requirements for approval of postcoital DES, specifying marketing, labeling, and conditions of use.[235] Even this belated action may have been prompted by outside pressure – the approval notice came just weeks before Senator Edward Kennedy was to hold yet another hearing on DES to consider Congressional action barring its use as a postcoital contraceptive.[236]

The FDA's proposed requirements contained a number of intended safeguards. The morning-after pill was to be prescribed "as an emergency treatment only" (although what constituted an emergency was left up to the physician's discretion), and only after a pregnancy test had been performed to rule out a prior pregnancy. "Repeated courses of therapy" were to be avoided. Should the treatment fail and pregnancy continue, "serious consideration [of] voluntary termination of pregnancy" was recommended in view of the possible risks to the fetus. To encourage completion of the full dosage regimen, postcoital 25-mg DES tablets were to be packaged in containers of ten, and were to contain leaflets for patients describing the drug's risks and the conditions necessary for safe use. Approval for all other 25-mg tablets was rescinded. On dosage strengths other than 25 mg, the package inserts for physicians – but not for patients – were to warn: "THIS DRUG PRODUCT SHOULD NOT BE USED AS A POSTCOITAL CONTRACEPTIVE."[237]

Whatever the merits of these safeguards, they applied only to manufacturers that sought FDA permission to market postcoital DES, and few companies had evinced any interest whatsoever. To try to entice companies into applying for approval, the FDA took another unprecedented step: It announced that it would accept an "abbreviated" new drug application in place of the standard application normally required to add new indications to an existing drug. With an abbreviated application, manufacturers did not have to provide proof of safety and efficacy but could rely on outside, independent research. Even so, most companies still had no interest in seeking FDA approval for the use of DES as a morning-after pill. Eli Lilly, already smarting from lawsuits over the prenatal use of DES, not only discontinued its own line of 25-mg DES tablets but sent letters to nearly 300,000 physicians and pharmacists in 1974 recommending against the use of DES for postcoital contraception.[238] The FDA did get one taker, however: In January 1974, a small New Jersey drug company called Tablicaps submitted the long-awaited abbreviated application to market postcoital DES.[239] It looked like the FDA's scheme for the morning-after pill might finally succeed.

But not for long. New opposition emerged within the agency

itself when the Director of the Generic Drug Division, Dr. Marvin Seife, objected to the use of an abbreviated application for this obviously controversial therapy and flatly refused to approve it. "In all my years with the FDA," Seife told an interested Senate Health Subcommittee in early 1975, "this is unique, this is the first time in my experience that a drug has been published for a new use in the Federal Register without any study, without any investigative new drug application for a totally new indication."[240] In 1976, Seife and his medical officer, Dr. Vincent Karusaitis, were called to testify before a Senate hearing.[241] With the blessing of Senator Kennedy and other influential members of Congress, these two officials simply refused to sign off on the Tablicaps application, despite the wishes of Crout and other higher FDA officials who favored approval. And they could not be coerced into compliance by their superiors because, as Karusaitis put it, "the mantle of security of the Senate rests upon our little heads."[242] Thus protected, Seife and Karusaitis continued to stonewall the Tablicaps application throughout the 1970s on whatever grounds they could find. It never was approved.

So the FDA's efforts to exert some control over the use of DES as a morning-after pill, controversial enough in their own right, finally came to naught. In 1978, the agency sent out another *Drug Bulletin* stating explicitly that no manufacturer had been granted approval to market DES as a postcoital contraceptive and reiterating the FDA's willingness to approve this indication under the specified conditions.[243] Despite this clarification and repeated warnings by DES Action and other consumer groups about the potential risks of the DES morning-after pill, confusion about its legal status persisted and use continued in college health clinics and hospital emergency rooms throughout the 1970s.[244] Meanwhile, anecdotal reports of babies being born following postcoital DES exposure began to surface.[245]

Use of DES as a morning-after pill did decline considerably in the 1980s as the drug's bad reputation finally began to catch up with it. As of 1985, however, as many as 15 percent of college and university health services around the country were still prescribing DES for postcoital contraception or referring women to clinics where they could get it.[246] Those that did not offer DES almost all used some other form of estrogen – generally Ovral (a combination estrogen–progestin birth control pill) – for the same purpose. Yet, like DES, these other drugs have never been approved by the FDA as postcoital contraceptives, thus their safety and efficacy remain unconfirmed.[247] Some critics fear that such alternatives could be as risky as DES to any fetuses that might survive if not to the women themselves; the evidence is just not in yet.

Once again, as in the battle against DES in meat, the victory was more symbolic than real. The words changed but the melody stayed the same.

Conclusion

The DES story has, for the most part, concluded. Yet, for many of the victims, its consequences are still unfolding. No one knows, for example, whether DES daughters will develop other tumors as they reach menopause, whether exposure to DES has affected other body systems besides the reproductive system, or whether additional reproductive disorders will yet emerge. Nor does anyone know what impact the morning-after pill's flash exposure might have on the women taking it or on any surviving fetuses. And no one can really say how trace amounts of DES and other chemicals in the food supply will affect the nation's health over the long run.

On the other hand, a great deal more is known now than a decade ago about the effects of prenatal DES exposure. By 1985, over 300 cases of clear-cell adenocarcinoma of the vagina or cervix in DES daughters had been identified.[248] For an individual DES daughter, the risk of developing such a cancer by age 35 is now estimated at 1 in 1,000; in other words, one out of every thousand exposed daughters will develop clear-cell cancer by her mid-30s. Some researchers (and most DES manufacturers) have questioned whether DES actually "caused" these cancers or instead serves as a marker for some other causal factor, such as the mothers' "problem pregnancies."[249] Although the answer to this question may never be known with absolute certainty, there is compelling evidence that DES played a primary role.[250]

Much more common than cancer among DES daughters are various other abnormalities of the reproductive tract, whose ultimate health impact is not yet fully understood. These abnormalities include adenosis, an apparently benign form of glandular growth in the vagina or cervix, and structural alterations – extra tissue or folds on the cervix called "hoods," "collars," or "cockscombs." Some of these alterations appear to recede as the DES daughter matures, but it is still uncertain whether this transition is complete.[251] In 1974, the National Cancer Institute launched a multicenter research effort, the National Collaborative DES Adenosis (DESAD) Project, to observe the natural history of vaginal and cervical changes in a representative cohort of DES daughters compared with a similar cohort of nonexposed women. By the early 1980s, the findings showed that 25–50 percent of DES daughters had structural anomalies of the cervix or vagina (versus 2 percent of the controls), and were roughly twice as likely as other women to have dysplasia and carcinoma in situ of the cervix or vagina.[252] (The latter are premalignant conditions, which can generally be successfully treated if promptly diagnosed.) Data from the DESAD and other studies indicate that many DES daughters have abnormalities of the upper genital tract as well, such as a "T-shaped" or constricted uterus, narrow cervical canal, or misshapen tubes.[253]

Not surprisingly, these various DES effects appear to interfere with reproductive function. A number of studies have now shown that DES daughters are at substantially higher risk than other women for a poor pregnancy outcome (miscarriage, stillbirth, ectopic pregnancy, premature delivery, etc.), and some evidence also indicates that they have more trouble becoming pregnant.[254] Such reports only confirm what DES daughters had been telling their doctors for some time, only to be reassured that "there was no evidence" to link their pregnancy problems with DES. They provide the first evidence that the drug's effects do not stop with the mothers, or their offspring, but extend to the third generation as well, in the form of the multiple risks associated with prematurity.

DES-exposed sons and mothers have had their own problems. Women who used DES during their pregnancies are now estimated to have a 40–50 percent greater chance than nonexposed women of developing breast cancer – which amounts to anywhere from 25,000 to 62,000 excess cases of breast cancer in American women, depending on assumed exposure levels.[255] DES sons, like daughters, are at risk for various structural abnormalities of the genital tract, including undersized penises and testes, epididymal cysts, and perhaps fertility problems due to lowered sperm counts and malformed sperm.[256] Studies to date have not found a higher rate of testicular cancer among DES sons as was initially feared, but the types of abnormalities they do have raise the possibility that tumors may develop later in life.

Since 1983, government support for research on DES has been cut sharply. Reduced funding forced the DESAD Project to abandon its detailed medical examinations of the cohorts of exposed women it had been following, cutting off the largest existing source of clinical data about the long-term effects of DES exposure on middle-aged women and about the drug's impact on the third generation.[257] DES is no longer a topic with which younger researchers are likely to establish their professional reputations. Some investigators, particularly clinicians, may have been discouraged from continued work on DES by the increasingly frequent requests to testify in DES lawsuits. Others may have found it distasteful to work in an area that so directly highlights a medical mistake, with its implicit criticisms of colleagues. The decline in funding and interest in DES-related health problems is unfortunate in light of the many medical unknowns that remain and the continuing use of other estrogens as growth promotants in livestock and as postcoital contraceptives in women.

The DES story raises a number of larger issues about medicine's role in society and the limits of regulatory safeguards. It reveals a pattern of deeply ingrained optimism about the benefits of medical science in solving perceived social needs, a cultural outlook shared by groups as diverse as doctors, farmers, scientists, and college coeds. So great was the optimism enveloping DES that it allowed an almost willful disregard of the abundant evidence of risk. The warnings offered by some concerned experts, and later by public interest activists, seemed simply not to penetrate.

Certainly this uncritical faith in DES was encouraged and reinforced by the drug's corporate sponsors. That pharmaceutical manufacturers played a primary role in the conception, staging, and denouement of this medical tragedy cannot be disputed. But companies could not have sold their deadly ware without a receptive market. Scientists, doctors, and farmers alike were only too ready to seize upon the latest innovation to help them ply their respective trades, and patients were eager to benefit from a new "wonder drug." And why should anyone have been skeptical? Here were most of the nation's leading medical and scientific experts, the highest authorities of modern technological society, assuring people that DES was not only without risk but would yield important benefits. Voices of dissent were easily drowned out amid the chorus of hope and praise sung by the drug's advocates.

It was, indeed, a powerful alliance: bold scientific experts, enthusiastic corporate vendors, trusting medical practitioners, eager livestock producers, and unsuspecting patients. Together, this constellation of forces propelled DES into almost every corner of American life with a fervor rarely matched in the annals of medical innovation. For the most part, the FDA was swept along in the pro-DES current by officials who genuinely believed in the drug's promised benefits, sympathized with industry interests, and discounted signs of peril. It was influential members of Congress – Delaney, Kefauver, Fountain, Kennedy, and others – who repeatedly blew the whistle, frequently at the urging of public interest groups. Only these powerful figures, apparently, had the intellectual and economic independence necessary to take the emerging danger signals seriously, and the political wherewithal to act on their convictions. And even they were often constrained by the political and financial realities of electoral politics.

One cannot look back at the history of DES without being struck by the consistent and often flagrant failure of regulatory agencies – notably the FDA and the USDA – to carry out their mandated responsibilities. To be sure, it was medical scientists and clinicians who developed and prescribed DES, and drug companies that promoted its use. Yet their behavior is hardly surprising. Pursuit of medical and technological innovation is a hallmark of Western civilization, even though it inevitably entails hazards. The role of private industry, in turn, is to translate scientific developments into marketable products, even though some of these products may pose risks to the public. Indeed, it is precisely to protect the public against untoward risks from unsafe drugs and chemicals that we have regulatory agencies such as the FDA.

That is why the story of DES, with its long string of regulatory failures, is so troubling: It reveals a regulatory system that functioned less as public protector than as a willing and cheerful ally in the pursuit of corporate profits through technological development. Whether this was because the agencies had been "captured" by the industries they were supposed to regulate or simply shared the same values and ideology as their corporate clients is immaterial; both were probably true. The result was that regulators tended to view the risks of DES through the same optimistic lens as did the drug's manufacturers, rather than from the vantage point of the potential victims. The absence of critical and objective regulatory scrutiny was an open invitation to the tragedy that occurred.

Even after the link with vaginal cancer was discovered, the government's response left much to be desired. Practical assistance for the victims of DES resulted largely from the political lobbying and grass-roots activities of women's groups and other consumer-oriented organizations – DES Action and the Health Research Group in particular. These groups prodded administrative and regulatory agencies into sometimes grudging action and helped stimulate scientific interest in a range of DES-related health problems. Moreover, DES Action, an organization comprised mainly of DES-exposed mothers and daughters, broke through the long tradition of medical paternalism reflected throughout the history of DES, working from the principle that patients can and should be informed about their own health problems. The success of DES Action's nationwide network of self-help groups in providing clinical information and emotional support for the DES-exposed offers an inspiring example of the ability of laypeople to meet the needs that professionals have been unable or unwilling to fulfill.

Could the DES story happen again? Without some major re-alignment of the forces that dominate medical and scientific in-novation, the chances seem very good – not in precisely the same form, to be sure, but with the same underlying causes and similarly poignant consequences. If the DES story is any guide, the best hope of avoiding such an event lies in strengthening government regulatory and enforcement mechanisms, in paying more attention to those whose warnings were ignored in the past – disinterested scientific critics, consumer advocates, and independent Congressional committees – and in increasing public awareness of the risks of medical and scientific innovation and the limits of present safeguards. Such steps are not foolproof, but they deserve to be tried.

4

The artificial heart

Thomas A. Preston

Neither the Jarvik-7 nor any of the several other total artificial hearts
being developed is yet ready to permanently replace a human heart,
even on a trial basis . . .
— Robert Jarvik, *Scientific American*, January 1981

Barney Clark expected that the operation would either kill him
quickly or cure him. Unfortunately, he was wrong. He could not
foresee his prolonged ordeal of dying. Nor did his doctors anticipate the incessant medical problems, tremendous cost, and trou-
blesome ethical issues they would generate in their rush to the
operating room. Moreover, most of all, the media reports did not
convey to the average person how the landmark operation would
change the nature of medical services for everyone.

When Dr. William DeVries exchanged Barney Clark's failing
heart for a permanent artificial one, he initiated a sequence of events
that will affect all patients. In time, the artificial heart program
will have an indirect but certain influence on mechanisms and levels
of medical insurance, the direction of medical research and the
funding for it, the federal health care budget, and more. How all
this works out will determine who gets what and how much
medical care, how we apportion our expenditures between medical
and nonmedical services, and whether individuals perceive the
distribution of services as just.

The history of the artificial heart

For centuries biomedical investigators have dreamed of the totally
implantable artificial heart in their search for mastery over the
fragility of man's corporeal existence. In pursuit of the dream,
investigators have reduced the human heart to a mechanical pump,
defrocked as the seat of the emotions and replaceable by a plastic

and Velcro® fabrication of ambitious men. The transition from dream to the reality of modern technology has taken place mostly since the mid-1950s, and, quite naturally, has paralleled and been dependent on other biotechnical advances.

In 1812, Julien-Jean-César Legallois wrote: "If one could substitute for the heart a kind of injection [of arterial blood], one would succeed easily in maintaining alive indefinitely any part of the body."[1] This statement foreshadowed attempts of medical investigators to provide artificial circulation for the entire body, although the goal of mechanical circulation for only one organ was modest. More than a century later, in 1928, H. H. Dale and E. H. J. Schuster in England constructed what was probably the first artificial pump intended to substitute for the function of both sides of the heart in an animal.[2] About the same time, Charles Lindbergh, who had made his solo transatlantic flight in 1927, became interested in heart replacement systems because of the serious heart ailment of his sister-in-law. In 1935, he and Alexis Carrel demonstrated the feasibility of total body perfusion with what journalists dubbed the "robot heart," or the "glass heart."[3] These experiments sustained and nourished the dream of the many investigators thinking about an artificial heart.

The dream turned to a more tangible reality in the 1950s. The seminal precursor was the heart–lung machine, introduced for clinical use in 1953 by John Gibbon.[4] This device enabled surgeons to divert blood from the heart while maintaining circulation to the rest of the body. With the application of heart–lung bypass to "open heart surgery," in several years investigators across the country and abroad were observing patients whose heart and lung function was replaced for hours by an external machine. This progress led several investigators almost simultaneously to begin serious laboratory work on a totally implantable artificial heart.[5]

Perhaps the most important of these early investigators was Willem Kolff, a Dutchman. Kolff designed the first successful artificial kidney machine while the Nazis occupied Holland during World War II. He emigrated to the United States in 1950, and continued his research at the Cleveland Clinic. In 1957, he and Tetsuze Akutsu took a large experimental step toward a totally implantable artificial heart by putting two compact blood pumps (one for each of the two pumping chambers of the heart) into the chest of a dog to replace its heart. The device was powered by an external source of compressed air, and the dog survived for about ninety minutes.[6] Over the next few years, Kolff's group in Cleveland experimented with other models of artificial hearts driven by electricity rather than by compressed air, and some dogs lived for up to six hours with these units.[7] Concomitantly, other investigators also constructed prototype hearts for experimentation in animals.[8]

These early investigations were carried out in academic and research centers, without federal sponsorship but occasionally with federal research grants from the National Institutes of Health (NIH). The early work pinpointed problems: Although some dogs survived for hours, they hardly regained consciousness. The prototype pumps had inadequate outputs, they caused destruction of blood cells (hemolysis), and they were too bulky to fit inside a closed chest.[9] Particularly, investigators recognized the difficulty of placing the blood-pumping mechanism and the motor or energy converter for it, plus a power supply, within the space available within the chest. There were other unsolved technical needs: blood-compatible materials able to withstand continual wear and flexion (over 100,000 times per day) while not forming clots on the surface; and a miniaturized control system sophisticated enough to synchronize the artificial heart to the body's physiological requirements.[10]

The federal connection

The other pressing need was funding. The capital necessary for long-term research on the several technical fronts was beyond the means of independent investigators and their institutions.[11] Individual investigators could apply to the NIH for grants from general funds, but they had to compete for these limited funds with investigators across the country working on different heart-related and other medical projects. Several investigators, therefore, began pressing their contacts at the NIH and Congress for special funding earmarked for the artificial heart.

The most notable of these investigators was Michael DeBakey, the noted heart surgeon from Houston. He also was well known in Congress for prior testimony as a lobbyist for NIH programs. In February 1963, Dr. DeBakey appeared before the Senate Subcommittee of the Committee on Appropriations. Senator Lister Hill, whose surgeon father named him after the famous physician Joseph Lister, presided and greeted DeBakey:

> Dr. DeBakey, you are an old friend of this committee. We
> have had no better friend and no friend who has helped us
> more and supported us with greater devotion.[12]

DeBakey, who was supporting the budget requests of the Department of Health, Education and Welfare, introduced to Congress for the first time the idea of specific funding for the artificial heart:

> Experimentally, it is possible to completely replace the heart
> with an artificial heart, and animals have been known to survive as long as 36 hours. This idea, I am sure, could be reached
> to full fruition if we had more funds to support more work,
> particularly in the bioengineering area.[13]

The NIH was on notice to do something, and shortly thereafter the matter was taken up by the Planning Committee of the National Advisory Heart Council, the National Heart Institute's top policy advisory body. In October 1963, this group recommended that the NIH pursue development of the artificial heart "with the highest level of priority rating," and a month later the full National Advisory Heart Council accepted and endorsed the artificial heart program as one of high priority.[14] The plan received special Congressional approval before the end of 1963.[15]

In February 1964, the director of the National Heart Institute called together an ad hoc advisory group, including a number of the leading investigators in the artificial heart field, to discuss the program.

To the surprise of no one, the group urged the NIH to pursue development of the artificial heart "with a sense of urgency," and recommended a special staff for the program. The group also recommended a contract program "to develop devices and materials and to explore new avenues of approach."[16] The first congressionally approved allocations for the artificial heart became available in July 1964, when an Artificial Heart Program Office was established.[17]

The political background

In the early and mid-1960s, the artificial heart program was attractive to virtually all segments of our society.[18] Public confidence in the ability of medical researchers to advance social progress was buttressed by recent remarkable achievements in improving and extending life: kidney dialysis and kidney transplants, heart pace-makers, open-heart surgery, and artificial heart valves. All these efforts were supported by grants from the NIH, which had gained increasing credit for coordinating America's medical research effort.

It was an era of unprecedented growth of medical research across the country, and for surgeons and physicians the artificial heart provided a glorious opportunity to pioneer new techniques, to conquer death, and to win fame and glory in the service of man-kind. The recent success in the space race gave wide support to the concept of federal funding for technological solutions to social and national problems. Also, the strength of the economy during the boom years following World War II allowed tremendous growth of medical research funding without jeopardy at that time to other federally funded programs.

Congress was favorably disposed to funding of medical re-search, and in most years appropriated more monies than the NIH requested.[19] In addition, both of the powerful chairmen of the two principal congressional committees had a personal interest in ex-panding medical research. Senator Lister Hill was a devoted pro-ponent of NIH activities, and Congressman John Fogarty, the prominent chairman of the House Appropriations Health Sub-committee, suffered from heart disease.[20] In hearings of his com-mittee in 1965, Fogarty said:

> We are spending millions and millions and millions of dollars
> in space and trying to get a man on the moon and on foreign
> aid. Here we are losing a million people who are dying every
> year because of some form of heart disease, and we quibble
> at trying to get a few million dollars to get going on this
> project.[21]

To be sure, this classic argument, based on exaggerations, as-sumptions and an appeal to humanitarianism, expressed the mood of the time. The artificial heart program seemed comparatively cheap; for the price of a bomber, Congress could save hundreds of thousands of lives and demonstrate its commitment to a popular and seemingly noncontroversial public program.

The NIH artificial heart program fit the political goals of the administration as well as the Congress. John Kennedy was pres-ident when the artificial heart program was being formulated in Congress. Many in his administration had become enamored of the coalition of scientific and industrial teams that had worked so well for national defense and the space effort. However, the massive expenditures for cold-war armaments were slowing down in the early 1960s. President Kennedy, who feared that decreased defense-related research and development would result in a loss of these research teams, formed the Committee on Possibilities and Policies for Industrial Conversion.[22]

The administration's goal was to use the researchers and engineers for other government programs to even out federal spending and maintain technological readiness for future needs. This coincided with the NIH's desire to maintain a working relationship with industrial teams. Industry certainly was not averse to the proposed partnership fueled with federal funds. Frank Hastings, the first director of the artificial heart program, explained the political rationale many years later: "The most important part of the... program is not to develop an artificial heart but to develop a biomedical capability in industry."[23]

When the artificial heart program got off the ground in 1964, Lyndon Johnson had become president, and it fit nicely into his promise of transformation of American life through the "Great Society." President Johnson, who also had heart disease, found no reason to oppose the program. In December 1964, the President's Commission on Heart Disease, Cancer and Stroke recommended the development of the artificial heart.

The federal agency most directly involved in the artificial heart program was, of course, the National Institutes of Health, which through the years preceding the artificial heart had forged a close alliance with the nation's leading medical research centers and their constituent investigators. This accommodated the political ideal of federal funding of private efforts. Through its program of grants the NIH was able to stimulate and coordinate research while maintaining the independence of investigators who were on the receiving end of a funding bonanza. In return, some of the most prominent investigators, such as Dr. DeBakey, became politically active in support of the NIH program.

Dr. James A. Shannon was the influential director of the NIH during the period of rapid growth of NIH-funded research, and during the initiation of the artificial heart program. Year after year, his success in expanding the NIH budget was the envy of lobbyists on Capitol Hill, and he later attributed his success to being able to bypass administrative budget cutters through his close working relationship with Senator Hill and Congressman Fogarty.[24]

In the expansive mood of the mid-1960s, no one social problem was perceived as in conflict with another, and the obvious solution was to throw more dollars into the budgetary pot. Medical research, through a partnership of science, universities, and private enterprise, was politically clean and ripe for political success, and the artificial heart program was just one of many medical research efforts enthusiastically embraced at all levels of government. Unfortunately, the political and economic good times of the 1960s did not compel those within the key congressional committees or within the NIH to scrutinize the assumptions underpinning the race to build an artificial heart.[25] Nor did it force them to apportion their allocations among basic research, preventive medicine, specific medical technologies, or even nonmedical projects. Much less did the planners understand the implications of their acts for later times when their successors would be forced to make such decisions.[26]

The NIH artificial heart program

As early as February 1964, the Ad Hoc Advisory Group of the National Heart Institute (one of the National Institutes of Health) recommended that the NIH enter into a development contract program as the approach to rapid development of the artificial heart. Although the first Congressional appropriations were not available until July 1964, in the spring of 1964 the director of the National Heart Institute assigned a coordinator for the program, who obtained temporary authorization from the secretary of HEW to consider contracts. A release of funds was obtained, and in June 1964, the NIH awarded through competitive bidding six contracts to private firms (mostly in the aerospace industry) for comprehensive analyses of different aspects of technical development and needs for the artificial heart. A seventh contract was awarded to a systems-oriented program contractor, to synthesize these analyses and present final recommendations.[27]

In November 1964, the National Advisory Heart Council endorsed "with confidence and commendation" the proposal that the program be approached "on a systems development basis," and that "the resources of industry be brought to bear on the entire development."[28] This recommendation, which would be instrumental in shaping the program for the next five years, arose from two assumptions. Both were stated succinctly in the NIH's proposed master plan, which referred to the "recognition that the development of the artificial heart is essentially an engineering problem and calls for greater participation and full utilization of industry's resources of manpower, facilities, know-how, and experience."[29]

The NIH planners assumed that construction of an artificial heart was predominantly an engineering problem.[30] They had become convinced by the statements of prominent investigators that sufficient basic biomedical information existed, and what they needed was just money and industry's technological expertise. Representative Fogarty, for instance, referred to a letter from Joshua Lederberg, a Nobel Laureate in medicine, who wrote, "I believe the problem is technically difficult but easily manageable within the framework of our present scientific knowledge and technical proficiency."[31] This, as subsequent events showed, was not the case.

The second assumption grew out of the successes of the systems approach in military and space programs. The idea was to use targeted contracts to private firms to develop component parts, such as energy systems, pump designs, and blood interface materials, all of which would be brought together for final integration into a clinically workable heart. As happens in industry, the plan also called for two or more groups to compete on each of the component systems.[32]

In 1965, Congressman Fogarty invited the NIH to "supply a statement for the record as to how much money you would need in 1966, in addition to your budget, to start a real planning program to develop an artificial heart."[33] The NIH was up to the challenge, and requested $40 million for fiscal years 1965–8. At the same time, the agency disclosed its "master plan" to achieve an artificial heart. The first phase, to be completed by the summer of 1966, was the contracted studies of needs and system specifications. The second phase, to terminate around the fall of 1967, was to design, test, and develop the entire system, including specifications for hardware, installation facilities, testing, and training facilities. During the third phase, to be completed by the fall of 1969, the actual design and development of a number of artificial heart prototypes would be accomplished by the winner of the second phase competition. The fourth and final phase, to be completed by 1970, would result in the "availability of specifications to which artificial hearts can be mass-produced, installed, maintained, and monitored."[34] In one version of their plan, the program's administrators promised to implant the first artificial heart on February 14, 1970 – the first Valentine's Day of the new decade.[35]

The summary report of the first-phase contract feasibility studies, prepared by Hittman Associates, a systems-oriented program contractor, was released in early 1966.[36] It called for development of a family of temporary and short-term emergency devices, long-term partial heart replacement devices, and total artificial hearts. More importantly, the report addressed certain medical, social, and economic questions: Who would need and receive the artificial heart? How well would it work, and what would be the consequences of it for those who received it and for society at large? How much would it cost?

The six subcontractors had each made estimates of the need for the artificial heart, ranging from between 10,000 and 500,000 new cases annually; the Hittman consultants threw out the highest estimate, averaged the others, and came up with a specific figure of 132,500 cases per year. When it came to calculating costs and benefits, they simply assumed that all implantations succeed without any operative deaths, and that recipients would return to normal and, on the average, live longer and healthier lives than other people because they would no longer be at risk of having heart disease. There was no talk of confinement to an intensive care unit, being dependent on a ventilator or enfeebled by a stroke; patients would have the operation, recover, and get up and walk out of the hospital. Thus rehabilitated, according to the report, these patients would return to the work force and add $19 billion to the GNP during the first decade of the artificial heart, and another $41 billion during the second decade. The Hittman consultants estimated that the average cost would be about $10,000 per patient, and optimistically said that the taxes paid by recipients of the artificial heart would more than pay for the entire federal program.[37]

The erosion of optimism

In the analysis of needs, costs, and outcomes the Hittman report mirrored the optimism and assumptions of those who enfranchised it. This view, first espoused by the heart surgeons and bioengineers who were its prime advocates, filtered through the alliance of governmental agencies and industrial consultants almost unchallenged, and, if anything, even embellished by the process. Up to the time of the Hittman report, which marked the end of the first phase of the NIH master plan, the artificial heart program had sailed along through various stages of unopposed advocacy, with no real scrutiny of its underlying assumptions or fantasies.

From the beginning, however, NIH Director Shannon was skeptical.[38] He was willing to support feasibility studies, but he thought a major effort to build an artificial heart was premature because of lack of basic knowledge and unsolved problems with biomaterials and an implantable energy source. Moreover, he also was concerned about the narrow base of advocacy for the artificial heart within the overall medical community, and the adverse effect a crash program might have on all other investigators who had to compete for research grants. In a letter to the surgeon general and to the secretary of the Department of Health, Education and Welfare, in October 1966, Shannon wrote:

> A rigorous review... of relevant knowledge and technology has led me to the conviction that total cardiac replacement is not a feasible program objective at the present time. Furthermore, the establishment, as a primary goal, of the implantable "artificial heart" tends to prejudice the development of program plans, priorities, and balance in an unsound manner as well as generate unreasonable expectations.[39]

He also informed key investigators and lobbyists of his conviction that the master plan was wildly optimistic, and "if too much money was authorized, we wouldn't spend it."[40] He quietly favored initial development of a partial artificial heart, known in the trade as an *assist device*. Because the natural heart is actually composed of two pumping chambers, a total artificial heart requires the integration and synchronization of two pumps. A single pump, to replace the major pumping chamber of the heart – the left ventricle – would have the advantage of decreased size and simplicity, and would probably suffice for the majority of patients who might need an artificial heart, and would serve as a "halfway house" on the treacherous technological journey to a total implantable artificial heart.

Although Shannon privately resisted full funding of the program, he publicly supported it, and avoided any public confrontation on the issue with the program's powerful allies. Accordingly, he did not inform key congressional members of his position, and when Congress increased appropriations to $10 million in 1967, Shannon convinced his colleagues in Congress that the artificial heart could be developed for less money than was appropriated. He renamed the program the "Artificial Heart–Myocardial Infarction Program," and through this strategem incorporated the funds into a broader umbrella program for research on heart disease. Approximately half the allocated monies were diverted to research on heart attacks.[41]

By 1967, the program was off and running, but it was running mostly in place. Congressional funding stabilized at about $10 million per year. After diversion of monies to other parts of the umbrella program, much of what was left was spent on atomically driven energy converters, leaving little for work on other components of the artificial heart.[42] Insoluble technical problems remained, foremost of which were biocompatible materials from which to design a blood pump, and an energy system suitable for implantation.[43] A nuclear power system, using plutonium–238, seemed to be the only feasible means of obtaining a totally implantable unit, free of the need of periodic energy resupply.[44] As a consequence, the Atomic Energy Commission (AEC) became interested in developing power sources and even total hearts, and much of the contracted nuclear system work was for a while conducted with joint funding by the NIH and the AEC.[45]

However, the two agencies could not agree on management jurisdiction and a uniform approach to development of the total heart, and in 1967 their collaborative relationship ended, with subsequent independent efforts and a decrease in exchange of information between the agencies. Until the search for a nuclear power source was aborted sometime after 1973, when the AEC dropped the program entirely, there were actually not one but two competing government programs to develop an artificial heart, with no central coordination by the Congress or the executive branch of the federal government.[46]

In addition, the NIH program floundered on its basic assumptions. Application of engineering principles was of no avail when the baseline knowledge required for a "systems" approach did not exist. Furthermore, the strategy of building competition into the program drastically reduced the flow of information necessary for a coordinated effort. Instead of the usual academic process of prompt publication of results and full interchange of information, the parceling of contracts to competitive private companies put a premium on secrecy that undoubtedly impeded the program.[47]

In fact, the NIH shift to targeted contracts rather than grants for basic research threatened to rupture the carefully nurtured relationship between the NIH and its constituent academic investigators. Whereas the NIH sought to forge an industrial base for medical research, the medical research community feared that the expansion of contract programs would lead to the reduction of support for basic research, to the curtailment of creative investigations, and eventually to the undermining of the medical research system itself.[48]

By the end of 1968, the program had stalled considerably under the weight of formidable technical problems. There were no striking technological advances to maintain the interest of Congress. Appropriations leveled off at about $10 million per year, which meant a slow decline when adjusted for inflation.[49] In 1967, Representative Fogarty died of a heart attack, and in 1968 Senator Hill retired. Dr. Shannon retired the same year. The unrestrained optimism of just three years before was wilting.

The redirection of the NIH program

In the late 1960s, two unforeseen events were instrumental in official changes of direction in the NIH artificial heart program. In 1967, Dr. Christiaan Barnard startled the world with the first human heart transplant,[50] and in no time there was a stampede of heart surgeons replicating the feat. Although the poor results of that initial surge of frenzied surgical activity culminated within a few years in a de facto moratorium on heart transplants which lasted for more than a decade,[51] the technical breakthrough of transplantation spawned unanticipated competition for the artificial heart program. Public attention and the energies and research activities of heart surgeons and other investigators interested in salvaging failing hearts turned to the dramatic new therapy which overnight produced walking and smiling testimonial triumphs. The artificial heart program, mired in the inevitable depression of setbacks and disappointments, faded from public notice and congressional attention in the euphoria of the latest clinical adventure.

In late 1968, the director of the National Heart Institute appointed a special advisory committee to study the broad issues presented by transplants and, potentially, by artificial hearts. Of the ten members of the committee, nine were drawn from the medical community. However, before publication of their report in August 1969, there was an unexpected and momentous event: the first implantation of an artificial heart in a human being. It was accomplished by Dr. Denton Cooley, renowned heart surgeon.[52]

Cooley, who was formerly a protégé of Michael DeBakey, had become a prominent rival of his senior colleague, and "one-upped" him by a highly questionable act. DeBakey had been working on models of artificial hearts since about 1960 and was the head of the Baylor–Rice Artificial Heart Program. He also was the most notable of the investigator-lobbyists for the federal program. Cooley was not active in research on artificial hearts, but also was associated with Baylor although he worked at a different hospital and in competition with DeBakey. He knew DeBakey's research team had fabricated an artificial heart and had begun experiments, although in all of seven cases the experimental animals had died within forty-four hours of receiving the prototype heart.[53]

In secret, Cooley contacted DeBakey's chief surgeon-investigator who was doing the animal work, and the engineer who had developed the power console. Cooley persuaded them to construct an artificial heart for him, modified for human use. All this was done without notifying DeBakey, and in violation of the regulations of NIH, which had funded DeBakey's research on the experimental heart. Two days after the equipment was delivered, on Good Friday, April 4, 1969, Cooley implanted the artificial heart into Mr. Haskell Karp. There was advance preparation for movies and photographs during the operation, and Dr. Cooley had told Mr. Karp that a transplant or temporary use of an artificial heart might be necessary if conventional open heart surgery was not successful. Three days later, after a dramatic nationwide plea for a donor heart, the artificial heart was replaced with a transplanted heart. Mr. Karp died on April 8.[54]

By all scientific standards, Cooley's use of the artificial heart was experimental at best. It was, to understate the case, ethically dubious, as there was no reasonable expectation of scientific advancement or benefit for the patient.[55] Cooley had not submitted a protocol for the experiment to the Baylor Committee on Research Involving Human Beings, as was required, and when called to account for his act he simply left the university.[56] The NIH could not discipline him for violation of their guidelines, for he had not himself received funding from them. Cooley summed up his position:

> I have done more heart surgery than anyone else in the world. Based on this experience, I believe I am qualified to judge what is right and proper for my patients. The permission I receive to do what I do I receive from my patients. It is not received from a governmental agency or one of my seniors.[57]

Cooley's defiance of norms of ethical and scientific behavior in this exploit raised many troubling questions about the uses of the artificial heart, such as who should get it, the fate of patients who do get it, informed consent for experimentation, and clinical reliability, among others. Partly because of its clinical failure, and partly because of the way it was done, Cooley's adventure cast a shadow over the entire artificial heart program, and in the opinion of many, set back the program.[58] Certainly it affirmed the thinking of many that an attempt to leapfrog from the base of known technical difficulties to human implantation was premature and counterproductive.

In August 1969, the Special Advisory Committee released its report, stressing the major technical obstacles to a totally implantable artificial heart – namely biomaterials and a power supply.[59] The committee recommended minor reforms: more experimental research at the NIH and closer surveillance of contractors, and more use of advisory groups. Also, they officially recommended a change in emphasis from the total artificial heart to partial assist devices, and, if funds became restricted, less attention to mechanical devices and more emphasis on preventive medicine.

The committee estimated an annual demand for artificial hearts of only about 32,000, downgraded from earlier predictions of 150,000. They also estimated a cost of up to $20,000 per patient, which would amount to a total national cost of $640 million. The committee asked who should or would pay, but offered no substantive answers. Finally, the committee urged that future planning should include psychological, social, ethical, legal, and religious concerns, in addition to medical issues.

Following Cooley's incursion, the artificial heart program again receded from public attention. Congress spent roughly two to three hours per year in reviewing budget requests,[60] and maintained appropriations at about $10–12 million per year.[61] The NIH did not emphasize the program in its annual requests, and most members of Congress were quietly content to keep it within overall appropriations. There were, along the way, some reported technical advances. In 1968, the NIH had entered into contracts with three companies to develop nuclear-powered hearts, and the Atomic Energy Commission financed a similar program through two other companies. The goal was a fully implantable artificial heart that would work for ten years.[62] In 1972, officials announced the first use of a nuclear-powered heart in an animal, and suggested that such a heart might be available to Americans by the end of the decade.[63]

In 1971 and 1972, Clifford Kwan-Gett, working with Kolff in Utah, achieved survivals of up to two weeks in animals fitted with a pump that was relatively free of problems with mechanical breakage and destruction of blood cells.[64] Early in 1973, Yukihiko Nosé, working in Cleveland, kept a calf alive for seventeen days with a new model, and in the same year an experimental animal survived twenty-five days at the University of Mississippi.[65]

Although the program lagged far behind schedule, concerns grew about the unforeseen effects of technology in general and the artificial heart in particular. Perhaps in part because of the Cooley incident and the recommendations of the 1969 advisory committee, but also because of the near-readiness of partial assist devices, in 1972 the NIH appointed another advisory group, the Artificial Heart Assessment Panel, to undertake "the difficult task of attempting to foresee and to come to grips with some of the societal implications of such a technological advance; to assess and to provide recommendations concerning the economic, ethical, legal, psychiatric, and social problems attending artificial heart development and use."[66]

The composition of this panel was remarkable for its difference from previous advisory committees; it consisted of two lawyers, two economists, three physicians (one a psychiatrist), a sociologist, a political scientist, and a priest-ethicist, none of whom had any significant previous knowledge of the artificial heart program or other medical technology.[67]

The panel's charge, however, was limited to an assessment of problems that might arise with a clinically workable heart.[68] Specifically, the panel was not asked to review technical issues or to consider whether a totally implantable artificial heart should be developed or whether there was reasonable likelihood that it would be. Nor was it asked whether the program should be entitled to priority over alternative programs for prevention or treatment of heart disease or other diseases or societal problems. As the panel said in its report, "Indeed, our advice on specific policies is not really sought at all, except on developmental matters and on possible measures that might help to prepare for the introduction and clinical application of the artificial heart."[69]

The panel was asked to work under the assumption of a device that would be totally implanted within the recipient's body, would contain a nuclear power source, would operate reliably and trouble-free without any dependence on external machinery for a period of at least ten years, and would allow the patient to be free to live an essentially normal and productive life. In short, the NIH restricted the panel to recommendations for the use of an already perfected device. In general, the panel concluded that if the optimistic assumptions were realized, the benefits to society of the artificial heart would be immense.[70] As we now know, these idealized circumstances bore no resemblance to the conditions of Barney Clark and those who followed him in the first wave of human implantations.

However, the panel took an expanded view of its charge, especially concerning recommendations for "desirable courses of development and use," which they interpreted as "opening up all of the non-technical problems inherent in the process of getting from here (the present state of the art) to there (the clinically accepted device)."[71] Unlike previous panels which were comprised largely of individuals who favored biotechnical solutions for medical–social problems, this panel probed into all the related "societal" issues, including the most imposing question of all: Should the artificial heart be developed? The panel called attention to the public's asserted right to "reject any particular technological development on the ground that, even though it may benefit some members of society, it is inconsistent with values that society as a whole considers paramount."[72]

The panel acknowledged that technology frequently has been misused for the purposes of prolonging life without regard for quality, and prophetically asked what would happen if the artificial heart fell below expectations and left patients alive but gravely impaired. Panel members raised the specter of patients maintained in states just short of death, and forecast that doctors might be reluctant to deny patients extended life at the cost of decreased quality.[73] They questioned the technological imperative and pointed out the difficulty in turning back from a technological fait accompli.

The panel also pointed out potential disadvantages to society as a whole, regardless of short-term benefit to individuals. Unrehabilitated patients might become a burden to their families and to the community as well. As the panel report poignantly put it: "Perhaps the worst possible outcome would be for the device to work just well enough to induce patients to want it – so that they could see their grandchildren grow up, for example – but not well enough to prevent the typical recipient from substantially burdening others."[74] The panel projected 50,000 implants per year, at an average cost of $15,000–25,000, or an aggregate annual cost of up to $1.25 billion in 1973 dollars.[75] Unless the federal government paid the bill for most patients, as the panel assumed it would, only the very rich could afford to have implants, and there would be violations of social justice. The panel also questioned whether such expenditures were justified when so many individuals in our society go without adequate routine medical care.[76]

The most lasting and specific accomplishment of the panel was its repudiation of a nuclear energy source,[77] on which most artificial heart investigators were pinning their hopes. In addition to the increased overall cost of making plutonium hearts, panel members feared that the recipient would be a walking target for kidnappers or graverobbers who would covet the implanted plutonium, valued at up to $16,000 in 1973.[78] They also were concerned about the hazard of irradiation of surrounding people, with possible induction of leukemia or other radiation-induced illnesses. They concluded that a spouse sleeping in the same bed as a recipient of a nuclear-powered artificial heart would have an increased probability of developing leukemia or some other type of cancer, and likely would be sterilized by radiation within 10–20 years.[79] The panel's strongly worded recommendation to reject a nuclear power source undoubtedly influenced official policy, although contracted work on nuclear energy sources continued until 1980.[80]

Lastly, the panel recommended establishment by the NIH of a permanent, broadly interdisciplinary, and representative group of public members to monitor further steps and to participate in the formulation of guidelines and policies for the artificial heart program.[81] To date the NIH policy directors have not followed this recommendation, and the 1973 panel report stands as the last balanced review of these issues.

By 1973 the focus of the NIH artificial heart program was narrowed to the partial heart, or the left ventricular assist device (LVAD).[82] There would be a rational, stepwise progression of orderly and careful development through bench testing and animal testing before any clinical experimentation would begin.[83] The first LVADs were for short-term use in patients whose hearts would not start up following open-heart surgery. There are about 1,000 such patients per year who die at the time of surgery, but who might survive, if given an LVAD for a period of hours or days during which the damaged natural heart has a chance to recover.[84] Pneumatic-powered devices were given approval for clinical evaluation for this purpose in 1975.[85]

The first implants of LVADs demonstrated their feasibility in supporting the natural heart for short periods, but through 1977 none of the seventeen patients who received the device lived long enough to be discharged from the hospital.[86] During the next few years, the record improved: Of 170 patients who received LVADs through March 1981, 66 were weaned from the device and 36 left the hospital alive.[87] These were bittersweet results; the trials demonstrated some true successes (virtually all the patients would have died without the LVADs), but there were technical problems with fibers chipping off the surface of the pumps, and the external pneumatic power supply remained an obstacle.[88]

Once again, in 1979, the NIH appointed a panel to assess the artificial heart program. This panel, made up of seven members drawn mostly from the medical community, generally avoided social issues, and in 1980 issued a fairly optimistic report endorsing LVADs.[89] However, technical problems remained, especially with an LVAD designed to last only two weeks. What if a patient was kept alive by the LVAD but could not be weaned from it? The panel judged pneumatically activated devices inadequate for long-term use because of quality-of-life considerations, and urged investigators to pursue a fully implantable system, which could be realized only with electrically energized LVADs.

Since 1980, the NIH program has been directed primarily toward development of partial mechanical heart systems capable of functioning for two years without mechanical failure, and free of tethering to bulky external power sources.[90] In most designs, the patient will wear a battery pack around the waist or in a vest, and electrical energy for the LVAD will be transmitted through transformer coils to the implanted energy converter and pump. Another system under development uses a thermal storage unit. Both the electrical and thermal systems require recharging several times a day.[91] Totally implanted systems, the goal of researchers from the beginning, remain elusive and a hope for the future.

The breakaway of the Utah heart

The federally directed artificial heart program extended across the country geographically, academically, and commercially. Until work on a nuclear energy source phased out about 1980, the Atomic Energy Commission was actively if not equally involved in a parallel effort. Over the decade after the beginning of the program in 1965, the NIH approved hundreds of targeted contracts, and later some nontargeted grants, to researchers, institutions, and commercial enterprises. The managers of the NIH program intended throughout to coordinate and direct the various dispersed activities and to achieve a stepwise development of devices through control of the processes of integration of component parts, animal testing, and ultimately clinical testing. That the most intriguing and perhaps significant chapter of the overall story of the artificial heart should emerge outside the aegis of the NIH is an ironic testimony to the politics of American medicine.

The development and clinical use of the Utah artificial heart (Jarvik-7) took place, for the most part, independent of and in many ways contrary to NIH policy. Nevertheless, because the Utah heart represents the social and medical introduction of this form of technology, it and not the NIH program gives us the parameters by which we may measure its importance to society.

The breakaway of the Utah heart is a case study of how personal and institutional incentives, politics, and high-level finances can influence the selection of medical technology. It is also the story of how, under the loosely knit American system of federal assistance to private medical researchers, one group of investigators was able to exploit the federal program to its own particular ends. It is the story of the uneasy marriage of commercialism with established scientific methods and governmental safeguards for the introduction of new and unproven medical technologies.

On December 1, 1982, at the University of Utah, Dr. William DeVries implanted a Jarvik-7 artificial heart into Dr. Barney Clark, a dentist from Seattle.[92] It was the first time a total artificial heart had been intended as a *permanent* replacement for a natural heart. Although school children may always learn of the event as the culmination of steady scientific progress, in the perspective of the development of the artificial heart the event was more accurately a deviation from the orderly course planned by NIH.

The Jarvik-7 artificial heart was developed at the University of Utah, and although it carries the name of the last of a series of investigators who worked on it, its real father, as mentioned previously, was the Dutch-born Dr. Willem Kolff, who moved from Cleveland to Utah in 1967.[93] Born in 1911, Kolff is a tall, straight-standing, white-haired man, fit and robust, who evokes immediate respect and liking. He remains fully involved in his work, is headstrong, demanding of his associates, and paternalistically generous. Above all, he is a single-minded devotee of technological solutions to medical problems. He deflects questions about social and eco-

nomic implications of artificial organs with parables of patients
with artificial kidneys who are able to take raft trips down the
Colorado River.[94] He considers any questioning of the benefit of
artificial hearts a display of ignorance and a subversion of science
and medicine. In his fathering of the artificial heart and his insular
dedication to technology, he is the J. Robert Oppenheimer of
medicine.

Kolff's move from Cleveland to the University of Utah was
greased by the politics of funding for medical research. During
the 1960s, the NIH was the chief source of support not just for
the artificial heart, but for virtually all medical research.[95] The
University of Utah at the time ranked near the bottom of American
medical schools in winning NIH grants, and Utah officials were
looking for "name" researchers who not only would bring their
own funding to the university, but who also would serve as a
magnet to attract NIH grants to other investigators within the
institution.[96] One person they recruited to accomplish this was
Kolff, who was a proven investigator and was internationally
known for his work on artificial hearts, kidneys and other organs.
In 1967, Kolff established the Division of Artificial Organs on the
University of Utah campus at Salt Lake City. By all standards,
Kolff was a brilliant success. Over the next sixteen years, his
institute garnered $23.4 million of research funds from outside
agencies, particularly NIH, for work on artificial kidneys, arms,
ears, and hearts.[97]

Under Kolff's paternalistic but lenient direction at Utah, a series
of investigators in turn fashioned improvements to the existing
artificial heart. In 1971, Robert Jarvik was 24 years old and had
just finished studying medicine for two years in Italy. He was
hired in Kolff's laboratory as a $100 a week design engineer.
Through Kolff's influence and financial assistance, Jarvik finished
medical school at Utah and began refining the artificial heart developed by his predecessors. Jarvik, who is trim and aggressive,
quickly became a formidable researcher. By 1972, he had modified
previous designs into a heart called, with Kolff's encouragement,
the Jarvik-3, which produced a modest increase in experimental
survival time in animals. After a series of further modifications,
in 1979, a Jarvik-7 model kept a calf alive for 221 days.[98]

By the mid-1970s, Kolff began having funding difficulties. An
unenthusiastic Congress had put a lid on appropriations for the
floundering program, and once again Kolff had to compete with
other leading investigators for support. Moreover, the policy shift

of the NIH, of deemphasizing work on the *total* artificial heart, and concentrating on *partial* artificial hearts, troubled Kolff in particular, because he had directed his efforts at Utah to the development of *total* hearts, and away from partial assist devices in which he had less interest.

Kolff saw that because he was working on total artificial hearts he could not count on renewal of his NIH grants after their expiration in a few years. As Kolff said, "The NIH put all their marbles in one basket, and it wasn't our basket."[99] For most of his career Kolff had been dogged by his dependence on benefactors for funding,[100] and once again he felt vulnerable. He decided to reduce that dependence by forming a company through which he could obtain long-term private funding.

Kolff knew he needed the approval and participation of his university research colleagues to form a new company to make the hearts they were working on. On March 11, 1976, he drove twenty-five miles from Salt Lake City, up into the snow of the Wasatch Mountains, to Big Cottonwood Canyon. There he met fifteen to twenty of his associates and co-workers whom he had invited to a Dutch-treat dinner at Silverfork Lodge. That evening, Kolff broached to the group his idea of a private company to manufacture artificial hearts. Jarvik, and most of the others, were skeptical of achieving a marketable heart in the foreseeable future, and initially rejected Kolff's offer of partnership.[101]

Five months later, however, Kolff prevailed and formed his company, Kolff Associates. At a penny a share, Kolff, Jarvik, and Dr. Donald Owen, a materials scientist, each bought 70,000 shares, a 20 percent interest for each. The other sixteen founding shareholders, all from the university's Institute for Biomedical Engineering, went in for 2,000–10,000 shares apiece. The total capital outlay was $3,500.[102] Kolff was chairman of the company, Owen was president, and Jarvik was vice-president for research and development. For the first few years, the company did a minimal consultation business, while its shareholders were working on development of the artificial heart at the university's institute, financed by public grants. The company attained a licensing agreement from the university to manufacture the hearts, and Kolff turned his energies to producing a better product.

However, Kolff and Jarvik did not get on well. Kolff set Owen to work on a different heart design in direct competition with the one Jarvik was developing. Jarvik was infuriated, and under threat of quitting and taking others with him, he forced Kolff to fire

Owen, the more senior investigator.[103] At company meetings the two men fought over policy decisions: Kolff was single-minded in the pursuit of research, whereas Jarvik, with his business orientation, was more concerned with the financial development of the company. As Kolff remarked, "It is not at all unusual that the son revolts against the father."[104] Kolff and Jarvik continued to draw apart in disruptive quibbling, and in 1979 Jarvik quit the board although he remained vice-president and consultant.

By 1980, the average survival time of the heart in animals was only two to three months, due in part to an unexpected and serious problem: Connective tissue, called *pannus*, grew uncontrolled along the suture lines, partially obstructing the flow of blood into and out of the device.[105] This phenomenon, however, was rarely seen in people with prosthetic heart valves, and Jarvik reasoned that it was related to the rapid growth of the experimental animals.

Meanwhile, Kolff feared he was losing momentum. His research grants were drying up, and the experimental results with the Jarvik-7 heart reached a plateau. His program needed a visible achievement. Thus Kolff decided on another gamble – a human implant. Jarvik and others again objected at first, but as before, Kolff soon prevailed.[106]

The hand-picked surgeon for the job was William DeVries, tall, slender, and sandy-haired. As a medical student in 1967, DeVries heard Kolff lecture about artificial organs. After the lecture he asked for a job, and Kolff asked him, "What's your name?" When DeVries told him, Kolff replied, "That's a good Dutch name, you're hired."[107] Thereafter, DeVries worked in Kolff's laboratory where he dreamed of using the artificial heart in human patients. After graduating from medical school, DeVries went to Duke University for his surgical residency, and returned to Utah in 1979 with implantation of the artificial heart as his primary goal. He is a dedicated and hard-working person who used to sit quietly by himself at medical meetings and social events before fame forced him into more polished public appearances. The artificial heart became his life's work and ambition. He was eager to progress to implantation in humans, and when Kolff gave the word, DeVries was ready to go.

The Utah team faced two hurdles. First, they needed approval of the University of Utah's Institutional Review Board (IRB), the local committee responsible for reviewing all human experimentation. Second, they needed approval of the Food and Drug Administration (FDA), which has jurisdiction over new medical devices. The Utah IRB proceeded deliberately. Its members placed restrictions on the type of patient eligible for an artificial heart, wanting to use it only for patients who otherwise would not survive open-heart surgery.[108] In January 1981, the IRB finally approved the team's application and announced it to the press.[109]

The NIH first heard of Utah's plans when a reporter from "NBC News" phoned John Watson, head of the Devices and Technology Branch of the NIH, asking his reaction to the intention of Kolff Associates to apply to the FDA for permission to implant a total artificial heart.[110] Watson, upon hearing the news, phoned Glenn Rahmoeller, director of cardiovascular devices at the FDA, to give him the story and to warn him that NBC would be calling him soon. Rahmoeller asked for the NIH's reaction to this report. Watson said they were not in favor of it, as they "did not believe the device was ready for clinical investigation."[111] Watson also called Utah, according to investigators there, and said "Don't do it." As Kolff put it, "They (the NIH) did everything they could to stop us."[112]

On February 3, NIH Director Fredrickson told the artificial heart program administrators to get involved in analyzing the issues regarding implantation of the Utah heart. On February 13, Peter Frommer of the NIH wrote Rahmoeller, recognizing the licensing authority of the FDA but arguing against approval of the Utah request. As Frommer noted in his letter:

> While we are told that the planned activity will be conducted independently of NIH funding, it is in part derivative of and related to past and current NIH supported activities at the University of Utah. Whether this subtlety is recognized by, or is important to, the general public is doubtful.[113]

Officials at both the NIH and the FDA were hamstrung by Utah's application. The NIH has no control over physician investigators who are not doing NIH-funded research. Dr. DeVries and his clinical team did not develop the artificial heart; they were only going to do the clinical implants, were not funded by the NIH, and therefore did not need direct approval of the NIH for their work. Nor did the NIH have control over the manufacturer of the device, Kolff Associates, which had not itself received NIH funding.[114] Kolff had foreseen the need to be independent of the NIH, which he achieved by transferring control of the device to a private company with no NIH strings attached. He had produced the circumstance in which the implant team could avoid substantive influence or control by the very federal agency that had provided major funding for the development of the artificial heart they were proposing to use.[115] The Utah team was set on a course independent of the NIH.

Nor could the FDA do the bidding of the NIH. The FDA has jurisdiction only over research involving an unapproved drug or medical device. Because the FDA regulates devices, and not investigators, it must deal with the manufacturer of the device, and not with the institution where it originated or with the physicians who propose to use it. Accordingly, whereas the NIH had no control over the manufacturer, by law the FDA could deal only with Kolff Associates and not with the surgical team, and on February 27 the breakaway company submitted its application to the FDA. The NIH, whose very program was at stake, was bypassed, and first received a copy of the application to the FDA from a member of the press who had received it from the University of Utah's public affairs office.

FDA officials, although sympathetic to their colleagues at the NIH, made it clear that their decision would be based on the narrow congressional mandates of "patient protection" and "the importance of the knowledge to be gained" – which under FDA regulations would be determined by the local (Utah) IRB – and would not be influenced by NIH policy or "societal issues."[116] Although the FDA rejected Utah's first application on a technicality, they invited a resubmission, and ultimately gave approval the following September for seven implants. However, this approval came after another astonishing episode, again not in any way a part of the NIH's "rational" program.

Once more, Denton Cooley made a sizable blip on the ragged progress curve of the artificial heart. After learning of the Utah team's plans and their pending application to the FDA – he had sounded out Utah surgeon DeVries at a medical meeting – Cooley was impelled to move first. In July 1981, he stunned the medical world a second time by implanting in one of his patients, ironically a Dutchman, an artificial heart designed by Tetsuze Akutsu, Kolff's former assistant at Cleveland. Two days later, the artificial heart in turn was replaced with a human heart transplant. The patient died a week later.[117]

Some observers saw Cooley's bold defiance of medical custom and the principles of sound investigation as an aberration of no long-term significance. However, his move highlighted and foreshadowed two important consequences of deviation from the planned program of the NIH. First, it produced immense publicity, as was undoubtedly intended. Reporters descended on Cooley from around the world, and the meaning of the act in terms of limited scientific knowledge gained through the use of an unproven and largely untested device was lost in the cover stories and television interviews.[118] It set a precedent as a medical spectacular. Cooley had beaten Kolff to the publicity, as fleeting as it was, and showed the world how to do it.

Second, Cooley spurned the customary procedure of gaining approval of the local Institutional Review Board before experimentation. Furthermore, since his previous artificial heart implant of 1969, the law had been changed, making it mandatory to gain FDA approval for the use of new medical devices. Cooley flouted the FDA by not doing so. He publicly stated his defiance, saying that his personal authority was enough to do what he needed to do for his patients.[119] The FDA authorities, clearly on the spot, informed Cooley and the Texas Heart Institute of their noncompliance with federal law, but did not pursue the matter beyond this effete slap on the wrist.[120] Their explanation was that the courts had never upheld FDA prosecution unless the FDA had given the offender at least one warning.[121] Surgeon Cooley had demonstrated the ability to circumvent the FDA under the rubric of "emergency therapy" for a dying patient.[122] It would not be the last time that someone used the artificial heart in this manner.[123]

The commercial ingredient

After the decision to proceed with a human implantation but before Barney Clark's operation, Kolff's company experienced tumultuous changes. In June 1981, because of a pressing need for capital, Kolff sought the assistance of a New York consulting firm which specializes in acquiring venture capital for fledgling health-care companies. The financial consultant, W. Edward Massey, probed the loose business structure and methods of the company, and concluded, as did others, that Kolff had neither the right temperament nor training to be an adequate business manager.[124] As a requisite to attract venture capital, Massey wanted Kolff replaced with someone more financially astute and aggressive, who would be a full-time president and could convince prospective investors of the company's promising future. To what extent the New York financier was influenced by Jarvik in this appraisal, no one has said, but Massey agreed to provide financing with a condition: appointment of Jarvik as president of a reorganized company.

Kolff was enraged at first, but finally relented for the survival of the company and so he could continue his research. Jarvik set to work, and his financial contact got in touch with a New York venture backing firm, Warburg–Pincus Capital Corporation. The combination of Warburg–Pincus's desire to invest in the health care field, and Jarvik's captivating analyses of the potential profits on sales of 15,000–50,000 hearts a year culminated in the purchase by Warburg–Pincus of a 20 percent share of the company for $1.2 million.[125]

A few months later, while Kolff was away from the store in Australia and New Zealand, Jarvik engineered a coup by forming a new management committee. Incredibly, Kolff lost control of his company, and also received an unfavorable allocation in a new incentive stock-option plan. As a result of the fracture, he and Jarvik no longer speak to each other, but Kolff refuses to acknowledge anger. As he said, "Do you think I should be bitter in my remaining years?"[126]

After the media success of Barney Clark's implantation, venture capitalists lined up, and in early 1983 a second package of $5 million credit was arranged for the young company with the dramatic product, including substantial investments by Hospital Corporation of America, American Hospital Supply Corporation, and Humana, Inc.[127] Along the way, surgeon DeVries picked up 27,000 shares of stock, whereas Jarvik amassed 712,500 shares.[128] At this point, the original shareholders still controlled 64 percent of the outstanding stock. In July 1983, the company went public, and offered 1.5 million shares at $12.50 per share. The issue sold out on the first day.[129] To obtain a better corporate image, and also to erase the Kolff legacy, Jarvik and his close associates came up with a new name for the company: Symbion, derived from *sym*-biosis and *bionic*.

The Barney Clark experiment

FDA authorities cite the second Cooley affair as coincidental in the approval of the Utah application less than two months later, but DeVries thought it was instrumental. Whatever the relevance, on September 10, 1981, the FDA granted permission to the Utah team to implant their total artificial heart, but only in a patient who could not be resuscitated after failure of the natural heart during open-heart surgery. In time, this proved to be an obstacle for the Utah surgical team, because of the near absence of patients who fit the FDA criteria. The open-heart surgical load at the University of Utah at that time was relatively small by comparison to many surgery centers, and the likelihood of getting such a patient was slim.

Months went by without an acceptable patient. In the animal laboratory, a calf named "Tennyson" lived 268 days with the total artificial heart, providing some publicity, but the average longevity in animals was still about 2½ months. DeVries and his surgical team needed to change the ground rules if they were to make progress. In May 1982, they submitted a revised application, seeking FDA permission to use the artificial heart in a person who was not dying at the moment, but who had chronic, severe heart disease with no hope of recovery. In less than a month the FDA approved Utah's revision, and the team began its search for the "perfect candidate."

Barney Clark, a Seattle dentist, first went to Salt Lake City for medical consultation in March 1982, following almost three years of slow deterioration, particularly weakness and shortness of breath.[130] While there receiving an experimental drug, Clark learned of the artificial heart program. Deterioration continued, and in October he again went to Salt Lake City for further testing. This time he talked to the surgical team about his chances and his potential candidacy for the artificial heart. He visited the animal laboratories and looked at calves standing in cages with air tubes running from compressed air consoles outside the cages to the artificial hearts within them. He heard the constant thumping of the pumps, and could see the clear plastic air hoses recoil with each mechanical pulsation. He expressed some concern about life after receiving the artificial heart. As he put it, "The sheep and cattle can't tell you" how they feel.[131] Nevertheless, the selection team, consisting of surgeon DeVries, two cardiologists, a psychiatrist, a nurse, and a social worker, considered him an excellent candidate.[132]

At first, Barney Clark rejected this solution fearing that his quality of life might be poor, but as his condition worsened he changed his mind.[133] At Thanksgiving Day dinner, by the account of his son, a head and neck surgeon:

His life was pretty much confined to a six-foot radius. He would have to sit for minutes on the couch just to muster up enough energy to walk to the dinner table. The next day, Friday, I called a cardiologist at Utah and said, "He's really in terrible condition." The Utah team decided then that Dad would be the first man to have the artificial heart.[134]

By all accounts, Barney Clark went into the experiment with his eyes open. He was not optimistic about the outcome. He seems to have been motivated by making "a small mark in this world,"[135] and in common with his family thought that he would die very quickly or would make a very quick recovery. As his son said after it was all over, "He never really thought the artificial heart would work *for him*."[136] Nevertheless, Barney Clark flew by commercial jet to Salt Lake City, and on the evening of December 1, 1982, was taken to the operating room for replacement of his struggling heart with a Jarvik-7 model, the first time an implanted artificial heart was intended to be permanent.

Barney Clark's life with the artificial heart was not as he or his family had predicted; why his physicians did not foresee his anguished course is less understandable. Forty-eight hours after the historic operation there was a tear in his lungs and air leaked into the tissue beneath his skin, requiring a half-hour operation for repair of the leak. On the sixth postoperative day, he suffered a series of seizures lasting two hours.[137] His son reported, "In my mind, Dad was never the same mentally after he suffered the seizures within a week of the implant. During the times I saw him, there would be short intervals of lucidity and much longer intervals of confusion. He was never quite together."[138]

On the thirteenth postoperative day the implanted heart began to fail: He had to have repeat "open-heart" surgery to replace a broken valve in the implanted heart. Only later did it come out that during animal testing the valve used in the Jarvik-7 had failed on several occasions, and the team had not specifically tested it for the long-term use that would be necessary in a human setting.[139] In the sixth postoperative week, Dr. Clark had severe nosebleeds, which subsequently required surgery to tie off the bleeding artery. During his life with the artificial heart, he had pneumonia, recurrent kidney disease, gout, epididymitis, and an intestinal ulcer. Finally he developed overwhelming infection and failure of his lungs, kidneys, and intestines, and after 112 days his heart was finally set at rest.

Marketing the operation

The Barney Clark case illustrates several issues relevant to the introduction of new medical technologies. For one thing, the Barney Clark saga was as much an experiment in the use of the public media to market a medical technology as it was a scientific experiment. No medical event had been so extensively and intensively covered as was this one, and never before was there so much debate over the reports in the media and the purposes for which they were used. If medicine was coming of age in public relations, the Barney Clark epic marked the beginning of the deliberate use of reporting to influence public acceptance of a medical "advance."

Once Barney Clark had his operation on December 2, 1982, the story shifted from the humdrum reporting of animal testing to the drama of the human patient with the bionic heart. For months the University of Utah public relations people had prepared for what they anticipated and created – a media blitz. Reporters (hundreds, notified in advance, came from all over the world) were each given a packet of background papers and articles describing the Utah heart and what it took to achieve it. Immediately after the operation, hospital spokesmen made regular announcements and were responsive and accommodating. The surgical team was available for questioning. The reporters dug in.

For the most part the public media reported the event as drama, not as medical experimentation. Understandably, the focus of the press and the Utah spokesmen was on the personal stories of the players in the drama. In the previous most notable medical media spectacular, swashbuckling surgeon Christiaan Barnard was the center of attention. This time the surgeon, DeVries, was too plain and reticent to make good copy.[140] The media turned to the patient who dared to sacrifice himself to the experiment. Barney Clark had character, and instantly schoolchildren across America knew about him and his devoted wife, Una Loy Clark. The operation was likened to landing on the moon, and Barney Clark, pioneer and explorer, was compared to Columbus.[141] Although some few reporters raised the hard questions about use of technology, cost, informed consent, and rationing of scarce resources, the event in general was presented to the media and duly reported to the public as a remarkable technological breakthrough, with all the trappings of a live soap opera.

The very first statement given to the army of reporters contained perhaps the most consequential claim to arise from the adventure. Barney Clark, they said, would have died within minutes had he not received the artificial heart.[142] There is no question about Barney Clark's condition when he went to surgery: He was very sick, with almost no chance for meaningful long-term recovery. However, no heart specialist can be sure when death is imminent for a patient with an ailing heart, unless that person has already died and been resuscitated. Yet, when the Utah surgeons made the claim of saving Barney Clark's life – a moot claim they repeated with subsequent implant patients – they in effect used the public media, rather than established scientific channels, to withdraw their act from the realm of experimentation and to establish it as therapy. The Utah team was sincere in its claim, and at the same time acknowledged the simultaneous experimental nature of their act, but the ex cathedra claim of therapy obviously had far-reaching implications in public acceptance of this technology, and set a precedent for other investigators.[143]

For the first few days after Barney Clark's implant, hospital spokesmen and the surgical team fed news to the assembled reporters. Spokesmen made statements such as "He's doing better, if anything, than a standard (heart) bypass patient."[144] DeVries said Barney Clark could leave the hospital "within a few weeks," and Jarvik said that tests of the artificial heart show it is capable of operating for at least 4½ years. Even Kolff, usually more cautious, made wild predictions that artificial heart recipients would be barred from marathons within fifteen years because they would be too strong.[145] These early statements, eagerly swallowed by a public hungry for good news, produced the first impressions of an uncertain, perilous experiment of a device that had sustained cloistered life an average of 2½ months in animals.

But the honeymoon with the press quickly faded. Clinical complications set in, and the press releases became less frequent and more controlled.[146] After a few days it was difficult for reporters to get to anyone but a designated spokesman who was getting his information secondhand. The issue of family privacy during a public event came under scrutiny. On the one hand, the family was hounded unmercifully; on the other, they bypassed an agreement to "pool" interviews by granting an exclusive interview to the Mormon-owned Salt Lake City *Desert News*, and also sold exclusive book rights to their stories.

The clinical team also had conflicts. Initially they were a leading part of the well-orchestrated press releases that planted the first seminal impressions in the minds of a public eager for medical breakthroughs. However, when publicity became adverse, they withdrew. The spokesmen who fronted the clinical team continued to emphasize positive events even as Clark's condition worsened. They mentioned the negative events, but frequently were not understood or were misinterpreted by the press, and they had trouble reversing gears to undo the success story they had crafted.

About ten days after the implant, they seemed to sour on the media, which had so willingly publicized the event, and began to perceive it as inimical to their interests. There was growing internal criticism of saying too much, and a collective desire of the team to turn off the story as it was being told. Spokesmen erroneously denied that Clark, a Mormon, was a long-time cigarette smoker. His doctors had not fully appreciated his chronic lung disease, which would be a clinical problem throughout his hospitalization. One assistant surgeon, when asked about it by a reporter, retorted, "Don't ask me about that." On the twelfth postoperative day, reporters were told that half of them would have to leave; what had begun as a media party was being called off.

The next day a valve broke in Barney Clark's implanted heart, and the reporters would not be shooed away. They stayed, but interviews with the surgical team became next to impossible. Nevertheless, as time went on, pressure grew to provide positive news about the nation's most talked about patient. Sometime in January or February, 1983, a reporter from the Salt Lake City public television channel did two separate videotape interviews of Barney Clark. During these interviews the confused patient did not look well, and had nothing good to say about his venture. The videotapes were never released or acknowledged to the press. [147] In a later interview, in March, the Utah team selected the best vignette from more than a half hour of probing for a good response. The nation watched on national television as DeVries asked his enfeebled and ashen patient, "You got any words of advice for somebody that's going through this again?" Clark replied, "Well, I'd tell them that it's worth it. If there's no alternative – they either die or they have it done." [148]

Utah spokesmen felt aggrieved by charges of press control. In fact, they often did tell both sides of the story, but reporters did not always listen to the negative side. [149] However, the selective withholding of information and the initial hype and unfounded claims of clinical success set the media on a predictable and irrevocable course. The public media undoubtedly contributed to the mistaken public conception of a medical miracle, but the collective failure of the Utah team to provide the interviews necessary for responsible investigative reporting made it difficult to change the story.

Only long after Barney Clark's demise did we learn how it really was for this medical pioneer. About a year and a half later, two psychiatrists who had followed the patient throughout the experiment detailed their observations in a psychiatric journal.[150] They reported medical impediments (an endotracheal tube needed for respiratory assistance, packing of the nose, and extreme fatigue) throughout the hospital course. He complained from time to time of a "pounding" feeling in his chest over which he expressed distress, but not pain. There were long periods of disorientation, and the last months of his life were marked by repeated episodes of "wanting to die" and trying to figure out means of accomplishing this. His wife sadly described him as being totally indifferent, as though she were "not there."[151]

Clark's surgeon son recalled:

> Dad was in some pain. There were interminable and unending medical procedures, simply an unbelievable number and all very necessary. Once, the news reports said that Dad had had a "darned good day." Well, if you'd been in the room with him as I had, you'd know that "darned good day" is a very subjective description. Sure, it was good in the sense that he wasn't nauseous or throwing up or having nosebleeds, but he was still confined to bed, unable to do anything.[152]

Barney Clark's lingering demise revealed more than clinical problems in marketing the artificial heart. The price tag for his care exceeded a quarter million dollars, dwarfing earlier estimates. Who should pay for expensive technology? Although this historic case did not answer that question, it did set some precedents for sources of payments. The University of Utah billed Medicare, which said it would not pay for the actual operation but allowed associated costs, and by one report paid more than $100,000. Clark's private insurance company – to which this writer also pays premiums – picked up $5,928 of the bill, and the remainder was covered by contributions to the university.[153] Public agencies already have set an example of paying for this unproven technology at a time when more than 10 percent of the nation's citizens are without insurance for basic medical care. Public expenditures on the artificial heart thereby threaten to deprive many persons of access to needed medical care.[154]

The corporate takeover

Although Barney Clark's experience was hailed by the press as "pioneering" and a clinical success, the surgical team met resistance in proceeding to the next implant. The Utah IRB was concerned about the safety of future patients and the process of informed consent.[155] Its members were not pleased that the investigators had not pretested the valve that broke thirteen days after implantation in Barney Clark,[156] and they also wondered why the team had not known of Clark's advanced lung disease. They also were concerned about DeVries's desire to recruit patients healthier than Barney Clark had been. The IRB requested more data about the Barney Clark case and about proposed changes in the protocol for selection of patients.[157] This produced delays, and the surgical team was further frustrated by the need to reapply to the FDA because of design changes in the heart. Although the FDA approved six more implants on June 19, 1984, the Utah IRB gave notice that it would approve only one at a time, and would require review after each implant. DeVries and his team had no assurance of being able to develop an ongoing program.

From the time of the first association of Symbion with its three proprietary hospital-chain investors, Jarvik had instituted programs for training surgeons from the corporations' hospitals, and from others as well, to do artificial heart implants.[158] The plan was to promote the product and gain acceptance through wide use of it, but now the connection took a more immediate turn. The idea of moving the primary implant team to operate in a hospital of one of Symbion's financial partners, with active promotion rather than constraint by the host institution, was compelling.

As DeVries became more and more vexed with the delays at Utah, he made clear his intention to leave for a more hospitable environment, if he was not given the opportunity to pursue his program of more implantations.[159] Officials of the Humana Heart Institute, in Louisville, Kentucky, pursued DeVries as ardently as university coaches recruit standout high school athletes. They offered him a free hand to do his research. At a dinner on the porch of Dr. Allan Lansing, director of Humana's heart program, Humana Chairman David Jones asked DeVries, "How many hearts do you need to find out if it works? Would ten be enough?" DeVries, not a born negotiator, allowed that ten would be just fine. "If ten's enough, we'll give you 100," replied Jones.[160] Humana officials also arranged for DeVries to have a private practice which would triple his income. The driveline of the artificial heart was now connected directly to corporate headquarters.

On July 3, 1984, DeVries applied to the Humana IRB in Louisville, a formality required by the move. Four weeks later, he announced his intention to move to Louisville, which he did before the Humana IRB gave official approval on September 10, 1984. The unanimous decision of the sixteen-member board to allow the surgical team to proceed with the six more implants already allowed by the FDA was remarkable in its contrast to the deliberations and divisions of the Utah IRB. The Humana IRB had no experience in dealing with this sort of issue, and when interviewed later, some members did not know they had acted differently from the Utah IRB. DeVries's move before official approval of the IRB, constituted by members who had an interest in promoting Humana, naturally raised the question among observers as to whether the Humana IRB had acted as a rubber stamp.[161]

Humana spokesmen denied profit-making as an incentive for the deal. President Wendell Cherry defended the use of the plastic heart with an ironic metaphor, "My heart is as ethical as anyone's heart."[162] Chairman Jones said they were doing it to gain "prestige" within the medical community, and, as he was widely quoted, "We hope to establish Humana as a brand name that stands for high-quality health care services at affordable prices."[163] The free media coverage Humana received from their first implant was probably worth more than all the television advertising time during the Super Bowl. As *Business Week* pointed out, after Humana's first implant the number of enrollees in the chain's health maintenance organization more than doubled in the next six months. They concluded: "Jones' shrewdest move to date – attracting the celebrated DeVries to Humana – has already paid off in spades."[164]

Proponents say the move to commercialism in the development of technology is necessary to infuse needed capital into projects too large for undertaking by nonprofit research institutions. They point to the value of "risk taking," the enhancement of innovation, and speedier marketing of beneficial medical products.[165] Of ten of the principals in the Utah experiment to whom I broached the subject of possible conflict of interest in the connection of their artificial heart program to the for-profit company making the hearts (in which the head surgeon held a sizable interest), only one, a social worker, expressed concern. As the chairman of the Utah IRB Subcommittee for the Artificial Heart pointed out, this sort of relationship between researchers and commercial producers is standard procedure across the country.[166]

Nevertheless, many of the leaders of American medicine are not so sure.[167] They fear diversion of precious resources from basic research to more profitable lines of work, to society's detriment. Dr. Arnold Relman, the editor of the *New England Journal of Medicine*, decried Humana's excessive interest in publicity, saying "Whatever they invest in paying for hospital care they certainly expect to gain back many times over. It's just like a commercial organization promoting the release of a new product."[168] The fear, whether from elitism or insight, is that commercial coopting of medical research will undermine the entire medical research structure, producing expensive therapies of limited value instead of fundamental medical advances. With regard to the artificial heart, the weight of capital seeking a salable product is almost certain to produce a course different from the originally planned progression of the artificial heart. The introduction of commercialism into the artificial heart creates the risk of premature dissemination of a device that may be harmful to its recipients and could lead to public disillusionment and rejection of the entire program.

Conclusion

Dr. DeVries implanted three more Jarvik-7 hearts in patients during a span of about six months after his move to Louisville. Another Jarvik-7 heart was implanted in a patient in Sweden. In all five cases, including Barney Clark, there were serious clinical problems, particularly with strokes. The quality of life in these patients was very poor. The Jarvik-7 model has not performed suitably for permanent human implantation, as predicted by Jarvik less than two years before it was first used in Barney Clark. The experience to date underlines the need for carefully controlled testing and development of new technology in order to avoid unacceptable clinical outcomes.

The basic problems of an implantable power supply and non-clotting surfaces remain unsolved. Nevertheless, there may be uses for the artificial heart short of permanent implantation. Investigators have shown the feasibility of using it as a temporary measure for patients awaiting a human transplanted heart. Moreover, the careful and measured program of the NIH continues. We may someday have a permanent artificial heart capable of prolonging useful life instead of extending death with devastating consequences. Still, we will have to ask if we want a device that benefits a relative few if we must buy it at a cost detrimental to the many. We will have to ask whether this device fits our broader human purpose of living useful, harmonious lives, rather than simply extending biological existence.

The central issue is whether we will utilize the artificial heart in accordance with the interests of a small group of medical investigators and their entrepreneurial backers, or in accordance with the interests of the society that has developed the technology and will have to live with its consequences. It is a classic conflict between the economic independence of individuals, and the determination of the public good in an increasingly interdependent society.

The existence of the artificial heart, with its potential for benefit and the insistence of the public for any new technology, forces decisions that are for the most part not medical, but social and political. How society chooses to handle the artificial heart will influence our lives much more than the few years of added corporeal existence the device may provide for some persons. The stakes are too high, and the consequences too great, to exclude the public from these decisions.

5

The swine flu immunization program

Sweet peace be his, who wipes the weeping eye,
And dries the tear of sobbing misery!
Still higher joys shall to his bosom flow,
Who saves the eye from tears, the heart from woe!
– A far, far greater honor he secures,
Who *stops the coming ill*, than he who cures.
> – Valentine Seaman to Samuel Scofield,
> letter dated August 15, 1809, in Samuel Scofield,
> *A Practical Treatise on Vaccinia or Cowpock*
> (New York: Southwick and Pelsue, 1810), p. v.[1]

When Private David Lewis, an Army recruit, reported for sick call at Fort Dix, New Jersey, in February 1976, complaining of mild, flulike symptoms, no one dreamed he would be dead within hours. Nor would anyone have imagined that his death would open an unprecedented episode in American public health: a federally funded campaign to immunize every man, woman, and child in the United States against a feared "killer" epidemic of swine flu.

At first the choice seemed clear: If the epidemic came, the nation would be protected; if it did not, the program would still have been a prudent investment. "Dollars for lives" became the byword of federal officials. Yet almost nothing seemed to go right. The program was mired in controversy from the outset and, beset with an unending string of problems, it fell farther and farther behind schedule. And the epidemic never came. Moreover, despite re-

---

The first part of this chapter (pp. 127–48) is based on a case study prepared for the John Fitzgerald Kennedy School of Government, Harvard University, by J. Bradley O'Connell and Thom Seymour under the supervision of Professor Laurence E. Lynn, Jr. and Dr. Harvey V. Fineberg of the Harvard University School of Public Health, edited by Diana B. Dutton, and supplemented with research conducted by the Stanford University project in Ethics and Values in Science and Technology (EVIST). The remainder of the chapter was written by Diana B. Dutton based on existing sources, historical documents, and interviews with participants obtained by the Stanford EVIST project.

peated assurance of safety, swine flu shots were later linked to Guillain–Barré syndrome, a paralyzing and sometimes fatal neurological condition. Over a decade later, lawsuits over alleged vaccine-induced injuries were still underway.

With all its setbacks and unanticipated obstacles, the swine flu story provides a rare opportunity to examine retrospectively the pivotal decisions and actions that led to these difficulties, and to speculate about how things might have been handled differently. Just how the program got started, and the momentum that carried it forward over seemingly insurmountable hurdles, provide a revealing picture of the way Washington and the medical establishment responded to a perceived public health crisis, and of the individuals and institutions that made it all happen.

The new flu

Swine flu was discovered in 1976 almost by accident. In January, when army recruits returned to Fort Dix, New Jersey, from their Christmas holidays, they began coming down with respiratory illnesses, as commonly happened at that time of year under the grueling and crowded conditions of boot camp. Army doctors assumed that the usual adenoviruses were responsible, such as those that cause the common cold. Nevertheless, they reported the outbreak to the county health officer, who in turn alerted Dr. Martin Goldfield, the chief epidemiologist for the New Jersey Public Health Department. Given the symptoms of the victims, the rapid spread of cases – hundreds of recruits were hospitalized – and other factors, Goldfield suspected that the outbreak was not adenovirus at all, but flu. He bet the health officer at Fort Dix that laboratory tests of throat washings from the sick soldiers would prove him right.[2]

Sure enough, of the nineteen specimens tested, eleven revealed the A/Victoria strain of flu, which since 1968 had been the most common type of human influenza. Among the other specimens, however, were flu isolates the state laboratory could not identify. These were sent to Atlanta, to the federal government's Center for Disease Control (CDC), for further investigation. Goldfield conjectured that if these were new strains of flu, to which most Americans would not have protective antibodies, they could be the harbinger of a major epidemic. Maybe, he jokingly speculated to a CDC virologist, the new strain would turn out to be the infamous "swine flu," the most virulent form of influenza known to modern medicine.[3]

Back at Fort Dix, on February 4, the same day that CDC received the mysterious specimens, Private David Lewis died. Sick for more than a week with a headache, sore throat, stuffy nose, and low-grade fever, Lewis had ignored doctors' orders to stay in the barracks for forty-eight hours and had joined fellow recruits on a strenuous five-mile march and training session in the snow. On the return trip, within a quarter mile of the barracks, he collapsed. He was rushed to the hospital, but was dead on arrival.[4] Throat and lung specimens from the dead recruit were immediately sent to CDC for identification.

By February 12, CDC had confirmed Goldfield's half-facetious speculation. The unidentified virus from the three initial specimens, plus Private Lewis, was indeed swine flu. Dr. David Sencer, director of CDC, hurriedly called a meeting for February 14 to discuss this disturbing finding. Attending the meeting in Atlanta were experts from the Food and Drug Administration's Bureau of Biologics, which sets standards for vaccine production, the National Institute for Allergy and Infectious Diseases, which sponsors influenza research, Goldfield, some military officials, and CDC staff. Most were astonished at what they heard, "I told you so," Goldfield gloated.[5]

Everyone recognized that the news was ominous. Although influenza is not normally a very serious illness, in 1918–19, a devastating swine flu pandemic – a worldwide epidemic – had swept the globe. Over 20 million people had died worldwide, some 500,000 of them Americans. It was the worst medical catastrophe in modern history. The pandemic had also infected and killed large numbers of hogs – hence the flu's nickname. Although the disease had died out among humans by the late 1920s, it continued to circulate among swine, emerging occasionally among humans in contact with pigs. Since then, there had been almost no cases of "human-to-human" transmission, such as had apparently occurred at Fort Dix. If the disease were now returning to humans as it appeared, no one under age 50 would have built up specific antibodies from previous infection. The toll in morbidity and mortality could be enormous.

There were, to be sure, several important qualifications. The virus responsible for the 1918–19 pandemic had never been identified with absolute certainty. It was not until 1931 that laboratory techniques were able to isolate the swine virus from that year's pig population. Based on the similarity between this virus and the antibodies still in the blood of people who had survived the 1918 pandemic, virologists concluded that it was the same virus that had caused the pandemic. Although most experts accepted this conclusion, a few remained doubtful.[6] Even if it were the same virus, however, it might not cause nearly as much damage as it had in 1918, since many of the deaths then had been due not to swine flu itself but to bacterial pneumonia brought on by the weakening of the respiratory system, which today could be prevented by antibiotics.

To most of the participants at the February 14 meeting, such considerations were negligible in light of all the other ingredients for a pandemic that seemed to exist. For one thing, all known prior pandemics had been preceded by a number of smaller, localized flu outbreaks, a process known as "seeding" the virus in the population. The 1918 swine flu pandemic itself had come in two waves, the first much milder than the second. The handful of swine flu cases at Fort Dix could be the precursor of a new pandemic the next flu season. Second, the Fort Dix virus had two key surface proteins, called antigens, which differed from those of the dominant strains of influenza viruses. Experts call this an "antigenic shift."[7] The prevailing view at the time was that because an antigenic shift left the population without immunity to the new strain of flu, it invariably led to a major pandemic. Third, previous pandemics seemed to come about every eleven years, having occurred in 1946, 1957, and 1968. Another one was due soon. Some experts had even predicted that the next pandemic would be caused by a swine virus, based on an influenza virus "recycling" theory then current, which postulated that the flu virus had a limited number of possible forms and hence that earlier forms would resurface when a large enough group of people without immunity had accumulated.[8]

The meeting's participants agreed that more data were needed to determine whether the outbreak at Fort Dix did indeed herald an emerging epidemic or was merely an isolated incident. They decided there should be a broad field investigation to determine the extent of the outbreak in and around Fort Dix. They also agreed that since the significance of the Fort Dix cases was still uncertain, there should be no publicity, which might prematurely and unnecessarily raise public concern. A few days later, however, fearful of uninformed press leaks, Sencer changed his mind, and CDC held a press conference on what was known about the Fort

Dix cases. The story made headlines. On February 20, 1976, the *New York Times*, in a front-page story, reported: "The possibility was raised today that the virus that caused the greatest world epidemic of influenza in modern history – the pandemic of 1918–19 – may have returned."[9]

That same day, Sencer convened a broader meeting at the Food and Drug Administration's Bureau of Biologics in Bethesda, including not only the people who had met at CDC the previous week but also scientists from state and local health departments, universities, and vaccine manufacturers. Research and epidemiological data on swine flu were reviewed. Thus far, intensive surveillance of influenza activity throughout the country had found only the A/Victoria virus. But the human-to-human transmission in the four swine flu cases at Fort Dix was confirmed: None of the diseased recruits had had any contact with pigs, and laboratory contamination of the cultures had been ruled out. This news evoked in many of the participants a sense of exhilaration, even excitement. Bureau of Biologics Director Harry Meyer described his own reaction.

> In the world I deal with every day, there are so many things you do that are not terribly interesting, but which are called "real chores." To have a challenge of something that is a real public health interest is really stimulating. So perhaps it is bad to have these things happen in one respect, but it is kind of stimulating to those of us who are in public health in another respect.[10]

Another participant, Dr. Maurice Hilleman, Vice President of Merck, Sharpe and Dohme Laboratories, a vaccine manufacturer, repeatedly sounded the theme of "heroism" in the ensuing discussion. "There [will] have to be some very heroic decision-making very soon," he told the group.[11]

During the following weeks no new swine flu cases were discovered at Fort Dix or in the surrounding community, elsewhere in the United States, or for that matter – according to the World Health Organization – anywhere in the world. There were plenty of new influenza cases at Fort Dix, but these were all caused by the Victoria strain. Had Goldfield made his bet just a week later, swine flu might never have been discovered.

Not enough was known about how flu epidemics spread to interpret with certainty the absence of further swine flu outbreaks. The swine virus might simply have "sunk" back into the pig population, or it might be spreading through the human population without causing clinical symptoms, only to erupt into a pandemic the next winter. Some scientists, including CDC's own chief virologist, felt that a swine virus so quickly dominated by the Victoria strain at Fort Dix was not apt to rise up later and sweep the world. But no one could say absolutely that it could not happen.

In the meantime, the surveillance effort at Fort Dix produced some troubling results. In addition to the four cases of swine flu identified initially by CDC through virus isolation, a fifth case was similarly established in a soldier who had been sick in early February. Another eight cases were identified on the basis of rising swine antibody levels in two successive blood tests, a highly reliable method of diagnosis. This brought the total number of proven "clinical" cases of swine flu to thirteen. In addition, single bloodtests from a large sample of recruits at Fort Dix revealed swine flu antibodies among some 500, suggesting that these men too may have been infected with swine flu at some time, with or without clinical symptoms.[12]

Sencer decides

David Sencer was known as a tough, effective, and dedicated administrator. Director of CDC for nine years, Sencer inspired considerable loyalty among his staff, despite some complaints that he held the reins too tightly. Outside CDC, he was regarded as something of an operator, a self-confident, brash, and sometimes "manipulative" bureaucrat not necessarily to be trusted.[13] He had always taken an active hand in most policy matters involving CDC, and so he did with swine flu.

On March 10 the group that had met February 14 reassembled at CDC and, with Sencer chairing, reviewed their findings with the Advisory Committee on Immunization Practices (ACIP), an external scientific advisory panel to CDC. In January, the ACIP had given drug manufacturers its recommendation for the 1976–7 flu season, namely, to produce enough Victoria flu vaccine – about 40 million doses – to immunize groups traditionally considered "high risk": the elderly and people with certain chronic diseases. The purpose of the March 10 meeting was to consider whether this recommendation should be revised in light of Fort

Dix. If any vaccine for swine flu was to be available in time for the 1976–7 flu season, CDC would have to act almost immediately so the manufacturers could begin production. The press was there; everyone understood that important decisions were at hand.[14]

No one at the meeting was willing to predict the occurrence of a pandemic in the next year, or even to estimate the chances. Privately, most seemed to consider it quite unlikely (although one participant, Dr. Edwin Kilbourne, a highly respected virologist, thought it "very likely.")[15] They all agreed that a pandemic was possible, however, and therefore, since everyone was potentially at risk, that steps should be taken to ensure the production of enough vaccine for the entire population. The group stopped short of actually endorsing mass immunization, but recommended that the necessary plans be developed.[16]

This recommendation was unprecedented in two respects. First, it expanded the scope of a federally sponsored vaccination program beyond traditional "high-risk groups," the targets of past influenza immunization efforts, to the population as a whole. Second, it gave the federal government a much more active role than it had ever had. CDC officials remembered keenly the lesson of the past two major influenza pandemics, in 1957 and 1968, when the federal government had been unable to mobilize immunization campaigns in time. In 1957, although Asian flu had been reported in China six months before it peaked in the United States, vaccine production and distribution, largely under private control, fell far short of needs, despite federal exhortations. In 1968, with almost five months of warning, an even smaller fraction of the population was immunized, again because of the limited supply and dissemination of vaccine. The 1957 pandemic was estimated to have caused 70,000 deaths, the 1968 pandemic about 28,000. So inadequate were both immunization drives that the federal officials responsible questioned whether they had had any detectable effect on either pandemic.[17] The message was clear: To assure a timely response to the next pandemic, the federal government would have to be directly involved in vaccine procurement and administration. Fort Dix seemed to present the perfect opportunity. "We had options for the first time," CDC virology head Dowdle recalled. "Up until then, it had all been frustrations."[18]

Legitimate concerns were also reinforced by the sense that here was a chance – finally – for preventive medicine to get its share

of the public spotlight. As Dr. Reuel Stallones, an ACIP member
and Dean of the Public Health School at the University of Texas,
later recalled:

> It was . . . an opportunity to strike a blow for epidemiology
> in the interest of humanity. The rewards have gone over-
> whelmingly to molecular biology which doesn't do much for
> humanity. Epidemiology ranks low in the hierarchy . . . . Yet
> it holds the key to reducing lots of human suffering.[19]

The principal advocate of delaying the decision on mass immunization was Dr. E. Russell Alexander, an ACIP member and
Professor of Public Health at the University of Washington. He
urged that the vaccine be produced but then "stockpiled" unless
there was another outbreak somewhere in the world. Alexander
was known for his cautious attitudes toward medical intervention.
As he later explained: "My general view is that you should be
conservative about putting foreign material into the human body.
That's always true . . . especially when you are talking about 200
million bodies. The need should be estimated conservatively. If
you don't need to give it, don't."[20] This was not, it should be
noted, the majority view. Most experts at the time, including the
other ACIP members, considered flu vaccines to be essentially free
of major risks – "just like water," in the words of Dr. Walter
Dowdle, head of CDC's virology division.[21]

Alexander also asked what would turn out to be a pivotal question: "At what point do we stop going on with our preparations
to immunize everybody and turn to stockpiling instead – what
point in terms both of progress of our preparation and progress
of the disease?"[22] Unfortunately, that question was never answered
at the meeting and was quickly forgotten. Had it been addressed,
then or later, the swine flu story might have turned out quite
differently.

Goldfield, not a member of the ACIP, also spoke in favor of
stockpiling, but was "squelched," he recalls, by Sencer. He described the atmosphere of the meeting as "unfriendly," speculating
that it discouraged several participants who had confided their
doubts to him privately, including the representative of the World
Health Organization, from speaking up at the meeting. Goldfield
came away from the meeting, he remembers, "sure that Sencer
would go for the big thing." (In fact Goldfield had had the same
reaction even after the February 14 meeting.)[23]

Sencer, as chair, had apparently decided that the question of stockpiling was not worth pursuing. He had discussed it with his staff the previous day, and they had concluded it was not feasible logistically. Inoculation took two weeks to provide immunity. If the virus reappeared, CDC staff warned, it could spread rapidly around the country with the aid of air travel ("jet spread"), gaining a foothold before the vaccine could be distributed, shots administered, and immunity built up. Besides, stockpiling made little sense if the vaccine really had no risks; better to store it in people's arms than on refrigerator shelves. Furthermore, there was the question of how stockpiling would be perceived if the pandemic occurred. As one official put it:

> Suppose . . . it comes out: "They had the opportunity to save life; they made the vaccine, they put it in the refrigerator. . . ." That translates to "they did nothing." And worse, "they didn't even recommend an immunization campaign to the Secretary."[24]

Sencer did not press for unanimity at the March 10 meeting but suggested that the group "go home and sleep on it," and he would call them within a few days. In any event, the ACIP's function was to offer medical recommendations, not to design administrative machinery. But, in a closing pun, he made his own position clear: "It looks like we're going to have to go whole hog."[25]

Two days later Sencer called four of the five ACIP members (he could not reach the fifth). Two were reportedly in favor of going ahead with immunization; two, including Alexander, favored vaccine production followed by watchful waiting for evidence of new outbreaks.[26] This left Sencer with the tie-breaking vote. There was little doubt as to how he would cast it.

Sencer immediately set the wheels in motion. He called his superior, Dr. Theodore Cooper, Assistant Secretary for Health in the Department of Health, Education and Welfare (HEW), and reported that the ACIP had agreed that the possibility of a major outbreak could not be dismissed and that an extraordinary federal response was therefore in order. Sencer added that he and his aides were preparing a more specific memorandum to that effect – in all likelihood recommending a national immunization drive – which he would bring to Washington that weekend.

Cooper asked what he called "the usual administrative questions," such as whether CDC had conferred with outside authorities and with the other relevant health agencies. Convinced of the seriousness and urgency of the situation, he arranged for Sencer's recommendations to receive expeditious consideration while he was away on an eight-day trip to Egypt. He instructed his Deputy Assistant Secretary for Health, James Dickson, who would be standing in for him, to pass Sencer's proposal on to HEW Secretary David Mathews and to arrange for Sencer to meet with Mathews. Cooper also mentioned to James Cavanaugh, Deputy Chief of Staff in the White House, that a flu immunization proposal was in the pipeline. According to Cavanaugh, Cooper said that he thought a full-scale immunization program might be necessary, but that he wanted to be certain first that CDC and the other line health agencies had adequately documented the need for and feasibility of such a program.

In a memo entitled "Swine Influenza – ACTION," Sencer summarized the situation for his superiors at HEW. The memo was addressed to HEW Secretary David Mathews, but as things turned out, it did not stop with Mathews but went on up to the Office of Management and Budget, to the Domestic Council, to the White House, and ultimately to President Ford, as *the* decision paper in the case. It was a forceful and persuasive document, aimed at convincing Congress to appropriate the necessary funds and at selling the program to Mathews. The seven "Facts" with which it began built the case for a swine flu epidemic in 1976–7 as a "strong possibility." Fact No. 2 was, so to speak, the killer:

> The virus [isolated at Fort Dix] is antigenically related to the influenza virus which has been implicated as the cause of the 1918–1919 pandemic which killed 450,000 people – more than 400 out of every 100,000 Americans.[27]

Sencer's position, as expounded in the memo, was that the only way a pandemic could be halted was through a program that would immunize the entire population. A half-hearted or more conservative vaccination effort would be little better than none at all (1957 and 1968 had proven that): "The magnitude of the challenge suggests that [HEW] must either be willing to take extraordinary steps or be willing to accept an approach to the problem that cannot succeed." This conclusion was based on a critical – and highly debatable – assumption: "The situation is one of 'go or no go'." The memo continued: "If extraordinary measures are to be undertaken there is barely enough time to assure adequate vaccine production and to mobilize the nation's health care delivery system." For mass immunization to be carried out during the fall of 1976, it concluded, a "decision must be made now."[28]

What this meant, in effect, was that two distinct and separable decisions – producing sufficient vaccine and embarking on mass immunization – were being rolled into a single "go or no-go" decision that had to be made by the end of March – that is, in the next two weeks.

The Sencer memo aimed two criticisms at the recommended approach, both of which tended to glance off it harmlessly: It would be expensive (the total cost for vaccine purchase and mass immunization was estimated at $134 million), and some people might be "needlessly re-immunized" (people over the age of fifty who might still have swine flu antibodies). Actually, to many politicians and government officials, the program's cost would prove to be one of its selling points. Compared with many of the major programs they dealt with, $134 million for a nationwide *anything* was cheap. And, lest anyone overlook the political significance of a swine flu epidemic prior to the upcoming presidential election, in which President Gerald Ford was seeking a second term in office, the memo noted pointedly that "the Administration can tolerate unnecessary health expenditures better than unnecessary death and illness."[29]

Cooper endorses

That weekend Sencer arrived in Washington, memo in hand. Cooper's Deputy, James Dickson, signed it on Cooper's behalf as instructed and arranged for Sencer to meet with HEW Secretary Mathews and other officials Monday morning, March 15.

Mathews was a relative newcomer to Washington and was viewed by many as a political lightweight. Formerly President of the University of Alabama, he had been appointed Secretary of HEW only the previous August and had little clout in the upper echelons of the federal bureaucracy. Mathews was suspected of being both uninformed about and uninterested in health matters – some health staffers called him "the phantom." Sencer considered him "notorious for not wanting to make decisions."[30]

So, taking no chances, Sencer pulled out all the stops in his meeting with Mathews. He pushed hard for a joint public/private program aimed at the entire population, the option recommended in his ACTION memo. He also hinted that Congress, in the person of Representative Daniel Flood, might act on its own and hold appropriations hearings on swine flu if no immunization initiative emerged from the Department.[31] Mathews's principal question, and the one that most frequently would be posed to him over the next ten days, was: "What is the probability of an epidemic?" Sencer, Dickson, and Bureau of Biologics head Meyer unanimously responded, "Unknown." To Mathews, that meant an epidemic *might* occur, and hence must be prepared for. He later recalled:

> The moment I heard Sencer and Dickson, I *knew* the "political system" would *have* to offer some response.... You can't face the electorate later, if it eventuates, and say well, the probability was so low we decided not to try, just two or five percent, you know, so why spend the money.[32]

According to Dickson, the example of 1918–19, where half a million lives had been lost, hung like a "ghastly vignette" over the discussion.[33]

There was also a sense of urgency. They felt, in Dickson's words, that they were in a "time-bind." Flu vaccine is made from killed virus, which is grown in eggs. In accordance with the ACIP's January recommendation, drug manufacturers had already committed all their existing egg supplies to the production of Victoria flu vaccine. If they were now to produce swine flu vaccine, on a scale ten times greater than usual, manufacturers would have to obtain a whole new batch of eggs, which would have to happen before the food companies made hash out of the roosters, as they ordinarily did each spring. Meyer believed that, with some difficulty, the eggs could be obtained and that vaccine could be ready for distribution by midsummer. The next major hurdle, also considered "do-able," was to complete mass inoculation before the onset of winter. Sencer thought this could be accomplished by sometime in November. Meyer thought "by Christmas" more realistic.[34] It was unclear whether HEW would require new authorization legislation to launch a swine flu vaccination program – that point would have to be explored with HEW lawyers – but a supplemental funding appropriation was definitely necessary. That meant the proposal would have to go through the president, and the Office of Management and Budget (OMB).

Mathews inquired about vaccine safety; Sencer's response was reassuring. Vaccines for other influenza strains had been in use for a quarter of a century, with about 20 million doses administered annually. Side effects were anticipated – many arms would be sore and some people would experience fever and chills for a couple of days – but no serious ones. In any case, the FDA would conduct extensive field tests with volunteers before any vaccine was administered to the general population. In addition, CDC would set up an elaborate epidemiological surveillance system to monitor both the spread of influenza throughout the season and the incidence of any major side effects.

Mathews did not announce a definite decision at the end of the meeting, but no one doubted what he would do. That same morning he wrote a note to James Lynn, director of OMB, strongly endorsing mass immunization:

> There is evidence there will be a major flu epidemic this coming fall . . . that we will see a return of the 1918 flu virus that is the most virulent form of flu . . . The projections are that this virus will kill one million Americans in 1976. To have adequate protection, industry would have to be advised now in order to have time to prepare the some 200 million doses of vaccine required for mass inoculation.[35]

Gone were all the caveats and qualifiers that even Sencer had included in his hard-hitting ACTION memo – let alone the guarded views expressed at the March 10 ACIP meeting only five days earlier. ACIP participants had considered an epidemic unlikely – odds in the range of 2–20 percent according to their later reports. Sencer's ACTION memo converted these (mostly unacknowledged) low odds into a "strong possibility"; Mathews's memorandum, in turn, translated that into a virtual certainty – "there *will* be a major flu epidemic." Projections of severity underwent similar escalation. The ACIP had explicitly stated that there was no way to predict whether a new swine flu epidemic would be as virulent as the 1918 pandemic.[36] Sencer's ACTION memo hinted at severity but in language that was carefully hedged: The new virus is "antigenically related" to the virus that "has been implicated" in the 1918 pandemic. Mathews took the identity and severity of the two viruses as a given and, since the population had subsequently doubled, simply doubled the casualty level of 1918 – ignoring the fact that antibiotics would in all likelihood prevent many of the concomitant deaths from pneumonia that had occurred in 1918. It is unclear whether Mathews deliberately overstated his case in order to impress OMB, whose job was to pass on the fiscal soundness of all budgetary items, or genuinely believed what he wrote. Without OMB's approval, however, there could be no mass immunization program.

As it turned out, the OMB people were, and remained, the most skeptical of all of the federal participants that an epidemic would occur. Even before receiving Mathews's March 15 memorandum, OMB staff were working on a swine flu memo of their own. Victor Zafra, chief of the health branch in OMB, had read the newspaper accounts in February about Fort Dix and the possible return of the 1918–19 virus. His reaction was open mistrust: "I didn't believe them." (Zafra commented later that he thought, in general, the "incentive system" in HEW and in most government bureaucracies discouraged "asking hard questions," whereas knocking the conventional wisdom was rewarded in places like OMB, which incubated skepticism.)[37]

Zafra and others at OMB and HEW met with Sencer and Meyer for a briefing later on Monday, March 15. Zafra thought that "they hadn't made their case." He was convinced that the Fort Dix virus had neither the spread nor virulence of the 1918 virus. First of all, the Fort Dix outbreak had occurred under unusual circumstances that increased susceptibility to infectious disease – crowded living quarters and a pool of recruits unaccustomed to the rigors of military life. Second, even under these conditions, only a handful of soldiers on the whole base had been stricken with swine flu; many others had apparently been exposed but had successfully resisted it.[38] OMB Director James Lynn reportedly told Zafra, "go ahead [and raise questions but] you'll never prevail."[39] And indeed, the tone of the internal OMB memorandum on the subject was cautious. It hinted at doubts, noting the absence of other identified outbreaks, the unknown probability of an epidemic, and the probably excessive funding requested – but did not argue directly against the program's approval. OMB staff had neither the time nor the personal connections to link up with other critics outside the federal government, such as Alexander and Goldfield. And without such technical backup, according to Zafra, they simply "did not know enough to say [a program] was definitely bad."[40]

OMB examiners did raise again the question of stockpiling the vaccine pending another outbreak of swine flu. Sencer and other HEW officials continued to insist that it was not feasible for both logistical ("jet spread") and medical reasons. By the end of the week of March 15, the OMB leadership had arrived at much the same view. On budgetary grounds, there was no reason to favor stockpiling over the preventive immunization approach; all but $8 million (for administration) of the requested $134 million would go for purchase of the vaccine in either case. Moreover, if stockpiling meant that not everyone who wanted the vaccine would get it in time, the administration might be held accountable. Even Zafra agreed that although an epidemic might be unlikely, the government was by this time boxed in: The swine flu program had become "the necessary political choice."[41]

Mathews's inquiries within his department had not turned up any opposition to mass immunization or any insurmountable operational obstacles. Also, Sencer, at Mathews's and Dickson's behest, had again polled the CDC's external advisory panel, the ACIP, and filled them in on the details of the vaccination proposal and the status of the federal decisionmaking. Sencer reported back to Mathews their unanimous concurrence. (He made no mention of Alexander's objections.[42]) Jack Young, Comptroller of HEW, advised Mathews, "In situations such as this, I see no alternative but to rely upon the advice of our health professionals."[43]

Upon returning from Egypt later that week, Dr. Theodore Cooper, the chief health official within HEW, gave his own emphatic endorsement to the program. As a strong advocate of preventive medicine, he was no doubt predisposed to be supportive. He regarded both Sencer and CDC as trustworthy and technically competent, and saw no reason to second-guess their conclusions. Neither his office nor other analysis-and-review operations within HEW were set up to undertake that type of medical and epidemiological investigation. As Cooper put it: "If you want to put layers of everything over everything to double-check everybody, then you might as well fire the whole goddamn thing – it ain't worth a damn. The technical expertise is down in the agencies."[44]

The seriousness of the swine flu threat – and its very unpredictability – weighed heavily on the nonmedical officials. All the technical experts' earlier words of caution were forgotten. A comment of William Taft, HEW General Counsel, indicates the impression that filtered up to those regions of HEW most distant from CDC's epidemiologists: "The chances seemed to be 1 in 2 that swine flu would come." An unknown probability had been translated into an even bet. Viewing the odds this way, and believing that the risks of the vaccine were negligible, government officials saw the program as politically inevitable under any circumstances, and especially so given the upcoming election. As an HEW Assistant Secretary put it:

> People at the top of the department came pretty quickly to a
> belief that inaction . . . was simply untenable. . . . And people
> were mindful of the fact that it was a presidential election
> year and that made the thing dreadfully more difficult in a
> sense – the consequences of doing nothing and having it later
> come to light.

If scientific concerns suggested that it could, perhaps should be
done, political concerns dictated that it would be done.[45]

It was assumed by everyone involved that the final decision
would have to come from the president, Gerald Ford. Simply as
a procedural matter, he would have to sign the request for a supplemental appropriation. More important, the decision could have
major implications for life and safety (the estimated half a million
to a million deaths from a pandemic) as well as policy (it would
set an important precedent for federal preventive medical programs). Paul O'Neill, the Deputy Director of OMB recalled:

> I guess it never occurred to me that, whoever the President
> might have been, he wouldn't have been deeply involved in
> this kind of a question. Because the national policy implica-
> tions of a threat of a major epidemic are not the kind of thing
> that, in my judgment, ought to be left to HEW and OMB
> to decide between themselves.[46]

Ford announces

All this came at a busy time for Ford and his advisors. Ford was
running for a full term after having assumed office following Nixon's resignation, and faced an uphill battle. He was not doing well
in the polls, and was widely viewed as inept, indecisive, and un-
inspiring. In the New Hampshire primary, Ford had only narrowly defeated Ronald Reagan. Although he won the next four
primaries, he was now facing a stiff challenge from Reagan in the
North Carolina primary, March 23. The swine flu program gave
him a chance to seize the initiative – to take the helm of a nationwide campaign in what experts said was a genuine public health
emergency. It looked like a political windfall.

At the White House, Domestic Policy staff members recognized
that the president would have to act promptly, even though some
had "real questions" about the whole immunization program after
their initial discussions with Mathews and Dickson. They therefore undertook a review of their own. This was the usual White

House response whenever a federal agency pressed for a new program marked with an "urgent timeframe." "There was no 'rush to judgment'," recalled James Cavanaugh, White House Deputy Chief of Staff. "We'd put the issue on a fast-track for decision but be damned sure we'd gotten a full staff review." As a former HEW man himself, Cavanaugh had contacts that he claimed reached "down into the bowels of the agencies," and he called on many of them to see what problems he could uncover. His inquiries turned up few if any significant criticisms. By the end of the week Cavanaugh, too, was resigned to the swine flu decision.[47]

By this time the press, having learned something of the decisions that were brewing, renewed its interest in Fort Dix and its implications. On Sunday, March 21, swine flu made its second appearance on the front page of the *New York Times*: "Flu Experts Soon to Rule on Need of New Vaccine." The article described the Fort Dix incident as "a single scream in the night and then silence." It went on to report that the government was expected to decide in favor of a mass immunization program, based on the recommendations of Drs. Sencer, Meyer, and other leading advisers. "It's a choice between gambling with money or gambling with lives," Meyer was quoted as saying.[48]

March 21 was also the day that Theodore Cooper returned from Egypt. He had not been in touch with Washington during his trip and was unaware of what had transpired during the week. His first piece of news upon arrival at John F. Kennedy Airport in New York was that he was to attend a meeting at the White House at 11:00 AM the next morning. Prior to the meeting Cooper satisfied himself that all the potential drawbacks of the proposed swine flu program had been adequately explored, and then became a forceful proponent of the position already generally supported by the other federal participants.

Attending the meeting with Ford were HEW Secretary Mathews and Dickson, a number of White House staff members and OMB Director Lynn and Deputy Director O'Neill. Various documents were distributed, including a list of "talking points" from OMB, Sencer's "ACTION memorandum," and an OMB attachment containing all of the questions and doubts Zafra and others still harbored (tactfully labeled "uncertainties"). None of these documents offered any serious objections to the mass inoculation program. Nevertheless, White House staff would probably not have agreed with the view expressed at the public health workshop the previous month – that dealing with the chain of events flowing from Fort Dix was exciting and "stimulating" rather than a "real chore."

The discussion at the meeting covered most of the questions and answers that had circulated through the agencies – the unknown probability of an epidemic, the expectation of few if any serious risks from the vaccine, the probability of minor side effects such as sore arms and fevers, the challenging but feasible production and distribution timetables, and other issues. The president reportedly maintained a "typical Ford" demeanor – "a conservative, quiet listener," asking a few questions.[49]

Some advisors were worried that the program might be a political liability; if swine flu did not spread, the president might be seen as a spendthrift and alarmist. One White House official recalled:

> It was going to cost a lot of money and a great inconvenience to people, and privately some of the political experts around thought that this might be very damaging to the President because people would have sore arms in October, just before the election.... That was never a serious consideration, but someone did raise it as a possibility.[50]

As a way of lessening the potential for political embarrassment, Mathews, and later Cooper, offered to announce the program to the public. The suggestion was left up in the air, although most of the White House advisers thought that convincing people to line up for flu shots would probably require an exercise of presidential leadership.

There was no devil's advocate per se at the March 22 meeting, although the OMB officials were still, according to Cooper, "very leery" of the program. Deputy OMB Director O'Neill stressed the importance of getting broad scientific input in view of the questions that remained unanswered. On this issue he struck a chord with the president. Ford asked, "How wide has the consultation been?" Cooper explained that CDC had followed the usual process of consulting with the other federal health agencies and putting the issue before its standing advisory panel, the ACIP, before bringing it to the HEW secretariat, but that no new scientific review body had been set up specifically to deal with swine flu. O'Neill argued that such a review should now take place at the presidential level. It would serve not only as a final opportunity to ferret out any scientific objections that had not been raised in HEW but, more important, as a way of extending responsibility for the decision beyond the federal bureaucracy and the administration. It would provide graphic public proof that the president was acting on the best scientific advice he could get.

Ford was enthusiastic about the idea of touching base directly with the "scientific community," and asked his aides to assemble a group of the "best" scientists and other experts, representing a "spectrum" of scientific views, to meet with him in two days. Although Ford did not announce a final decision at the March 22 meeting, the participants emerged convinced that the mass vaccination program was now a near certainty.

The task of putting the panel of experts together fell to White House Deputy Chief of Staff Cavanaugh, who relied mainly on lists drawn up by Cooper, Sencer, and the other line agency heads. Cavanaugh ultimately lined up about thirty nongovernmental experts for the meeting. Among them were several people, such as Maurice Hilleman of Merck, Sharpe and Dohme, Reuel Stallones of the University of Texas, and Edwin Kilbourne of Mt. Sinai Medical School, who had participated in either the March ACIP meeting or earlier swine flu meetings, plus representatives from state and local health departments, the AMA, and sundry other public figures. Also invited were both Jonas Salk and Albert Sabin, developers respectively of the killed- and live-virus polio vaccines, whose names were, to the public, almost synonymous with vaccination. They were not only outside the ACIP circle, which Cavanaugh thought would increase the group's diversity, but were well known to be adversaries, both personally and on most scientific issues. Since Sabin had already indicated his support of a mass immunization program, Cavanaugh expected Salk to spot any faults the proposal might have.

Notably *not* on Cavanaugh's invitation list, despite the request for a spectrum of views, were critics such as Alexander and Goldfield, both of whom had expressed doubts at the March ACIP meeting. Cavanaugh did not know them; the others did not propose them. Also not invited was another "inside" critic, Anthony Morris, an FDA bacteriologist.[51] Nor was anyone from Capitol Hill asked to attend. The White House aides thought it would have seemed either "political" or "weak" for Ford to have brought in members of Congress, especially Democrats, to advise him on *his* decision.[52]

On the day of the meeting, March 24, the invited experts assembled at the White House. Only serious criticism from some of them would stop the program now, and that, based on the advance checking White House aides had done, seemed extremely unlikely. President Ford evidently thought so as well. Just before entering the meeting, he left the Oval Office wearing a white shirt. When he reappeared minutes later, his shirt was television-blue.[53]

The meeting began at 3:30 PM in the Cabinet room. Ford welcomed the group, and Sencer summarized the background and facts. The president then turned to Salk, who gave a prepared presentation strongly urging the president to mount a mass immunization campaign such as Sencer had outlined. Sabin followed with his own statement of support. Ford asked for opinions from the other experts attending, but only a few participated very actively. Everyone agreed that no figure could be placed on the probability of an epidemic.[54] The 1918 disaster was another recurring topic. None of those who spoke up had a disparaging word for the immunization proposal. When Ford asked how many were in favor, all hands went up. He also asked whether anyone present had any reservations about this course of action. (Bureau of Biologics Director Meyer later compared this to asking those at a wedding to "speak now, or forever hold your peace.")[55] There was a long silence. Ford repeated the question a few minutes later, adding: "If anyone has any doubt about this [and] would like to speak to me privately about this, I would like him to do so. I will be in my office for the next ten minutes if anyone wants to come in." He told an aide, "You make sure that they come in."[56]

Ford adjourned the meeting, and retired to the Oval Office to await any dissenters. None arrived. This was hardly surprising, for by then the consensus had been established. Anyone who got up and marched into the Oval Office would have been seen not only as breaking rank with fellow experts but also as opposing a program the president apparently supported. Those who had private reservations kept them to themselves. "Later," one participant acknowledged, "I regretted not having spoken up and said, 'Mr. President, this may not be proper for me to say, but I believe we should not go ahead with immunization until we are sure this is a real threat.' "[57]

Some of the experts contended subsequently that the whole March 24 meeting had been "staged," "orchestrated" to produce the desired decision, which had already been made. Sencer himself described it as a "rubberstamp" intended in part to enhance public credibility.[58] And Sencer had indeed called some of the scientists the night before to tell them when to speak and what the President would ask. White House aides claimed that any coaching by Sencer would have been outweighed by Cavanaugh's statements to the scientists, when inviting them to the meeting, that the president was seeking their independent counsel, and by Ford's declarations during the meeting itself. The point may be moot. Given the choice of those who would attend, the conclusion was all but foreordained.

Indeed, so confident were White House staff members that the meeting would produce a "go" decision that they set in motion ahead of time the machinery for announcing the program. Before the scientists had even arrived, a top White House aide sent the congressional liaison aide a short memo stating that the president would "brief the press at the conclusion of the meeting to announce his decision to give the go-ahead to pharmaceutical manufacturers to produce enough vaccine to immunize every American, at least 200 million doses."[59] And, while the president was still consulting with the scientists in the Cabinet room, government press offices had received swine flu "Fact Sheets" ("embargoed for release" until 5:00 PM) bearing the message he would deliver at the meeting's conclusion.[60]

Around 4:50 PM Ford appeared in the White House press room with Salk and Sabin on either side. Secretary Mathews and Drs. Cooper and Sencer stood respectfully in the background. The president said that the federal health officials and the "very outstanding technicians" who had just met with him had advised that a swine flu epidemic the following year was a "very real possibility."[61] Consequently, he continued,

> I am asking the Congress to appropriate $135 million prior to their April recess . . . to inoculate every man, woman and child in the United States. . . . I am asking each and every American to make certain he or she receives an inoculation this fall.[62]

Ford was no doubt pleased. He had launched an ambitious and seemingly noncontroversial program that he was persuaded was unquestionably in the public interest. And he might have gained some political mileage in the process. Congressional aides thought he had.[63]

His pleasure was short-lived. No sooner had the announcement been made than controversy erupted. That night on the Cronkite show, Robert Pierpoint of CBS News reported that "some doctors and public health officials . . . believe that such a massive program is premature and unwise, that there is not enough proof of the need for it. . . . But because President Ford and others are endorsing the program, those who oppose it privately are afraid to say so in public."[64] A local CBS reporter in Atlanta following the story had called sources inside CDC, and had been told on "*deep background*" that, based on present evidence, nationwide immunization was unjustified, "a crazy program" or words to that effect.[65]

The next day, all three networks aired criticisms from various sources, most notably Dr. Sidney Wolfe of Nader's Health Research Group, a frequent critic of the medical establishment. Cronkite reported that the World Health Organization had "expressed surprise at the president's decision. A WHO spokesman said there is no evidence of an epidemic and no plans in other countries for massive inoculations."[66] A CBS insider later revealed that what they learned from CDC convinced them early on, before they got on to Wolfe, that it "was a rotten program, rotten to the core. We thought it was politically inspired . . . it certainly was awful in technical terms . . . unwarranted . . . unnecessary."[67]

This barrage of criticism apparently took Cavanaugh, in the White House, by surprise. He had been assured by both Sencer and Meyer that there was unanimous support among the scientists they had consulted. To be sure, most medical organizations, including the American Medical Association, did support the program. But obviously there were some experts who felt otherwise. Cavanaugh asked Cooper himself to check again at CDC for internal dissent. A day later, Cooper reported that he had found none, that CBS was wrong.[68] In any event, the wheels were now in motion.

A shaky start

Congress responded speedily to the president's call for funds. An appropriation of $135 million was added to a pending supplemental bill by an accommodating Senate Appropriations Committee, and the two key Congressional figures in health matters, Edward Kennedy, Chairman of the Senate's Subcommittee on Health, and Paul Rogers, his House counterpart, promptly scheduled hearings on the proposal. Both men were favorably disposed toward the program, being firm believers in the value of preventive medicine and convinced that the threat of swine flu was real. Kennedy also saw the program as a means of promoting children's vaccination against other, potentially more serious diseases such as measles and rubella.[69]

Each subcommittee held a one-day hearing and invited the usual array of government officials (Sencer, Cooper, Meyer, etc.), drug company representatives, and various other experts all generally supportive of the president's proposal. Yet many of the problems that were to plague the swine flu program over the coming months were raised – if not fully appreciated – at these hearings. The epidemic's uncertainty was acknowledged (Cooper called it a "strong possibility") but not pursued. C. Joseph Stetler, President of the Pharmaceutical Manufacturers Association (PMA), a trade organization representing the drug industry, warned that the "probabilities" were that drug companies could not produce enough vaccine to inoculate all Americans – 213 million doses – by the target date of October or November.[70] Subcommittee members asked about the vaccine's safety and efficacy, and whether there would be side effects. Cooper was reassuring; the vaccine would be 70–90 percent effective, he said, and side effects would be minimal.[71]

In fact there was, and still is, considerable controversy about the effectiveness of all influenza vaccines, including swine flu. According to a Government Accounting Office report, the estimated effectiveness of past flu vaccines ranges anywhere from 20 to 90 percent. In 1968–9, based on one of the few well-controlled clinical trials to test the protective effect of immunization during an actual flu epidemic, the head of CDC's virology division and colleagues concluded that "optimally constituted influenza vaccines at standard dosage levels have little, if any, effectiveness and that even very large doses of vaccine do not approach the high degrees of effectiveness that have been achieved with other virus vaccines."[72] Graphic evidence of the low effectiveness of influenza vaccine was available at Fort Dix, where many of the recruits had come down with A/Victoria flu even though all had been vaccinated against it. A common reason for ineffectiveness is a slight mismatch between the vaccine and the current virus strain, since flu viruses are always undergoing minor but continuous changes ("antigenic drift"). Privately at least, federal officials feared that such changes might also render swine flu vaccine ineffective.[73]

Another concern raised at the hearings that presaged later problems was how to inform vaccine recipients of the potential risks. The President of the American Academy of Pediatrics testified that it was "questionable whether adequate informed consent is possible, indeed practical, in a mass immunization program of this magnitude."[74] The drug companies were worried about legal liability. PMA President Stetler warned that there were "major product liability problems associated with mass immunization programs," and called for government indemnification of vaccine manufacturers. "Quite frankly," the PMA Vice-President told the Senate subcommittee, "the liability is so enormous here, we doubt whether we could obtain the necessary insurance coverage."[75] This startling statement evoked only mild interest from the subcommittee.

Despite these indications of brewing trouble, most members of Congress were in favor of the swine flu program. Believing the experts that a pandemic could be potentially devastating, they viewed the proposed immunization campaign as a wise investment. Like Ford, they undoubtedly feared what would happen in the November elections if swine flu became rampant in October and Congress had denied the president's request for funds. The supplemental appropriation was approved handily in the House and Senate and signed into law April 15.

Meanwhile, planning efforts were getting underway. On April 2, a week after the president's announcement, CDC held a giant meeting in Atlanta to acquaint state health officials and private physicians with the program's goals and to coordinate local planning. To the chagrin of administration and CDC leaders, many of those in attendance apparently viewed the whole enterprise with suspicion if not outright hostility. One sore point was that the requested federal funding would pay only about twelve cents per shot toward the costs of administering the vaccine, a process that usually cost from fifty to seventy cents; state and local governments would end up paying the difference.[76] Massachusetts Commissioner of Public Health Jonathan Fielding did not favor mass immunization because, as he recalled, "I thought the risk of an epidemic small and I didn't want to divert resources from other programs."[77] Doubts and apprehensions were evident in the comments and questions from the audience. "Many emphasized that the government reacted too hastily with only one outbreak at Fort Dix . . . as evidence," the Associated Press reported.[78] Several states urged that the vaccine be stockpiled for all but high-risk people until further sign of an epidemic, prompting the only sustained applause of the conference, according to one participant.[79]

Dr. Martin Goldfield, the New Jersey epidemiologist whose hunch had uncovered swine flu at Fort Dix, expressed his opposition to the program in no uncertain terms, focusing especially on the risks to pregnant women. "When we talk emotionally about gambling with lives, we must also remember that we are gambling with lives, health and welfare if we throw around 200 million doses of the vaccine," he warned.[80] Although specialists discounted Goldfield's concern about pregnant women, all the network TV news shows that evening gave him feature coverage. "There are as many dangers to going ahead with immunizing the population as there are [in] withholding," he declared on national television. "We can soberly estimate that approximately fifteen percent of the entire population will suffer disability reaction."[81]

Some experts were glad that someone had finally given voice to the doubts they had secretly been nursing. One CDC official told Goldfield confidentially, "Marty, keep it up. I can't say anything."[82] Most scientists, however, were dismayed at what they saw as a breach of professional conduct; such public candor ran very much against the grain of accepted scientific and political behavior. One senior epidemiologist chastised Goldfield privately: "Marty, you have some good points. I agree with much of what you say. But the decision's made. Now is the time to close ranks. You are wrong to go public."[83] By all accounts, including his own, Goldfield was never forgiven for resisting peer pressure and expressing his doubts in public.

Washington professor and ACIP member Russell Alexander, who shared many of Goldfield's views, was more circumspect. Not having attended the Atlanta meeting, Alexander wrote Sencer a tactful note pointing out that he, for one, did not agree that vaccine administration "need necessarily be carried out, unless there was another swine outbreak." Noting that stockpiling was a well-established military strategy, Alexander urged similar contingency planning for the swine flu program. "There still seems to be time to be cautious if there is no further evidence of significant swine outbreaks by September," he concluded, signing the letter, "With personal regards from your 'half-a-hog' colleague...."[84] Two months later, in June, Alexander would finally break ranks too and join the by-then growing list of defectors in openly opposing mass inoculation.

On April 6, four days after this rather unsettling meeting in Atlanta, CDC officials got another jolt, this time from a biting editorial in the *New York Times*. Challenging each of four key assumptions underlying the program – that the swine flu threat was real, that vaccine production and administration could be completed in time, that the benefits of the vaccine would outweigh its medical and financial costs, and that the vaccine would be effective – the editorial concluded:

> The President's medical advisers seem to have panicked and to have talked him into a decision based on the worst assumptions about the still poorly known virus and the best assumptions about the vaccine.... A convincing case for the President's proposal... cannot be made until those who support it debate publicly with the medical and scientific skeptics who are already voicing their doubts.[85]

This editorial was written by Harry Schwartz, a longstanding foe of public medicine. Schwartz was convinced that the program was without scientific merit and that Ford's endorsement was pure "politics." It was a view that would appear repeatedly in *New York Times* editorials throughout the coming summer. The majority of newspapers, however, supported the program. In late April, an HEW press analysis showed that nearly 90 percent of major newspapers around the country had responded favorably, although by May the figure had already dropped sharply to only 66 percent favorable.[86]

The reaction in other countries to the swine flu threat was cautious and rather skeptical. In early April, the World Health Organization held a meeting of international experts in Geneva to consider appropriate responses. Noting that "extensive investigations in the United States have revealed no further infections since [Fort Dix]," the meeting concluded that it was "entirely possible that this may have been a unique event in a military recruit population and will not lead to widespread epidemics."[87] Its recommendations were thus comparatively modest: increased surveillance, and production of vaccine for stockpiling or administration depending on a country's resources and priorities. The British elected to add swine virus to existing vaccines and distribute them to "high-risk" groups (the elderly, disabled, and pregnant women), along with a small stockpile for the general population. In Canada, the government recommended flu shots for roughly half the population considered to be at greatest risk (this excluded healthy children and most adults between 40 and 65), but did not give the provinces any additional funds for this purpose. U.S. experts contended that the main reason other countries were not responding as aggressively as we were was that they lacked the funds and facilities. Most foreign observers, by contrast, claimed that the United States was overreacting.[88]

That other countries were choosing to stockpile vaccine rather than immunize prospectively did bother some American officials, especially at OMB. In early June, OMB staff again sounded a warning signal, urging Cooper and the White House aides to "rethink" the swine flu program and seriously consider a retreat to stockpiling. "The main reason for a possible change in approach," one OMB health examiner wrote, "is that there have not been *any* further cases of swine flu reported *anywhere* in the world since the 12 Fort Dix cases in February.... There is no available evidence to demonstrate that the 1976 version of swine flu is any more virulent than other current strains of flu."[89] Even Dowdle, the head of virology at CDC, had begun to lean toward stockpiling, although he never said so publicly.[90] The White House rejected the OMB recommendation. This decision was undoubtedly less President Ford's than that of his advisers, principally Sencer. As Sencer himself later put it, "The White House was not rigid. Ford would listen to Cavanaugh and Cavanaugh would listen to me. Through Cooper, Mathews, and Cavanaugh, we could have brought about a change if it was deemed necessary."[91]

More problems and mounting opposition

The field trials of swine flu vaccine began on April 21. They were the largest ever in the history of influenza, involving more than 5,000 people carefully divided into different age groups receiving vaccines from different manufacturers. Each person was to get a single shot. Although it was well known that children, who lacked long exposure to related viruses, often required more than one dose in order to achieve immunity, planners did not include tests of two doses for children in the field trials, mainly because, as one of them later admitted, "We just didn't think of it."[92] This was a regrettable oversight, as would soon become clear.

Vaccine production had meanwhile begun, but manufacturers were running into problems. On June 2, Cooper announced that one of the four companies producing swine flu vaccine, Parke–Davis, had somehow used the wrong virus – a strain that was slightly different from the one isolated at Fort Dix – in making several million doses of vaccine. These all had to be discarded, a setback that would delay the start of immunization by four to six weeks.[93] Other companies were having their own troubles. Production was behind schedule across the board because the eggs were yielding roughly one dose of vaccine per egg instead of the expected two doses. (According to some critics, this low yield indicated that the swine flu virus was not very virulent and hence an unlikely agent of a pandemic, but little attention was paid to this new suggestion casting doubt on the swine flu threat.[94]) In mid-June, vaccine manufacturers announced that the first 80 million doses would probably not be ready until October, with the remainder to follow in December and thereafter. This was well behind the original timetable, which had full-scale immunization starting in July and being substantially completed by October.[95]

The results of the field trials were reported on June 21 and 22, exactly two months after the trials had begun, at a huge meeting in Bethesda held by the National Institute of Allergy and Infectious Diseases, a branch of the NIH. All the relevant government agencies were represented – CDC, FDA, ACIP, and also state health departments. Sabin had been invited, at his own request. Also attending were numerous observers and critics, including Sidney Wolfe of the Health Research Group, and a full contingent of the press. All told, about 200 people were present.

The news was mixed at best. Although the vaccine appeared to be effective in adults based on antibody responses, and to have few side effects, it worked poorly in children. "Whole" vaccine produced immunity in children, but caused many adverse reactions ranging from sore arms to high fever; "split" vaccine, on the other hand, caused few side effects, but failed to confer immunity. ("Whole" and "split" vaccines refer to two different methods of preparing killed-virus vaccine.)[96] The solution seemed to be to give children two half-strength doses of vaccine, some weeks apart. But this regimen had not been field tested. So the experts reluctantly concluded that another set of field trials was needed to assess the results of a two-dose regimen for children, which would take another two months to complete. By then, however, producing twice as much half-strength vaccine as originally ordered in time for the flu season would be impossible.

What this meant, although it was not publicly acknowledged at the time, was that children were out of the program unless an epidemic erupted. If it did, they would get single doses of whole vaccine, adverse reactions notwithstanding; if the epidemic were serious enough, mothers presumably would not mind their children's discomforts. The exclusion of children could not be publicly announced, officials felt, because it would sound crazy to most Americans, who, after polio, considered shots for kids the heart of preventive immunization. Children are in fact generally considered to be the most important spreaders of influenza through their contacts in schools, summer camps, and other facilities. How could any self-respecting national campaign leave out the nation's children?

The second day of the meeting, June 22, was devoted largely to the question of stockpiling, a topic Sabin, surprisingly enough, had insisted be added to the agenda. With no trace of swine flu anywhere in the world, including the southern hemisphere where flu season was approaching its peak, Sabin had come to view the situation quite differently than he had back in March. He no longer supported prospective immunization but now favored watchful waiting, for many of the reasons Alexander and Goldfield had given earlier. The likely exclusion of children from the program only strengthened the case. Sabin maintained that with proper planning and preparation, we could still inoculate in time if signs of a major epidemic were to appear, by utilizing brigades of local volunteers to conduct assembly-line inoculations in each community. Such an approach, he insisted, fueled by a sense of national emergency, could drastically reduce the time required for immunization. Alexander, also at the meeting, strongly supported Sabin's appeal.

Sencer and other CDC officials responded with the standard arguments against stockpiling, buttressed by a brief CDC staff feasibility study. By CDC estimates, it would take nine to twelve weeks to complete inoculation once an outbreak had been identified and another two weeks for the vaccine to provide immunity – too long to prevent the flu from spreading and to avoid unnecessary deaths. (These estimates did not take into account the kinds of emergency measures Sabin was proposing, of course.)[97] Stockpiling was also more expensive. Furthermore, state plans were now well underway and could not easily be revised without a serious loss of momentum. To some participants, even those initially sympathetic to stockpiling, the latter was a key point.[98] Besides, if the vaccine was in fact perfectly safe as most experts believed, then stockpiling made even less sense. "Remember," one CDC official recalled, "we thought that the vaccine had no risks."[99] For whatever reasons, a clear majority of the advisory committees at the meeting reaffirmed their commitment to mass immunization.

That evening, all three TV network news shows featured the debate over stockpiling, and two offered wry comments about the ambiguous status of children in the program. Sabin led the dissenters, and was joined by others, including Alexander, who had finally decided to go public with his doubts. "As time goes on," Alexander told the nation, "most people think that the probability is there will not be an epidemic in the 1976–77 season due to swine influenza."[100] The *New York Times* renewed its attack on the program.[101]

Studies questioning the program also began to appear in the medical literature. A paper in the respected British journal *Lancet* reported that six volunteers exposed to swine flu virus had shown only "mild" symptoms, implying that the virus was less virulent than many other flu viruses and was unlikely to cause widespread disease. Another article in the same issue, by the leading British influenza expert, concluded that preparing for mass immunization was "highly questionable," even in the United States, "until the shape of things to come can be seen more clearly."[102] In late July, the Assistant Director of CDC acknowledged publicly that there had been no reason to "raise the specter of 1918" in connection with Fort Dix. "We have nothing on which to base a similarity of behavior between the two viruses," he admitted.[103] Sydney Wolfe of the Health Research Group, citing doubts about the likelihood of an epidemic and the vaccine's potential side effects, offered an eerily prescient warning to Congress. "The major disease in the U.S. this year related to Fort Dix will not be swine flu," he predicted, "but, rather, swine flu vaccine disease."[104]

CDC officials were not the only ones smarting under these attacks on the program. Touchy pork producers complained that all the talk about "swine flu" might give the industry a bad name, and suggested that the disease be renamed "New Jersey flu." (New Jersey officials politely declined the honor). The pork industry did convince federal officials not to consider a mass immunization campaign for the nation's pigs, which one veterinary expert had proposed.[105]

The growing criticism, and especially the defection of key scientific supporters like Sabin, was beginning to worry many members of Congress. Their concerns were soon eclipsed by news of a new blow to the program: The insurance industry refused to insure the manufacturers of swine flu vaccine.

The impasse over liability

On June 25, Leslie Cheek, head of the American Insurance Association's Washington office called CDC and the White House to announce that the manufacturers of swine flu vaccine would not get liability coverage. The insurers, which the AIA represented, were simply too worried about the potential liability of a nationwide immunization program to be willing to underwrite vaccine producers, Cheek said. Existing coverage would terminate June 30. And the manufacturers would not bottle or release the vaccine without insurance.

This issue had been looming for some time, although Cheek's call still came as a shock to government officials. Ever since the initial Congressional hearings on the program, vaccine manufacturers had been talking darkly about liability problems. Merrell, one of the smaller manufacturers, had apparently been warned in early April that it might not be able to obtain liability coverage. On May 24, Merrell had announced that it would not provide its share of vaccine – a quarter of the total – unless the government insured manufacturers for most prospective legal costs resulting from the program. Other companies soon followed Merrell's lead, declaring that they would not proceed with production without liability insurance. Now Cheek was saying that *no* vaccine manufacturers would have insurance coverage beyond June 30. The government was trapped; if it wanted the program to go forward, it would have to indemnify the manufacturers. Reluctantly, HEW lawyers began drafting an indemnification bill which, still more reluctantly, OMB approved.

In defense of their position, insurers pointed to the side effects that inevitably accompany all influenza inoculations, and the possible breakdown of quality control under the pressures of a "crash" program. But what they feared most of all, they stressed, were the costs of defending against all the damage claims – groundless as well as valid – that would be entailed in a program the size of swine flu. The number of frivolous claims alone could be enormous, and their defense expensive, quite apart from any damage awards. To make matters worse, two recent court cases had held a polio vaccine manufacturer liable for injuries suffered by vaccine recipients who subsequently contracted polio on the grounds that the manufacturer had failed in its "duty to warn" the individuals being vaccinated of the potential risks – even though the company *had* included in the shipping cartons a printed statement containing adequate warning, and experts had testified that the polio had probably not even been caused by the vaccine.[106]

Manufacturers were understandably alarmed at the prospect of having to assume this newly precedented duty to warn in an immunization program that would cover the entire population. Even Cooper agreed that it was reasonable for the government to assume this part of the program's legal burden, and HEW lawyers had tried to incorporate the necessary language in the contracts with manufacturers to accomplish this. But assumption of the duty to warn was not the same, to Cooper or anyone else, as reimbursement for all legal costs, let alone compensation for all vaccine-induced injuries – which was what, it seemed, the manufacturers were now demanding.

On June 16, HEW's indemnification bill arrived in Congress, where it met a cool reception. The House Health Subcommittee held a hearing on it on June 28 and heard testimony from representatives of HEW, the drug companies, and the insurance industry, including Cheek of the American Insurance Association. Skeptical subcommittee members pressed hard. How could insurers refuse to cover a vaccine said by leading experts to be perfectly safe – and indeed shown to be safe in the largest field test ever conducted? Moreover, insurers themselves admitted that they had been held liable in only *two* flu vaccine suits in the past seven years.[107] It was blackmail, Congressman Henry Waxman charged. Insurers were holding the program hostage to force the government to absorb costs that were rightfully theirs, a step that could set a dangerous legal precedent for other areas of product liability. Chairman Rogers suggested that the whole thing was merely an elaborate way to dramatize the insurance industry's concerns about adverse court decisions and large personal injury awards, and threatened that Congress might move toward imposing federal licensing requirements on the insurance industry if the industry was unwilling to meet public needs.[108] "The subcommittee did not understand," Cheek recalled, "why insurers were reluctant to insure a vaccine whose medical risks appeared minimal; they certainly did not sympathize with the industry, or with our argument that the incalculable number of spurious claims and new liability doctrines made the manufacturers uninsurable at any price.... They just didn't believe us."[109]

With legislation stalled, HEW officials renewed their efforts to solve the liability problem through special language in the vaccine procurement contracts. Negotiations dragged on through much of July, getting essentially nowhere. Meanwhile, some influential members of Congress were having growing doubts about the need for mass immunization.[110] Flu season in the southern hemisphere was at its height, with major outbreaks of A/Victoria flu reported on virtually every southern continent, but no sign of swine flu. HEW and White House staff again raised with President Ford the possibility of abandoning the foundering program. Ford wanted to know whether his advisers still considered the pandemic possible. They did. Although many experts believed the odds were declining with each passing week, the likelihood remained "unknown," hence "possible."[111] This Ford apparently viewed as decisive, and he turned his energies toward persuading Congress, due to recess shortly, to act. "There is no excuse," Ford declared in a letter to Chairman Rogers read aloud at a July hearing of the House Health Subcommittee, "to let this program, a program that could affect the lives of many, many Americans, bog down in petty wrangling. Let's work together to get on with the job."[112] The letter served to calm some tempers and voices, but at the end of the session the fundamental issues were no nearer resolution.

Caught in a deadlock between Congress and the insurance industry, with vaccine production on hold and scientific doubts growing, the swine flu program was in real jeopardy. Then, by fluke, outside events intervened. On August 1 the press reported an outbreak of a mysterious respiratory ailment among people who had attended an American Legion Convention in Pennsylvania. The next day, at least eight were reported to have died of what was called, for want of a better name, Legionnaires' disease.[113] For a few days, swine flu seemed a possible culprit. This was ruled out on August 5, when CDC announced that laboratory tests had shown that the infectious agent, although still unidentified, was definitely not swine flu. The scare nevertheless had a big impact on Congress, and President Ford seized the opportunity to do some more arm-twisting. "These tragic deaths were not the result of swine flu," he told Congress at a nationally televised press conference:

> But let us remember one thing: They could have been. The threat of swine flu outbreak this year is still very, very genuine . . . . I am frankly very dumbfounded to know that the Congress . . . has failed to act to protect 215 million Americans from the threat of swine flu . . . . Further delay in this urgently needed legislation is unconscionable.[114]

With an election coming up shortly, this was more pressure than most members of Congress could stand. Both Houses bowed to the president's wishes and hastily drafted a new swine flu liability bill, passing it on the eve of the convention recess. Many legislators still resented giving in to what they saw as the avarice of the insurance industry, and were uneasy about getting the government into the insurance business. "I hate this bill," Senator Ted Kennedy was quoted as saying, "but suppose there is a swine flu epidemic? They'll blame me."[115]

The bill did not formally indemnify manufacturers but, using the Federal Tort Claims Act as a model, assigned legal liability to the federal government for everything except negligence. HEW was to draft a written consent form informing vaccine recipients of the risks and benefits. Manufacturers would be freed from the duty to warn and from the costs of defending against baseless suits. All claims would be filed against the government, which could sue manufacturers to recover damages resulting from negligence; manufacturers, in turn, could collect from their insurers. (As it turned out, recoveries for negligence were minimal, so insurers ended up pocketing nearly all of the roughly $8.6 million in premiums manufacturers had paid as pure profit.)[116] Ford signed the bill into law (PL 94–380) on August 12.

The swine flu program had been snatched from the jaws of defeat. President Ford, Sencer, and other top program leaders were jubilant.[117] Some lower-level officials, on the other hand, were secretly sorry. By late summer, with no sign of swine flu anywhere in the world, some people were pursuing the program only half-heartedly; they felt, CDC's chief virologist recalled, that "there wasn't much point in going on."[118] Now they had no choice. It was their job to resurrect a program that was by then hopelessly behind schedule, beset by legal and logistical uncertainties, castigated by the news media, and for which the need was becoming less apparent daily. Their reaction was, "Oh shit."[119]

Immunization begins – finally

The new swine flu law itself caused further delay. The law did not go into effect until October 1, the start of the new fiscal year, and manufacturers and their insurers refused to permit the vaccine to be used before then. Meanwhile, there were more problems with production. It was discovered that all of the vaccines being produced lacked a key component (the surface protein neuraminidase) that normally contributes to vaccine effectiveness, although how significant this deficiency was for swine flu vaccine was unclear.[120] Production was also falling farther and farther behind schedule. Manufacturers were now projecting that only 20 million doses could be packaged and delivered by October – a quarter of what they had promised by that date back in June. Even by December, vaccine supplies would still fall well short of the 143 million people in the population over the age of 18 for whom flu shots had been recommended (children's dosage had still not been finalized).[121]

Projected shortfalls in supply were accompanied by declining public demand. According to a Gallup Poll at the end of August, 93 percent of all Americans had heard of the swine flu program, but only 53 percent intended to get shots.[122] The program's numerous problems had also taken their toll on many state and local health agencies, whose carefully developed distribution plans had been thrown into disarray by the endless confusion and uncertainty created by the summer-long struggle over liability legislation. To make matters worse, CDC's arrangement with the Advertising Council, a nonprofit organization representing large advertising agencies, which had agreed to help with public relations efforts, also fell apart over the summer. CDC finally reached a new agreement with the Council on October 4, four days after immunization had begun, but it was not until late November that the Council's promotional materials were finally distributed to the states and news media.[123]

Problems with the informed consent forms threatened to cause additional delays. Earlier in the summer CDC had prepared what it called a "Registration Form" briefly describing the risks and benefits of the vaccine.[124] Some 60 million copies, dated July 15, had been printed and distributed nationally. But the August legislation included a provision (at the Kennedy Health Subcommittee's insistence) requiring the National Commission for the Protection of Human Subjects, an independent ethics advisory panel, to review and approve the consent forms. The National Commission had various criticisms: The forms gave no hint that the likelihood of a swine flu epidemic was in any doubt; they stated flatly that the vaccine "can be taken safely during pregnancy," yet it had not been tested on pregnant women; they gave no information on recovery rights against the government in the event of vaccine-related injuries; and finally, the title "Registration Form" did not make it clear that, in signing the forms, vaccine recipients were signifying their informed consent. Several members of the Commission suggested discarding the original forms and starting over.[125]

Sencer and his people at CDC were furious. Here was yet another detour, and one which they regarded as largely unnecessary. Their solution was a hasty compromise: a cover sheet, which included some of the Commission's recommended changes and ignored others, to be stapled to the original form. The result pleased almost no one. The two-page form was hard to follow, some of the language was technical and difficult to understand, the instructions for children were left vague, and the lack of data on use by pregnant women remained unclear (the form stated that "there are no data specifically to contraindicate vaccination . . . in pregnancy," failing to say that this was because no studies had been done – and the original statement of safety during pregnancy remained).[126]

The problem for CDC was how, in one official's words, "to protect without discouraging."[127] It was a delicate balance to strike, but CDC's consent form didn't come very close. By omitting any mention of the epidemic's uncertainty, or of uncertainties about vaccine efficacy, the forms subtly overstated the likely benefits. Conversely, risks were minimized, not only for specific population groups such as children and pregnant women but also for the population at large. The forms did warn that "As with any vaccine or drug, the possibility of severe or potentially fatal reaction exists," but went on to say that this was "highly unlikely" and that most people would have no side effects at all, or only minor ones such as "fever, chills, headache or muscle aches."

Noticeably absent from either the original form or its appended cover sheet was any mention of the possibility of neurological complications – even though all four manufacturers had included such a warning on the package inserts they provided with the vaccine.[128] The association between flu vaccines and neurologic illness was well known, albeit considered to be rare. In a July statement, the ACIP had listed neurologic disorders as one of three possible types of adverse reactions to swine flu vaccinations.[129] Dr. Michael Hattwick, the head of the swine flu surveillance program at CDC, said he had alerted his superiors to the possibility of neurological complications. Sencer denied, on national television, ever receiving any such warning. "That's nonsense," was Hattwick's response. "I can't believe that they would say that they did not know that there were neurological illnesses associated with influenza vaccination. That simply is not true. We did know that."[130] Certainly some of the swine flu officials knew. They just didn't want to mention it on the consent form for fear of frightening an already wary public.

On October 1, mass immunization finally got underway in the states that had vaccine; others joined in as they received supplies. In the first ten days, more than a million adult Americans got swine flu shots (children were still on hold).

Then the program was hit with more bad news. On October 11, three elderly people, all suffering from heart disease, dropped dead shortly after receiving swine flu shots from the same clinic in Pittsburgh, Pennsylvania. UPI picked up the story and it quickly made national headlines. The same batch of Parke–Davis vaccine turned out to be involved in all of the deaths, and the local coroner suggested that the batch might be bad. Many local health officials panicked. The Allegheny County (Pittsburgh) Health Department suspended the program, and nine states immediately followed suit. The *New York Times* fired off another searing editorial, calling for President Ford to "order a halt in the vaccination program," until there had been "a second hard look at the costs and benefits of what is being done to forestall the disease that isn't there."[131] CDC called the Advertising Council and told them to stop everything.[132]

For three days, the news media featured scare stories on vaccine recipients around the country who had died, regardless of cause. Sencer and others at CDC did their best to reassure an anxious nation that these deaths were in all likelihood simply coincidental – "temporally related deaths" occurring soon after the flu shots but not caused by them. On October 14, the panic abruptly subsided. Cooper reported laboratory results exonerating swine flu vaccine in the Pittsburgh deaths, and Ford and his family got their flu shots on national TV. Allegheny County and five states announced resumption of mass immunization, and the other states soon followed.[133] Federal officials heaved a collective sigh of relief. CDC gave the Advertising Council the go-ahead again, but suggested that one tag line be dropped from the planned advertising: "The swine flu shot. Get it. Before it gets you."[134]

"Temporally related deaths," it now appears, were another problem that program planners had foreseen but decided not to publicize, fearing that they would be misunderstood by the public and would discourage people from getting shots.[135] In fact, the panic and misunderstanding caused by the Pittsburgh deaths was undoubtedly much greater than it would have been had adequate information been provided in advance.

In spite of all the negative publicity, 2.4 million people were vaccinated during "Pittsburgh week." Inoculation rates continued to rise as state plans got underway, reaching a peak of 6.4 million people in mid-November. After that they fell, as the sense of national emergency receded with swine flu still nowhere in sight.[136] On November 16 federal officials finally announced appropriate dosages for children (two shots of split vaccine, a month apart), but disclosed that available vaccine supplies would cover only 7 percent of eligible children (roughly 2 percent ended up getting shots).[137]

Vaccination rates varied widely among geographic areas. Delaware, at the high end, immunized almost 90 percent of its population, Louisiana only 12 percent.[138] There were also large differences among population groups. Even though influenza is almost twice as common among poor adults as the more affluent, with death rates that are 50 percent higher among minorities, such groups were least likely to get swine flu shots – the standard pattern for most forms of preventive care. The government targeted urban ghettos and other poverty areas with special promotional campaigns, but residents remained both skeptical about the need and fearful of the shots.[139] Altogether, more than 45 million people received flu shots between October 1 and December 16 – twice the number ever immunized before in a single season, but less than a quarter of the total population.[140]

In early November, Republican Gerald Ford lost the presidential election to Democrat Jimmy Carter. Ironically, after all the earlier charges of "politics," the swine flu program played little if any role in the campaign. It was neither the political bonanza Ford might have hoped for, nor a political liability the Democrats could take advantage of (this would have been risky as long as any possibility of a swine flu epidemic remained). Swine flu officials, in the meantime, kept up a brave front. "Reestablish the pre-Pittsburgh enthusiasm," one CDC staff member exhorted his colleagues. "Extinguish defeatist attitudes...."[141] Another adviser said he would "call AMA to see if he can generate a happening and get out a press release."[142] Privately, however, morale was sagging. At a staff meeting one CDC official suggested sardonically that "the Flu Program should be considered the 'Vietnam of the [Public Health Service],' and that at some point victory should be claimed and we should pull out."[143]

In view of the final, fatal blow about to befall the program, this might not have been such a bad idea.

Guillain–Barré syndrome and the program's demise

It began routinely enough – a call in mid-November to CDC from Minnesota reporting an unusual health problem, an ascending paralysis called Guillain–Barré syndrome (GBS), in someone who had recently received a swine flu shot. At first this drew little notice. All fall CDC's surveillance center had been investigating many such reports of serious reactions and fatalities following flu shots. These had all been judged to be, like the Pittsburgh deaths, chance events unrelated to the shots. But within a week, three more cases of GBS (one fatal) were reported from Minnesota, plus three from Alabama and one from New Jersey. At this point CDC decided further investigation was required. An active search for GBS cases in these three states plus one other was initiated by contacting neurologists and other providers likely to treat patients with the syndrome. The goal was to see if GBS was occurring at a higher rate among the vaccinated than the unvaccinated; if it was, this would imply that GBS was being triggered in some way by the flu shot.

The data that began to accumulate were disturbing, though not conclusive. Because GBS was a rare disease and the medical literature on it sparse, scientists were unsure of the normal incidence rate in the United States, and the small number of cases on which incidence rates were based made statistical comparisons less reliable. Nevertheless, trends toward higher rates of GBS among the vaccinees were unmistakable even in the preliminary data. On December 10, Dr. Lawrence Schonberger, who had been asked to direct CDC's investigation of GBS, reported the initial findings at an internal CDC meeting. Although most of those attending were dubious that the association was real, they were nevertheless alarmed, and agreed that surveillance efforts should be expanded and Sencer informed; meanwhile, reports of GBS continued to come in.[144] Sencer turned for advice to his usual sources on the ACIP and at CDC, FDA, and NIH. Many of them, too, were not convinced that swine flu shots were actually *causing* the observed GBS cases, yet the mere hint of such a risk, after nine long months of accumulated doubts and problems, was enough to create concern among even the staunchest program advocate. For some embattled officials, it was the final straw.[145]

On December 14, after a conference call with outside experts, CDC announced publicly that it was investigating a possible association between GBS and swine flu shots, but stressed that "there was no evidence to link the reported cases to vaccination."[146] Two days later, after another conference call, CDC reversed its position based on revised estimates and recommended suspension of the program pending a nationwide investigation of GBS.[147] Sencer informed Assistant Secretary Cooper, who immediately conferred with Secretary Mathews and also Salk, in Paris. All agreed that inoculations should be halted until more was known about the connection with GBS. They told Ford, who, sighing, concurred.

That afternoon Cooper announced the suspension of the swine flu program, saying that he was acting "in the interests of safety of the public . . . of credibility, and . . . of the practice of good medicine."[148] He emphasized that this was only a temporary step while further investigation of the GBS data was being conducted. But most program officials, already battle-scarred and weary, knew in their hearts it was "the death knell for the program."[149]

They were right. The epidemic never came, and the mass immunization campaign was never reinstated.

But the swine flu story was not quite over. In January 1977, Joseph Califano, newly appointed Secretary of HEW under Carter, was informed after little more than a week in office that an outbreak of A/Victoria flu had erupted in a nursing home in Florida causing three deaths, and that the bivalent A/Victoria–swine flu vaccine was the only thing available to prevent further spread. The ACIP had recommended that it be released from suspension, the risk of GBS notwithstanding, since A/Victoria flu generally poses a much greater danger to high-risk groups such as the frail elderly. But how would the public react to the news that this much-maligned vaccine was once again to be put to use? It was a potentially treacherous situation, especially for a political newcomer. Califano decided to confront the issue head on, by consulting with a wide range of impartial advisors and holding key discussions in full public view. He told President Jimmy Carter that he planned to "ask other experts to join the [ACIP] so that we will have as broad and objective a base as possible." He wanted people outside CDC and the ACIP, he later explained, since both of these agencies had "a strong interest in promoting immunization programs and in vindicating their earlier judgments on this one."[150] The president himself was not to be directly involved.

The group met for an all-day session February 7 in front of TV cameras and a roomful of spectators, including known critics and members of public interest groups. Califano himself was present for much of the discussion. Many were expecting a real donny-brook, given the divergent views and perspectives of the various participants. What ensued instead was, in the words of a *Washington Post* editorial, "a full day of reasoned, wide-ranging and extraordinarily comprehensive discussion, and ultimately a consensus."[151] At the end of the day, the group recommended the release of bivalent swine flu vaccine and voluntary inoculation of groups at high risk for A/Victoria flu. The next day the Secretary announced his acceptance of those recommendations.

In fact relatively few people chose to get the bivalent swine flu shots, no doubt for fear of GBS.[152] Nevertheless, there was scarcely a murmur of public criticism of Califano's decision. Even veteran opponent Harry Schwartz of the *New York Times* wrote approvingly: "The Government stands now where it should have stood all along: focused on high-risk individuals and poised to do more, but only if necessary."[153] An editorial in the *Washington Post* praised not only the decision but the decisionmaking process: "What struck us almost as forcefully was the wide-open way that it was made – the 'sunshine' approach, if you will." An issue so acutely sensitive and fraught with controversy, the *Post* emphasized, "cried out for a large measure of public understanding, and . . . a broad base of support among scientists and physicians – which, in this case, is precisely what it received."[154] The need had been no less, of course, throughout the program's beleaguered existence. That such support was significantly lacking, particularly toward the end, had as much to do with the way decisions had been made as with the decisions themselves.

Legacies and lessons

The swine flu program was finally over, but a number of issues remained unresolved. First was the matter of GBS, on which data were still accumulating. The national surveillance effort at CDC continued through the change in administrations in January, and by February 1977 over 1,000 cases of GBS had been reported. Detailed epidemiological analyses of these data yielded an inescapable conclusion: Although the risk of vaccine-related GBS in the adult population was extremely low (less than 1 case per 100,000 vaccinations), during the six-week period after vaccination vaccinees were significantly more likely to have had GBS (about seven times more likely) than the unvaccinated.[155] Subsequent crit-

icisms of CDC's methods and data led to a detailed reevaluation
of this conclusion, which basically reconfirmed the original findings.[156] The pathological mechanisms that link GBS and swine flu
shots, however, remain a mystery.

Another major issue left in the wake of the swine flu program
was the question of liability. Under the August law, the federal
government had assumed responsibility for all liability claims resulting from the program. Claims began to come in almost as
soon as immunizations ended in mid-December; within two weeks
HEW had received thirty-one claims totalling $1.2 million.
Roughly a year later, 1,241 claims had been filed for damages
totaling over $608 million.[157] Government lawyers were swamped,
and the processing of claims moved slowly. Injured vaccine recipients complained about government foot-dragging. To try to
expedite the process, and "to provide just compensation for those
who contracted Guillain–Barré as a result of the Swine Flu Program," HEW Secretary Califano announced in June 1978 that GBS
victims would not have to prove negligence in order to receive
federal compensation. Claimants would have to show only "that
they in fact developed Guillain–Barré as a result of a Swine Flu
vaccination and suffered the alleged damages as a result of that
condition."[158]

This proved more difficult than it sounded. Altogether, more
than 4,000 claims were filed, of which HEW denied about three-
quarters. The government also won more than 80 percent of the
claims that were denied and subsequently filed as lawsuits, mostly
on the grounds of the plaintiff's failure to prove either the diagnosis
of GBS or the injury's causal relationship to the swine flu shot.
Even so, the government paid out over $90 million in total dam-
ages, and of course foots the bill for all the legal expenses involved
in adjudication and settlement.[159] Chapter 8 examines the performance of the swine flu compensation program in greater detail
and its lessons for compensation of other medically related injuries.

Besides legal expenses and damages, the swine flu program had
other costs, both financial and nonfinancial. Considering state and
local government expenditures for administration and publicity,
the time spent by federal employees analyzing the program and
its consequences, and the countless hours of professional and volunteer time devoted to the immunization effort – all on top of the
originally budgeted $135 million – the whole enterprise probably
cost the government – and taxpayers – well over $400 million for
direct outlays alone.[160] And this is not to mention the costs of lost

opportunities for more profitable uses of the funds, or the human costs borne by victims of vaccine-related GBS and other side effects, or the disruption to careers and lives experienced by many of the program leaders.[161] As an insurance policy, the swine flu program proved considerably more expensive than most had expected.

Many public health authorities feared that the swine flu program's sorry demise might erode public trust in preventive medicine, leading to reduced participation in future immunization programs. This does not appear to have happened, at least based on one small opinion survey.[162] Whether damage was done to more general attitudes about CDC's credibility, or that of the medical establishment, remains unknown. However, the swine flu program may well have reduced subsequent immunization rates in other ways. Immunization levels among children had begun falling during the early 1970s, and by 1976 more than a third of all children were without adequate protection against preventable childhood diseases such as polio, measles, rubella, and whooping cough.[163] The all-out effort required for swine flu forced many states to divert resources away from these other immunization activities (as well as other public health programs), leading to further declines. Presumably as a result, the number of rubella, measles, and whooping cough cases nationally rose in 1977 by anywhere from 39 to 115 percent.[164] Accordingly, the Carter administration mounted a major immunization initiative and by 1979 childhood immunization levels had risen to between 80 and 90 percent. In 1981, the Reagan administration reversed this policy, cutting grants for immunization programs by 25 percent.[165]

Certainly no one was entirely happy with the course of events in the swine flu program. Critics have called it the "most misguided vaccination program ever attempted," a "fiasco," a "debacle."[166] Suppose a swine flu epidemic *had* come. Would the program then have been judged a success? The answer must be, for the most part, no. For, as we have seen, much of what went wrong would have been even more serious had the epidemic come. The program suffered a seemingly endless series of logistical, legal, and political problems for which there had been little if any contingency planning: lagging production, the need for double dosages for children, the occurrence of temporally related deaths, inadequate informed consent forms, and, most of all, the impasse over liability. As a result, its future was continually in doubt, undermining public confidence, disrupting state plans, and causing major shortfalls in vaccine production and distribution. Because of such problems, less than a quarter of the population (and almost no children) had been vaccinated by mid-December – in all likelihood too few people to provide significant protection against an oncoming epidemic. Had swine flu erupted that fall or winter, the program would surely have been judged at least as harshly on these grounds.

The epidemic did not come, however, and the main question that has haunted the swine flu program is whether it was an unnecessary and avoidable waste of resources or, rather, a prudent insurance policy against a threat that just happened never to have materialized. The initial response following the Fort Dix events seems reasonable. All the ingredients for a pandemic appeared to exist: a double antigenic shift, timing compatible with the eleven-year cycle thought to characterize influenza epidemics, a virus similar to that implicated in the deadly 1918 pandemic, a susceptible population, and human-to-human transmission, including one fatality. In view of these circumstances, it made sense to arrange for the necessary vaccine to be obtained from manufacturers and to begin planning for its distribution and administration – but *not* to commit irrevocably to a mass immunization campaign. Several members of the ACIP itself, in March, were reluctant to endorse immunization without further evidence of an epidemic. Sencer, in his ACTION memo, compressed these two separable steps – preparation and immunization – into a single decision, "go or no-go." Once tied together, they became harder and harder to separate.

If a commitment to mass immunization in March seems premature, the persistent refusal to back away from that commitment during the summer is even harder to defend. The more time that passed with no sign of swine flu, the less likely an epidemic became. But program officials seemed fixated on the fact that an epidemic was still possible, no matter how remote the chances or how much trouble the program was in. They were apparently persuaded that stockpiling was not feasible, although other experts, notably Alexander and Sabin, disagreed. Sabin and others argued forcefully for stockpiling at the June meeting in Bethesda, to no avail. After that came the summer-long contest of wills with Congress over liability. By the fall, the risks of going ahead with immunization had probably begun to outweigh the risks of doing nothing, or at least of retreating to stockpiling. Yet although many might have welcomed such a step, there was by then too much face to be saved – and too much potential political fallout – for it to be taken.

The people involved were all intelligent, well-motivated, and highly competent individuals. They were all sincerely trying to do their jobs well and to do what was right, as they saw it. Indeed, many of the participants stoutly defended their actions and judgments, based on the information available at the time. "Placed in a similar position again," Sencer maintained in 1978, "I would certainly have made the same recommendations as I did then." Even HEW Secretary Califano confessed that he was "not sure that anybody would have made any different decisions [from those] the prior Administration made on swine flu."[167]

Is there, then, nothing to be learned from this unfortunate saga? Was it really not a tale of "failure" after all but, like the classic Greek tragedy, simply the playing out of inexorable events over which the actors, whoever they may have been, had no control? If not, how and why did things go wrong?

One is struck, first of all, by the consistently optimistic assumptions upon which planning was based. The vaccine would be effective, it would cause few if any side effects, production would occur on schedule, federal–state planning would go smoothly, there would be no unusual liability problems, vaccinations would be completed in record time, and so forth. Plainly such optimism was not justified. Worst-case thinking, rather than best-case, might have helped ferret out many of the lurking problems in advance, permitting development of contingency plans. As one senior health official put it:

> Hell, the thing that was needed in planning the swine flu program was a day around the table brainstorming Murphy's Law: "If anything can go wrong it will"; and all the permutations anyone could think of. That would have done it. It certainly would have caught a lot of things that went wrong – they weren't so hard to think of, after all.[168]

The excessive optimism went largely unchecked for at least three important reasons. One was the substantial influence of medical scientists in aspects of the program that fell outside their expertise. Few of these experts had had much experience with the practical problems of trying to immunize 200 million people under crisis conditions – in fairness, almost no one had. They might have obtained valuable insights, however, by setting up a broad trouble-shooting advisory group including, say, liability insurers, Congress, state health departments, organized medicine, the news media, the drug industry, and perhaps even civil defense or the military for their experience in mass programs and emergency preparedness.[169] Such a group, especially if asked to imagine all the things that could go wrong, would undoubtedly have come up with at least some of the problems encountered. The decision might still have been to go ahead, but at least it would have been based on a more realistic view of what lay in store. Forewarned is forearmed.

The second reason why optimism prevailed over realism undoubtedly had to do with Ford's role. Once the prestige of the presidency had been thrown behind the program, officials had substantially less room to modify their course of action as problems multiplied and the chances of an epidemic dwindled. "There's no doubt that Ford's role complicated things," Dowdle, CDC's chief virologist, recalled. "When the president says something, you don't want to back out of it."[170] Had the president not made this "his" program, health officials might well have let it quietly die during the long battle over liability. Also, there was a natural tendency for those higher up in the political hierarchy, most of them neither scientists nor physicians, to gloss over or ignore the technical qualifications and cautionary footnotes the experts had carefully woven into their recommendations. Indeed, such simplifications and overstatements often made it easier to promote the campaign. For a more balanced view of these scientific issues, what decisionmakers needed was direct and ongoing dialogue between proponents and critics about the various technical issues in dispute. President Ford's decision may have been partly political, but it was a choice pressed upon him by his most trusted scientific advisers.

The third and perhaps most important reason for the excessive optimism that pervaded swine flu planning was its insulation from the probing scrutiny of public discussion and criticism. Throughout, critics complained that decisions were being made by a small clique of government officials and their hand-picked scientific advisers, and that these leaders refused to debate publicly with challengers either inside or outside the government. The resulting decisions were suspect, critics charged, because all of the major participants, in one way or another, had something to gain from the program: Government bureaucrats would get an infusion of funds to their agencies and some yearned-for public limelight; scientists would be able to test immunization theories and develop new vaccines; the drug industry would reap essentially risk-free profits and some useful publicity about its liability problems; private physicians would have the chance to expand their patient clienteles; and government officials could claim to be champions of public health.[171] Few of these participants had much incentive to question the proposed program, or its comfortable acceptance of medical intervention as the route to disease prevention. Critics, by contrast, insisted that any medical intervention, even the lowly flu shot, involved risks, in this case side effects without corresponding benefit. Public dialogue might have provided the needed filter for uncritical and seemingly myopic expert advice.

Indeed, the real problem was not so much the decision itself as the way it was made. By refusing to confront openly and candidly the charges of the critics, decisionmakers left themselves open to precisely the kinds of attacks and suspicions that arose. Had they done as Califano did – encouraged full public debate and invited critics and the press into the inner sanctum to participate in policy discussions – things might have turned out quite differently. More important, however they turned out, there might have been less rancor and suspicion surrounding the program.

In short, the swine flu story was *not* inevitable. There were alternative actions and decisions that might have produced a happier ending. Many were steps that critics tried in vain to bring about. Their failure highlights the danger of relying too heavily on the views of technical experts, whose unswerving confidence in the safety and efficacy of medical intervention (even in the name of prevention) seemed to blind them to impending problems, and of insulating national health policy from public scrutiny. The lessons seem clear in hindsight. Harder will be knowing how and when to apply them next time.

6

Genetic engineering: science and social responsibility

Diana B. Dutton and Nancy E. Pfund

No recent scientific accomplishment has given rise to such wildly enthusiastic expectations – and apocalyptic warnings – as genetic engineering. Emerging from an esoteric area of molecular biology, genetic engineering – known in its early years mainly as recombinant DNA research and, now, as biotechnology or simply gene-splicing – has opened up possibilities once relegated to science fiction. These include an essentially unlimited supply of previously scarce vaccines and hormones; the creation of protein-rich food-stuffs from waste products; plentiful new stores of "clean" energy from organic "biomass"; plants bioengineered to be self-fertilizing; animals genetically programmed with the traits of other species – the list is virtually endless. The ultimate and most controversial achievement, of course, is human genetic engineering – actually altering our own genetic makeup.

The implications boggle the mind. Companies of all types are rushing to add specialized new facilities and personnel, and industry analysts spare few superlatives in describing biotechnology's prospects. According to one investment analyst, "We are sitting at the edge of a technological breakthrough that could be as important as electricity, splitting the atom, or going back to the invention of the wheel or discovery of fire."[1] Noted science writer Lewis Thomas called the current biological revolution "unquestionably the greatest upheaval of biology and medicine ever."[2]

Even its critics describe gene-splicing in revolutionary terms. Activist Jeremy Rifkin, head of the nonprofit Foundation on Economic Trends and a vocal opponent of the most controversial uses of genetic engineering, predicts that in the broad sweep of history, this single technology will "fundamentally alter humanity's entire relationship to the globe."[3] But critics insist that such changes also have a dark side. Giving scientists the unprecedented power to engineer the internal biology of living organisms, they warn, opens up a Pandora's box of social and ethical questions that society is simply not prepared to handle. We may gain certain medical and economic benefits, but at what cost? In seeking mastery over nature, will we ultimately trade away our humanity?

The enormity of the powers unleashed by genetic engineering was not lost upon the scientists who helped to develop the technique. James Watson, who in 1953, along with Francis Crick, had discovered the double helix structure of DNA, for which they were awarded a Nobel Prize, recalls fearing that scientists might create a genetic combination in the test tube that would "rise up like the genie from Aladdin's lamp and multiply without control, eventually replacing preexisting plants and animals, if not man himself."[4] Scientists were sufficiently alarmed in the early 1970s to call for a voluntary moratorium on much of the research while hazards were investigated. The matter did not stop there, for the controversy had raised more fundamental questions about who should be making the potentially fateful decisions required. Public interest advocates argued that elected representatives and other lay groups had a right to participate in these decisions, since the public was potentially at risk and since tax funds had supported much of the research leading to the technique's development. Scientists countered that public intervention would threaten scientific freedom and disrupt research; that society would be best served by letting knowledgeable experts determine appropriate safeguards and set research priorities.

As initial fears about risks subsided, the challenge to scientific autonomy largely collapsed. By 1980, any lingering concerns were overshadowed by mounting enthusiasm over the unfolding array of promised commercial applications. Yet questions about the public's right to a voice in science and technology policy were never fully resolved. Civic, environmental, and religious groups continued to raise concerns about such issues as human genetic engineering, the deliberate release of genetically engineered organisms into the environment, and commercial priorities for biotechnology. Few scientific advances better illuminate the ambivalent and often conflicting attitudes toward technological progress that exist in our society, and the ongoing struggle to define an appropriate and productive relationship between science and its public sponsors.

The early years: scientists' fears

The story really began in 1871, with the discovery of DNA (deoxyribonucleic acid) in the sperm of trout from the Rhine River.[5]

But many other developments were necessary before scientists came to understand the genetic functions performed by DNA and how they could be manipulated. Not until roughly 100 years later had all the pieces of the biological puzzle been assembled that would allow scientists to create, in the laboratory, hybrid organisms containing the genes of two different species – for example, a mouse and a bacterium. In 1973, biologists Stanley Cohen of Stanford, Herbert Boyer of the University of California at San Francisco, and colleagues devised a simple yet ingenious procedure to produce "recombinant" organisms: They used special enzymes to snip apart the DNA of different species, then spliced these DNA fragments together in new, hybrid DNA molecules, which were in turn introduced into bacterial hosts such as the ubiquitous bacterium *Escherichia coli,* and, later, into animal hosts. To select out the recombinant bacteria, they linked the new DNA fragments to genes for antibiotic resistance and then exposed all the bacteria to antibiotics. Only the antibiotic-resistant bacteria, many of which carried recombinant molecules, survived.

Because DNA contains the all-important hereditary information necessary for generating life, the possibility of untangling its mysteries with this sophisticated new technique opened exciting vistas in molecular genetics. According to James Watson, most scientists' first reaction was one of "pure joy."[6] With such a powerful tool, they could gain new understanding of the structure of genes and the mechanisms responsible for turning on and off the production of proteins, the building blocks of life. The new technique might also help solve the puzzle of cell differentiation which, on the one hand, may control abnormal cell development associated with cancer, and on the other, governs the creation of organisms ranging in complexity from the simple amoeba to human beings.

But the initial excitement was soon tempered by the specter of unsuspected dangers. In Watson's words: "We began to ask whether in the process of possibly discovering the power of 'unlimited good' we might simultaneously be setting the stage for discovering the power of 'unlimited bad.' "[7]

Concerns first surfaced over a set of experiments being planned in 1970 by Paul Berg, a Stanford biochemist (Berg would later receive a Nobel Prize for research in this field). The experiments entailed the use of an animal tumor virus (SV40) as a vehicle to transport genetic information between cells. A graduate student in Berg's laboratory, troubled about being involved in what she correctly perceived would be the forerunner of human genetic engineering, told biologist Robert Pollack at Cold Spring Harbor Laboratory on Long Island about the work she and Berg were doing. Pollack was dumbfounded. Had they considered the dangers of replicating large quantities of a tumor virus in *E. coli,* a bacterium that commonly makes its home in the human intestine? Fearing that Berg might be "making something that in ten years would give everybody in Stanford a gut cancer," Pollack phoned him and urged him not to do the experiment.[8] At first, Berg resisted, but after consulting with more colleagues who also expressed doubts, he too became convinced of the possible risks and decided to postpone the experiment.

Berg's action was in keeping with the environmentally oriented mood of the 1970s. Many scientists, sensitized to the harmful effects of science and technology by the antiwar and environmental movements, were already worried about the risks of research involving potentially serious pathogens, which was becoming increasingly common in molecular biology.[9] Berg, in fact, helped organize a conference in 1973 on the risks of research using animal viruses, one of a series of efforts within the scientific community to confront the question of research safety.

The views of many leading scientists at that time bore a striking resemblance to those that would later be expressed by various lay groups (although by then the weight of scientific opinion had shifted considerably). At the 1973 conference on biohazards, for example, participants gave voice to three important themes that public interest advocates would echo in the years to come: the desirability of caution in the presence of unknown dangers; the right of laboratory workers and others exposed to research risks to participate in decisions about safeguards; and the need for appropriate health monitoring and risk assessment to detect unforeseen harms. An NIH scientist cited the case of prenatal DES to illustrate the pervasive uncertainty of all risk projections.[10] The meeting concluded with a unanimous recommendation for the National Institutes of Health (NIH) to conduct epidemiological studies focusing especially on laboratory workers, who were presumably at greatest risk. NIH's response was also a precursor of things to come: The recommendations were largely ignored.

Meanwhile, concerns about biohazards continued to mount. Robert Pollack, the Cold Spring Harbor microbiologist who had first called Berg, was quoted in the influential journal *Science,* the official journal of the American Association for the Advancement of Science, as saying: "We're in a pre-Hiroshima situation. It would be a real disaster if one of the agents now being handled in research should in fact be a real human cancer agent."[11]

It was at the 1973 Gordon Conference on Nucleic Acids, however, a meeting normally devoted entirely to scientific issues, that the full import of the new technique really dawned. One of the participants remarked innocently, "Well, now we can put together any DNA we want to." The meeting hall began to buzz as the meaning sank in.[12] Several participants convinced the meeting's co-chairpersons to take a half hour of the previously set program for a discussion of the potential hazards that might accompany this newfound capability. Following this discussion, a majority of the 142 scientists in attendance voted to write the President of the National Academy of Science (NAS) stating their concerns and requesting that the NAS set up a committee to make recommendations. To draw attention to the issue, the participants agreed to publish their letter in *Science*.[13] Little did they know just how much attention it would soon get.

Berg was asked to head the NAS committee, and in April 1974 he met with Watson and other prominent scientists to consider a response to the *Science* letter. The group quickly agreed that a larger meeting should be held to discuss the situation and decided to organize an international conference the following February at the Asilomar Conference Center in California. But what to do in the meantime? According to Berg, it was Norton Zinder, a microbiologist at Rockefeller University, who suggested, "If we had any guts at all, we'd tell people not to do these experiments." "That came as a shock," recalls Berg. "It seemed rather radical."[14]

Radical or not, that is what they did. Berg drafted a letter issuing this unprecedented request of fellow scientists, and it appeared in July 1974 in *Science* and its British counterpart, *Nature*.[15] Cosigning the letter along with Berg were ten of the most distinguished scientists in molecular biology, all members of the NAS committee. Although noting that scientific concerns were based on "potential rather than demonstrated risk," the letter called for a worldwide, voluntary moratorium on recombinant DNA experiments believed to pose the greatest dangers: those that would confer antibiotic resistance, oncogenicity (the ability to cause cancer), or toxin-producing capacities to bacterial strains that do not normally have these traits. The letter also requested the NIH to establish its own advisory committee to assess the risks more thoroughly and devise appropriate safeguards.

The controversy goes public

The moratorium letter caused a sensation in the public press as well as the scientific community. Never before had scientists acted collectively to restrict their highly prized "freedom of inquiry" for purely conjectural dangers. Newspaper headlines featured such titles as "The Scientific Conscience" and "Possible Danger Halts Gene Tests."[16] Berg and the other scientists, viewing the issue in terms of technical risks and safeguards, were surprised at all the interest. "We hadn't thought that the public would be in on it," Berg later commented, "and I must admit I didn't give much thought at all to the press."[17]

Despite some grumbling, most researchers around the world complied with the request – reassured, perhaps, that the matter would be reopened shortly at the upcoming meeting at Asilomar. In Great Britain, which was particularly sensitive to the problem of biohazards because of the recent escape of a laboratory-bred smallpox virus which had killed several citizens, the government immediately appointed a parliamentary group to investigate the dangers posed by recombinant DNA research.[18] Most other countries were content to wait and see what the international gathering of scientists at Asilomar would produce.

As the months passed, many molecular biologists began to chafe at their self-imposed moratorium. By the time of the Asilomar meeting in February 1975, even the conference organizers, although still concerned about how little was known about possible risks, were hoping to get on with the research by designing suitable safeguards.[19] With that goal in mind, they invited nearly 150 of the world's most respected biological scientists in academe, government, and industry to meet at Asilomar.

Deliberations at Asilomar were intense and frequently got bogged down in highly technical disputes. There was considerable jockeying over which areas of research were hazardous; as one participant put it, "People tend to draw lines just north of themselves."[20] Several senior scientists, notably Nobel Laureates James Watson and Joshua Lederberg, argued forcefully against imposing any constraints at all. By the end of the conference, however, the majority of the participants had agreed on a rough categorization of experiments based on the assumed level of hazard and a general plan for controlling risks through physical and biological safeguards.[21] Asimolar had achieved its purpose: The research could go forward with the agreed-on controls.

The only nonscientists at the meeting were selected members of the press and four law professors who had been invited to talk about the legal and ethical responsibilities of scientists. The lawyers' comments were quite unsettling to many of the assembled researchers. "Academic freedom is limited," declared University of Pennsylvania law professor Alex Capron. "Freedom of thought does not extend to the causing of harm." He continued: "This group is not competent to judge the risk–benefit ratio of experiments. That is a social decision, as is the judgment on benefit itself."[22] Another speaker addressed a legal issue closer to home: the likelihood of massive personal injury lawsuits if any harms did materialize. These talks had a chastening effect on those present. Previously rejected safeguards found new support. When one scientist was congratulated for having been so persistent in arguing for restraints, he replied that he didn't have much clout, "but the lawyers had a tremendous influence."[23]

Still, most of those gathered at Asilomar remained convinced that the main questions involved technical biohazards and that responsible self-regulation was an appropriate and sufficient response to these uncertainties. Indeed, in the opening session, future Nobel Prize winner David Baltimore had announced that the larger ethical issue raised by human applications of genetic engineering or its possible military uses were outside the scope of the meeting.[24] Thus was established a pattern that would prevail throughout the controversy: Scientists defined the questions as technical matters requiring scientific data, and thereby secured for themselves the dominant role in decisionmaking. Social, ethical, and political concerns, which entail different forms of expertise, were declared off limits from the outset. The very framework of the debate served to define who could participate legitimately in decisionmaking, and who could not.

After the Asilomar conference, the job of translating the agreed-on controls into specific guidelines fell to the Recombinant DNA Molecule Program Advisory Committee ("RAC"), an NIH advisory committee appointed by HEW in response to the Berg committee's request. Initially, RAC was comprised entirely of scientists, although two nonscientists were quickly added for symbolic reasons at the behest of RAC itself.[25] Coming up with workable guidelines was a herculean task, given the wide range of opinions about risks and the absence of reliable data. It took sixteen months of wrangling – and one false start that was widely criticized as being too lax[26] – for the group to agree on containment strategies which, in its view, struck an acceptable balance between risk-reduction and progress with the research. In February 1976, the NIH finally released a draft version of the research guidelines for public comment. These guidelines helped to define worldwide standards. Virtually every country where recombinant DNA research was occurring enacted some sort of regulations, most borrowing heavily from the NIH guidelines.[27]

To many observers, the proposed guidelines seemed arbitrary, and byzantine in their complexity. One NIH official, referring to the often bitter struggle over their formulation, described them as 90 percent emotional and only 10 percent scientific.[28] Each experiment was classified according to its presumed hazard and assigned corresponding levels of physical and biological containment, with a provision for annual review and modification as new evidence became available. The least restrictive level of physical containment, designated P1, involved little more than standard microbiological practices, while the most restrictive level, P4, required special protective clothing, double air-locked doors, and negative air pressure inside the laboratory (so that all air currents that might carry germs would flow into, rather than out of, the laboratory). The highest-risk experiments, including nearly all work with viruses, were banned altogether.

Biological containment involved the use of special strains of enfeebled experimental organisms that could not survive outside the laboratory. Most recombinant experiments in those days used a laboratory-bred strain of *E. coli* bacteria known as K12. There had been considerable controversy, both before and after Asilomar, over the use of such an ubiquitous bacterium, especially in view of its well-known proclivity for the human intestine. But *E. coli* was favored over other organisms because of scientists' intimate knowledge of its behavior gained from decades of previous research. Furthermore, it was believed that the specially bred K12 strain, having adapted itself to life in a laboratory petri dish, would not survive very long under the harsher circumstances of life in the wild. Because the survival of *E. coli* K12 under natural conditions had not been systematically studied, however, one of the major recommendations at Asilomar had been to create a special strain of *E. coli* K12 that was guaranteed to be incapable of surviving outside of a highly artificial laboratory environment. Roy Curtiss, a microbiologist at the University of Alabama, undertook the task and developed a bug whose survival rate was close to 0 (1 out of 100 million escaping bacteria). In honor of the

Bicentennial Year, he named it Chi 1776. The Chi 1776 strain represented the safest level of biological containment, followed by *E. coli* K12. The lowest level of protection involved so-called wild-type *E. coli*.

To implement the NIH guidelines, each institution receiving federal funds was required to have a local "biohazards" (later renamed "biosafety") committee. These local committees were to review researchers' proposals detailing how their experiments would meet containment requirements and forward them to NIH. The NIH, advised by RAC, was responsible for approving researchers' proposed containment measures and for revising the guidelines as appropriate. As the guidelines underwent successive revision, the local biosafety committees assumed more and more responsibility, to the point where they were eventually operating almost entirely independently of NIH.

The guidelines were, strictly speaking, voluntary, although violation carried the threat of withdrawal of research funds. This setup, although imposing constraints, was congenial to the academic community because it involved professionally defined standards rather than externally dictated laws. And biologists could look to the NIH, their most generous benefactor, as a friendly enforcement agency. The NIH could not dictate safety procedures to institutions that were not receiving federal funds, however, which meant that, by and large, private industry remained outside the scope of the guidelines. At the time, commercial involvement in biotechnology was extremely limited, and few were troubled by what was to become a very large loophole.

The breaking of consensus: the emergence of dissident scientists

Throughout their self-imposed moratorium in 1974, the scientific community maintained a reasonably united front. But as the view of most mainstream scientists swung back toward resumption of the research, dissent began to emerge. Among the new critics were well-established scientists troubled by the increasing power of science as well as younger researchers with a history of political activism. They questioned both the adequacy of the proposed safety measures as well as the process by which research policies were being decided.

One of the earliest voices of dissent was Science for the People, a national group formed during the Vietnam War to protest the use of defense-related science in the war effort. The Boston chapter included three respected biologists – Jonathan King and Ethan Signer of MIT, and Jonathan Beckwith of Harvard – who had become increasingly uneasy about the implications of recombinant DNA research. The group outlined its concerns in an open letter to each participant at Asilomar, urging that recombinant DNA not be added to the list of prior technological disasters such as radium, asbestos, and thalidomide. Scientists, the letter emphasized, especially molecular biologists who were actively engaged in developing the techniques, had no right to make these decisions alone; this was "like asking the tobacco industry to limit the manufacture of cigarettes." The letter continued: "Since the risks and danger of these technologies are borne by the society at large, and not just scientists, the general public must be directly involved in the decision making process."[29] This letter was never formally considered at the Asilomar meeting, and only one person ever wrote the group in response.[30] Nevertheless, it is clear that most of the participants at Asilomar were adamantly opposed to the idea of sharing control over science policy with lay citizens.

While Science for the People was generally viewed as the political fringe, others raising similar questions were squarely in the professional mainstream. Indeed, some of the new critics were senior scientists who commanded great respect, including Robert Sinsheimer, then chairman of Biology at California Institute of Technology, and Erwin Chargaff, emeritus chairman of Biochemistry at Columbia University Medical School and recipient of the National Medal of Science, the United States' highest tribute to a scientist. Both men argued that recombinant DNA research had far broader implications than just its possible laboratory biohazards. Sinsheimer, for instance, warned that splicing genes from unlike hosts could wreak havoc on barriers between species, barriers created by a delicate process of evolutionary checks and balances and set asunder only with unpredictable consequences. Using genetic engineering on humankind, Sinsheimer wrote "would in the end make human design responsible for human nature. It is a responsibility to give pause . . ."[31]

Chargaff painted a more ominous picture: "Is there anything

more far-reaching than the creation of new forms of life?" he asked in a letter to *Science*. The letter concluded with a deeply foreboding forecast:

> An irreversible attack on the biosphere is something so unheard of, so unthinkable to previous generations, that I could only wish that mine had not been guilty of it . . . Have we the right to counteract, irreversibly, the evolutionary wisdom of millions of years, in order to satisfy the ambition and curiosity of a few scientists? The world is given to us on loan . . . My generation, or perhaps the one preceding mine, has been the first to engage, under the leadership of the exact sciences, in a destructive colonial warfare against nature. The future will curse us for it.[32]

At February 1976 hearings on the draft guidelines, critics stressed how little was known about even possible short-run hazards. Several scientists from Science for the People contended that even if the enfeebled Chi 1776 bug itself could not survive outside the laboratory, it might still be able to transfer pathogenic capabilities to other, "wild" *E. coli*. This could occur, they suggested,

> on lab surfaces . . . in someone's body following inhalation or ingestion . . . in "safe" cultures which are contaminated. Perhaps, above all else, prolonged survival and genetic exchange could occur in special natural environmental conditions sometimes encountered by *E. coli*, that we in our ignorance of this bacterium's specialized habitats, know nothing about.[33]

To assess the risks, critics called for a series of experiments that would answer such questions as how long recombinant bugs could survive in different environments; whether cancer-causing genes spliced into *E. coli* would be "expressed" (i.e., cause cancer); and whether recombinant bacteria would transfer their genetic load to other "wild" strains of bacteria that might travel outside the laboratory. Without such answers, they argued, safety – even with the use of the Chi 1776 strain – could not be ensured. They also urged systematic health monitoring of laboratory technicians and other exposed workers.

Most of the critics were dubious that their concerns had much impact. Richard Goldstein, then an assistant professor of microbiology at Harvard Medical School, expressed his dissatisfaction with the guidelines and their development in a letter to NIH Director Donald Fredrickson. Despite the appearance of public discussion, Goldstein wrote, "I was struck . . . that America is once again, as in the case of atomic power, opting only for development rather than development, leadership, and control of a new technology."[34]

For Goldstein, who would later be appointed to the Recombinant DNA Advisory Committee, the frustration had only begun. As the debate evolved, dissident scientists became increasingly estranged from the mainstream scientific community. For their efforts to make science more publicly accountable, critics earned the intense and enduring antipathy of many of their peers.[35]

After the February hearings, the guidelines went through one more revision before their formal release in June 1976. Meanwhile, the debate between the various protagonists and antagonists of the research was attracting increasing attention from the press. The brewing conflict did not go unnoticed in Congress.

Recombinant DNA goes to Congress

Congress had for some time been interested in the social implications of scientific advances. In 1971, in fact, James Watson had testified that developments in molecular biology were "far too important to be left solely in the hands of the scientific and medical communities."

> The belief that . . . science always moves forward represents a form of laissez-faire nonsense dismally reminiscent of the credo that American business if left to itself will solve everybody's problems. Just as the success of a corporate body in making money need not set the human condition ahead, neither does every scientific advance automatically make our lives more "meaningful."[36]

These views (which Watson later sharply repudiated) were similar to those of Senator Edward Kennedy, then chairman of the Senate's Health Subcommittee. For Kennedy, who had a long-standing interest in science and medicine, the genetic engineering controversy raised important questions about the role of science in society. In April 1975, a few months after Asilomar, he opened a series of hearings on recombinant DNA research.[37] The House followed suit, under the direction of Health and Environment Subcommittee chairman Paul Rogers. All told, between 1975 and 1978, recombinant DNA generated nine separate sets of congressional hearings.[38]

In contrast to many political cartoons about genetic engineering, often depicting rogue *E. coli* and microscopic monsters, the initial Senate hearings had a distinctly academic flavor. Molecular geneticists, like Stanley Cohen and Herbert Boyer, the developers of the basic recombinant technique, tried to convey the importance of their research in language senators could understand, and to convince them that the NIH guidelines provided adequate protection. Bioethicists, health policy specialists, and a few dissident scientists presented opposing views, criticizing the guidelines and urging broader public participation in decisionmaking. As support for their position, critics pointed to the nation's increasing commitment to protection of human subjects in research, and to the patient's right to "informed consent" in medical treatment. In the case of recombinant DNA, the "patient" was society itself.[39]

By the summer of 1976, warnings of dissident scientists and rumors of federal legislation combined to intensify the controversy. As the lines of debate hardened, the tone of Congressional hearings, even disputes over "scientific" issues, became increasingly political. Environmentalists, including the Natural Resources Defense Council, the Environmental Defense Fund, and Friends of the Earth, joined forces with scientific critics to form the "Coalition for Responsible Genetic Research."[40] In press releases and Congressional testimony, Coalition members warned that the creation of self-replicating, potentially hazardous recombinant organisms could lead to a "molecular meltdown." Friends of the Earth called for a complete moratorium on the research pending further investigation of risks and "implementation of democratic procedures that will insure open discussion and public decision..."[41] In July 1976, shortly after the guidelines were released, Senator Edward Kennedy, along with Senator Javits of New York, wrote President Gerald Ford urging federal control of all recombinant DNA research.[42]

The environmentalists' position on recombinant DNA, unlike some of their other goals, did not make an adversary out of organized labor. Indeed, in an important undoing of the familiar environmentalist–labor antagonism, health and safety officials from the AFL-CIO and the Oil, Chemical, and Atomic Workers Union (OCAW) echoed the sentiments of groups like Friends of the Earth. In House hearings in March 1977, OCAW official Tony Mazzocchi warned that it was one thing to issue the NIH guidelines, and quite another to see that they were followed. "Guidelines?" he asked skeptically. "I do not even want to dignify any discussion around guidelines because no one generally pays attention to guidelines nor less, the law."[43]

Mazzocchi's remarks proved prophetic. A few months later, a science writer described her ninety-five-day visit to Boyer's laboratory and the disturbingly cavalier attitudes toward the NIH guidelines she observed there. "Among the young graduate students and postdoctorates," she wrote, "it seemed almost chic not to know the NIH rules."[44] This article drew vehement protests from Boyer and other senior scientists that the laboratory's safety procedures were perfectly adequate, but later events revealed that the guidelines had in fact been violated.[45] In the months to follow, violations also occurred at Harvard Medical School, and, in 1980, at the University of California at San Diego – although none of these incidents, NIH officials stressed, posed any danger to the public health.[46]

By 1977, organized religion joined environmentalists and labor in the debate. In May of that year, the National Conference of Catholic Bishops issued a plea that "urgency in the formulation of guidelines . . . not be allowed to short-circuit reflection on: the purpose and implications of these forms of DNA modification; the effect of this type of genetic research on our understanding of ourselves and of our relation to nature; and the correlation between the scientific advance possible through recombinant DNA research and human progress as judged by a variety of criteria."[47] Such concerns were to be reiterated many times by the country's major religious groups in the years to come, as they watched the pace of the research quicken and the opportunities for meaningful social choice fade.

One of the sharpest critics in the recombinant DNA debate, then as now, was Jeremy Rifkin, at the time director of the Peoples Business Commission. The real question, Rifkin declared, was not biohazards but whether the research should be done at all, given its potential for bringing us into the age of human genetic engineering and steps closer to a Brave New World.[48] For scientists to avoid dealing with the ultimate consequences of their work, he claimed, was like an atomic physicist in the Manhattan Project saying, "I am only splitting the atom and you are talking about nuclear holocaust."[49]

It was precisely such long-run consequences that most scientists refused to address. This was one reason the issues could not be joined: The biologists and their critics were really talking about different things. This may partly explain why legislators were not immediately convinced by the reassurances they heard from molecular biologists. Allusions to nuclear research probably struck home with many members of Congress, who had learned, painfully and dramatically, that science could be used for evil as well as good. Furthermore, while the process of developing the guidelines reflected noble intentions on the part of the scientific community, from the standpoint of politics it was elitist and undemocratic. And the questions raised by critics seemed legitimate. Were voluntary controls, enforced only by the threat of withdrawn funding, adequate? What about the lack of any controls for industry-based research, where the risks were likely to be greatest? To most observers, some sort of legislation seemed almost inevitable.

Throughout 1976 and 1977, both houses, led by Kennedy and Rogers, considered some sixteen bills between them to regulate recombinant DNA research. Most of these bills extended the NIH guidelines to all recombinant DNA research, public or private, and required licensing and inspection of research facilities. The major Senate bill, drafted by Kennedy, proposed to give regulatory authority to a commission involving a majority of nonscientists, whereas the principal House bill, framed by Rogers, took a more traditional approach, lodging responsibility in HEW.[50] The House and Senate bills also differed on the issue of federal preemption – that is, whether state and local governments could set their own regulatory requirements. Kennedy was impressed by what had been happening in places like Cambridge, Massachusetts, his own backyard, and so was inclined to let communities decide for themselves whether to adopt requirements that went beyond the federal guidelines. Rogers, in the House, believed that uniform federal standards were preferable, and so placed greater constraints on local authority.[51]

Most scientists strongly preferred the House bill. As local activity increased, in fact, many in the scientific community supported preemptive federal legislation as the best way to prevent what they feared would be a hodgepodge of inconsistent state and local regulations. Their hope, as expressed by Philip Handler, then president of the National Academy of Sciences, was to terminate what Handler called the "feckless debate which has offered outlets for anti-intellectualism and opportunity for political misbehavior while making dreadful inroads on the energies of the most productive scientists in the field."[52]

Feckless or otherwise, the surge of lay participation in the once off-limits area of science policy, especially at the grass-roots level, remains one of the most vivid chapters in the recombinant DNA story.

Taking it to the streets: recombinant DNA at the local level

In communities where the research was occurring, people were becoming increasingly alarmed by continuing press reports that scientists could not agree about the potential risks. From early 1976 to mid-1977, recombinant DNA showed up on the agenda of city council hearings, citizen review committees, town meetings, city and county health department investigations, and countless public forums on science and society in places like Ann Arbor, Princeton, San Diego, and, most notably, Cambridge, Massachusetts. In February 1977, Cambridge passed the country's first local ordinance regulating recombinant DNA research; within the next year, five other cities and towns followed suit. Seven state assemblies considered bills, although only two, New York and Maryland, ultimately passed state legislation.[53]

Nowhere was grass-roots involvement greater – or more colorful – than in Cambridge.[54] It began in June 1976, when a local newspaper disclosed that Harvard was planning to install moderate-risk (P3) containment facilities in a fourth-floor laboratory of an old building near Harvard Square. The story noted that the building's electricity was unreliable; that the pipes often broke causing flooding; and that the building was plagued with an apparently ineradicable infestation of ants and roaches. To many townsfolk, it seemed a poor choice indeed for a laboratory intended to provide special protection. Moreover, Harvard had apparently done little, if anything, to involve city officials in discussions about public safety.

When Cambridge Mayor Alfred Vellucci learned of all this, he was furious. Vellucci was famous for his outspoken hostility to Harvard in a community with fierce town–gown antagonisms. With the support of two distinguished Harvard biologists, Ruth Hubbard and her husband, Nobel Laureate George Wald, Vellucci scheduled an open meeting of the city council and led it off with a chorus of "This land is your land" by the local high school choir. The discussions were heated. Vellucci referred to stories about "monsters" and "creatures," reminding everyone present that it was his duty as a public servant "to make sure nothing is being done in the private or public laboratories that may be injurious to the health of the people of the city."[55] Scientists from Harvard and MIT, along with a special delegation from the NIH, tried in vain to allay his fears. The outcome, eventually, was a compromise: Scientists would refrain from all P3 research while a specially appointed citizen group, the Cambridge Experimentation Review Board, reviewed the hazards of gene-splicing experiments.

So began a precedent-setting experiment in public participation in science policy. The Review Board included no practicing scientists although the members included two physicians and a nurse. Everyone, particularly members without professional training, had difficulty understanding the highly technical issues being discussed. They depended on scientists to define some of the basic terms and to validate underlying assumptions. What transpired, according to one observer, amounted to "a debate between two scientific camps slugging it out in public."[56]

Still, the Board had to make its own judgments about the merits of the arguments it heard. After four months of intense work and meetings, the group finally decided that the research should proceed, but recommended various precautions that went beyond the NIH guidelines: mandatory health monitoring of laboratory personnel, additional laboratory safeguards, and a special city biohazards committee with oversight over all recombinant DNA research. Cambridge's local ordinance subsequently incorporated all of these recommendations.

Members of the Cambridge Review Board decided at the outset not to address the ethical implications of the research, but their final report alluded to the broader philosophical context of the debate. And, the report confronted head-on the prickly question of free inquiry:

> Knowledge, whether for its own sake or for its potential benefits to humankind, cannot serve as a justification for introducing risks to the public unless an informed citizenry is willing to accept those risks. Decisions regarding the appropriate course between the risks and benefits of a potentially dangerous scientific inquiry must not be adjudicated within the inner circles of the scientific establishment.[57]

This historic confrontation in Cambridge between the body politic and science – the "little people versus the big people," as

Mayor Vellucci put it[58] – was not without its lighter moments. In May 1977, Vellucci, who delighted in needling his opponents, wrote National Academy of Sciences President Philip Handler asking him to investigate reported sightings of "a strange, orange-eyed creature" and a "hairy, nine foot creature" to determine whether they might be "in any way connected to recombinant DNA experiments taking place in the New England area." Handler wrote back:

> I have been assured by scientists from research centers in the area that the careful records of their experiments with recombinant DNA indicate that none of their creations have escaped and that all are accounted for. Presumably, the bizarre creatures in the Cambridge area are among the rarer specimens of *Academicus americanus, Cambridgiensis.* They are usually identified by their green rucksacks and bear close watching as they may be wolves with sheepskins.[59]

All this was much to the chagrin of recombinant DNA researchers and science policy professionals, who warned that the NIH regulatory model they had so carefully constructed might be jeopardized by such homespun antics and fears. As public involvement in the recombinant DNA controversy spread from Cambridge to other cities and states around the country, many scientists were openly dismayed, arguing that allowing local governments to make decisions about research safeguards would be an invitation to poorly informed, special interest, and, ultimately, debilitating restrictions on scientific progress.[60]

In the end, however, local and state regulations invariably used the NIH guidelines as the stepping-off point, adding only modest additional requirements. The most significant impact of these ordinances was to make the NIH guidelines mandatory rather than voluntary and to extend them to commercial as well as federally funded research. Typically, they gave state or local governments somewhat greater control over regulation and enforcement. A few, as in Cambridge, set up special community-based biohazards committees to provide additional oversight, or required mandatory health monitoring of workers in recombinant DNA laboratories to assess short-term health effects and to develop baseline information for interpreting long-term health trends.[61]

In 1979 and 1980, growing commercial interest in recombinant DNA technology prompted renewed concern about the adequacy of existing regulations, which in most places were still only voluntary for industry. In turn, as international competition in the field mounted, many industry-based scientists worried that local restrictions might put them at a disadvantage. Once again, Cambridge and surrounding communities, where a number of the new biotechnology companies were locating, debated the pros and cons of recombinant DNA in public hearings and city council meetings, focusing this time specifically on the hazards that might be associated with commercial biotechnology. Cambridge, Boston, Somerville, Newton, and Waltham, Massachusetts all passed laws regulating industrial applications of the research.[62] But this second wave of local activity never matched the "David and Goliath" aura that marked the early forays into a streetwise science policy.

Exit research controversy

By the spring of 1977, at least six bills were pending before Congress, and states and communities around the country appeared increasingly eager to pass laws of their own. Meanwhile, scientific evidence was accumulating that cast doubt on many of the initial concerns about risks. Indeed, most leading molecular biologists, including Berg and the cosigners of the original moratorium letter, had come to believe that the research was probably quite benign. This new attitude toward risks was articulated in a "Consensus Agreement" prepared by Sherwood Gorbach, chairman of an NIH-sponsored, invitation-only workshop on risk assessment in Falmouth, Massachusetts, in June 1977. There was "unanimous agreement," Gorbach wrote, "that *E. coli* K12 cannot be converted into an epidemic pathogen by laboratory manipulations with DNA inserts . . . [and] does not implant in the intestinal tract of man." "Extensive scientific evidence . . . provides assurance that *E. coli* K12 is inherently enfeebled and not capable of pathogenic transformation by DNA insertions."[63]

Although Gorbach may have been stretching things a bit, this statement clearly reflected the growing confidence of most mainstream scientists in the safety of the research. The more confident they became, the more they resented the prospect of regulatory legislation. James Watson referred balefully to the "regulatory Frankenstein" molecular biologists had unwittingly unleashed in calling for a moratorium back in 1974, and urged fellow scientists to help contain it before it threatened all of biological research.[64] The key, most agreed, was to defuse the risk issue.

Scientists duly set about trying to quell fears about risks. Watson suggested that even those who had called for a moratorium had never really been worried. "To my knowledge," he wrote in May of 1977, "none of us then was deeply concerned, but since others had expressed worry, we thought the responsible course was to inform the public..."[65] Even accumulating evidence that there were, indeed, risks was interpreted in a positive light. A striking example, according to some participants at the Falmouth workshop, was Gorbach's highly reassuring rendition of the meeting's conclusions, which they claimed was misleadingly unequivocal. Richard Goldstein, one of the workshop organizers, wrote NIH Director Fredrickson that "though there was general consensus that the conversion of *E. coli* K12 itself to an epidemic strain is unlikely (though not impossible)...there was *not* consensus that transfer to wild strains is unlikely. On the contrary, the evidence presented indicated that this is a serious concern."[66] Several other participants wrote concurring letters.[67] But it was Gorbach's optimistic summary, not the reservations expressed by Goldstein and others, that drew attention on Capitol Hill and in the media.[68]

Armed with antirisk arguments, leading investigators and professional societies like the American Society for Microbiology and the National Academy of Science embarked on a major lobbying campaign against legislation they considered oppressive, especially the Kennedy bill.[69] Universities also played a major role, through a lobbying group called "Friends of DNA" representing prominent scientists along with the president and senior administrators of some of the most prestigious American academic institutions.[70] In letters, conference statements, and speeches, scientists charged that the legislation being considered imposed unjustifiable controls over many important areas of scientific inquiry, and, pointing to the damage done to Russian genetics by Lysenkoism, could lead to political repression of research.[71] Two professional lobbyists hired by Harvard helped spread the group's antilegislation gospel. At Stanford University, the vice president for public affairs urged resident "DNA fans" to send Congress "gentle and persuasive letters."[72] Norton Zinder, the Rockefeller microbiologist who had first proposed the idea of a moratorium in 1974, urged his colleagues to "lobby like crazy" with the Congressmen from their states.[73] "Health staff on the Hill said they had never seen anything like it," an environmental lobbyist recalled.[74]

Later that summer, results from a then-unpublished experiment by Stanford's Stanley Cohen created an even greater stir in Congress. Cohen sent a manuscript describing his results to Fredrickson, explaining that he was taking this "unusual step" because the findings had "policy, as well as scientific, importance with regard to the regulation of recombinant DNA."[75] Bootlegged copies of the manuscript and Cohen's letter were soon circulating widely. Cohen's findings indicated that genetic recombination between different organisms, once thought to be the exclusive province of laboratory manipulation, also occurred naturally in the cell's own processes. This constituted "compelling evidence," Cohen wrote, that recombinant DNA molecules artificially synthesized with the help of particular enzymes "simply represent selected instances of a process that occurs by natural means." The implication was that recombinant DNA research was not so special after all and thus presented little, if any, risk.

Again, dissident scientists argued that the case had been overstated. Richard Novick, a noted plasmid biologist, contended that "the conditions under which this interpretation took place are so extremely artificial that there is essentially no chance of their occurring in nature."[76] Cohen's results, MIT's Jonathan King testified in Congress, show only "that by the use of the very potent recombinant DNA technologies, one can generate certain rare events in bacteria in the lab." To infer from these results that genetic recombination between unrelated organisms has been going on in nature, King said, was "equivalent to saying that since we can construct typewriters in factories, typewriters must assemble themselves spontaneously in nature."[77]

While critics of the research challenged both the relevance of Cohen's conclusions and their release prior to publication, they nonetheless began to see the burden of proof concerning risks shift from the proponents' camp to their own. As pressure from the science lobby mounted, the Coalition for Responsible Genetic Research organized a counterconsensus statement in support of legislation, signed by almost 100 "concerned citizens," including anthropologist Margaret Mead and scientist/activist Barry Commoner as well as a number of dissident scientists. This statement expressed strong disagreement with the "misrepresentation and exaggeration of recent data purporting to show the safety of re-

combinant DNA research," and protested that scientists were using scientific data for their own political purposes.[78]

Such objections proved no match for those of the scientific
establishment, which had directed its considerable professional and
technical clout into a full-blown lobbying effort. Critics, in contrast, formed only lean and scattered ranks: Environmentalists had
Alaska wilderness and other legislative battles to fight; labor unions
were overwhelmed with known toxic hazards; and dissident scientists had to endure increasingly overt professional ostracism. In
one widely quoted interview, James Watson referred to critics of
the research as "kooks, shits, and incompetents."[79] It was especially difficult for younger faculty members without tenure to
withstand hostility and intimidation from senior colleagues, and
many withdrew from the controversy, fearing for their careers.

In late September 1977, the science lobby witnessed its first
major victory. Senator Kennedy withdrew support from his own
bill, citing changes in scientific evidence on risks and Cohen's
results in particular.[80] As a replacement, Kennedy proposed a one-
year bill to extend the NIH guidelines to industry while a commission studied the issue. Although Kennedy reaffirmed his continuing support for public participation in science policy, in fact
this announcement choked off most of the momentum toward
legislation in either the House or Senate, and bills in both chambers
died in the ensuing months.

Scientists had won their political battle, but in the process lost
some of their innocence. As Norton Zinder confided to Paul Berg
in 1977, "I've been busy so long calculating the results of moves
– did I push too soon? too late? were the right people contacted?
. . . how far can I stretch 'truth' without lying? – that I may have
lost all perspective."[81]

NIH reform: form versus content

As talk of legislation abated, the focus of controversy shifted increasingly to the NIH, the only remaining source of governmental
control over research safeguards. RAC had begun working on
modifying the NIH guidelines not long after their release in 1976.
Some eighteen months later, in December 1978, the first proposed
revisions of the guidelines were released.[82]

Numerous meetings and hearings were held in the course of

these revisions in which critics and proponents presented sharply divergent views.[83] Most scientists argued that containment requirements should be greatly reduced in light of evidence, from the Falmouth conference and various risk assessment meetings in Europe, that risks were less than initially feared. They warned that the original guidelines were making it difficult for American biologists to keep pace with their foreign counterparts in Western and Eastern Europe, where regulations were less codified and hence, in some cases, more permissive.[84] Critics, by contrast, opposed relaxation of containment requirements on the grounds that systematic risk assessment had not yet occurred. In addition, they accused the scientific establishment of once again making most of its decisions about the guidelines in private, with little outside scrutiny. Friends of the Earth threatened to sue the NIH for failing to prepare the required Environmental Impact Statement, and for failing to allow public access to relevant information leading to the proposed changes.[85]

The revised guidelines issued in 1978 struck a compromise between the complaints and suggestions of critics and the pleas of scientists to "quietly dismantle the whole hateful [regulatory] artifice."[86] Containment levels were substantially relaxed, and the NIH shifted most of the responsibility for enforcement to the local institutional biosafety committees. In response to critics' charges that molecular biologists had monopolized policy decisions, the NIH broadened the RAC to include more individuals from fields outside the biomedical sciences and required each biosafety committee to include at least two members from outside the institution to represent community interests. The 1978 guidelines also mandated more comprehensive risk assessment.

The procedural reforms achieved by critics proved, ultimately, to be more symbol than substance. Dissidents could no longer claim they were excluded from the policymaking process, yet a few critical voices on the RAC were easily overridden by the solid majority of proponents of the research. Tufts University professor Sheldon Krimsky, formerly a member of Cambridge's citizen Review Board and one of the new nonscientists on the RAC, recalled: "The scientists still dominated the committee – as much by the extreme technicality of the deliberations as by their numerical advantage – but the critics were coopted."[87] On the local level, the presence of lay members on biosafety committees tended to enlarge the scope of the committees' concerns and to encourage more critical review of research proposals, although again scientists generally played the major role in decisionmaking on most committees. In some biosafety committees, despite the obvious intent of the 1978 reforms, biological scientists from other institutions, not lay citizens from the community, were brought in as the outside members.[88]

Meanwhile, the no-risk arguments were gathering momentum. Scientists at an NIH-sponsored meeting in Ascot, England, in January 1978 pronounced the likelihood of hazards from viral recombinant DNA research "so small as to be of no practical consequences."[89] Similarly, the Committee on Genetic Experimentation (COGENE), an international group of scientists, held a series of conferences in Europe in which participants discussed the latest scientific results and went home with more reassuring evidence.[90] By the spring of 1979, COGENE urged the NIH to deal the guidelines a speedy coup de grace.[91]

At its September 1979 meeting, RAC came close to doing just that. Over the objections of the recently added critics, the committee voted to recommend exempting from the guidelines most experiments involving *E. coli* K12, which comprised 80–85 percent of all recombinant DNA research.[92] "I do not know of a single piece of new data," Wallace Rowe, an NIH scientist and coauthor of the proposal, said, "that has indicated that K12 recombinant DNA research could generate a biohazard."[93]

Another blatant exaggeration, critics charged. Rowe's own risk assessment results, they argued, gave reason for concern. Rowe, along with another NIH researcher, Malcolm Martin, had undertaken several "worst case" experiments recommended by RAC to investigate the effects of genetically engineered tumor viruses on mammals. In these experiments they spliced the tumor-causing polyoma virus into *E. coli* K12, and then fed the recombinant bacteria to mice and hamsters. They found that recombinant polyoma virus caused fewer tumors than did the original (unrecombined) virus, a result they termed "highly reassuring with respect to the safety of cloning viral genomes in *E. coli*."[94]

Looked at another way, these results were anything but reassuring. Jonathan King pointed out that in transforming intestine-residing bacteria into tumor-causing agents, even weak ones, researchers were creating a new vehicle for polyoma to infect humans. Until the recombined bacteria approach came along, polyoma had no such ticket to the human body.[95] Other dissident scientists agreed with King's interpretation, and raised additional issues, including the possibility of a recombinant bug triggering an autoimmune disease (i.e., a potentially damaging reaction of the body to itself), worker health and safety, and the unexplored hazards of large-scale industrial biotechnology. To many critics, the evidence of risks seemed to be growing, not shrinking as the scientific community claimed.

Other studies began to appear which reinforced these concerns. For example, researchers at the University of Texas found that enfeebled *E. coli* K12 had entered the sewer system, despite provisions in the guidelines requiring chemical or physical disinfection of all experimental cultures before disposal, and that many of the organisms survived a sewage treatment plant in viable condition.[96] Another study reported that enfeebled K12 could survive for days on laboratory surfaces.[97] Scientists at Tufts University found that the survival rate of the enfeebled strain Chi 1776 in humans was increased greatly when it carried a commonly used cloning plasmid.[98] Equally worrisome were research results showing that K12 could colonize in mice that had been given certain antibiotics.[99]

Such evidence received little media attention or consideration by the larger scientific community. The NIH was deluged with letters from researchers supporting the proposed blanket exemption of K12 experiments from the guidelines.[100] Again, environmental groups and scientist-critics objected strenuously. Yet not all the objections came from those who wore the label of "critic." Roy Curtiss, the originator of the enfeebled strain of *E. coli* K12, Chi 1776, also opposed the exemption on the grounds that little new information had accumulated since the guidelines were last revised in 1978. Moreover, Curtiss noted,

> essentially all data obtained since 1977 by NIH contractors conducting risk assessment experiments to evaluate the safety of the *E. coli* K12 systems have indicated that host strains and . . . vectors survive better in many environments than previously believed . . . As a consequence, transmission of recombinant DNA from K12 to other microbes is a more probable event than was previously believed.

He spelled out various unanswered questions regarding risks, and, citing DDT and thalidomide, warned that "there must remain uncertainty since none of us is totally clairvoyant." The proposed exemption Curtiss concluded, "was based more on the politics of science than on the data of science."[101]

The NIH's final decision essentially confirmed Curtiss's assessment. In November 1979, Fredrickson announced that experiments using *E. coli* K12, although not formally exempted from the guidelines, would henceforth be slated for P1 containment, a level achieved by almost any standard microbiological lab, and that review would take place at the local biosafety committee level – in many cases, after an experiment had already begun.[102] In 1981, the RAC considered removing the withdrawal of funding as an enforcement measure for the guidelines but, in part to avoid re-raising the question of legislation, left this provision intact, while again reducing the range of experiments covered by the guidelines and delegating still more responsibility to institutional biosafety committees.[103] These trends have continued. Today, most recombinant DNA research falls entirely outside the guidelines, and there is virtually no meaningful NIH oversight of local biosafety committees.

Scientists had finally conquered their "regulatory Frankenstein." They succeeded in "burying Asilomar," as Watson put it, by confining the debate to laboratory biohazards, about which they had the greatest technical knowledge, and then by emphasizing the positive evidence and discounting or ignoring the negative evidence.[104] Armored with the shield of technical expertise, they won the political battle for control.

Only time will tell whether this was wise. So far, no known health problems have resulted from genetically engineered organisms, but some experiments are still considered potentially hazardous, and disagreement persists as to whether the widening array of uses and methods of genetic engineering pose long-term risks to health or the environment.[105] That no evidence of harm has yet become apparent is obviously no guarantee that, as in the case of DES, problems will not emerge later on.

DNA hits Wall Street

As the debate over biohazards was dying down, increasingly optimistic talk of the hoped-for benefits of genetic engineering began to take its place. The first glimpse of the technique's vast productive possibilities came in late 1977, when NAS President Handler announced, at a Congressional hearing on the regulation of recombinant DNA research, that Dr. Herbert Boyer and other scientists at the University of California at San Francisco had succeeded in getting bacteria to synthesize somatostatin, a scarce and expensive brain hormone with significant commercial potential.[106]

Although the hearing was held to consider the possible need for federal legislation, and included fiery questions to researchers who had violated the guidelines, news coverage was dominated by Handler's announcement. The public relations value of this disclosure did not go unnoticed. "I do not believe that a political debate is an appropriate forum for scientific announcements," *San Francisco Chronicle*'s science writer David Perlman wrote in *Science*. "The propriety of Dr. Handler's testimony," he added, "however politically useful, should, I believe, be widely discussed."[107]

Such shifts in the accepted code of scientific practice, although noted here and there in 1977, were only the precursors of more dramatic convulsions that would occur as the technique's full commercial potential began to unfold. The first harbinger of change was the formation, by a few scientists at the cutting edge of molecular biology, of their own small private biotechnology companies to develop marketable products, with backing by venture capitalists and contracts from large pharmaceutical and petrochemical companies. UCSF's Boyer was among the earliest of these scientist-entrepreneurs. In 1976, along with a partner, Robert Swanson, Boyer founded a small California company called Genentech, which would go on to become one of the leaders in commercial genetic engineering.

The claims, even then, were bold. Promotional literature from Cetus Corporation, a small company across the San Francisco Bay from Genentech, was promising in 1975 that recombinant technology would revolutionize pharmaceutical production processes:

> What we have done in the past, while exciting and profitable, pales compared to what we can accomplish in the future . . . *We propose to do no less than to stitch, into the DNA of industrial microorganisms, the genes to render them capable of producing vast quantities of vitally-needed human proteins* . . . This concept is so truly revolutionary to the biomedical sciences that we of Cetus predict that by the year 2000 virtually all the major human diseases will regularly succumb to treatment by disease-specific artificial proteins produced by specialized hybrid microorganisms.[108]

Still, money was not always easy to find. Many Wall Street investors, accustomed to overblown promises, dismissed genetic engineering as just another fad that would run its course and sink from sight. Ronald Cape, a cofounder of Cetus, recalled that in the mid-1970s, when he tried to attract venture capital for the company's visions, "the very same people who today are evincing such wonderful excitement were telling us that we were crazy."[109] Through late 1979, there were still only a handful of small companies in the biotechnology business, all with leading molecular biologists among their founders or advisors.

But the number of commercial applications of genetic engineering was expanding at a dizzying pace. By 1980, the extraordinary opportunities offered by industrial genetic engineering had gotten front-page coverage in nearly every major periodical.[110] Researchers promised a cornucopia of scarce and potentially valuable new drugs and hormones, and a wide range of useful applications in the agricultural, chemical, food processing, and energy industries.

The time was ripe to cash in on some of this promise if the fledgling companies were to expand and grow. In 1980, Genentech issued a public stock offering that made Wall Street history when the price per share was bid up from $35 to $89 in the first twenty minutes of trading. At the close of the day, the initial investment of $500 each put up by Boyer and his partner was valued at some $529 million.[111] The molecular goldrush was on. More than eighty new biotechnology companies were formed during 1980 and 1981 alone.[112] Petrochemical and pharmaceutical corporations such as Du Pont, Dow, Monsanto, and Exxon, many of whom were among the first investors in the early biotechnology firms, began gearing up their own in-house capabilities in genetic engineering. Industry spent more than $250 million in 1980 on biotechnology research, up fivefold from 1979.[113]

Biotechnology's commercial future was insured in June 1980, when, in the long-awaited *Chakrabarty* ruling, the U.S. Supreme Court decided that genetically altered life forms could be patented.[114] A patent entitles the patent holder to sell licenses to those who wish access to a patented creation, thus assuring some return on the investment that went into its discovery. Although proprietary interests can also be protected by keeping key scientific procedures "trade secrets," this is not always possible when so many research groups are working on similar problems using closely related methods. The Supreme Court decision was therefore an important green light to the commercial development of genetic engineering. Patent applications mushroomed, as did the number of patents issued. "Before the *Chakrabarty* decision," commented one patent lawyer, "the Patent Office was just sitting on the applications. Then all hell broke loose."[115]

The Supreme Court ruling raised anew some of the earlier ethical questions surrounding recombinant DNA research. The *Chakrabarty* case involved a microorganism. Would patents also be allowed on higher forms of life? The Court offered no guidance on where, or whether, to draw the line. Indeed, the Court held that the very distinction between living microorganisms and inert chemicals was "without legal significance," a judgment some considered an affront to prevailing humanistic and theological views about life and living creatures.[116]

The Court's decision drew strong protests from religious groups. In July 1980, the leaders of Protestant, Catholic, and Jewish national organizations issued a joint letter questioning the patent decision and warning of the dangers posed by the rapid growth of genetic engineering. "Who shall determine how human good is best served when new life forms are being engineered?" the letter asked. "When the products are new life forms, with all the risks entailed, shouldn't there be broader criteria than profit for determining their use and distribution? . . . Do we have the right to let experimentation and ownership of new life forms move ahead without public regulation?"[117]

No one was addressing these "moral, ethical and religious questions," the religious leaders charged, and called on the president and Congress to set up a mechanism that would enable "a broad spectrum of society" to play a role. This task was assigned to the President's Commission for the Study of Ethical Problems in Medicine and Biomedical and Behavioral Research, a government advisory panel appointed in 1978, whose members included prominent figures in law, ethics, medicine, and other fields. The commission's report, issued in 1982, was reassuring about many aspects of the research, but concluded nevertheless that the time had come "to broaden the area under scrutiny to include issues raised by the intended uses of the technique rather than solely the unintended exposure from laboratory experiments." Such scrutiny, the commission noted, should be by a group that is "broadly based and not dominated by geneticists or other scientists . . . "[118]

The *Chakrabarty* ruling evoked practical as well as ethical concerns. Some academicians worried that patents would inhibit the free exchange of scientific information, as individuals sought to keep their work secret in order to protect their eventual patent rights. On a more global level, critics warned that the *Chakrabarty* ruling would exacerbate the trend toward declining genetic diversity already evident in the plant world, which they attributed to the increasing concentrations of control over the world's food supply by a few multinational corporations.[119] Genetic diversity is essential to successful plant breeding and serves as protection against damage from epidemic pests and diseases. While the extent of genetic erosion is difficult to measure, it is generally believed that several hundred plant species become extinct every year.[120]

Further controversy over the patenting of new life-forms erupted in 1987, when the U.S. Patent and Trademark Office ruled that genetically engineered animals could be patented. This decision produced an outcry from animal welfare agencies, farm groups, religious denominations, and assorted other public interest organizations, and legislation was introduced in Congress to prohibit such patents for two years while Congress examined the issues. Never before in the 197-year history of the American patent system had legislation been proposed to block a patent policy that had already gone into effect. This unprecedented step reflected, as one bioethicist put it, "the reluctant recognition that human beings have discovered how to deliberately change and alter biological evolution."[121]

The impact of patenting plants and animals on society's moral values, or on the earth's genetic diversity, will never be fully known, for the effects of patents are hopelessly intertwined with a multitude of other social, economic, and ecological changes. The effects of patents also mingle with other aspects of commercialization in influencing patterns of scientific communication and collaboration. Although the specific causes remain unclear, there was nevertheless general agreement within the scientific community that communication was less open than it had been, a trend that was generally ascribed to the increasing rivalry of the new scientist-entrepreneurs. "No longer do you have this free flow of ideas," Stanford biochemist Paul Berg said in 1979. "You go to scientific meetings and people whisper to each other about their company's products. It's like a secret society."[122]

In the years that followed, claims to private ownership of biotechnology products expanded steadily, and concerns about the open exchange of information and materials persisted. These concerns reached new heights when, in 1987, a private company announced plans to lay claim to the discovery of the entire human genome – a map of all genes in human beings, which researchers believe will have major scientific value as well as many potential clinical applications. Many scientists objected strongly to the idea of anyone "owning" such a fundamental social resource and exploiting it for private gain.[123]

Even as the Supreme Court was coming to its landmark decision in the *Chakrabarty* case, the first of what was to become a long series of bruising battles over patent rights was taking shape. David Golde, a hematologist at the University of California in Los Angeles, with a colleague, took some cells from a dying leukemia victim and, as a scientific courtesy, sent some of them to a longtime research colleague. The colleague noticed that the cells produced the antiviral agent interferon and passed them on to an interferon researcher, Sidney Pestka at the Roche Institute, a research arm of the drug firm Hoffmann–La Roche. After several months of work, Pestka was able greatly to enhance the interferon-producing capability of the cells, resulting in a cell line potentially worth fortunes. Roche contracted with Genentech to clone a bacterially made version of Pestka's interferon gene. In June 1980, Genentech accomplished this feat. The timing was fortuitous; shortly thereafter, Genentech announced plans to issue a million shares of public stock.[124]

Roche then applied for a patent on its new product. But was the gene Roche's to patent? The University of California claimed *it* should have first rights to any profits from the now-lucrative gene, because U.C. researchers had isolated it, and Golde denied having given permission for it to be distributed. It was Pestka's manipulations, Roche rejoined, not the initial isolation, that had made the cell line commercially attractive. Hoffman–La Roche and the University of California avoided the embarrassment of a full-scale public brawl over the right to cells developed under government funding by reaching an out-of-court settlement; Roche paid the university an undisclosed sum in return for its claim to the cell line. But some nasty blows were exchanged. Roche sued Golde for defamation, claiming that Golde had called Roche employees "crooks" and "thieves" who had "stolen" or "hijacked" the cell line. For his part, Golde called the whole episode a very "bitter" experience.[125] One conclusion from it all was inescapable: The rules of scientific collaboration were being rewritten with a clear eye on the bottom line.

Of boomtown and gown

Biologists-turned-businessmen and major industries were not the only ones interested in the fortunes to be made from biotechnology. University administrators, faced with research costs that were outpacing federal funding, suddenly found themselves sitting on a molecular goldmine. It was university faculty members, after all, who were the field's original prospectors, relying on the accumulated wisdom of decades of academic research. Now that they had finally struck a prolific vein, weren't the universities that had nurtured them also entitled to a share?

However logical the reasoning, the first tentative academic foray into the business world touched off a storm of controversy. Shortly after Genentech sold stock publicly, Harvard President Derek Bok announced that his university was considering setting up its own corporation to develop and market products from its genetic research. As well as making money for Harvard, Bok pointed out, this arrangement would speed the transfer of technology to the market. "The potential benefits to society and to Harvard," stated a memorandum describing the proposal, "seem to warrant careful innovation and some prudent risk taking..."[126]

Debate among the faculty was vigorous, but the weight of opinion was negative. Ten of the seventeen members of the Harvard Biology Department cosigned a letter urging the administration to drop the plan. The author of the letter, Woodland Hastings, wrote:

> The whole matter violates the role of the university in our society so extensively and so terribly that I don't see how anything can come of it. The university would no longer be a nonprofit organization. It would mean that in everything we do, in our laboratories, in our scholarship, we are joining with the University to make a profit.[127]

A member of the faculty council warned that the plan would seriously compromise the university's perceived independence and autonomy:

> When we speak out for or against such things as nuclear power or air pollution we are listened to, in part, because people see us as members of an institution which is impartial. I think this technology will have good and bad impacts on society. If I speak about it, will people believe me? They will say "Universities are just like industry, they have an interest in it."[128]

Reactions in the press were also largely critical. The *Washington Post* titled its discussion of the proposal, "Ivy-Covered Capitalism."[129] The *Washington Star* wondered whether Harvard would now be known as Harvard, Inc., and suggested that its motto might be changed from "Veritas" to "Cupiditas."[130] Cambridge ex-Mayor Alfred Vellucci, a veteran pungent critic of Harvard, suggested, with characteristic irreverence, that Harvard's new motto should be the dollar sign, and its school color green instead of crimson.[131]

Stung by all the criticism, Harvard eventually decided against the plan. Other kinds of ties with industry were not long in developing, however, not only at Harvard but at virtually every major university across the nation. By this time, most of the country's leading faculty in genetic engineering had already formed consulting arrangements with industry, and growing numbers had equity (ownership) positions in corporate ventures. Industry's access to university-based genetic engineering programs was also steadily expanding as both industries and universities came to appreciate the mutual advantages of collaboration. Joint ventures proliferated. In 1981, Hoechst AG, the giant West German chemical company, gave Massachusetts General Hospital, a Harvard Medical School teaching facility, $70 million over a ten-year period to create an entirely new Department of Molecular Biology; in return, Hoechst was promised exclusive license to any patentable discoveries that emerged from the research it sponsored. Du Pont and other companies, in a similar arrangement with Harvard Medical School, gave $6 million over five years to form a new Genetics Department. At Stanford and the University of California at Berkeley, six major corporations contributed $2.5 million over a four-year period to create a for-profit company, Engenics, which was jointly owned by line officers and consultants, the six corporations, and a non-profit institute created at the same time, and which was assured licensing rights to any commercially patentable products emerging from the university-based research the institute sponsored. The list goes on and on.[132] In 1984, nearly half of all biotechnology companies in the United States provided some kind of funding for university-based genetic engineering research.[133]

While these new collaborative relationships brought a welcome influx of capital for universities, some in the academic community were troubled about the control that was being relinquished in the process. When industrialist Edwin Whitehead offered MIT more than $125 million in 1981 to set up an independent research institute affiliated with, but separate from, the university, for example, a number of MIT faculty objected to the proposed institute's separate-but-equal status, which bypassed usual university governance procedures. "If [Whitehead] really wants to support research," Jonathan King observed, "we have an excellent mechanism; he can establish the institute within MIT."[134] Other critics questioned the propriety of allowing a private institute to exploit what was, in their eyes, public property. "Are MIT's resources and reputation, created largely at public expense, really MIT's to sell?" asked historian David Noble.[135] Despite these and other objections, the faculty finally voted in favor of Whitehead's original proposal, although expressing "serious concern" about some of the implications.[136]

As industry–university ties multiplied, such concerns drew increasing attention. Incidents had been reported in which graduate students had been assigned to work on projects for a faculty member's company, or required to subordinate their own research to the company's priorities. (In one case, the situation was so objectionable that most of the students working with a particular professor, who had formed his own biotechnology company, asked to be transferred out of his laboratory.)[137] The image of academic science seemed to be getting a bit tarnished. The U.S. solicitor general, in his reply brief for the *Chakrabarty* patent decision, dismissed an argument made by a prominent molecular biologist who had set up a company, explaining that he was "hardly an impartial observer in the debate" over genetic engineering.[138]

In 1981 and 1982, then-Congressman Albert Gore held a series of hearings to explore what the commercialization of academic biomedical research was doing to the nation's research climate and academic values. While many witnesses testified to the benefits of closer ties between industry and academe, a number described examples of colleagues refusing to provide special strains of bacteria or other biological material, traditionally considered routine scientific etiquette. Stanford University President Donald Kennedy reported various instances in which a researcher at scientific meetings "actually refused on questioning to divulge some detail of technique, claiming that, in fact, it was a proprietary matter and that he was not free to communicate it."[139] Such secrecy, Kennedy pointed out, prevents other scientists from replicating experiments and thereby verifying existing theories. This is not a trivial issue. Replication lies at the heart of the scientific method; it is the way science moves forward.

To try to develop appropriate guidelines for industry–university interaction, Kennedy organized a national conference of university and business leaders at Pajaro Dunes, not far from Asilomar on the California coast. Like the 1975 meeting at Asilomar, which had launched the earlier recombinant DNA debate, Pajaro Dunes, too, was closed to the public. The press, students, and public interest organizations were excluded, Kennedy explained, in order to ensure "full and frank" discussions.[140]

It was no small task that the participants at Pajaro Dunes faced. For, at the root of all the controversy over industry–university collaboration lay profound – some would say irreconcilable – differences in the aims, methods, and values of the two enterprises. At best, the new alliance between biology and business required compromise and mutual accommodation; at worst the quest for knowledge was simply incompatible with the pursuit of profit. Berkeley immunology professor Leon Wofsy expressed the basic conflict as follows:

> The business of business is to make money, to beat the competition, and the mode is secrecy, a proprietary control of information and the fruits of research. The motive force of the University is the pursuit of knowledge, and the mode is open exchange of ideas and unrestricted publication of the results of research.[141]

The president of Schering–Plough, a major pharmaceutical company, speaking at the opening of his company's $20 million biotechnology facility, put it more bluntly. "Schering–Plough is not in business to do research," he said. "It's in research to do business."[142]

The participants at Pajaro Dunes clearly did not consider the goals of business and academe irreconcilable. The group concluded, as did another national conference on the same subject in 1983, that university–industry collaboration was basically good – good for universities, good for business, and good, by promoting technology transfer, for the public as well.[143] The Pajaro Dunes conference outlined general principles that should guide industry–university interaction, as it had set out to do; critics complained that they were too vague to have any real impact, and furthermore, implementation was left to each institution's discretion.[144] Many contentious issues were in fact left unresolved. For example, no mechanisms were adopted to ensure that a faculty member's involvement in commercial activities did not conflict with performance of academic responsibilities. Nor were any safeguards proposed to protect students against exploitation by faculty members involved in corporate work.

In allowing, indeed encouraging, the convergence of industry and academe, the participants at Pajaro Dunes were simply reinforcing a trend that was already well underway. Science itself had undergone a major transformation in the postwar period, and its values and practices had changed accordingly. Changing attitudes toward "intellectual property" illustrate the extent of this transformation. Describing the ideal norms of science in the 1940s, sociologist Robert Merton emphasized the collective nature of scientific discovery, and the importance of full and open communication of findings. "The substantive findings of science are a product of social collaboration and are assigned to the community," Merton wrote. "The scientist's claim to 'his' intellectual 'property' is limited to that of recognition and esteem . . . ." With respect to communication, he wrote: "The institutional conception of science as part of the public domain is linked with the imperative for communication of findings. Secrecy is the antithesis of this norm; full and open communication its enactment."[145] Today, the notion of "intellectual property" is routinely accepted in science. The "rights" to such property, in the form of patents, are assigned to scientists or their institutions. The need for secrecy prior to obtaining patents is taken for granted. In short, scientific knowledge is no longer considered a collective good but, rather, private property that can be owned and traded for profit.

Even when the knowledge is produced under a government grant – that is, has been supported by taxpayers' money – researchers and their institutions may still claim "ownership" rights, entitling them, in some cases, to sizable profits. The basic recombinant DNA technique, developed in 1973 by Stanford's Stanley Cohen and Herbert Boyer of the University of California, was patented by the two universities, even though both investigators were at the time supported by an NIH grant, and their accomplishment was the result, in Cohen's words, of "multiple discoveries carried out by many individuals over a period of time."[146] Both Boyer and Cohen have turned over their personal share of patent royalties to their respective universities, which have profited handsomely. By 1985, licensing fees paid to the two universities had already totaled $3.5 million, and industry analysts predict that the Cohen–Boyer patents could be worth more than $1 billion over their lifetime – many times what the two universities have earned to date on all their other patents combined.[147]

The dangers of treating knowledge as a commodity that can be owned and sold are subtle, but real. They threaten the basic norms

and values of science that Merton described, which have traditionally served to promote the communication and advancement
of knowledge while protecting the intellectual integrity of the
scientific enterprise. In the case of genetic engineering, the risk is
magnified by the unprecedented scale of commercial involvement;
never has an entire academic discipline, or at least the large majority of its leading practitioners been so closely and extensively
involved with private industry.[148] Furthermore, genetic engineering, unlike many applied disciplines, contains no built-in buffer
separating the process of basic discovery from the production of
salable commodities. These dangers have been widely discussed;
the trends continue. Like earlier concerns about safety, they, too,
have for the most part been shunted aside by the glittering prospects of biotechnology's bright commercial future.

Biotechnology: ℞ for the economy

During the early 1980s, the biotechnology industry grew almost
overnight from an embryonic venture by academic scientists into
a major industry with immense economic significance. It suffered
some growing pains along the way. In 1982, pressure from Wall
Street to start delivering on all the promises, combined with an
economic downturn, forced many of the early research-oriented
startup companies to cut back on staffing and redirect their efforts
toward products that could be brought to market more quickly.
Some of the early genetic engineering companies were acquired
by larger corporations, although only a handful actually failed.[149]
Since then, the industry has recovered much of its early dynamism.
In late 1984, over 300 companies were involved in biotechnology
in the United States.[150] Industry investment in biotechnology has
also mushroomed, reaching $2.5 billion by mid-1985.[151]

All this investment has yet to yield much in the way of marketable products. By 1986, only a few genetically engineered products for humans had been approved for marketing in the United
States, including recombinant DNA-derived human insulin, a type
of interferon, human growth hormone, and a vaccine for hepatitis
B.[152] Diagnostic techniques based on related methods such as mon-
oclonal antibodies were also being marketed. Some other biotechnology products had been approved for clinical testing, and
many more are in the research pipeline or await federal approval.
The major emphasis so far has been on pharmaceutical and health

products, because of their high "value-added" (i.e., profit margin), and also because this was where the first scientific breakthroughs occurred.[153] Down the road, industry analysts foresee a broad array of applications in the chemical, agricultural, and energy industries as well.

Forecasting the future of such a pathbreaking technology is a tricky business given the many unknowns, but everyone agrees that its likely market potential is large. The Department of Commerce estimates that the worldwide market for products of biotechnology in the year 2000 will range between $15 billion and $100 billion.[154] Domestic sales of biotechnology-based pharmaceutical and agricultural products are expected to jump from $200 million in 1985 to $1.5 billion by 1990.[155] With economic stakes of this magnitude, international competition in the field has, not surprisingly, been fierce. Although the United States is still the acknowledged world leader, the Congressional Office of Technology Assessment (OTA), in an extensive study in 1984, warned that continued preeminence is by no means assured.[156] Japan is a leading competitor because of its wide experience in traditional bioprocess technology, the OTA report said, followed by several European countries. In explaining America's vulnerability to foreign competition, the report cited, among other things, "the relatively low level of U.S. government funding for generic applied research in biotechnology" (research on general problems associated with the application of biotechnology), estimated at $6.4 million in fiscal 1983.[157] Yet as the report also pointed out, government support of "basic" biotechnology research in the United States is orders of magnitude greater than that of any other country. In fiscal 1983, for example, the U.S. Government spent $517 million on basic and generic applied biotechnology research combined – almost nine times as much as the $60 million apiece spent by Japan and most European countries.[158] Moreover, U.S. government funding for this field has been rising sharply, increasing by about 50 percent every year since 1978.[159]

Business also supports commercially oriented R&D, of course, and is reportedly boosting its outlays for biotechnology in response to foreign competition.[160] But the federal government still provides the bulk of funding for scientific research and development, even under the Reagan administration, which officially disdains government interference in the marketplace. "Not only is basic research an essential investment in the nation's long-term welfare," explained George Keyworth, President Reagan's science advisor, "but it is largely a federal responsibility because its benefits are so broadly distributed. Quite simply, basic research is a vital underpinning for our national well-being."[161]

Viewing biotechnology as one of the "key sources of future growth and productivity for the U.S. economy," the Reagan administration has done what it could to minimize regulatory interference with commercial development. The President's Office of Science and Technology Policy set up an interagency group representing the Departments of Health and Human Services, Commerce, Agriculture, State, and others, as well as the major regulatory agencies, to assure that federal agencies would "speak with one voice" on matters of biotechnology research and regulation.[162] Starting from the general premise that genetic engineering was not inherently dangerous, this group concluded that no new legislation was needed to regulate either the conduct or products of biotechnology. Instead, it developed a regulatory framework that delineated the responsibilities of existing federal agencies for each point in the research, development, marketing, shipment, use, and disposal of genetically engineered products.[163] Under this scheme, the RAC will continue to review certain NIH-funded gene-splicing experiments, and four new committees, consisting primarily of scientists experienced in biotechnology, will be created to perform a similar function at the Environmental Protection Agency, the Food and Drug Administration, the Department of Agriculture, and the National Science Foundation, for products that fall within each agency's purview. Government officials hope that this new framework will unsnarl biotechnology's regulatory tangle, smoothing the way toward rapid commercial development.

Not everyone is convinced that existing legislation is adequate to meet the needs of biotechnology. Jeremy Rifkin filed a lawsuit asserting that the new regulatory framework for biotechnology did not provide sufficient safeguards for protecting the public health and the environment. (The suit was subsequently dismissed.)[164] In 1985 hearings on the proposed plan, Congressman John Dingell, chairman of the Oversight Subcommittee of the House Committee on Energy and Commerce, noted that Congress was "aware that the existing legislation was not drafted with biotechnology in mind – either to promote its development or to protect against its associated risks. Thus, even the most appropriate and intelligent operation of current programs may not suffice." He stated that Congress would be closely monitoring the actions of regulatory agencies in carrying out their assigned responsibilities, and that his Committee would continue to hold public hearings on biotechnology to "advance our understanding and the public debate about the role of government."[165]

Genetic futures

Biotechnology, born amid controversy, continues to provoke fresh debate as scientists' journey of discovery carries us steadily onward into new and unexplored territories. But it is no longer primarily scientific curiosity that propels the journey. It is the dazzling array of new commercial products and possibilities.

Most of the immediate applications will be in the biomedical arena. Using recombinant techniques, scientists are trying to replicate substances that form the body's natural defenses. Among their goals are hormones that can control blood pressure, dissolve blood clots in heart-attack victims, and single out and destroy cancer cells.[166] Research is also underway to develop genetically engineered vaccines for such often intractable infectious diseases as AIDS, malaria, gonorrhea, and herpes. At least thirty different companies have been testing types of interferon in treating cancer and other diseases.[167] Biotechnology may also permit, for the first time, direct comparison of the DNA of normal and diseased people, with the possibility of identifying genetic "markers" of latent disease. Safer vaccines, synthetic brain hormones, cheaper drugs, and, eventually, direct human gene therapy for genetic diseases are just a few of the other points on the biomedical horizon.

Biotechnology also has a host of applications in agriculture and animal husbandry, some of which are already on the market. Genetic engineering permits crop plants to be adapted to the environment, rather than, as had traditionally been the case, trying to adapt the environment to plants through fertilization, irrigation, and pest control. Current research is directed at improving the yield and nutritional content of food crops, increasing their resistance to disease, enabling them to obtain nitrogen directly from the soil rather than from fertilizers, and developing plant lines with greater tolerance for frosts, droughts, and pesticides. Genetic engineering has a number of possible uses in animal breeding – to increase milk and meat production, for example – as well as yielding a wide range of new veterinary drugs.[168] These efforts, if successful, could transform much of modern agriculture. Yet they may also have less desirable consequences. A 1986 report by the Office of Technology Assessment predicted that genetic engineering and other emerging technologies, far from being cost-cutting tools that will help save small and medium-sized farms, could instead raise costs and cut the number of American farms in half.[169]

In the long run, biotechnology could have an even greater impact on the petrochemical and energy industries. These industries were developed in a period when oil was cheap. When oil is more expensive, organic production of chemicals from genetically engineered "biomass" – biological materials such as cheap grains or wood pulp – might prove more efficient and, some researchers say, might yield fewer toxic byproducts. In addition, biotechnology may allow creation of alternative sources of energy like ethanol, the primary ingredient in gasohol, or more efficient utilization of existing energy. (For instance, microorganisms might be able to liquify the highly viscous crude oil that gets left behind in oil wells.) Recombinant microorganisms might also be used to enhance the effectiveness of naturally occurring microbes now used as "organic" pollution control devices, by degrading hazardous wastes, or eating up oil spills. Again, however, the social and environmental consequences of such developments are problematic.[170]

Many critical questions remain unanswered about all of these potential uses of biotechnology. The answers will determine biotechnology's future: what uses are given priority, and what risks are taken in the process. Although it is impossible to foresee that future in detail, certain trends seem clear.

First of all, the profit potential of each of the many possible applications of biotechnology will play a decisive role in determining whether they reach maturity. Because of the extensive involvement of industry in all phases of this technology, including the basic research done in universities, priority will go to lines of research that have the best commercial prospects. As David Baltimore, a Nobel Laureate associated with one of the biotechnology companies, put it, "the desire of people to seek profits determines what research gets turned into useful products for the general economy. If they are not even seen as potentially profitable, they are not pursued or produced."[171] Recent history provides ample evidence that molecular biologists are not lacking in profit-seeking desires.

The quest for profit has unquestionably led to many important technological discoveries. But even the most ardent supporters of the private marketplace acknowledge that there are limits to what we can expect from commercial incentives. One issue, for example, is the short-run orientation that many smaller, less heavily capitalized companies have had to take to survive. Yet another Nobel Prize winner, Walter Gilbert, a former Harvard professor and founder of Biogen, one of the early biotechnology firms, stated flatly that his company undertook only work "that has direct social value next year, or three years from now, not something of social value twenty years from now." "We cannot take that long time-frame," he explained; "we have to focus on the immediate."[172] The efforts of many biotechnology companies to ensure an early return on investment means that more fundamental – and ultimately more significant – questions may not receive adequate attention. "The most important long-term goal of biomedical research," MIT's Jonathan King testified at the 1981 Gore committee hearings, "is to discover the *causes* of disease in order to *prevent* disease."[173]

The short-term orientation of much commercial biotechnology poses a special threat to academic research. Speaking at the same Gore committee hearings as King, MIT President Paul Gray described the appropriate role of university scientists as "oriented toward exploring and expanding the frontiers of knowledge, rather than producing products."[174] Yet with the massive infusion of corporate capital into universities, as noted previously, these priorities appear to be changing. Collaboration between industry and academe requires that interests converge, and they apparently do. A government study of over 100 cooperative university–industry research projects found that industry and university scientists both ranked development of "patentable products" as the most important goal of their joint research.[175] As ties with industry proliferate, some observers fear that the focus on product development will impede more basic research, gradually eroding the very base of fundamental scientific knowledge that made biotechnology possible.

Also likely to be neglected by private market incentives are the needs of the world's poor. According to the World Health Organization, the diseases that ravage Third World countries, including the so-called enteric diseases such as bacterial and ameobic dysentery, cholera, and typhoid, and parasitic diseases such as malaria and leishmaniasis, account for well over 80 percent of illness worldwide; enteric diseases alone cause 20 million deaths annually.[176] Most biotechnology companies have not made a major research commitment to such diseases but have concentrated instead on medical and other products for the more affluent markets of industrialized countries.

Representative John Dingell, in a speech about the dilemmas posed by biotechnology, cited malaria vaccine as an example of a socially valuable use of biotechnology that might not be commercially feasible. "Over 150,000,000 people get malaria each year," he observed, "but they are among the poorest people in the world. What commercial incentive exists to invest development money if the potential consumers have no buying power? Has this dilemma already slowed the development of malaria vaccine?"[177]

This last question was undoubtedly in reference to the withdrawal of Genentech Corporation, in 1983, from a collaborative project to develop a malaria vaccine. Genentech backed out when the World Health Organization, which was funding the project, denied the company exclusive rights to manufacture the vaccine. In explaining this action, Genentech's vice-president noted that the company's products had been carefully selected to yield a significant return on investment, and that "development of a malaria vaccine would not be compatible with Genentech's business strategy." (This project was subsequently rescued when another company, Hoffmann–La Roche, agreed to manufacture the vaccine and distribute it to the World Health Organization at a reduced price.)[178]

Private business strategies, in turn, may not always be compatible with society's interests. One example is the effort by some petrochemical companies to develop crops that are resistant to weed-killing herbicides so that farmers can use more herbicides without endangering their crops. As critics have pointed out, crops may be spared, but heavier use of toxic chemicals will only increase chemical pollution in the rest of the ecosystem.[179] Nor will such bioengineered seeds and chemicals necessarily benefit small landowners, the bulk of the world's farmers. Under the control of agribusiness, biotechnology may be no more successful than the first "green revolution," where higher-yielding crop varieties and fertilizers were supposed to solve Third World hunger but sometimes served instead to squeeze out small farmers, disrupt local economies, and reinforce the dependence of developing countries on multinational agribusiness. Since 1970, Africa has slipped from food exporter to food importer, and today is the only region of the world where per capita food production is falling.[180]

Ideally, the new genetic technologies would be directed toward trying to eradicate critical social problems such as world hunger, epidemic diseases, and environmental pollution. However, for many of these problems, the limiting factors are social and economic, not technical, and some critics fear that high-tech approaches like gene-splicing will have little impact. Testifying before Congress in 1977, MIT scientist Ethan Singer stated it well:

> Right now even the rich have a low standard of medical care, and the poor, of course, a much lower one. Any benefits of recombinant DNA will fit right into this pattern. Miracle cures won't suddenly find their way to rural or ghetto hospitals, and drug companies won't suddenly put health before profit. We are stalled in medical care, but not for lack of recombinant DNA. And all this focus on recombinant DNA is going to draw even more resources away from what we really need. If we are worried about our people's health, let our main course of action be to give them what we can already. Then we can have the pie in the sky for dessert . . . [181]

Continuing controversies: hubris and humility

No use of biotechnology has prompted more metaphysical soul-searching than human genetic engineering. In 1980, a researcher at UCLA, Martin Cline, made an unsuccessful attempt to perform gene therapy on humans, an effort which was widely criticized as premature.[182] Now, many experts believe that enough of the technical problems have been solved for clinical testing of human gene therapy to begin in the near future.[183] The NIH approved elaborate guidelines for such experiments in 1985, and has received a number of applications from investigators planning to use gene-splicing techniques on humans.[184]

In the early 1980s, nearly two-thirds of the American public opposed the creation of "new life-forms," and almost a third feared that genetic engineering might result in more harm than benefit.[185] Since then, these fears have abated somewhat under the influence of a steady stream of favorable publicity from the scientific community, yet the prospect of modifying the human genetic makeup still evokes substantial moral and personal qualms among many

scientists and nonscientists alike. In a 1986 survey, 42 percent of
the respondents said they thought that genetic alteration of human
cells to treat disease was morally wrong, regardless of the purpose.[186] The National Council of Churches, reporting the results
of its two-year investigation of genetic engineering, speaks of "the
folly of arrogance, of the presumption that we are wise enough
to remake ourselves." "When we lack sufficient wisdom to do,"
the report urges, "wisdom consists in not doing. Caution, restraint, delay, abstention are what this second-best (and, perhaps,
only) wisdom dictates with respect to the technology for human
engineering."[187]

Most scientists directly involved with human gene therapy think
whatever risks may be involved are worth taking. In congressional
hearings held by the Gore subcommittee in 1982, genetic researchers as well as some medical ethicists expressed confidence that the
benefits of human genetic engineering would greatly outweigh the
risks. Public fears, several suggested, were due to ignorance or
misinformation. Gore was apparently not persuaded. "I think science would be ill-advised to dismiss the concerns of the American
people," he said at one point; "they may ... be grasping some
truths that the specialists are not."[188] Following the hearings, Gore
introduced a bill to establish a presidential advisory commission
consisting of a majority of nonscientists to monitor genetic engineering in humans. A compromise bill was passed by Congress,
but subsequently vetoed by President Reagan.[189]

The most controversial form of human genetic engineering, all
agree, would involve the so-called germline cells, human reproductive cells whose traits are passed on to future generations. In
June 1983, prompted by a ten-page letter circulated by Jeremy
Rifkin, over forty religious leaders, representing almost every major church group in the United States, signed a resolution calling
for an outright ban on genetic engineering of human reproductive
cells.[190] Government-sponsored reports, although more moderate
in tone, have also raised concerns about germline therapy. A 1984
OTA report concluded, for instance, that there is "no agreement
about the need, technical feasibility, or ethical acceptability of gene
therapy that leads to inherited changes."[191] Nonetheless, some
researchers do not rule out the possibility of germline therapy at
some point. Nor can they guarantee that even gene therapy of
noninherited ("somatic") cells will not inadvertently affect germline cells.

Another prospect that troubles some observers is the U.S. Department of Agriculture's research effort to develop "transgenic" animals by splicing genes from other species or humans into the animal's natural genes. One immediate goal is to produce "super" sheep and pigs, perhaps twice the size of normal livestock, by transplanting human growth hormone genes into their germline cells. Scientists have succeeded in producing giant mice and fish, as noted in Chapter 2, and are working on techniques to introduce other commercially desirable traits into farm animals.[192] Rifkin calls such research "morally reprehensible." "There is nothing we can do to animals that is more cruel than to rob them of their genetic uniqueness," he says. "It is a violation of the moral and ethical canons of civilization."[193] Many researchers scoff at such objections, pointing out that we have been crossbreeding plants and animals to produce new species for thousands of years. The enhanced food production that could come from larger or faster-growing animals, they argue, more than justifies whatever philosophical problems might be raised by cross-species gene-splicing. Unimpressed, Rifkin, along with the Humane Society, sued the USDA to stop the experiments on the grounds that an environmental impact statement had not been filed. In 1986, a District Court judge ruled against Rifkin, and the case is under appeal.

A more immediate risk is posed by the deliberate release of recombinant organisms into the environment, which many ecologists believe could conceivably do irreversible harm to the existing ecological balance. There are numerous past examples of disruption caused by the introduction of nonnative organisms into new ecosystems: Gypsy moths and starlings, brought to the United States from abroad, multiplied rapidly and periodically destroy trees and crops; kudzu and other foreign plants now dominate native flora in some areas; and microorganisms from abroad have killed large numbers of chestnut and elm trees.[194] Testifying at congressional hearings on the environmental effects of genetic engineering held by the Gore subcommittee in 1983, ecologists warned that under the right circumstances, recombinant microbes could result in similar damage to the environment. Unfortunately, no one knows what those circumstances are, but as the subcommittee noted, although the likelihood might be low, the damage that could occur is great.

Risks or no, researchers working on plant and animal experiments were eager to begin testing some of their laboratory-bred products under natural conditions, and in May of 1983 the NIH officially approved the first open-air field test of genetically engineered organisms. In the proposed experiment, University of California at Berkeley scientist Stephen Lindow and colleagues planned to spray a 200-foot row of potato plants with genetically altered ("ice-minus") bacteria to see if it would prevent frost from damaging the plants. Those plans came to an abrupt halt when a coalition of environmental groups led by Rifkin filed suit, charging that the NIH had not adequately evaluated the ecological consequences of the release it had approved. "Ecological roulette," Rifkin called it.[195] A federal district court ruled in Rifkin's favor, and the NIH and U.C. promptly appealed. In February 1985, a federal appeals court upheld the ruling, stating: "We emphatically agree ... that NIH has not yet displayed the rigorous attention to environmental concerns demanded by law, and that the deficiency rests in NIH's complete failure to consider the possibility of various environment effects."[196]

The appeals court granted permission for Lindow's experiment and other deliberate releases to proceed when appropriate environmental assessments were completed, which the NIH had already agreed to perform in response to Rifkin's suit. In the meantime, the Environmental Protection Agency adopted its own regulations prohibiting companies from testing genetically altered substances outside of a "contained facility" – such as a greenhouse or laboratory – without first obtaining EPA's approval.[197] The EPA approved Lindow's proposed release in May 1986, but by then, mounting local public opposition had convinced university officials to withdraw permission for the experiment. In April 1987, the field test finally took place but was immediately vandalized by a radical environmental group.[198]

The environmental debut of genetically engineered organisms that finally occurred was nothing short of a fiasco. In November 1985, the EPA had granted a California biotechnology company, Advanced Genetic Sciences, Inc. (AGS) a permit to conduct outdoor tests of another genetically altered frost-retarding microbe on 2,400 strawberry plants in Salinas Valley, a fertile farming area in California. Rifkin tried unsuccessfully to obtain a court injunction blocking this release, and local residents jammed a public meeting of the county board of supervisors to express their fears and suspicions.[199] The county imposed a temporary ban on the tests, and the company agreed to seek a new location. Then, to everyone's astonishment, a *Washington Post* story (obtained on a tip from Rifkin) disclosed that a deliberate release had already occurred, unbeknownst to the EPA or local officials! AGS had, months earlier, gone ahead and injected genetically engineered microbes in trees growing on the company's roof, without informing federal or state officials or seeking EPA approval. One EPA official called the news "a real shock."[200] AGS claimed that it had not violated EPA regulations because the bacteria had been injected rather than sprayed. The EPA disagreed. "We do not believe a tree is a contained facility," a high-ranking official stated tersely.[201]

The EPA charged AGS with deliberately falsifying key scientific data, suspended its testing permit pending additional studies, and fined the company $20,000. "It's unbelievable," said Rifkin. "Here is the first deliberate release experiment. The whole world is watching. And the company is violating the public trust, Federal rules, and accepted safety standards."[202] Later, with the world still watching, the EPA's own position softened. The penalty imposed on AGS was reduced to $13,000, and the charge of "falsification" was changed to "failure to report." Rifkin called these charges "a clear signal that EPA is succumbing to pressure by industry and the White House not to properly enforce environmental standards."[203]

No sooner had the furor over AGS's misdeeds begun to die down when an even earlier, unannounced environmental release came to light. This time the culprit was the government itself. In late 1984, the Department of Agriculture's Animal and Plant Health Inspection Service had quietly authorized a Nebraska-based company, Biologics Corporation, to field-test a genetically engineered vaccine intended to prevent pseudorabies in swine, without consulting either the Recombinant DNA Advisory Committee or its own internal biotechnology review committee. In January 1986, the Inspection Service issued a license allowing Biologics Corporation to begin marketing the vaccine. These actions did not become public until April 1986, when the peripatetic Rifkin learned of the matter, and charged the Department of Agriculture (USDA) with running roughshod over federal policies for releasing recombinant organisms into the environment. Rifkin threatened to sue if the company's license was not revoked within ten days. Again, scientists in and out of government expressed shock and dismay that this historic step had been taken in such an unhistoric fashion. "In my view Rifkin is exactly right," said one top USDA scientist. "We had no knowledge whatever that this decision had been made until we got it through the grapevine . . . there's no question in my mind that it violated our trust and the public's trust."[204] Agriculture officials defended their actions as "scientifically sound," but, admitting that they had not followed the prescribed procedures, revoked the company's license. Sales of the vaccine then resumed two weeks after the USDA concluded the vaccine was safe and effective.[205]

These two episodes cast real doubt on the adequacy of regulatory mechanisms governing the environmental release of genetically engineered products, and the willingness of private industry to follow those mechanisms. In a scathing editorial titled "A Novel Strain of Recklessness," the New York Times accused both AGS and the Agriculture Department of arrogance, deceit, and irresponsibility, and gave credit to Rifkin's organization for exposing the "ragged" federal system for regulating this new technology.[206]

Reckless or not, the era of genetically altered plants and microbes was definitely underway. In May 1986, the EPA approved another deliberate release, this time involving gene-altered tobacco plants. Many other field tests of genetically engineered plants, microbes, and pesticides await federal approval, and thousands more redesigned plants and organisms are being developed. Violations of federal regulatory procedures for environmental releases continued to occur. Meanwhile, the regulatory process itself was embroiled in controversy in both the environmental community and Congress, where five subcommittee chairmen in the House sided with ecologists who favor close federal scrutiny of proposed releases.[207] Nevertheless, in November 1986, responding to complaints by the research community, the USDA abandoned its previously announced rules governing the conduct of biotechnology research, which were to have been part of the Reagan administration's coordinated strategy for regulating biotechnology research and products.[208] Until the NIH's RAC comes up with revised rules, existing NIH guidelines will be used in conjunction with established USDA criteria to evaluate most proposed deliberate release experiments. Whether these will prove adequate, only time will tell. It is too soon to judge what harvest we will reap from the man-made seeds of life now being sown.

A final concern, deeply disturbing to many, is that biotechnology will be used for biological warfare. Some scientists and military analysts believe that by reducing the potential cost of biological weapons development as well as the potential risks of detection, advances in biotechnology have dramatically increased the threat of germ warfare.[209] Adding to the worry is the significant expansion of the Defense Department's biological research program since 1980. An international treaty signed by the United States and other industrialized nations in 1972 prohibits development and use of biological weapons, but permits research for "defensive" purposes. This is a major loophole, critics contend, since knowledge about defensive measures is often directly applicable to offensive measures. In 1982, RAC members Richard Goldstein and Richard Novick, in an effort to strengthen the 1972 treaty, proposed that NIH guidelines formally prohibit use of recombinant DNA methods to develop biological weapons. The overwhelming majority of letters to NIH about this proposal favored it, but the RAC turned it down on the grounds that it was redundant with the 1972 treaty.[210] Rifkin also tried, unsuccessfully, to block RAC approval of biotechnology research by the military, but did succeed in stalling the construction of a controversial new biowarfare laboratory pending investigation of its environmental impact.[211]

Each of these areas of controversy involves complex technical issues and conflicting interpretations of scientific data. In each, a plausible rationale can be constructed for the choices that have been made. Nevertheless, taken as a whole, the pattern is unmistakable: In virtually every arena, public concerns about ethical and environmental issues have been overridden by the imperative of scientific progress on the one hand and larger economic or military goals on the other. Efforts to establish regulatory safeguards for genetic engineering and its products beyond those routinely required have largely failed. Nor is there any special vehicle, such as the predominantly lay advisory commission proposed by Congressman Gore, for any form of direct public oversight and influence. In short, this extraordinary technology is being treated in a quite ordinary fashion.

Many observers – the President's Commission, the National Council of Churches, Gore, Dingell, and others in Congress, and a host of academic critics and environmental groups – have argued that biotechnology is *not* like other technologies in key respects, and that these differences demand greater public accountability. They cite three critical differences. First, the basic discoveries in biotechnology emerged directly from federally funded university research, whereas the development phase of most previous technologies occurred in industrial settings. In the case of semiconductors, for example, government support of electronic engineering during the 1950s and 1960s served primarily to establish a research and training infrastructure for the electronic industry; few semiconductor innovations resulted directly from federally funded university research.[212] Having funded the development of the recombinant method, it is argued, the government is now entitled to a major role in directing its use. Second, there are valuable applications of biotechnology that private industry will not find profitable to pursue. For the world's needy to share in the fruits of genetic technology, and for broader values and concerns to play a role in charting its future course, public bodies, whether national or international, will have to guide and supplement the activities of the private sector.[213] Third, gene-splicing carries awesome risks and responsibilities for all of society, which only a publicly accountable body can properly assume. Many believe that public priorities, not just private interests, should determine biotechnology's overall impact on society.

Governmental bodies are in a position to assess the most pressing problems for biotechnology to address and to develop coordinated, long-range planning based on enunciated social priorities. The question is which goals and concerns are given precedence. The government is the steward of the public interest. If it ignores the public's wishes and fears, citizens have, for all practical purposes, nowhere to turn. This, in the end, may be biotechnology's most serious threat. Echoing sentiments heard frequently during the earlier debates over recombinant DNA research, Congressman John Dingell declared:

> We probably face greater danger from creating an exclusive policymaking process than from any direct consequence of genetic engineering. We cannot be faulted for not having all the answers now, but we will be guilty of serious negligence if we fail to ask questions about the opportunities and consequences of genetic engineering and biotechnology. And we will lose the public's trust if we fail to create a process that it can understand and in which it can participate.[214]

Conclusion

The genetic engineering story closes, then, on the same note with which it began. The issues have changed, many of the actors are new, but the underlying question is essentially the same: Who has the right to control this powerful new technology which will affect all of our lives?

No other modern technology scrapes so bluntly at the raw nerve of our being, DNA, the very essence of our heredity. Precisely because it embodies our greatest hopes for medical and economic progress as well as our deepest fears, genetic engineering is, in a real sense, the flagship of modern technology. Currents and events that might pass unnoticed with lesser technologies are writ large in the history of genetic engineering, marked with milestones as unique as the technique itself. A voluntary scientific moratorium over speculative risks, serious legislative efforts to regulate a technology that had not yet caused any harm, and fiery debates about science policy measures which, in some communities, literally raged in public meeting halls – all these events speak of a deepening mistrust of scientific "progress" among lay citizens and a growing disillusion with experts and government officials in representing what many citizens perceive as the public interest. Equally remarkable, especially in light of the profound doubts and fears generated by this technology in its early years, was the ripple of commercial interest in the mid-1970s that, by the mid-1980s, had swelled into a tidal wave engulfing nearly an entire field of basic research. These are the landmarks of no ordinary technology.

Yet underlying these precedent-setting events are the same basic questions that arise with almost every major scientific or technological advance. How are risks defined? How much effort should go into exploration of possible, but unproven, dangers? How should risks and benefits be balanced, particularly when the groups that benefit most directly are not the same as those most directly at risk? And, as always, whose answers should we be listening to?

The evolving controversies over genetic engineering illustrate the groping efforts of citizens and lay groups to confront these questions. Their answers have been, for the most part, ignored. The lack of any mechanism to funnel and respond to expressed public concerns, a pattern established first by scientists and later reinforced by the involvement of private industry, is one of the most striking features of the history of genetic engineering.

It is also one of the most disturbing. For, in discovering how to alter the genetic structure of living organisms, scientists have begun a journey that will transform not only the relation between science and society but, very likely, the genetic heritage of society itself. We are all entitled to a voice in its course.

PART III

Lessons, questions, and challenges

7

Risks and rights

At times, human ingenuity seems unable to control fully the forces
it creates... with Hamlet, it is sometimes better "to bear those ills
we have than fly to others that we know not of."
                                    – Chief Justice Burger,
                           referring to arguments against
                     patenting genetically engineered bacteria;
                              Diamond v. *Chakrabarty*,
                           447 U.S. 303, 316 (1980)

All medical advances involve risks. Nothing ventured, nothing
gained, as the saying goes. But that does not mean all risks are
worth taking, or that we would all make the same choices. In at
least two of the four case studies, DES and swine flu (and some
might add the artificial heart), the risks were definitely *not* worth
running, a judgment some observers made at the time. The periodic public controversies over genetic engineering also underscore the wide differences in people's responses to, and even
definitions of, risks.

   This chapter looks at how we view risks in general and the risks
of medical innovation in particular. What makes some risks "acceptable" and others not? Why are patients, like most of those
who took DES, often not told about the uncertainties and hazards
of treatments? More broadly, what is or reasonably could be done
to protect individuals and society against the potential dangers of
new medical and scientific technologies? In the picture that
emerges, the central problem appears to be the lack of concern
about possible hazards on the part of those making decisions on
the one hand, and the lack of communication of either dangers or
doubts to those at risk on the other. These failures explain much
about why the warnings of scientific and lay critics in the four
cases had so little impact and why even predicted problems were
not avoided.

Varying perceptions of risks (or, The risk you see depends on where you sit)

When, in 1971, the link between DES and clear-cell vaginal cancer among DES daughters was discovered, many were outraged. Feminist health critics called it a major medical scandal, an example of corporate greed, governmental collusion, and professional irresponsibility. Drug company defenders countered that there was a real need for a synthetic estrogen, that the possibility of fetal risk was unforeseeable, and that the significance of the studies by Dieckmann and others in the early 1950s indicating that DES did not prevent miscarriages had not been fully appreciated. Researchers also pointed out that the actual risk of clear-cell adenocarcinoma among DES daughters is very small (1 case per 1,000) – far less than the risks associated with smoking or driving a car. Why such an outcry, some wondered?

One important difference between injuries caused by DES and those caused by smoking or driving is the individual's awareness of risk. Smoking and driving entail dangers that people presumably understand and accept, whereas the risks of DES were largely unknown to doctors and entirely unknown to most patients. Furthermore, the harms resulting from DES were iatrogenic – caused by medical treatment – and, in the eyes of many victims, warrant compensation just like other personal injuries inflicted by someone else.

The dictionary defines risk as "the chance of injury, damage, or loss." Strictly speaking, it refers to the objective likelihood that something bad will happen. But perceptions of risks are anything but objective. Reactions vary depending on whether the risk is felt to be under individual control, as in the case of smoking, or imposed by some outside agent, as in the case of DES and many other modern technological developments such as genetic engineering. Responses also depend heavily on the context and manner in which the risks are presented and on underlying value differences concerning the type of "injury, damage or loss" involved. People are likely to exaggerate the perceived likelihood of events they consider particularly undesirable and to minimize the likelihood of events that cause little concern.[1]

Perceptions of risks also differ between experts in a given field and nonexperts. Studies by psychologist Paul Slovic and colleagues suggest that both groups make systematic, albeit different, errors.

Nonexperts tend to overestimate the risk of infrequent events (e.g., death from botulism) and to underestimate the risk of common events (e.g., death from heart disease). Experts, in contrast, tend to minimize the likelihood that something will go wrong, a bias that Slovic and colleagues attribute to overconfidence in current scientific knowledge, inadequate appreciation of how complex systems function as a whole, and insensitivity to human errors.[2] Experts are also more likely to suffer from what has been called "specialized blindness": the tendency to define issues in the narrow and technical terms of their own specialty and to ignore related nontechnical problems.

The case studies bear witness to the prevalence of expert optimism. Enthusiastic administrators proclaimed that a successful artificial heart would reach mass production within five years of the program's inception in the mid-1960s. The medical profession eagerly embraced prenatal DES as a treatment to prevent miscarriage and improve pregnancy outcomes, and many physicians continued to prescribe it long after repeated clinical trials had failed to find it effective. Similarly, even as the swine flu program was mired in public controversy and practical problems, embattled government officials staunchly maintained that a nationwide immunization campaign would be safe, effective, and manageable. To be sure, reliable information on risks and benefits was limited and incomplete in all of these cases. It is in just such conditions that professional optimism flourishes. Studies of particular clinical innovations have shown that the poorer the data and the less rigorous the evaluation, the more exaggerated the claimed benefits tend to be.[3] Such uncertainties, commonplace in the early stages of medical innovation, provide a built-in bias toward further development.

Not all expert optimism is the product of unconscious biases, of course. By definition, professionals have a long-standing commitment to their field, which is often reinforced by personal, professional, or financial ties.[4] People's backgrounds and occupations can strongly influence the way they perceive given "risks." In a 1980 survey on attitudes about regulatory safeguards, for example, only 5 percent of the business executives agreed that "the overall safety of society will be jeopardized significantly during the next twenty years unless technological development is restrained," compared with 56 percent of the general public. Likewise, a 1980 Harris poll found that only 19 percent of top corporate executives agreed that society "has only perceived the tip of the iceberg with regard to the risks associated with modern technology," compared with 62 percent of the general public.[5]

Professional and financial ties can also lead to deliberate skewing of information. Drug company detail men played down data on the risks of DES and the evidence that it did not prevent miscarriages. Scientist-entrepreneurs involved in biotechnology companies held strategically timed press conferences to publicize their latest scientific breakthroughs. The news media were also used to garner early support for the artificial heart program. Shortly before the initial, critical funding decisions were to be made at the National Heart Institute in 1966, heart surgeon Michael DeBakey, the program's chief proponent, performed a dramatic operation using a left ventricular assist device, the forerunner of the artificial heart, and alerted TV, radio, wire service, and newspaper reporters.[6] This surgical media event was hardly geared toward an evenhanded treatment of the risks and benefits of the new procedure.

All of these factors enter into perceptions of risk. People respond differently to uncertainty, and their responses are colored by past experiences, the information currently available to them, and what they believe is at stake. This means, among other things, that it matters a great deal *whose* views are considered in deciding whether given risks are acceptable or not.

Of risks and doctors

Decisions about risks run throughout every aspect of medicine. For patients with serious or fatal illness, the treatments themselves may be life-threatening; yet it may be even more risky to do nothing. Modern pharmaceuticals, medicine's principal weapon in the war against disease, often injure as well as cure: Adverse reactions to prescription drugs result in an estimated one million hospital admissions per year, and more than 130,000 fatalities.[7] Inevitably, mishaps occur. In one study of a university hospital's general medical service, more than one-third of the patients were found to have some kind of iatrogenic illness.[8] Even some forms of disease prevention recommended for healthy people – immunizations and jogging, for example – pose hazards, however minimal. Critics such as Ivan Illich have charged that the iatrogenic injuries caused by modern medicine outweigh its benefits. Yet the value of many drugs and medical procedures, when used wisely, is undeniable. Unfortunately, as with DES, they are *not* always used wisely: Our society is often described as "overmedicated."[9] Although many factors contribute to this problem, the medical profession must bear a major share of the responsibility.

All too frequently, doctors decide unilaterally what treatments are best for their patients – sometimes, perhaps often, without even informing them of the risks involved. However well-intentioned, many physicians do not believe that patients are capable of comprehending the intricacies of medical matters. Sometimes consciously, sometimes not, they slant the information presented, to lead patients to the "right" decision. Often this means minimizing the possible dangers. Dr. Theodore Cooper, Assistant Secretary for Health at the time of the swine flu program, explained the tendency of physicians to play down risks. "If health care were represented to the American people as faulty, unreliable, and hazardous," Cooper wrote in 1976, "the people would certainly tend to make less and less use of the services... Because we in the health field recognize that, to some undefined degree, health care is better than no health care, we encourage people to avail themselves of it. And we do this by extolling its virtues and accomplishments."[10]

This was obviously the spirit behind the government's campaign in 1976 to convince a dubious public that it had little to fear and much to gain from participating in the swine flu immunization program. In accepting liability for vaccine-related injuries, Congress had insisted that written consent be obtained from all those vaccinated. This requirement was at odds with the goal of immunizing as many people as possible, however, and the consent procedures actually employed left much unsaid. The initial documents, labeled "registration" forms, made no mention of the substantial uncertainty about whether a swine flu pandemic would in fact occur, of scientific disagreement about vaccine efficacy, or of the possibility of serious neurological reactions. The National Commission for the Protection of Human Subjects, directed by the swine flu legislation to review the forms, insisted on some revisions, but in general the contents remained strongly tilted toward the benefits of vaccination.

As it turned out, the swine flu consent procedure achieved neither an informed, immunized public, nor a legally reliable consent document. People signed the forms after they had come in to be vaccinated and had, presumably, already made up their minds. A

majority of those who received the shots, when questioned subsequently, were unaware of the information provided on the consent form (indeed, 15 percent said they did not even remember
signing the form at all).[11]

Such failures of communication are, unfortunately, not uncommon. Many physicians avoid disclosing the risks of medical procedures to patients for fear of producing imaginary symptoms
among "suggestible" patients, or of undermining faith in the physician's curative powers.[12] Even less likely to be communicated
are the numerous uncertainties involved in many medical procedures. Physicians themselves have a hard time coming to terms
with this pervasive uncertainty, and are often reluctant to share it
with patients.[13] Yet at least one well-known psychologist, Irving
Janis, contends that full disclosure, far from being harmful, allows
patients to engage in "the work of worrying," a process he claims
is vital to their ability to cope with their disease and treatment.[14]
Surveys have shown that patients would generally prefer more
detailed information about their diagnosis and proposed treatment
than physicians provide, and they are more likely than physicians
to believe that the final decision about therapy should rest with
the patient.[15]

The reluctance of physicians to disclose risks and uncertainties
to patients dates back to ancient times. Hippocrates counseled
physicians:

> Perform [your duties] calmly and adroitly, concealing most
> things from the patient while you are attending to him. Give
> necessary orders with cheerfulness and sincerity, turning his
> attention away from what is being done to him; sometimes
> reprove sharply and emphatically, and sometimes comfort
> with solicitude and attention, revealing nothing of the pa-
> tient's future or present condition . . .[16]

This paternalistic attitude is still evident today in many aspects
of medical training and practice. Beginning in medical school,
students are discouraged from giving patients a major role in treatment decisions. The arcane technical vocabulary they acquire further impedes communication. Medical training, which generally
emphasizes rare and serious illnesses, stresses the value of therapeutic intervention – typically an attempt to cure rather than prevent – causing many physicians to overrate the benefits of the
therapy they provide and to underestimate its risks. The highly
specialized and bureaucratic settings in which modern medical care

is delivered can further obstruct patients' access to comprehensive data on risks, benefits, and alternatives.[17] Small wonder that patients often end up with little real understanding of the risks of recommended treatments.

The patient's right to informed consent

The exclusion of patients from decisions about their own medical care runs counter to one of the fundamental tenets of the American legal tradition: the individual's right to self-determination, or autonomy. Autonomy, as viewed by the courts, entitles patients to the information necessary for them to exercise ultimate control over what happens to their bodies.[18] Over the last several decades, legal doctrine has taken an increasingly broad view of what that entails. The law has long recognized the importance of a patient's consent to medical procedures, but this meant only that physicians had to disclose the nature of the intended procedure.[19] In the late 1950s, some courts began to expand the concept of consent and to hold physicians negligent if they failed to provide information on risks or other issues essential to the patient's informed decision.[20]

Generally, the adequacy of disclosure was determined by the prevailing practice of other physicians (see, for example, *Canterbury* v. *Spence*, 1972). Recognizing that such a standard bore no necessary relationship to the patient's need for information, a few courts adopted a standard based on what the "reasonable man" would require (*Wilkinson* v. *Vesey*, 1972).[21] One precedent-setting case (*Cobbs* v. *Grant*, 1972) went even further, judging adequate disclosure in terms of the information needed by that particular patient to make an informed choice.[22]

These patient-oriented court decisions remain the minority, however; in fact, a number of states have subsequently moved in the opposite direction, passing legislation that mandates standards based on prevailing professional practices.[23] And the resistance of the medical community to informed consent remains strong. Even though physicians are now significantly more likely than in the past to disclose threatening information, such as a diagnosis of cancer, many continue to believe that fully informed consent is at best unattainable, at worst downright harmful.[24] As law professor Jay Katz has observed, "the doctrine's appealing name – informed consent – promises much more than its construction in case law has delivered."[25] After all is said and done, Katz concludes, the courts imposed primarily a duty-to-warn on physicians, focusing on the information disclosed rather than on the adequacy of the communication process or on what the patient actually understood.

Informed consent and medical experimentation

Patients undergoing experimental medical treatments have an even greater need for information than do those receiving standard care, since the risks are often higher and there are many more unknowns. Most people consider informed consent essential in clinical research. In one public survey that described various medical experiments, consistently large majorities of the respondents said that informed consent should be required in virtually all of the experiments, even if this made it difficult to conduct research with high potential benefit.[26] Informed consent has been part of the American Medical Association's code of ethics since the mid-1940s and was included among the "Ten Commandments of Medical Research Involving Human Subjects" developed for the Nuremberg war trials.

Abundant evidence over the years indicates, nevertheless, that patients in experimental settings are often as poorly informed as those undergoing routine treatment. Patsy Mink and other women in the 1952 Dieckmann study of prenatal DES reported later that in order to safeguard the study design, the investigators had told them that they were taking "vitamins."[27] Public outrage over the revelation of abuses by medical researchers, such as the experimental administration of cancer cells to elderly patients in the Brooklyn Jewish Hospital, led the federal government in 1966 to adopt guidelines intended to protect the rights and welfare of subjects of all federally funded human experimentation.[28] These guidelines require investigators to "obtain and keep documentary evidence of informed consent..." Prospective patients must be informed of the possible risks and benefits of the treatment, of alternative treatments available, and of their right to withdraw from the experiment at any time.[29] Food and Drug Administration regulations include similar guidelines, requiring that patients give informed consent when receiving experimental drugs and medical devices (e.g., the artificial heart).

Experimentation with DES as a morning-after pill to prevent

pregnancy, which began in the late 1960s, clearly fell within these federal regulations. Yet subsequent surveys of women participating in the early research at the University of Michigan, reported by health activists Kay Weiss and Belita Cowan in congressional testimony, indicated that the majority of women said they had not been aware that the DES morning-after pill was experimental.[30] Despite the adoption of federal guidelines, such failures of informed consent appear to be commonplace. Even research institutions that scrupulously follow the Federal guidelines have tended to focus mainly on the technical aspects of the consent forms used rather than on the goal of improving patient/subject comprehension.[31] One survey of women in a study of labor-inducing drugs found that twenty of fifty-one "interviewed after the drug administration had begun did not know that they were participating in a research study, despite the fact that all had signed a consent form."[32]

Lack of informed consent was also the charge brought against Dr. Denton Cooley by the widow of Haskell Karp, the recipient of the first total artificial heart implant in 1969. Mr. Karp had signed a consent form, but very reluctantly, his widow testified after his death. She brought suit against Cooley, charging that her husband's consent had been fraudulently obtained and was therefore not "informed." Her husband was not aware, she claimed, that the mechanical heart "had not been tested adequately even on animals, [and] had never been tested on a human being."[33]

Dr. Cooley had not, in fact, sought approval from the hospital's human subjects committee for a clinical experiment. He maintained that the artificial heart was implanted in an emergency effort to save Karp's life and hence was not a medical "experiment." The court accepted this argument, holding that the distinction between therapy and experiment was irrelevant if the physicians, whatever their scientific curiosity, had also been motivated by a therapeutic intent to save Mr. Karp's life.[34] The one physician who might have been expected to support Mrs. Karp's allegation that her husband was the unwitting victim of human experimentation was Dr. Michael DeBakey. It was from DeBakey's laboratory that the artificial heart had been taken, and he was widely known to consider it unready for clinical testing. DeBakey refused to testify, however, and in the absence of expert supporting testimony, the judge found Cooley innocent of any wrongdoing.

Does this verdict mean that the judge believed Cooley had followed proper procedures and that informed consent had been achieved? Certainly not. It means only that Mrs. Karp could not produce an expert witness, a physician, who was willing to support her allegation. The reluctance of physicians to testify against each other has been a frequent difficulty in medical litigation.

The Cooley–Karp case shows the blurred line between medical practice and clinical experimentation, and the broad discretion physicians have in drawing it. Simply by claiming therapeutic intent, Dr. Cooley exempted himself from normal experimental protocol. (Later investigators in the artificial heart program would claim the same privilege.) Another difficulty in the Cooley–Karp case was how to define and obtain meaningful informed consent from a patient facing imminent death and with no alternative treatment available. Such patients are, understandably, desperate for any hope offered. This situation presents a potential dilemma for the physician-investigator, who must reconcile obligations to patient care with the need for willing research subjects. Ideally, the dual responsibilities of the physician-investigator are complementary, but they may sometimes come into conflict, especially with regard to disclosure of risks and alternatives.

Evidence of such a conflict occurred not only in the first human artificial heart implant but also in the first human application of genetic engineering. In 1980, Dr. Martin Cline, a specialist in blood disorders at the University of California in Los Angeles, used the new genetic technique to treat two patients with a fatal blood disease (beta-thalassemia), by attempting to fuse normal bone marrow cells with the defective cells. This attempt, which was unsuccessful, met with strong criticism from the research community.[35] Many experts felt that given the limitations of previous animal research, experimentation on humans was scientifically premature and might lead to public distrust of the nascent field of gene therapy. Some of Cline's colleagues blamed his misjudgment on the desire for personal recognition as the first to apply genetic engineering to humans.

Much of the criticism focused on the issue of informed consent. Cline performed the experiments in Italy and Israel, after having been denied permission by the UCLA Human Subjects Use Committee. The committee had concluded that the risks to patients were too high and had recommended further work with animals. Although the patients signed consent forms, some observers questioned how meaningful such consent was in view of the lack of scientific evidence for appraising risks and benefits. The patients could not help but be misled, charged one scientist, since testing the technique on humans implied more hope for success than was warranted from animal research.[36] Ultimately, Cline was asked to resign his post as department head at UCLA and lost much of his federal research funding.[37] In the judgment of many of his peers, he had violated the accepted ethical standards of both clinical research and medical practice.

The exploits of medical pioneers such as Cline and Cooley provide vivid examples of the paternalistic tradition of medicine, in which patients' rights depend heavily on doctors' judgments. Informed consent seeks to give patients more of a voice in defining their own rights, and is part of the general trend in society toward greater egalitarianism, as well as a reaction to the disclosure of abuses.

To date, the major emphasis has been on giving people more medical information. This was the goal of the Carter administration's plan to include so-called patient package inserts in prescription drugs, a plan that was abandoned under Reagan despite the efforts of consumer groups, and it was the intent of a landmark 1981 California law requiring doctors to inform breast cancer patients about all options for treatment.[38] Informed consent procedures themselves have become considerably more elaborate. By 1985, the informed consent form for the artificial heart was seventeen pages long. However, more information does not always mean greater understanding. One study found that three out of every four informed consent forms required a college education to comprehend. In some cases, too much detailed information can actually impede understanding.[39]

As construed by the courts, the goal of informed consent is not merely to protect patients from unknown injury but also to promote patient autonomy. Provision of information is necessary but not sufficient; autonomy requires that patients have a chance to use this information in medical decisions. Ideally, autonomy would foster effective two-way communication between patients and physicians and a partnership relation in clinical decisionmaking. Achieving that goal will require major changes in both professional attitudes toward patients and public deference to expertise.

Risks to society

Informed consent, even if successfully implemented, can protect individuals only against unknowing exposure to the hazards of personal medical care. But the risks of medical innovation involve more than just the sum of each individual's personal risks. There are also risks to society as a whole – what we might call "collective risks." These include large-scale health and environmental hazards, economic investment costs, damage to moral and spiritual values, and loss of health benefits from alternative medical advances. By their nature, collective risks must be dealt with at the regional or national level. They typically involve large-scale effects, or society's responsibility to future generations. Judgments about such risks are the responsibility of regulatory agencies, with oversight by Congress and the courts. The Food and Drug Administration is charged with ensuring the safety and efficacy of drugs and medical devices, the Centers for Disease Control with controlling the spread of infectious disease, the Occupational Safety and Health Administration with reducing occupational injuries, and the Environmental Protection Agency with preventing unintended damage to the environment. There is also a growing national effort to monitor doctors' performance more rigorously by physician-run licensing boards as well as state or federal review agencies.[40]

Developments in genetic engineering illustrate various types of collective risks. One is the possibility of ecological damage, a long-standing fear that resurfaced in 1983 when scientists sought permission to release genetically engineered microorganisms into the environment. Critics, calling such planned releases "ecological roulette," warned that foreign organisms can have unpredictable, and sometimes devastating, effects on the existing ecological balance. By some estimates, nearly half of the major pest insects in the United States originated abroad.[41]

Just what are the dangers? As discussed in Chapter 6, no one really knows. Although many experts maintain that genetically altered organisms present no unique ecological hazards, not everyone agrees. Gene-altered "ice-minus" bacteria prompted speculation that weather patterns might be disrupted. Some meteorologists contend that there is at least circumstantial evidence that a substantial reduction in the number of unaltered ice-forming bacteria could decrease total rainfall.[42] Genetically engineered vaccines have also raised questions. Most animal viruses do not infect people, yet a few do; the AIDS virus, for instance, is known to infect green monkeys in Africa. According to the scientific director of the Humane Society, inoculation of animals with live viral vaccines could infect other animals, or even people. "There could be a human danger," he said. "When we're filling farm animals with weakened live viruses it could become a big health problem because they can transfer to the human host."[43] Most experts believe that the most likely form of transfer would be through airborne transmission or animal secretions, not through eating the meat. Of course, experts also did not expect DES given to poultry and cattle as a growth stimulant to end up in the meat eaten by humans, but it did.

Plants and seeds genetically engineered to be resistant to specific brands of herbicides pose an obvious danger to the environment. A number of major companies are developing plant lines that can survive megadoses of toxic pesticides and weedkillers. Typical of these efforts is Ciba–Geigy's arrangement with its subsidiary, Funk, to produce seeds that are chemically treated for special tolerance to one of Ciba–Geigy's herbicides.[44] Such plant lines will allow higher doses of pesticides and herbicides to be used in agriculture, increasing the sales of petrochemical companies but at the same time exacerbating the growing problem of environmental and groundwater pollution.

Even organisms genetically engineered for beneficent purposes could end up doing more harm than good. The creation of a microbe that eats toxic chemicals, for example, led to some disturbing speculations about the problems that could arise. In 1981, Dr. A. M. Chakrabarty, the scientist whose landmark patent on a genetically engineered organism opened the way for the commercial exploitation of biotechnology, produced a microbe that lives on toxic chemicals such as the herbicide 2,4,5-T. After a spraying of 2,4,5-T, the microbe could be applied to clean it up. The problem is that although most of the microbes would die when they ran out of the herbicide to eat, they would not necessarily all die. The next time 2,4,5-T was sprayed, the surviving microbes would experience an enormous population boom as they devoured the 2,4,5-T, reducing its effectiveness as an herbicide. Therefore, as Chakrabarty put it, although the microbes would be able to degrade the herbicide: "You might have to use more of the chemicals to kill the weeds. The chemical companies might sell more of the chemical."[45] The ultimate winner in this curious contest between microbes and chemicals is unclear, but it will probably not be the environment or public health.

Another concern is that genetic engineering will compromise the integrity of the gene pool, the genetic inheritance of future generations. By eliminating "defective" genes, the new technology might gradually reduce genetic diversity. It is this diversity that has allowed our species to evolve and meet environmental challenges. Moreover, because our knowledge of how genes function is still quite primitive, there is always the danger that genetic manipulations will be based on incomplete or faulty understanding. We now know, for example, that sickle cell trait, long considered an undesirable genetic abnormality, also bestows a genetic advantage on carriers in tropical climates by making them more resistant to malaria.

Genetic manipulations of so-called somatic cells, which do not affect genetic inheritance, could also cause problems. Criticizing Cline's experimental use of gene therapy on humans, Paul Berg, a pioneer of the genetic technique and sponsor of the Asilomar Conference, stated: "I don't think any of us know enough about the fine molecular structure of genes, how they operate and how they are regulated, to be planning and carrying out that kind of operation...I don't know what to predict. I don't think anyone can predict. We ought not to take chances."[46] Cline himself later acknowledged that unexpected reactions were possible, although in his view highly unlikely. "When you insert new genes into cells," he admitted, "it is conceivable that they would be so aberrantly controlled that they would change the nature of the cells, make the cells defective, or even possibly malignant."[47]

Finally, genetic engineering poses a variety of risks to traditional social and moral values. Especially troubling to many is the image of scientists "playing God" – using genetic manipulations to create new types of organisms or to "correct" physical or mental traits deemed deficient. The ethical implications of the new technology have also evoked concern within the religious community. Pope John Paul II has warned against setting science above "the right of individuals to their physical and spiritual life."[48] With increasing military involvement in biotechnology research, the specter of new and more dangerous forms of biological warfare adds yet another threat to our physical and spiritual lives.

However intangible and difficult to measure, such collective risks to society are real, and they elude individual safeguards such as informed consent. Most require protective actions at the national or even international level. The likelihood of such actions depends on how much is known about the risks, whether there are mechanisms requiring or encouraging early disclosure, and how seriously the threats are taken.

Changing attitudes toward collective risks

Attitudes toward risks have fluctuated widely over the last several decades, evolving through spirals of optimism and disenchantment, risk and recoil. The 1940s and 1950s witnessed a wave of major new drug discoveries: penicillin and other antibiotics, cortisone, the first tranquilizer, oral antidiabetics, and others. This period of unprecedented accomplishments was followed by a tightening of drug regulations. In 1958, due largely to the efforts of Congressman James Delaney, an influential member of the House Rules Committee, Congress passed the Delaney Amendment, as it came to be called, banning any food additive shown to be carcinogenic. In 1959, the Kefauver committee in the Senate held hearings on the high prices and market power of the drug industry. And in 1962, in the wake of the thalidomide tragedy, Congress substantially expanded the Food and Drug laws, enlarging FDA's jurisdiction over experimental drugs and requiring proof of efficacy, as well as safety, in the drug review process.

These legislative actions belied the generally positive outlook toward science and medicine in the early 1960s. These were boom years, in which scientific successes were extending America's power up into space and, with the discovery of DNA structure, down into the building blocks of life. Science and technology were on our side, it seemed, and powerful chemicals were fast becoming the present-day staff of life. In this heady climate, manufacturers of DES, along with the cattle industry, were able to convince Congress in 1962 to grant a special exemption to the Delaney Amendment that allowed continued use of DES, a known carcinogen, as an animal growth promotant, provided residues were not detected in meat. It was also around this time that medical researchers were beginning to experiment with the use of DES as a morning-after pill contraceptive, the known risks of all estrogens again notwithstanding. And this was the period in which the artificial heart program was launched amid wildly optimistic projections of speedy accomplishment and net financial benefit, with only the briefest nod to the dangers of a nuclear-powered heart.

By the early 1970s, the mood had shifted considerably. Vietnam, Watergate, and a growing list of technological failures – pollution, toxic chemical dumps, acid rain, radioactive wastes, and others – all contributed to declining public confidence in social institutions in general and in science and technology in particular. Reflecting this new mood of concern, scientists declared a moratorium on the more hazardous recombinant DNA experiments while risks were investigated. Problems were emerging in medicine as well. There was growing evidence of adverse drug effects and unnecessary surgery. Congress held hearings on abuses in the medical technology industry. In 1971, Herbst discovered the transgenerational association between DES and clear-cell cancer. Reviewing the artificial heart program in 1973, an outside assessment panel deemed the risks of a nuclear-powered artificial heart unacceptable, whereas less than a decade earlier they had been essentially ignored. In Congress, this mood of caution was reflected in a spate of laws to protect human health and the environment: the National Environmental Policy Act, the Toxic Substances Control Act, the Occupational Safety and Health Act, the Consumer Product Safety Commission, the Clean Air Act, the Technology Assessment Act, and the Medical Devices Amendments, among others.

In the late 1970s, with the economy beginning to slump, the mood changed again, as national leaders sounded the battle cry for a rebirth of the freewheeling, risk-taking spirit of the frontier days. Risks were the necessary price of progress, they exhorted, and science and technology were again heralded as saviors of a society plagued by continuing social and economic problems. Those who pointed to the dangers of modern technology were mocked as hopeless utopians – antiprogress Chicken Littles in search of a risk-free society. "We are hyperconcerned about risk," declared Philip Handler, then President of the National Academy of Sciences. "The world has never been safe," he stated. "There will be no advances without risks."[49] Others claimed that it was the fear of risks rather than the risks themselves that posed the greatest danger. In the effort to avoid risks, the head of General Electric charged, "we are in danger of risking everything: our liberty, our hard-won standard of living, our range of choices, our great and venturesome system of private enterprise."[50] A new theme had emerged, one that would dominate much of the 1980s. In the words of political scientist Aaron Wildavsky, "No risk is the highest risk of all."[51]

The politics of risk

While many factors underlie these shifting attitudes toward risk, perhaps the single most important one has been the state of the nation's economy. In good times, we can afford to be cautious; when the economy is lagging, we are told to take more risks to stimulate innovation and productivity. The push to promote risk-taking in the 1980s can be seen as a direct response to the nation's economic difficulties.

What is not made clear in the drive to legitimize risks, however, is exactly *who* will bear *what* risks. And with good reason. For while it is private industry that benefits most directly from a re-laxation of safeguards, it is consumers and the general public that bear the brunt, should harms occur. Furthermore, private com-panies frequently pass on the costs of whatever risks they face by raising the price of products or by turning to public agencies for help. That was precisely what insurance companies did, for in-stance, in forcing Congress to provide liability insurance for swine flu vaccine producers. A report to Congress later estimated that the insurance firms had earned roughly $8.7 million in premiums from the vaccine producers, most of which represented windfall profits at the taxpayers' expense since the government had assumed almost all liability costs.[52] Similarly, DES manufacturers have waged a lengthy legal battle to avoid paying damages for DES-related injuries. Many biotechnology companies – unable to get liability insurance because of the essentially unknown risks of large-scale commercial genetic engineering – have nonetheless pushed ahead with development and marketing.[53] If serious health or environmental damage were to occur, many of these companies would probably have to declare bankruptcy, as the Johns–Manville Corporation did when faced with several billion dollars in asbestos claims. If so, any major harm caused by this new technology would be borne by the potential victims.

The smaller the perceived risks, the more likely they will be accepted by society. So industry has been trying hard to convince us that we have nothing to fear. The Chemical Manufacturers Association, a trade group in Washington, hired the J. Walter

Thompson Advertising Agency to compose a $5 million media package with the message that life depends on chemicals and risks are just a normal part of life.[54] Dow Chemical advertisements state reassuringly: "The chemicals we make are no different from the ones God makes. . . . There is essential unity between chemicals created by God and chemicals created by humans. . . . Birds, for example, are extraordinarily beautiful products produced by God."[55]

Scientists and engineers have been particularly useful in industry's campaign to legitimize risk. Experts are seen as having less of an "ax to grind" than management, explained the *Wall Street Journal,* and so are more likely to be trusted by the public and the press. The Chemical Manufacturers Association spent $2.7 million in 1981 on advertisements featuring company scientists praising corporate efforts to clean up the environment. In what Dow called its "Visible Scientists" program, industry toxicologists and environmental specialists were assigned specifically to the task of public relations.[56]

Also helpful in refurbishing ideas about risks have been the growing ties between industry and academic institutions. According to a 1979 survey, private support of universities was the highest since the survey began around 1920, although still only a fraction of governmental funding.[57] Some of this support has gone to individual scientists and departments; some has been used to set up interdisciplinary "risk institutes" for the study of quantitative risk assessment, environmental toxicology, and other topics of joint interest to industry and academics; and some has been used to sponsor conferences or symposia sympathetic to industry views. Such ventures have given the private sector a growing foothold in academic research and increasing opportunities to steer it toward commercial interests.

Corporate sponsorship sometimes has strings, but often the effects are more subtle, reflecting nothing more than the academician's desire for continuing support. In a book entitled *The Regulation Game: Strategic Use of the Administrative Process,* the authors, professors of business and economics, instruct would-be executives about how to exploit this desire. University faculty can be "coopted," they suggest, "by identifying the leading experts in each relevant field and hiring them as consultants or advisors, or giving them research grants and the like. This activity requires a modicum of finesse . . . for the experts themselves must not recognize that they have lost their objectivity and freedom of action.[58]

Even the courts have been a target of industry's efforts to influence concepts of risk and regulatory policy. One example is the Law and Economics Center (LEC), a privately financed institute formerly affiliated with the University of Miami Law School and now located at Emory University in Atlanta. Since 1974, the LEC has been schooling federal judges in what it calls "a straightforward treatment of market economics" – that is, the "free-market" brand of economics of conservatives such as Milton Friedman. Over 200 judges, a third of the entire federal judiciary, have attended at least one of the LEC's courses. These courses are designed not only to familiarize judges with economic and business topics but also to change the outlook of those who attend – to convince them, as one introductory speaker put it, "that what you thought was bad really may not be . . ."[59] Commented a participant in another one of LEC's programs, "Perhaps such analyses can help us retard the imposition of unduly restrictive governmental regulations by showing the true cost of such regulations . . ."[60]

Industry's role in shaping governmental policy on risks has been considerably more direct. The government is responsible for ensuring that the economy prospers, and must weigh industry's complaints about health and environmental regulations against the goal of protecting society against untoward rights. The balance struck varies according to the state of the economy and the political and ideological disposition of the administration in power. The Reagan administration has been unusually friendly toward business interests, undertaking massive deregulation of federal agencies responsible for consumer health and safety. One of the victims of deregulation was the National Center for Health Care Technology, a small agency created by Congress in 1978 to monitor and evaluate emerging medical technologies. Despite its efforts to appease business and professional interests, the Center came under increasing fire. In 1981, it was explicitly instructed not to "unreasonably inhibit the innovation of new technologies," and more industry representatives were added to the governing council.[61] A year later the agency was dismantled entirely.[62]

Reagan's solicitude of the private sector is perhaps unusual in degree, but certainly not unprecedented. Government, science, and industry have long made common cause. The FDA was a willing and helpful partner to pharmaceutical manufacturers in their efforts to obtain approval of DES in the 1940s, and also took the lead in trying to get DES approved for use as a morning-after pill. When federal regulation of recombinant DNA research was proposed in the mid-1970s, most officials at the National Institutes of Health were sympathetic to the onslaught of protest from the scientific community. Later, even though many of these same scientists had formed close ties with biotechnology companies, they continued to serve as trusted advisors to the NIH on matters of governmental regulatory policy.

A striking example of the symbiotic relation between government and industry occurred in late 1984, when the Agriculture Department, without consulting either the NIH's Recombinant DNA Advisory Committee or its own internal biotechnology review committee, quietly authorized field testing of a genetically engineered swine vaccine, and then issued a license to Biologics Corporation for commercial sales of the vaccine. When these actions finally came to light, they were roundly assailed and led to a congressional investigation. Existing federal procedures had obviously failed. "We do not have any strong controls to prevent somebody from going out and putting [gene-altered organisms] into the environment on his own," said Representative Harold Volkmer at a House subcommittee hearing. "What I'm seeing is you've got a loophole here where you can drive a big truck."[63] Even the Government Accounting Office, asked by Congress to review the Agriculture Department's monitoring of deliberate releases, concluded that it "lacked authority and direction." The GAO report noted that the department did "not want to impose cumbersome regulations that might stifle growth in biotechnology," and that "the timing and pace of its actions have been influenced by the White House's Office of Science and Technology Policy."[64] Rifkin, never one to mince words, called the Agriculture Department's oversight of genetic engineering "a joke."[65]

The FDA, too, has demonstrated its faith in biotechnology. In 1981, brushing aside concerns about the unknown properties of recombinant organisms, the FDA designated genetically engineered drugs for "fast-track" review, a shortcut intended for important new drugs with excellent therapeutic potential. As with the morning-after pill, the FDA acted more as active proponent in this decision than as passive regulator. According to then-Commissioner Jere Goyan, designating the fast track for genetically engineered drugs "reflects an FDA not content with waiting for a [New Drug Application] to show up but rather one reaching out to the very edge... to find out if there might be potential difficulties and to resolve such difficulties far in advance."[66]

Given the forces at work, it is not surprising that business interests have tended to dominate notions of appropriate and acceptable risk. A symposium of academic and industrial experts sponsored by General Motors in 1979 revealed the kinds of attitudes this has produced. There was much discussion of unrealistic public fears ("based on infirm and dubious facts"), but little mention of the narrow, and arguably lopsided, range of risks being considered. One of the symposium participants, sociologist Charles Perrow, observed that in every case, "the risk-acceptance position is the one that favors private business and industry, and the risk-aversive position would harm it." What about other types of risk, he asked (rhetorically): experimentation with income-maintenance programs, unilateral arms reduction, highly progressive tax reforms, a crash program for solar energy, or legislation facilitating class action lawsuits or freedom of information suits? He concluded sardonically: "No amount of chanting that risk is an accepted part of life, that without risk there is no progress, that willingness to take risks is a condition of biological success, that without risks there can be no benefits, etc. would induce most of [the symposium] participants to take *these* risks."[67] The real issue, in short, is not how many risks we should take, but rather what kinds and for what social and economic ends.

Risk assessment: analysis and ideology

The growing attention to risks has heightened interest in analytical techniques for risk assessment and management. Although these techniques seem objective and rigorous enough, in fact many contain subtle biases that work in favor of taking risks and making money. Their aim is simple: to compare the costs, or risks, versus the benefits of alternative policies in order to identify the one with the maximum net benefit. Nothing else is straightforward, however. The first step is deciding which costs, risks, and benefits to include. These choices, which are crucial to the outcome of the analysis, are inevitably arbitrary, and necessarily reflect a particular set of values and interests; for the results to be valid, these values must represent, and reconcile, the full spectrum of social priorities. Then, costs, risks, and benefits are translated into a common denominator; for costs and benefits, this is usually dollars. Yet many of our most treasured values – human life and dignity, the pleasure of fulfilling communication with other people, the aesthetic appreciation of a clean environment, the sense of personal control over bodily health – cannot meaningfully be reduced to dollars (or to any other quantitative terms for that matter). Typically, such values are omitted from the analysis, and often ignored entirely. Quantified effects are the currency of risk assessment and cost–benefit analysis, even though the unquantifiable effects may be understood to be more important.

In addition to problems of measurement and quantification, risk assessment has two other critical limitations. First, it is basically a static technique. It typically deals only with the end results of policy choices, while ignoring potentially vital questions about the process used in identifying or choosing among the alternative options.[68] The second limitation is more fundamental: It is impossible to weigh risks that are unsuspected. Even randomized clinical trials may fail to detect problems that are very rare or that remain latent for some time. Swine flu vaccine was tested on more than 5,000 people, the largest number in vaccine history, yet these trials still failed to discover the very low incidence of Guillain–Barré syndrome associated with the vaccine. Also difficult to assess are risks that are extremely unlikely but catastrophic if they occur, such as the potentially irreversible effects of genetically engineered organisms on the environment. Environmental risk analyses rest on a series of shaky, and often untestable, assumptions about how to calculate exposure, how to extrapolate from high-dose effects on animals to the risks of low doses for humans, and about the disease mechanisms themselves.

These various limitations tend to give quantitative cost–benefit and risk analyses an optimistic, protechnology bias. The benefits of a new technology are generally relatively clearcut and easily quantifiable, especially those that can be measured in dollars; risks, by definition, are usually remote and speculative, especially when the technology is new or when reliable epidemiological data are not available.[69] As a means of weighing the pros and cons of technological innovation, mathematical cost–benefit analyses are, as one critic put it, "about as neutral as voter literacy tests in the Old South."[70]

A cost–benefit analysis of the swine flu program published in the *New England Journal of Medicine* in September 1976, illustrates some of the pitfalls. It was characteristically optimistic, concluding that immunization for all adults over age twenty-five would be economically justifiable if vaccination rates exceeded 59 percent. The scope of the analysis was narrow; it compared total dollar benefits of immunization, including the presumed savings in medical care costs and worker productivity, to the total dollar costs of mass immunizations, including the costs of program administration, vaccine supplies, and sick days due to adverse reactions. As usual, key intangible factors that could not readily be assigned a dollar value were omitted, as the authors pointed out, including "pain, suffering, and grief associated with the illness or vaccination," the "disruption of community services" that an epidemic might cause, and problems associated with liability insurance.[71] But the main reason for the overly optimistic conclusion – that immunization would be cost-effective under certain conditions – was simply the failure to account for the enormous uncertainty about whether the epidemic would in fact occur. By September 1976, doubts were widespread; opinions varied widely, but most experts abroad, and a number in the United States, considered an epidemic quite unlikely. The authors assumed a likelihood of 10 percent, a conservative estimate relative to some expert predictions (which ran as high as 40%), but did not seriously consider the possibility of no epidemic at all. Not surprisingly, faulty assumptions produced faulty conclusions.

Similar problems and uncertainties confound almost all quantitative cost– and risk–benefit studies. The Office of Technology Assessment, in an exhaustive review of cost–benefit analysis published in 1980, concluded that in light of the method's limitations, it "is useful for assisting in many decisions, [but] should not be the sole or prime determinant of a decision."[72] Nevertheless, the following year, President Reagan issued an Executive Order directing that all federal regulations be evaluated quantitatively for their "cost-effectiveness," and only those enacted that showed a positive ratio.[73] This directive brought much regulatory activity to a screeching halt, no doubt exactly what was intended. The staff scientist for the House Subcommittee on Health had warned back in 1978 that cost-effectiveness and risk–benefit analyses had been adopted by "those who do not wish to regulate, or to be regulated . . . to avoid taking action which is necessary or desirable in order to protect the health of the public or the integrity of the environment."[74] Despite – or perhaps because of – their limitations, quantitative analyses have evidently become a strategic new weapon in the arsenal of federal health policy.

Beyond risks and benefits

Risks and uncertainty are an inescapable part of modern medical science; they are the price we pay, in a sense, for betting on progress. Lest we forget this price in our enthusiasm over medicine's spectacular achievements, we have DES, thalidomide, and other medical horror stories to remind us that the bet does not always pay. Today's catalog of risks includes a host of man-made dangers as well as natural disasters. Our increasingly synthetic environment and our growing reliance on chemical and mechanical technologies have introduced an array of hazards unknown in simpler times. With increasing technological complexity and interdependency has come a greater likelihood of accidents that cannot be prevented through standard safeguards or improvements in system components – what Charles Perrow has called "system" or, to stress their increasing frequency, "normal" accidents. Traditional calculations of risk for these new hazards are likely to be too low, Perrow argues, because they are based primarily on the expected frequency of component failures, with only limited attention to interactive possibilities and almost none to unanticipated interactions.[75] This is a sobering view. It suggests that advanced medical and scientific technologies are now so complex that breakdown is essentially inevitable and that advance prediction and prevention are often impossible.

The growing emphasis on informed consent in personal medical encounters offers some protection against individual medical risks. At a minimum, informed consent seeks to ensure that patients understand the risks and benefits of proposed treatments and the therapeutic alternatives. More broadly, the legal doctrine of informed consent, derived from the principle of self-determination, aims to promote the autonomy of patients. Its goal is not only to improve communication but also to increase patients' involvement in medical decisions.

For larger-scale risks, we need a social analogue of informed consent for the general public. This requires adequate information on the full range of costs, risks, and benefits associated with medical and scientific innovations, data that are frequently lacking. More importantly, it requires that we give more of a voice in governmental policy decisions to those likely to bear the major burden of the risks – both the groups affected by particular medical innovations, and also the larger public. The ultimate goal of societal informed consent, like its individual analogue, is not merely protection against risks but promotion of autonomy: social control over the welfare of the body politic. The 1973 Artificial Heart Association Panel saw itself as a vehicle for providing "societal informed consent" in the quest for an artificial heart, although one of the panel members objected that the term "consent" implied a more "passive and supine" role than the panel had played, and which might properly be expected for the public.[76]

Social choices about risk will never be easy. The issues are often highly technical, and invariably involve conflicting goals and competing priorities. A basic question is whether the burden of proof should rest with those who demand evidence of safety or with those who demand evidence of hazard. Should we assume that a new technology or medical procedure is safe until proven harmful, or potentially harmful until proven safe? Put another way, when both risks and benefits are uncertain, should we take risks in hopes of hastening technological progress, or should we give up some of the benefits of such progress in order to avoid possible dangers?

In general, this country has opted for risk over security, economic growth over social stability. Yet psychological research indicates that most people, given the choice, are risk averse — that is, they prefer to forego benefits rather than risk losses in order to obtain benefits.[77] The existence of a multi-million-dollar insurance industry bears witness to the phenomenon of risk aversion, as do many of our regulations for medical and consumer products. Furthermore, we may already be taking more medical risks than we realize. A 1982 report by the Congressional Office of Technology Assessment warned that the public, accustomed to governmental protection against the hazards of nonmedical products, may assume "that it is likewise protected in undergoing any medical or surgical procedure recommended."[78] That, the report noted, is not necessarily the case.

In the antiregulatory climate of the 1980s, the desire for protective regulations has often been derided as overly cautious. But a little unnecessary caution is not necessarily such a bad thing. Harold Green, chair of the 1973 Artificial Heart Assessment Panel, has argued that unnecessary caution "is precisely the kind of inadequacy — or indeed, error — that society is best able to tolerate. Where benefits are underweighted or risks overweighted, the result may be a decision that bars or postpones a benefit; but this means only that society is deprived of an opportunity for improvement. On the other hand, the opposite approach may expose society to actual injury."[79] The same theme is found in the ancient medical maxim *"primum non nocere"*: above all else, do no harm.

Risks, whether unknown or unknowable, assessed or ignored, broadly conceptualized or narrowly quantified, inevitably accompany medical innovation. There can be no "scientific" answer about how to deal with these risks, only a political one. In a democracy such as ours, it is the process by which political choices are resolved that we look to for protection. Decisions about risk cannot meaningfully be isolated from the social, political, and economic context in which the risk occurs, nor, indeed, from the measurement process itself. Dangers of very different types, affecting different population groups, must be weighed and compared, either explicitly or implicitly, and decisions made as to who will bear what risks. These decisions are matters of public policy as well as personal concern. They are, or should be, part of the process of democratic choice.

8

Compensating the injuries of medical innovation

In 1,600 cases scattered across America, people who got swine flu shots in 1976 have sued the United States Government charging that the shots caused Guillain–Barré and other injuries. Total damages claimed exceeded $2.2 billion. More than a decade later, nearly thirty of these suits were still pending in the courts. Of those that had come to trial, the government, represented by the Justice Department, won in more than eight out of ten cases.[1]

DES victims have also turned to the courts for relief, although only a handful have received awards of any magnitude. Many DES daughters have been unable to sue because they cannot identify the particular brand of DES their mothers took. Others have been barred because their injuries came to light after their state's statute of limitations had expired. Those not blocked by legal restrictions face the time, expense, and trauma of bitter court battles, and the likelihood of lengthy appeals by manufacturers. Eli Lilly, the largest manufacturer of DES, has already spent several million dollars fighting DES lawsuits, with no end yet in sight. Although somewhere between 500,000 and six million people were exposed to DES, only about 1,000 DES suits had been filed in the United States as of 1987.[2]

Amid periodic insurance "crises" and publicity aimed at limiting manufacturers' liability, the plight of injured victims often gets lost. This chapter discusses some of the difficulties victims of medical and technological injury face in obtaining compensation, and the efforts courts have made to adapt traditional legal doctrine to changing patterns of risk and new concepts of social responsibility. In addressing these issues, it is useful to consider not only the strengths and weaknesses of the American tort liability system but also the approaches used in other countries.

Injury compensation: goals and problems

Compensation is the payment to an injured person for losses caused by the injury: medical expenses, lost income, and in some cases, pain and suffering. The largest of these losses is usually income. For both medical malpractice and serious automobile accident claims, lost wages make up about three-quarters of total injury losses, while less than a quarter are medical expenses.[3] In the United States, depending on the nature of the injury and the individual's circumstances, compensation for such losses might come not only from court awards but from state or federal disability insurance; workers compensation; veterans benefits; private accident, disability, or health insurance; or public assistance. Each of these programs covers only a limited range of people, however, and recoveries for personal injury are uneven and often pitifully inadequate. More than 38 million Americans, 19 percent of the population, do not even have basic health insurance.[4] No other industrialized country gives people who have been injured so little assistance. Every industrial nation but the United States provides its citizens with some form of national health insurance covering medical expenses, and most also have much more generous social and disability benefits. For many Americans who are injured, the courts may offer the only hope of redress.

The ideal compensation system would provide payments to a relatively large fraction of those injured, compensate similar injuries similarly, and ensure that the amount is commensurate with the degree of injury. To the extent possible, it would make the victim "whole" again. Fair compensation is, or should be, a matter of justice.

Justice also requires an effort to minimize or prevent future injuries. In the legal system, deterrence is usually sought by making those who commit careless or wrongful acts, called "torts," bear the costs of resulting injuries. (Regulatory mechanisms, of course, seek to deter injuries more directly.)

In addition to providing fair compensation and promoting deterrence, a compensation system should be efficient. Reparations should occur quickly, and the system should channel most of the funds to injured individuals, not to insurance companies, courts, and lawyers.

The priority given to each of these goals depends upon society's values, as well as on the interests and power of the parties involved. Although no one system can accomplish them all, the American system of tort law is increasingly perceived as unsatisfactory on all counts, especially in dealing with the product-related injuries of modern consumer society. Most personal injury cases never go to trial, and of those that do, the awards may bear little relation to the degree of loss. Although a few large damage awards have attracted a great deal of attention, fewer than half of all people seeking damages for medical malpractice recover anything at all.[5] Highly adversarial trials can become performances in which theatrical skill and striking visual effects have a dramatic impact on the jury, sometimes overshadowing the basic issues. On similar facts, one victim may recover nothing because of varying legal requirements or an unsympathetic jury, while another gets full damages. Many DES cancer cases have been rejected on legal technicalities; in the few that have come to trial, DES daughters have received anywhere from \$150,000 to \$1.75 million for comparable injuries.[6] As a DES lawyer who has handled cases in many states observed, "Cancer might have a very different price in a small town in Iowa than in Manhattan, New York City."[7]

The legal system is also highly inefficient. Legal fees and costs typically consume a third of the damages awarded, and in complex cases may run as high as two-thirds.[8] In general, injured victims end up with less of the total benefits than insurers, law offices, and courts. The product manufacturer faced with a lawsuit alleging a defective product has a powerful incentive, and usually the resources, to fight a lengthy court battle, which can generate staggering costs. A federal study of products liability cases found that it took sixty-seven cents in legal and other costs to provide fifty-eight cents of compensation to the injured party.[9] Changes in the current process, such as arbitration, or judicial reforms to speed up the trial system, may save some money, but as a method of compensation, the tort liability system will never be very efficient.

One of the fundamental reasons for many of these problems, especially from the point of view of equitable compensation, is that tort law was developed to deter wrongdoers by making them pay injury costs; it was not designed to compensate the injured. Underlying the whole body of fault-based tort law is the principle that the need for compensation alone is not a sufficient basis for an award. In theory, compensation requires proof of fault – a wrongdoing that resulted in injury. Even though this fault principle has been eroded significantly in recent years, its imprint on legal doctrine remains pervasive.[10]

Another source of dysfunction in the legal system stems from the central role of the insurance industry. Few manufacturers or physicians these days are without some form of liability insurance. Indeed, the large majority of personal injury cases are resolved not by courts but by insurance negotiators. Like tort law, the private insurance industry did not develop with the goal of providing fair compensation. The insurer's interest is in defeating as many claims as possible. As one insurance company defense attorney put it, "Frankly, we are in business to wear out the plaintiffs. . . . Companies are smart to do this. This is big business; it's dirty business. We're not a charity out to protect the plaintiff's welfare."[11] Delays drive up total costs while also putting pressure on plaintiffs' attorneys to settle out of court for smaller sums than may be deserved. And, by insulating manufacturers from financial responsibility, liability insurance also tends to weaken the deterrent effect of tort claims.[12]

To the lucky few who make it through the thicket of legal restrictions and overcome the arguments and courtroom theatrics of defense lawyers, the American legal system can be very generous. Damage awards in personal injury cases are often considerably higher than in other countries. Entrusting legal verdicts to juries of our peers is consonant with our democratic tradition, and has helped protect judicial decisions against political and economic pressures. Nevertheless, in the view of some legal scholars, tort litigation is more akin to a lottery than a system of reparative justice.[13] One British observer characterized the American tort-liability system as "inefficient, slow, arbitrary, expensive, almost incomprehensible, and in many ways unjust," and concluded "it just does not seem possible to stretch tort law into an efficient and humane compensation system."[14]

DES lawsuits: who pays when drugs cause injuries?

These problems are at their worst in cases involving large-scale exposure to risk, uncertainty as to the specific manufacturer or product responsible, and harms that remain latent or have multiple contributory factors. Lawsuits over DES and other so-called "toxic time bombs" – asbestos, Agent Orange, vinyl chloride, and other toxic chemicals – all share these features, and are flooding the legal system.[15]

Most drug injury litigation involves the drug manufacturer

rather than the prescribing physician. So it has been with DES, even though many physicians prescribed DES prenatally long after controlled studies had shown it to be ineffective in preventing miscarriages. Many of these physicians are dead or retired, and the relevant records are unavailable. Also, to prove malpractice, DES victims would have to show that physicians prescribing DES had departed from "acceptable medical practice." This usually requires the testimony of other physicians, which might be difficult to obtain, given the well-known reluctance of doctors to testify against their colleagues.[16] Furthermore, until 1971 at least, when the link between DES and clear-cell cancer was discovered, prenatal prescription of DES was still common, if not accepted, practice. It is easier, and usually more financially rewarding, for DES victims to sue the manufacturer.

Drug injury suits may argue that the manufacturer was negligent, as a number of DES suits have, or, under the doctrine of "strict liability," that the product was defective.[17] What constitutes a defect is often hotly disputed. With a drug, the occurrence of injury is generally not sufficient. All drugs have risks, many of them known; for life-threatening diseases, even potentially fatal side effects may be tolerated. Food and Drug Administration approval of a drug means that its benefits are deemed to justify its risks, not that the drug will never harm anyone. (The use of DES in pregnant women, of course, turned out to have no benefits, only risks.) The law generally treats drugs as "unavoidably unsafe" products that are not defective if the manufacturer has taken reasonable steps to test them and has issued proper warnings.[18] As a result, much drug litigation hinges on the adequacy of the testing and the nature of the manufacturer's warning.

Risks that are not detected with proper testing are considered "unforeseeable," and manufacturers are generally not held liable for resulting injuries.[19] Foreseeability is the line drawn by courts between negligence and fate. But what is deemed foreseeable at any given time is inevitably a matter of judgment; it depends on the level of scientific knowledge, the testing procedures available, and the extent of investigation. There are often bitter courtroom battles over what facts were known, or could have been learned, and when. Strict liability may encourage manufacturers to perform more extensive testing and research to detect harms that might otherwise have been deemed unforeseeable.[20] In short, the line between negligence and fate is neither as clear as traditional legal theory would imply nor as immutable.

The issue of foreseeability plays an important role in many DES suits. For example, when Joyce Bichler, a DES daughter who developed clear-cell vaginal cancer and underwent a hysterectomy and vaginectomy, sued Eli Lilly for $500,000 in damages, the jury was asked to decide whether Lilly had been negligent in failing to test DES for its effects on the fetus.[21] Lilly argued that the trans-placental effect of DES could not have been foreseen, and hence the company should not be held liable for injuries to offspring. Testing for intergenerational effects was not standard practice until after the thalidomide tragedy of 1962, and DES manufacturers introduced evidence showing that none of the major producers had tested DES for its effects on the unborn child. Yet Lilly conceded at trial that it was aware by 1947 that prenatal DES posed the threat of cancer to pregnant women, that DES had been shown to cross the placental barrier, and that DES had caused malformations in the offspring of pregnant mice to which the drug had been administered. Although Lilly contended it had no reason to suspect that DES would have similar effects on humans, the jury did not find this claim credible. Lilly appealed, but the judgment was upheld by New York's highest court, which concluded that the evidence presented showed that "cancer to the offspring was a foreseeable risk of ingestion of DES during pregnancy."[22]

A manufacturer has a duty to warn physicians of any drug hazards that it should reasonably know about, including any new evidence that appears. Unfortunately, this does not always happen. Cases of deliberate suppression of known risks, even fatalities, are periodically reported.[23] DES manufacturers, too, apparently revealed less to the FDA than they knew. Even though one of the studies cited by Lilly in its supplemental application for FDA approval of prenatal DES specifically questioned whether DES might have carcinogenic effects on pregnant women or cause glandular problems in the unborn child, these concerns were noted in Lilly's application to the FDA.[24]

Even the warnings that are issued may have little impact in the face of manufacturers' promotional efforts.[25] The prescription drug industry has become big business, with a market value of more than $15 billion in 1983 and profit levels almost double those of other manufacturing industries.[26] Drug industry sales efforts are massive. Companies advertise by direct mail to physicians, in medical journals, and by sending salesmen, called "detail men," to hospitals and doctors' offices. In 1978, the average physician received over $4,000 worth of direct drug advertising.[27] These promotional efforts pay off. Studies have shown that physicians get most of their information about drugs from drug company detail men or advertisements.

In theory, the law's insistence on warnings should help to curtail inappropriate or excessive drug use; in practice it may do just the opposite. To protect themselves against liability, manufacturers have listed growing numbers of adverse effects in product literature while still highlighting the drug's benefits. The more adverse effects listed, the more a company can promote a drug while retaining some protection against legal liability – and the less likely the warnings are to be read and understood. Advertisements typically contain long columns of warnings in almost illegible print amid eye-catching pictures and bold claims in large type.[28] Even if read carefully, these elaborate warnings may not be very helpful since they rarely tell much about relative risks or predisposing circumstances. More comprehensive information may be available in medical journals, but with roughly 400 new drugs introduced each year, few physicians have the time or training to stay abreast of all the latest data. In the case of DES, manufacturers' brochures minimized the strong evidence that it was ineffective in preventing miscarriages, focusing instead on the older and less reliable positive evidence. Physicians continued to prescribe DES prenatally for more than a decade after six controlled studies had been published showing DES to be worthless in preventing miscarriages.[29]

A new view of medical risks

The harm wrought by DES and various toxic substances can be devastating, whether the risks were foreseeable or not. Over the past several decades, growing concern for the needs of injured victims and changing perceptions of risks have gradually reshaped traditional legal notions of negligence and foreseeability. The principle of foreseeability harks back to an earlier age when risks were considered either known or unknown. Unknown risks were viewed as the luck of the draw. "Fate" was responsible, and the unlucky victims had to pay the costs. We have learned that the issue is not so simple. No amount of advance testing will reveal all of a drug's possible effects. Some may be too rare to be detected in clinical trials. Guillain–Barré syndrome was not discovered despite the largest clinical testing effort in vaccine history. Other drugs may have delayed-onset or cumulative effects that take years to develop or result from the drug's interaction with other substances.

Modern medical practice suggests the need for a new view of risks. Particular harms may not be predictable, but the likelihood of some injuries in the aggregate is fairly certain. Taking this view, the legal notion of foreseeability simply places the costs of these purportedly unforeseeable injuries on the victims rather than the manufacturer. Yet these costs can arguably be considered part of the price of doing business and, therefore, most appropriately borne by the manufacturer. Moreover, as already noted, holding manufacturers liable for damages may result in safer drugs. If a company knows that it must pay injury costs, it will have a greater incentive to discover imperfections, improve its products, and provide *effective* warnings.

As the courts have gradually expanded drug manufacturers' liability and devised alternative ways to compensate injured individuals, manufacturers and their insurers have lobbied for statutory "reforms" to limit liability. Nearly every state passed laws during the 1970s limiting malpractice claimants' access to the courts or the size of awards. Many of these laws created new hurdles for DES suits. Some states precluded liability for injuries not detectable by "state-of-the-art" testing at the time of the product's marketing, thus possibly barring recovery by many DES victims. Others shortened, or at least failed to lengthen, the statute of limitations (the time period during which lawsuits may be filed), leaving DES offspring and other victims of injuries with long latency periods with no legal recourse.[30] Until 1986, for instance, New York's statute of limitations ran from the time the injury occurred. Under that rule, a New York attorney observed, "DES daughters have to get cancer by the time they're 21. If they get it when they're 22 or 25, forget it."[31] Many states, including New York, have sought to ease the statute of limitations problem by adopting the so-called discovery rule, which allows suits to be filed within a limited number of years (e.g., one to three) from the time an injury is discovered rather than from the time it was first caused. These discovery statutes are essential for litigation of DES suits, although they by no means remove all the barriers to recovery.[32]

Breaking new legal ground

Some of the most formidable obstacles in DES cases arise from the misfit between legal doctrine and the production patterns of modern industrial society. The law traditionally requires that suits name a specific party as defendant – in this case a specific DES manufacturer – which many DES victims cannot do. Over 200 manufacturers produced DES using a standard chemical formula (although five companies accounted for roughly 90 percent of all the DES sold).[33] This formula was never patented, and DES was usually prescribed by its generic name, diethylstilbestrol. Even women who were given a brand name may not remember it now, some 15–30 years later, and the records of pharmacies or physicians that might help to track it down have often been lost or destroyed. Consequently, many DES victims have tried to persuade the courts to allow suits against producers of DES collectively.

Until 1980, they were unsuccessful. Many DES cases were summarily dismissed, while others stalled in the courts. Then, in the 1980 *Sindell* decision, the California Supreme Court introduced a novel theory, called "market-share liability," which allowed the major manufacturers of DES to be sued jointly, with damages being apportioned acording to each company's market share at the time the injury occurred.[34] This new theory was needed, the court said, in order not to deprive injured victims of legal remedy when technological advances and complex distribution systems result in the production of many identical products that cannot readily be traced to a particular producer.

The U.S. Supreme Court refused to review the drug companies' appeal of this decision, so the ruling stood and DES trials in California could proceed. By mid-1986, courts in at least four other states (New York, Michigan, Washington, and Wisconsin) had issued their own rulings obviating the need to identify a specific DES manufacturer, either by fashioning new theories of liability or by reinterpreting older theories. Although the legal reasoning varied, these decisions generally placed the burden of proof on the manufacturers to prove that their DES did *not* cause the victim's injury instead of on the victims to prove that it did.[35] The Wisconsin court explained that it chose to deviate from traditional notions of tort law in "the interests of justice and fundamental fairness." "The drug company can either insure itself against liability, absorb the damage award, or pass the cost along to the consuming public as a cost of doing business," the court noted. "We conclude that it is better to have drug companies or consumers share the cost of the injury than to place the burden solely on the innocent plaintiff. Finally, the cost of damage awards will act as an incentive for drug companies to test adequately the drugs they place on the market for general medical use."[36] Commented a prominent DES lawyer about the Michigan decision: "We are beginning to understand the time-bomb effect of some drugs and this court has shown there is a way the law can meet changing technology."[37]

This trend has ramifications for many other products besides DES and has created considerable legal turmoil. Assessments of the various rulings to date range from "an epic victory for consumers" to "a disaster for the legal system."[38] The strongest opposition, not surprisingly, has come from pharmaceutical and insurance companies, which have sponsored both state and federal legislation to overturn the rulings. In California, largely as a result of the lobbying efforts of the consumer group DES Action and allies, industry-sponsored bills were defeated in 1980 and again in 1982.[39] Federal legislation limiting liability has also been defeated in Congress, but manufacturers and insurers have continued to push for a major overhaul of federal product liability law.

Another dilemma presented by cases of large-scale injury such as DES is the inefficiency of repetitive, case-by-case adjudication of similar facts. Class action suits, in which all those with comparable injuries seek reparations in a single court action, would avoid much needless repetition. In 1979, in another pioneering DES decision, a federal judge in Massachusetts provisionally allowed a class action suit by DES daughters without cancer, and he too cited the changing social context of drug injuries. "Traditional models of litigation, pitting one plaintiff against one defendant, were not designed to, and cannot, deal with the potential for injury to numerous and geographically dispersed persons that mass marketing presents."[40] This class action was subsequently vacated, however, and most subsequent efforts by victims of DES-related injuries to initiate class action suits have been rejected for failing to meet federal requirements for class certification.[41] Some of the inefficiency of individual trials has been eliminated in New York, where the New York Court of Appeals ruled that DES daughters did not have to relitigate many of the facts that were decided in the *Bichler* case – for example, whether DES causes cancer, and whether the manufacturer should have tested for intergenerational effects.[42] This ruling set another important precedent for DES cases.

DES also presents courts with difficult questions concerning harms other than cancer. In a 1985 Massachusetts case, a DES daughter received $50,000 for fertility problems related to DES that left her unable to have children.[43] But most DES daughters have been unsuccessful in obtaining damages for vaginal adenosis, an abnormal but apparently nonmalignant condition, and for the costs of ongoing medical monitoring, which can average several hundred dollars a year (although some have gotten out-of-court settlements for such claims). If courts granted all DES mothers and their offspring compensation for medical costs, this could easily triple or quadruple total damage awards.[44] Courts sometimes award damages for future costs that can be reasonably well projected, such as lost earnings after being disabled, but they have generally been reluctant to award damages on the basis of future risks.

However the courts ultimately resolve the various legal issues raised by DES and other "toxic time bomb" cases, they are slowly reshaping traditional tort doctrine to reconcile it with changing social needs. Unfortunately, legal restrictions still prevent most DES victims, even those with cancer, from getting any relief through the courts.

Liability for immunization injuries: lessons from the swine flu program

It was in this environment of expanding and rather unpredictable liability that vaccine producers were asked, in early 1976, to provide record quantities of vaccine for the swine flu immunization program. They had reasons for balking. Suits against vaccine producers had been increasing during the 1960s, and the potential liability of the swine flu program, with its intended scope of 200 million people, was orders of magnitude greater than any prior immunization effort. Two precedent-setting cases by polio victims against vaccine manufacturers gave added reason for concern. In *Davis* v. *Wyeth* (1968) and *Reyes* v. *Wyeth* (1974), the courts had held manufacturers responsible for warning *individuals* in mass immunization clinics of known risks, not just the doctors or health agency.[45] Individual warnings were needed, the courts held, because of the increased danger of vaccine-induced injuries in the absence of the patient's regular doctor and personal medical records.[46] It did not matter to the jury how remote the risks (the chance of contracting polio in the *Reyes* case was estimated at 1 in 1.4 million) or how likely it was that the vaccine had caused the injury (in the *Reyes* case, it probably had not). The Reyes child had gotten polio, and the parents had not been warned. The jury awarded them $200,000.

For vaccine manufacturers, the law seemed to be heading beyond strict liability toward absolute liability – that is, liability for any and all vaccine-related injuries, regardless of fault, warnings issued, or foreseeability of risks.[47] Insurers feared that potential liability in an immunization effort the size of the swine flu program, could prove disastrous. So, following sound business practices, they refused to take the gamble. After all, as one study observed, the insurers "were in business to spread risk, not to take it. What they couldn't calculate could not be spread on any terms they cared for."[48] If President Ford wanted the swine flu program, the government could underwrite the risks.

This was not an altogether unreasonable position. Almost all vaccines produce a small number of adverse reactions, no matter how careful the production, testing, and administration. There is a plausible rationale for expecting the government to assume liability for these statistically rare, but sometimes serious, injuries. The government regulates vaccine development and is heavily involved in the funding and administration of most immunization programs. Moreover, society has a collective interest in widespread immunization in order to increase the resistance of the entire community, and it is arguably the government's responsibility to compensate people who are injured in pursuit of the public good.[49] Furthermore, in the swine flu program, the government had done the testing of the vaccine and claimed that side effects would be minimal.

Still, many members of Congress were convinced that the insurance industry was overreacting. To be sure, liability insurers had suffered some financial setbacks during the early 1970s, leading to a contraction in available insurance coverage and substantial rate increases. Insurance premiums for drug companies had risen 613 percent between 1971 and 1976, contributing to a sharp drop in the number of vaccine manufacturers.[50] The insurers blamed the rise in premiums on the increasing volume of claims and expanding liability. More objective observers were not so sure. A subsequent government investigation mandated by the swine flu legislation concluded that it was "not possible to determine whether the premium increases reflect the actual increase in liability exposure."[51]

One thing was clear, however: The Swine Flu Act, drafted in haste and rushed through Congress in early August, was a good deal for the insurance companies. The government took responsibility for defending against all claims and suits and assumed all liability arising from the manufacturers' newly precedented "duty to warn" individual patients. Vaccine producers were liable only for negligent injuries, a risk insurers were quite willing to bear since producers' negligence had rarely been a factor in prior vaccine litigation (nor has it been in most swine flu suits). Nevertheless, liability insurers charged the vaccine manufacturers – and ultimately taxpayers – nearly $8.7 million for insurance coverage, and ended up pocketing almost all of it.[52]

Some angry members of Congress likened these financial arrangements to corporate blackmail.[53] In reality, the problem lay at a deeper level, in the uneasy mix of private insurance and public legal justice. The financial interests of the insurance industry may not always jibe with perceptions of individual equity, or, as in the swine flu program, with collective action in the public interest. The refusal of private insurers to cover vaccine manufacturers simply exposed the latent conflict between public and private interests that underlies the entire tort-liability system.[54]

With proper planning and anticipation, the swine flu program could have furnished an ideal opportunity to design a model compensation program for a limited purpose that avoided the pitfalls of tort law – a kind of national experiment in social insurance. The stated intent of the law Congress passed was promising: The program was "to assure an orderly procedure for the prompt and equitable handling of any claim for personal injury or death arising out of the administration of [swine flu] vaccine." What emerged, however, was a hybrid approach which gained few of the advantages of social insurance while retaining all of the delays, costs, and moral ambiguities of tort law.

Windfall profits for insurers were only the beginning of the swine flu program's liability problems. According to the swine flu law, all claims had to be filed first with the government, which reviewed them administratively. Claims that were denied could then be filed as lawsuits against the U.S. Government in a federal court. There they faced several important procedural constraints: no juries, no punitive damages, and strict limits on lawyers' fees. Such restrictions were intended to discourage frivolous lawsuits and curtail the size of damage awards.[55]

Claims came in faster and in larger numbers than anyone was expecting, and soon became enmeshed in procedural complexities. In the six months following HEW Secretary Califano's announcement in 1978 that the government would accept liability for Guillain–Barré syndrome without proof of negligence, the number of Guillain–Barré claims nearly tripled.[56] Altogether, over 4,000 claims were filed, of which the government denied about three-quarters. How many of the denied claims were valid will never be known with certainty, but one can assume that a good number at least did not seem provable under legal standards of proof, as fewer than half were subsequently filed as lawsuits. Of the suits that were filed, the government won more than eight out of ten as already noted, mostly on the grounds of diagnosis or causation issues.[57] Federal judges are less likely than juries to stretch legal doctrine out of sympathy for injured victims. Although the total damages requested in the initial claims exceeded $3 billion, the amount awarded in out-of-court settlements plus court judgments came to roughly $90 million as of mid-1987, less than 3 percent of the total amount sought.[58]

Ethically, swine flu suits put the government in a ticklish position, since it was the government's own much-touted program that was being charged with causing the alleged injuries. Yet the Justice Department has not been inclined to pay off trivial claims along with the serious ones. As Jeffrey Axelrad, the government attorney heading swine flu litigation, explained, "You can't expect us to give somebody one million dollars because he was sick for a month."[59] One can just hear the cries of boondoggle, especially for claims that appear dubious. Attorneys representing swine flu victims have characterized the government's legal battles against "crippled American citizens," as one plaintiff's attorney called them, as "hard-nosed," and lacking "any real sense of humanity." Locked into an adversary position, they say, the government may spend more to defend these cases than it would cost to settle them out of court.[60] For the injured individual, taking on the Justice Department may seem a formidable, if not foolish, undertaking.

Swine flu damage awards, as in most personal injury cases, have ranged widely. In relying on the courts to resolve disputed claims, the drafters of the swine flu law in effect guaranteed that judgments would be inconsistent, varying according to the court and lawyers. The family of one man who died from vaccine-induced Guillain–Barré agreed to a settlement of $285,000, while another family received damages of $5 million.[61] In most cases, people who were injured but not killed by vaccine-related reactions have received much more than the families of those killed outright, a standard pattern in personal injury litigation. The largest injury award so far is $3.9 million, awarded to a woman who was paralyzed from the waist down after she received an inoculation. Like many other vaccine victims, she had gotten the shot because, in her words, "the president [Ford] got on TV and made it seem like my patriotic duty."[62] The government appealed this judgment, but it was upheld.

Finally, contrary to the program's stated goal, compensation of swine flu claims has been anything but "prompt." Predictably, case-by-case litigation has proven to be a slow and inefficient way to handle generic questions of fact. The government began to receive claims in 1977, and trials have been underway since 1979. Yet it will take a decade or more for all of the cases, including appeals, to be wrapped up. Meanwhile, the delays have created hardships for injured victims. They must pay medical bills and support themselves at least until their cases end (and forever if the verdict is unfavorable). They are the real losers.

The swine flu compensation program's problems – delayed awards, uneven amounts, and huge legal costs – are characteristic of tort litigation. Far from representing a model injury compensation system, the swine flu program is widely viewed as exemplifying the problems of handling vaccine-related injuries through the courts.[63] Built into its very design was the fundamental incompatibility between tort law doctrine, with its emphasis on fault and foreseeability, and the view that society has a moral duty to compensate victims of vaccine injuries for harms suffered for the sake of a collective goal. That duty is increasingly recognized both in this country and abroad. At least six nations provide publicly funded "no-fault" compensation for vaccine injuries.[64] In the United States, a few states established compensation programs in the late 1970s covering legally mandated children's immunizations.[65]

The number of companies willing to manufacture vaccines has been declining steadily since the early 1960s, a trend companies attribute in large part to their expanding liability. By the mid-1980s, with only one or two firms still producing some vaccines, concern over continued vaccine supply led to a spate of proposals to limit manufacturers' liability in various ways while also assuring injured vaccine victims of no-fault compensation. In 1986, Congress finally passed a bill establishing a no-fault national compensation system for children injured by mandatory vaccines which would be run by a special judicial panel with awards limited to a maximum of $500,000. By mid-1987, however, it had still not gone into effect because of disputes over the system's funding, and President Reagan was seeking to rewrite the law to remove the government from the compensation process.[66] Whatever the outcome, there is clearly growing public support for providing people harmed by vaccines with the option of guaranteed compensation without litigation.[67]

Injuries due to medical experimentation

Many of the arguments for no-fault compensation of vaccine injuries also apply to injuries caused by medical research. As with immunization, some harms are inevitable even with the most careful precautions. Indeed, the very purpose of research is to explore the unknown risks and benefits of new therapies, and society as a whole stands to benefit from the knowledge gained. If the subjects are healthy volunteers, they gain nothing personally, other than the satisfaction of contributing to medical knowledge. Many research subjects do hope to benefit personally, of course, such as those who are desperately ill or who have not been helped by conventional therapy. Yet even these subjects have to accept some extra risks for the sake of society's interest in gaining new knowledge: They must submit to formal research requirements that may not be in their own personal interest – as, for instance, when "control group" patients are denied access to the therapy being tested. By accepting these risks, the research subject makes a gift to the larger society.[68] Society, in turn, has a special moral obligation to compensate those who suffer serious harm.

Existing information suggests that serious research injuries are rare. In a 1976 national survey, investigators reported that among 132,615 subjects, 3 percent experienced trivial adverse effects and 0.7 percent suffered temporarily disabling injuries; fewer than 0.1 percent were permanently disabled or died. More recent, smaller-scale studies indicate similar frequencies.[69] Yet, the few people who do suffer major injury may receive neither compensation nor care, especially if they have extended medical needs. Less than half of all research institutions and private companies nationally report providing treatment or compensation for research-related injuries (although many investigators and institutions may informally provide emergency and short-term medical care when complications occur).[70] Nor are the courts a very likely source of relief for people harmed in medical experiments. There have been relatively few claims and even fewer actual lawsuits against investigators for research-related injuries.[71]

The four case studies illustrate some of the problems facing such lawsuits. Patsy Mink and two other women filed a $77.7 million lawsuit on behalf of themselves and the other thousand Dieckmann study subjects, charging the University of Chicago, where the study took place, with "conducting a medical experiment on them without their knowledge or consent," and with "intentionally [concealing] the fact of the experiment and information concerning the relationship between DES and cancer."[72] The women contended that they had not been told they were part of an experiment, and remained unaware of any danger until the university finally informed them, in 1976, of their participation and the possible risks of DES – some five years after the link with clear-cell cancer had been discovered. They sought compensation for "severe mental anxiety and emotional distress due to the increased risk to their children of contracting cancer and other abnormalities," although neither they nor their children had yet suffered any direct physical harm.

Their class action suit was denied, and the court refused to award damages for emotional suffering or the increased health risks of DES offspring. The court did allow the women to pursue the charge of lack of informed consent, and litigation dragged on for five years. The three DES mothers finally won a bittersweet victory in 1982, when the university agreed, in an out-of-court settlement, to pay the women $75,000 each and to provide free medical care for the children of study participants – quite a step down from the original $77.7 million suit, but better than what most DES victims get. Was it adequate compensation for the trauma of being an unwitting human guinea pig and the subsequent risks and uncertainties? "There is no amount of money that can ever pay me back," said Patsy Mink. "The anxieties really relate to the future. We just don't know . . ."[73] Of course, money can never really make any injured victim "whole" again; it can only repay dollar losses and offer some financial compensation for non-financial harms. What such harms are worth, whether emotional or physical, is always a matter of arbitrary judgment.

Even when medical experiments lead to major injuries or death, lawsuits may be unsuccessful for various reasons. Mrs. Karp's suit against Dr. Cooley, in which she charged that her husband had not been given adequate information about the risks of the artificial heart, failed mainly because she could not obtain the necessary expert testimony. Many research subjects may be reluctant to bring suit because, if they were already sick, it may be difficult or impossible to prove that the therapy rather than the underlying illness caused an adverse event. Many therapies for life-threatening conditions entail extremely serious known risks which are tolerated in hopes of an overall improvement. Unless the investigator has been grossly negligent, recovery through the courts is unlikely.

During the 1970s, no less than five separate governmental review bodies urged some form of no-fault compensation for people injured by medical research.[74] Most comprehensive was the report of a 1977 HEW Task Force, which recommended that institutions provide compensation for "physical, psychological, or social injury proximately caused by [government conducted or sponsored] research" if the harm "on balance exceeds that reasonably associated with [the subject's] illness" and usual forms of treatment.[75] The amount of compensation should be "commensurate with the 'excess' injury," defined as indicated in the report, and should be available to all research subjects, including those who were sick and hoped to benefit from the experimental procedure as well as healthy volunteers.

The 1977 Task Force recommendations, although never enacted, were influential, and in 1978, HEW announced that research institutions would henceforth be required to tell prospective research subjects whether or not compensation was available for research-related injuries.[76] Many saw this as a move to embarrass institutions into offering compensation, and predicted a variety of dire consequences. Yet the regulation went into effect and survived a significant reduction in federal research regulations two years later.[77] Its impact remains uncertain, however, as no data have subsequently been collected on the number of institutions providing compensation.[78] It may have led institutions to distinguish more sharply between negligent and nonnegligent injuries, providing reimbursement where they might be legally liable but explicitly disavowing responsibility for nonnegligent consequences.[79]

The 1982 report of the President's Commission for the Study of Ethical Problems in Medicine and Biomedical and Behavioral Research backed away somewhat from the position taken by the 1977 Task Force. Although agreeing it would be "ethically desirable" to compensate people harmed by medical research, the President's Commission concluded that the federal government did not have "an ethical obligation to establish or to require a formal compensation program." The commission questioned the need for such a program as well as its potential cost, noting that current limitations on federal funding for biomedical research offer "more than the usual reasons to question any suggested expenditures in new areas." If compensation were to be provided, the commission recommended limiting coverage primarily to "nontherapeutic" research, in which individual subjects were not expected to benefit personally.[80]

Both the President's Commission and the 1977 Task Force excluded injuries caused by "innovative therapy" – informal clinical experimentation that is not part of a formal research project. They reasoned that such departures from standard practice were intended primarily to benefit the patient rather than to obtain new knowledge for society; hence society had no obligation to compensate the patient for unforeseen problems. Yet in cases where "innovative therapy" really amounts to therapeutic research, as when physicians began using DES as a morning-after pill contraceptive in the mid-1960s, patients may be exposed to greater risks than in formal clinical experiments. For one thing, they are deprived of federally mandated research safeguards like informed consent and review by institutional ethics committees. Furthermore, without standardized research procedures, treatment results are inevitably more difficult to compare and interpret – and the risks correspondingly more difficult to justify. Innovation in the absence of good experimental methods is apt to be poor research – "fooling around with people," in the words of three noted evaluation experts.[81] In such situations, the unwitting subjects of informal medical experimentation would seem to be at least as deserving of compensation as those injured during formal research. Of course, given the wide variability of regular medical practice, it may often be difficult to distinguish informal clinical experimentation from standard care. Adding injuries of innovative therapy would clearly expand the scope of any compensation program, but might be feasible within a no-fault system.

Because society as a whole benefits from medical research, the President's Commission and its predecessors recommended that the costs of compensating research injuries be borne largely by the federal government. Some analysts have argued that these costs should also be shared by researchers and research institutions in order to motivate them to scrutinize risks more carefully and reduce the incidence of injuries.[82] But researchers might then have as much incentive to avoid admitting responsibility for injuries as to avoid causing them. Deterrence and compensation, unless uncoupled, are generally at odds.[83]

Other countries provide compensation for research injuries in many different ways, though few rely on local institutions as a major source of funding. Sweden has had a national insurance program for people injured by biomedical research since the mid-1970s. In 1976, West Germany passed legislation requiring all firms conducting pharmaceutical research to contribute to a central fund that is used to compensate anyone injured in clinical investigations.[84] In 1983 the British Pharmaceutical Industry Association urged member companies to compensate patients injured in drug trials without regard to negligence, and in 1984, a royal commission in Great Britain recommended "strict liability" for medical research injuries.[85] Likewise, the World Health Organization has considered guidelines that would entitle all research participants to compensation.

Despite their diversity, these various programs suggest that no-fault compensation for research injuries is not unduly expensive, does not stimulate excessive claims, and is infrequently abused. Although similar schemes in this country might not be equally successful, the evidence from American research institutions with compensation programs is encouraging. Based on studies of two such institutions, the President's Commission concluded that their compensation programs did "not stimulate excessive or unmerited claims of injury."[86]

New problems, new solutions

Tort law was conceived for simple situations in which one person's act injured another, as with a punch in the nose, or leaving a sidewalk slippery. But now we live in a complex technological world where advances in science and industry have pushed the legal system to the breaking point, leaving some industries and institutions unable to obtain liability insurance at prices they can afford.[87] Courts are poorly equipped to resolve scientific disputes, and traditional legal notions of fault and responsibility are often irrelevant to the inevitable hazards of medical progress. Significant changes have already occurred as the law adapts to the needs of modern society, and a great many more, including sweeping reforms, have been proposed. Not surprisingly, there is growing interest in the way other countries handle compensation and legal liability.

Joint litigation

One problem is the sheer volume of litigation. Cases involving "mass torts," especially long-term injuries due to drugs and toxic substances, threaten to overwhelm the already overburdened courts. So many "long-cause" cases, including DES, were awaiting trial in the San Francisco Superior Court in 1984 that the presiding judge warned that the civil courts might literally be forced to close down.[88] Several methods of consolidating similar cases have been devised which allow at least some aspects to be handled jointly. Class action suits are one approach which, although still the exception in product liability cases, has been attempted in suits involving DES and various toxic substances. Courts have adopted several other hybrid efficiency measures as well in some DES and other mass tort cases, deciding common issues in one "master trial" and leaving only the specifics for individual trials.[89]

Also increasingly common are mass settlement funds: lump-sum, out-of-court settlements by defendant companies to be split among multiple plaintiffs. The largest fund as of 1987 was the $2.5 billion settlement offered by the Manville Corporation to settle asbestos claims. Smaller, but still sizable, funds have also been set up to settle Dalkon Shield and Agent Orange cases.[90] These settlements avoid the enormous legal costs of protracted trials, but require controversial decisions about which injuries are compensable and how to divide the money among deserving claims. They involve a trade-off for both sides. For manufacturers, they resolve massive product liability suits in a quick, if expensive, stroke; for injured victims, they allow more people to be compensated more efficiently, but, if additional claims emerge, the original settlement may prove to be unfairly low. In the Agent Orange case, the number of claims grew from 5,000 initially to over 200,000 a year later.[91] In contrast, some legal experts say that the Manville settlement may have been larger than the company would have had to pay if it had fought each claim separately in the courts. DES manufacturers have also offered mass settlements in some cases. In Michigan, a group of drug companies settled

240 DES daughter cases, most not involving cancer, for an undisclosed sum.[92]

Toxic torts

Joint litigation and settlement make sense in mass tort cases. Handling large-scale injuries in the traditional case-by-case manner is neither efficient nor always appropriate. The traditional tort law definition of causation, in which a particular action is shown to lead to a specific harm, is less suitable for cases involving large numbers of people, especially when the harms may not be immediately apparent.[93] In so-called toxic tort cases – litigation over the actual or feared dangers of things like asbestos, radiation, pesticides, and toxic chemicals – causative mechanisms are often ambiguous and complex. The high rate of association between Guillain–Barré syndrome and swine flu vaccine – about six times that observed with previous vaccines – suggests causation, for example, but scientists do not know why, or what other factors– say, an earlier illness – might also be involved. With large-scale epidemiological data, associations can often be seen, whereas causation is impossible to prove in an individual case. But probabilities may or may not satisfy legal standards of proof. Incomplete evidence and unresolved scientific questions can lead to additional uncertainty.

Developments in toxic tort cases point the way toward a new framework for injury compensation which, according to some legal authorities, may gradually replace present tort litigation.[94] The courts are apt to play a minimal role in such a framework, with claims being handled instead through some kind of administrative mechanism. Possible models exist, ranging from government-funded programs like the Black Lung Act of 1969 for disabled coal miners, to industry-financed mass settlement funds, to jointly financed programs like the "Superfund" program created by Congress in 1980 to clean up hazardous waste sites (although Superfund does not include compensation for personal injuries).[95] Handling claims through a single administrative agency rather than in separate court decisions facilitates collection of aggregate statistical data for establishing causation, the central problem in both toxic torts and most large-scale medical injuries.

An innovative way of handling the causation issue was proposed for compensating atomic testing fallout victims, and could conceivably be applied in other situations where causal links are tenuous. In this scheme, people with fallout-related cancers are entitled to compensation for actual damages, prorated by the estimated probability that their cancer was caused by radiation exposure.[96] Although this "probability-of-causation" approach rests on a number of statistical assumptions, it may nevertheless provide the fairest basis for awarding damages in many toxic tort and medical injury cases, where direct proof of causation is impossible. Like mass settlement funds, it should allow more deserving victims to be compensated, even though each one might receive only partial damages. This scheme was given an additional boost when a federal judge in Utah, relying on epidemiological evidence, ruled that government negligence was responsible for causing cancer among citizens exposed to nuclear fallout. Although the ruling technically involved only ten cancer cases, its acceptance of statistical evidence of causation was expected to have far-reaching legal and economic repercussions.[97]

Broader manufacturer liability

Another important trend has been the gradual expansion of manufacturers' legal liability. The primary rationale for this trend has been equitable allocation of injury costs. An injury can be financially devastating to an individual, whereas the manufacturer can spread the costs over all the product's users by raising prices to cover the expense of liability insurance. A spokesman for Eli Lilly, for example, commented that the company was well enough insured that the hundreds of DES lawsuits filed against Lilly would "not have an adverse effect on the company's financial condition."[98] Even companies that have declared bankruptcy because of lawsuits, such as the Manville Corporation and A. H. Robins, the maker of Dalkon Shields, have emerged, reorganized and refortified, and continued to operate. The chief executive of Robins declared that the company's reorganization "will benefit all concerned – the company, its employees and shareholders, creditors and customers, as well as those wishing to assert Dalkon Shield claims."[99]

Broader manufacturer liability should also, in theory, encourage greater caution in the introduction and promotion of drugs and new medical products.[100] The manufacturer has the most knowledge about the product, controls research and testing, and determines the nature of warnings and sales efforts, and so is in a position to minimize risks. If the financial liability is large enough, it should give companies an incentive to study potential adverse effects more carefully before marketing, to develop drugs and other products with the widest margin of benefits over risks, and to educate physicians in *appropriate* prescribing rather than simply promoting sales. (The liability would have to be greater than the cost of safety precautions and any revenues lost, however, as the Ford Pinto case made clear.)

Yet the continuing expansion of manufacturers' liability is not without problems. Fears have been expressed that it will lead to an excessive rise in drug prices, or stifle research, robbing consumers of valuable new drugs or delaying their availability. These fears reached a peak in the mid-1970s, with the "crisis" in malpractice insurance, and again in the mid-1980s, when virtually all liability insurance rates skyrocketed and some industries and cities were left without coverage. In part because of such rate increases, the majority of vaccine manufacturers stopped production during the 1970s. Insurance companies have blamed the precipitous increases in premium costs on the growing volume and size of personal injury damage awards, and accuse lawyers of manipulating juries to win huge judgments – and fat contingency fees. Industry arguments are difficult to assess, as detailed financial data on insurance companies – the only major financial service industry that is not federally regulated – are often not available. Trial lawyers and consumer groups accuse the insurance industry of raising premiums to cover previous unsound investment practices and then lobbying for liability "reforms" that would limit recovery by injured victims and erode corporate accountability for unsafe or harmful products.

Despite several major government investigations stressing the inadequacy of compensation for injured victims, nearly every state passed laws limiting malpractice claims during the 1970s, as noted previously. The 1980s brought a similar wave of limitations in product liability laws. At least 1,400 "liability reform" bills were submitted by state legislators during the first half of 1986 alone, and more than forty states passed laws that, variously, limited the size of court awards and attorneys' fees, reduced recovery for noneconomic damages such as pain and suffering, restricted liability to fault-based and foreseeable injuries, and made it harder to recover full damages from a party only partly at fault. By 1987, however, the liability insurance crisis seemed to be ending as abruptly as it began, although many of the rate increases remained in effect.[101]

Restrictions on tort law recovery may ease manufacturers' liability problems, but they will only exacerbate the problems of injured victims. On the other hand, the expansion of manufacturers' liability cannot continue indefinitely. At some point, it makes sense to consider more fundamental kinds of legal reform.

Social insurance

One such reform is social insurance. Some countries, such as New Zealand and Sweden, offer social insurance for a broad range of injuries, including those caused by medical products and treatment.[102] In these systems, fault and other principles of tort law play no role. All injuries falling within the scope of the system are compensated. The governments pay the costs, since society as a whole benefits from the risky activities that harm an unlucky few. In the words of a New Zealand commission: "The toll of personal injury is one of the disastrous incidents of social progress, and the statistically inevitable victims are entitled to receive a coordinated response from the nation as a whole."[103]

New Zealand's program, instituted in 1974, includes all persons who suffer any kind of "personal injury by accident" and is funded from tax revenues. Reimbursement covers medical expenses, partial lost wages, pain and suffering, and dependents' benefits. Although the chief goal is compensation rather than deterrence, the collection of centralized data on injuries facilitates other forms of prevention. So far, New Zealand's plan has been judged quite successful by most observers. Program beneficiaries are generally satisfied, claims have been processed fairly rapidly, and the need for lawyers has been minimal. There has been some difficulty in identifying "accidents" caused by drugs or medical treatment, but at least partial solutions seem to be evolving. Most other problems involve technicalities rather than the program's basic principles.[104]

The experiences of New Zealand and Sweden suggest that, under certain conditions, government-funded no-fault insurance can be both more equitable and also more efficient than tort litigation. Whether such an approach would work in the United States is unclear. Two of the largest government insurance programs, the state-run workers' compensation system and the federal Social

Security Disability Insurance program, have been widely criticized for a variety of shortcomings, including inadequate benefits, inordinate delays, and inconsistent and overlapping coverage.[105] A more comprehensive system, similar to New Zealand's, might eliminate some of these problems. With a unified plan, people would look to a single agency for relief instead of having to deal with a bewildering array of separate programs. The need for personal injury lawsuits would be greatly reduced, eliminating expensive legal fees and channeling a substantially greater fraction of the total damages to victims. The major risk of such a program, if workers' compensation is any indication, is that benefit levels would be inadequate. Government-run programs are always vulnerable to reductions in scope and coverage during times of fiscal austerity; the courts are less susceptible to the lobbying of special interest groups. It is the courts, after all, that have been moving steadily toward more generous compensation over the past several decades, while state legislatures, under pressure from manufacturers and insurers, have moved in the opposite direction.

Private no-fault insurance is another alternative. No-fault auto insurance laws were passed in twenty-four states during the early 1970s, and have apparently functioned very well. A major review in 1977 by the U.S. Department of Transportation concluded: "No-fault plans of sharply varying objectives and character are widely seen as successes. No problem has arisen in the implementation of no-fault for which there does not appear to be a readily available and feasible solution, given the political will to make the necessary change. No-fault insurance works."[106] Analogous no-fault schemes have been proposed for medical providers and product manufacturers. In most of these plans, professionals or manufacturers would purchase no-fault insurance covering injuries resulting from their products or services. Some plans would be strictly elective, others would be implemented through legislation. In general, these schemes would avoid, or at least reduce, costly litigation over fault determinations, and would limit providers' or manufacturers' liability in return for guaranteed compensation of all accident victims. Most plans would reimburse only for verifiable losses – medical expenses and income losses – and not for "pain and suffering," which is hard to measure in dollars and often extremely expensive. For medically related injuries, the primary problem would be the usual difficulty in distinguishing compensable "injuries" from the underlying disease.[107]

Social insurance with liability funding

Social insurance systems are concerned mainly with compensation, leaving prevention of future injuries to regulatory agencies. Certainly for injuries that "blend uncertainty and untraceability," like those caused by DES and some toxic chemicals, governmental regulation is essential for public protection, regardless of any financial incentives created by legal liability.[108] But no-fault insurance is not necessarily incompatible with financial deterrence, providing the two functions are kept separate. For example, with elective no-fault insurance for health providers, patients would be guaranteed compensation for specified injuries regardless of fault, while the premiums paid by providers could be based on damage awards paid out, giving providers an incentive to prevent injuries. Some no-fault plans would retain an element of financial deterrence by allowing no-fault insurers to collect from other insurance companies for injuries involving fault.[109]

Another intriguing model that combines social insurance with liability funding is Japan's system of compensation for pollution-related health injury. In this system, established in 1973, the government identifies "pollution-induced" diseases, thereby assuming the considerable burden of proving causation. It also designates official pollution zones, and certifies that victims in fact have one of the diseases eligible for compensation. The polluting industries are charged "pollution levies" based on the volume of health-damaging pollutants they emit, which are collected by the government and used to pay compensation claims. This system offers the advantage of guaranteed reimbursement to injured victims, while giving companies a direct financial incentive to reduce injuries by limiting toxic substance emissions.[110] Although rising pollution levies have led to mounting opposition from Japanese industry, victims of pollution-related disease still have a much better chance of obtaining at least some recompense through this system than under tort law.[111]

In principle, the idea of taxing a manufacturer to raise revenues for those injured by its products could be applied to drugs or medical devices. Determining which injuries to reimburse would undoubtedly be difficult and to some degree arbitrary, although for large-scale injuries, harms could often be identified more readily in aggregate statistical data than in individual cases. Although the financing of such a system would promote deterrence, its central goal would be to provide quick, efficient, and equitable no-fault compensation to injured victims. In exchange for the large payments for pain and suffering that tort litigation sometimes provides, victims would be assured of greater certainty of recovery, without long delays and exorbitant legal costs. Public opinion studies suggest that most people would opt for such a trade-off. In Michigan, after experience with the most comprehensive no-fault auto insurance law in the nation, citizens said, by a 53 to 18 percent margin, that they would relinquish their own rights to sue for payment for pain and suffering "in exchange for prompt and complete payment of medical bills and lost wages."[12]

The notion of combining social insurance with liability funding is not new, although it has yet to find much acceptance in the United States outside academic circles. A number of legal scholars have advocated this general approach, and have described in some detail schemes for providing no-fault compensation for medical injuries or other accidents that would also incorporate elements of deterrence.[13] Most of these plans would be run by the government and financed with a combination of general tax revenues and contributions from manufacturers or providers. One proposal outlines a compensation fund for DES victims, to be administered by a state agency and financed primarily by DES manufacturers based on their market share. In effect, this fund would provide comprehensive medical insurance for victims, reimbursing all medical expenses resulting from DES exposure, including both past losses and any future problems that might emerge (although not pain and suffering). This would solve one of the major problems DES victims face in tort litigation, given the reluctance of courts to grant damages for potential future injuries. Moreover, it would force drug manufacturers to bear "the cost of suffering inflicted by their product."[14] DES victims are not alone in deserving swift and dependable compensation when they suffer harm at the hands of the healing enterprise. Similar principles could, and should, be extended to all forms of medical misadventure.

Conclusion

In combatting the ravages of disease and disability, medical advances entail their own risks: harms caused by human interventions – drugs, medical procedures, new technologies – rather than by natural disasters or acts of fate. Many of these injuries are neither negligent nor malicious; they are simply an unavoidable part of progress through science and technology. It seems unjust to expect individuals to bear these costs without compensation, since they are often unaware of the risks they are taking and have little voice in deciding societal policies on innovation.

This is not to deny the importance of providing needed medical care and economic compensation for *all* people who suffer suffer illness or accidents, whatever the cause. The personal suffering and medical needs of the woman whose cancer has no identifiable cause are no less deserving of appropriate care than those of the DES daughter. Ideally, our society would assure all citizens of reasonable health and disability benefits when they are sick or injured. But there is an additional reason for compensating victims of human intervention. Society as a whole has a stake in assigning to those responsible the costs of injuries inflicted through human action, since the resulting deterrent effects theoretically produce a more socially efficient, as well as equitable, allocation of resources.

There has in fact been a trend worldwide toward compensating the victims of medical and technological injury regardless of fault. Determining causation remains the most redoubtable obstacle in all programs that reimburse only injuries of a particular origin, such as vaccine-induced injuries or those due to medical research. This problem can be avoided only by providing comprehensive health and disability benefits for all people suffering illness, regardless of cause or of who, if anyone, was at fault. On the other hand, determining causation cannot be avoided in any system that seeks to deter injuries by assigning their costs to those responsible for them.

With appropriate safeguards and adequate public accountability, no-fault compensation run by the government and funded at least in part by medical manufacturers or providers might, ideally, allow more equitable reparations for injured victims, while separate aggregate decisions about cost allocation would preserve the deterrent effect of financial responsibility. Such an approach would avoid the costly court battles of the tort system and assure injured individuals of reasonable, albeit more modest, compensation. For such a system to be fair, however, the determination of compensation awards would have to be protected against political and economic pressure from manufacturers or professionals seeking to avoid liability. If this could be done, the results might ultimately prove both more just and less expensive for many types of medical injury than the present tort system.

That the cost of injuries caused by medical and technological innovation should be borne by those who promote or profit from them is rather widely accepted in theory. The challenge is translating this principle into practice.

9

What is fair? Medical innovation and justice

In October 1981, Dale Lot, a thirty-seven-year-old veteran working as a fireman in Florida, was forced to quit because of severe congestive heart failure. His case was terminal, doctors told him, and transplant surgery was impossible. The artificial heart was his only hope. He thus began a desperate – and highly publicized – campaign to persuade the University of Utah to let him be their first "human guinea pig." His appeal, backed by the fire fighters' union and orchestrated by a flamboyant trial attorney, included repeated attacks on the "murdering bureaucrats" at Utah who were blocking his request, a barrage of publicity, and finally a public telegram to the White House. The University of Utah was already trying to broaden patient eligibility, and Lot's campaign probably hastened their efforts. But Lot finally abandoned his personal quest for an artificial heart when it became clear that he would be refused on the grounds that his heart condition was accompanied by diabetes and hypertension, and was linked to long-term alcoholism.

Dr. Barney Clark, the man finally selected as the first recipient, was, by most accounts at the time, an ideal candidate. He was plucky, and he had strong support from his wife and children. He seemed, in the words of a University of Utah social worker, "a classic choice."[1] Yet he was sixty-one years old, considerably older than Lot and many other applicants (too old for a heart transplant in fact) and his longtime two-pack-a-day smoking habit had seriously damaged his lungs as well as his heart. Six months after his death, several members of the Utah team were reported to consider Clark a "poor choice" because of his lung condition.[2] In turn, the experiment may not have been such a good choice for Barney Clark. Dale Lot, denied the heart, survived for nearly a year after Clark's death, despite doctors' earlier predictions that he had but days or weeks to live.[3]

The bill for Barney Clark's implant ran well over $250,000, and the University of Utah subsidized much of his care. Medicare – that is, federal taxpayers – paid more than $100,000 of Clark's bill.[4] Humana, Inc. Hospital, where four of the subsequent implants were performed, promised to subsidize 100 implants. Eventually, however, patients will presumably be expected to pay for artificial hearts themselves, unless the government were to provide coverage, which at present seems unlikely. *Should* government programs cover the latest treatments and technologies, or has this goal now become economically prohibitive?

This is only one of the many perplexing ethical questions raised by modern medicine's extraordinary scientific and technological achievements. With steadily expanding technical capabilities for medical intervention, society faces unprecedented decisions about when to use these new powers, and how to pay for them. Escalating medical costs have led to major cutbacks in publicly funded programs providing basic care for many of the disadvantaged, while high-cost new technologies continue to proliferate. Federal and state agencies grapple daily with poignant pleas for reimbursement of expensive therapies, which are sometimes patients' last hope. Even private companies are being drawn into these painful choices through employee benefit plans.

Here we consider these issues from the standpoint of justice – what is fair for individuals and also for society as a whole. We begin with the selection of candidates for scarce and potentially life-saving technologies. Because scarcity depends, at least in part, on the resources invested, the chapter then examines the relationship between medical resources and health needs, and the consequences of the private sector's expanding role in medical research and health care delivery. Medical innovation appears to be on a collision course with cost containment. Present responses to this problem – reductions in care for the disadvantaged, particularly disadvantaged children – are incompatible with any reasonable notion of justice, and are an inefficient use of society's resources.

What is "justice" in medical innovation?

Since Aristotle, equality, treating like cases alike, has been the central principle of social justice. But "alike" in what respects? There are many possible answers. For most goods and services, we accept merit or social worth, as reflected in income, as a tolerably "just" basis of distribution. Health care, however, is often viewed differently. Egalitarian theories of justice generally hold that health care should be distributed according to medical need, not merit.[5] For example, philosopher Robert Veatch argues that every human being should have an equal claim on health – a "right," in effect, to equal health. That claim, he contends, entitles people to "the amount of health care needed to provide a level of health equal, insofar as possible, to other persons' health."[6]

Philosopher Norman Daniels offers a compelling argument for trying to equalize health.[7] For social worth to be a fair basis for the allotment of other goods and services, Daniels points out, everyone must have an equal opportunity for self-development and accomplishment, the preconditions for achieving social worth. That requires equal health. A just distribution of health care therefore cannot assume a fundamental background of equality of opportunity but rather must work to promote it. In short, adequate health care is essential to ensure that people have fair equality of opportunity in all of the other arenas of life. Other philosophers have argued, similarly, that medical resources should be distributed according to need in order to maximize personal autonomy, which depends on decent health.[8]

The notion of a right to health appears not only in philosophical theory but in various governmental documents and declarations. For example, in passing the Comprehensive Health Planning Act of 1966, Congress proclaimed rather grandly that the "fulfillment of our national purpose depends on promoting and assuring the highest level of health attainable for every person, in an environment which contributes positively to healthful individual and family living . . . "[9] The preamble to the World Health Organization's (WHO) constitution states: "The enjoyment of the highest attainable standard of health is one of the fundamental rights of every human being without distinction of race, religion, political belief, economic or social condition."[10] In 1978, the 134 member countries of WHO reiterated this right, pledging support for a global effort to ensure "Health for all by the year 2000."[11]

Obviously, such statements speak of ideals, not reality. Neither in the United States nor elsewhere in the world is the population's health status the highest attainable, although inequalities are less in some European countries than in the United States.[12] Indeed there is no realistic way to ensure equal health. But it is still useful to consider equal health as a goal, since this helps to define society's responsibilities in distributing medical resources.[13] Medical care alone cannot eliminate existing inequalities in health; unequal social and economic circumstances are a major cause of differential morbidity and mortality.[14] Nevertheless, medical care can help to redress such inequalities. To do so, resources must be distributed according to medical need, however measured. To promote greater equality of health, therefore, access to care should be unequal, with the amount of care based on the extent of medical need.

That was exactly what the 1978 National Commission for the Protection of Human Subjects, a blue-ribbon advisory panel on medical ethics, recommended. "Justice," the commission stated, "requires that both access to health services and the costs of these services be fairly distributed and that those most in need receive the most benefits."[15] The National Commission's successor, the 1982 President's Commission for the Study of Ethical Problems in Medicine and Biomedical and Behavioral Research, was somewhat more circumspect, recommending "equitable access to health care" for all Americans, defined as "an adequate level of care without excessive burdens."[16] The President's Commission pointed out that the concept of "medical need" was both difficult to measure and subject to varying interpretations. However, there are numerous ways to quantify health needs which, although not without limitations, could help guide health priorities.[17] Many ethical principles are extremely difficult to act upon, yet this does not diminish their value as moral precepts.

Were equality of health the goal, resources would be directed toward groups with a disproportionate share of disease, disability, and early death. By most indicators, medical needs tend to be greatest among traditionally disadvantaged poor and minority groups. Such groups have higher rates of chronic and severe acute illness, more disability, higher age-adjusted death and hospitalization rates, and shorter average life expectancies. Chronic illness takes a particularly heavy toll on the disadvantaged. Heart disease, the nation's leading cause of death, is nearly three times as common among the poor as the affluent (11.4% vs. 3.5%).[18] Age-adjusted death rates from heart disease are also higher among the poor and minorities, as are death rates from cancer, stroke, and diabetes (all among the top five causes of death). Restricted activity and other measures of disability reveal comparable disparities. An egalitarian distribution of medical resources would ensure that certain basic needs of all individuals were met, and then would allocate the remaining resources so as to bring everyone up to the highest level of health possible.

The total amount of society's resources that should be devoted to health is a social judgment. Surveys indicate that the majority of the public would like government spending on health to increase, and only 21 percent of the population would restrict opportunities to use expensive technology.[19] Based on present trends, some analysts predict tht health care expenditures will consume as much as 15 percent of the GNP by the year 2000.[20] Still, public and private spending on health cannot continue to rise indefinitely. At some point, resources are better used for other purposes.

As we approach that limit, wherever it is set, it becomes increasingly important to be sure we are getting a reasonable benefit from our health care dollars – that the resources invested are "cost-effective." Resources should go not simply to people in need, but to those who are also likely to be helped by them. Without considering the potential fruitfulness, or cost-effectiveness, of the care provided, trying to equalize health could mean channeling the bulk of resources to people who are desperately ill but have little chance of recovery – for instance, people on their deathbeds or with irreparable damage to several organ systems. (This is in fact exactly what is already happening, as we shall see.) In practice, the fruitfulness or cost-effectiveness of medical intervention, like need, is often difficult to measure precisely. Yet even crude comparisons of current health care investments reveal large differences in cost-effectiveness – differences that are both inequitable and inefficient. If our limited health resources are to be used wisely as well as justly, priorities must reflect cost-effectiveness as well as need.

To achieve the highest, and most equal, level of health possible, prevention of disease must be medicine's paramount aim, and maintenance of people with debilitating disease the lowest priority.[21] There are many different ways to prevent disease, of course, including nonmedical as well as medical actions. Prevention can also occur at different stages of a disease: *Primary* prevention seeks to avoid the occurrence of the disease entirely; *secondary* prevention involves early detection and intervention; and *tertiary* prevention attempts to arrest deterioration and reduce damage from an established condition through rehabilitation. To minimize the amount of unnecessary illness, prevention efforts, like curative medicine, must be targeted toward populations most vulnerable to disease and disability as early in life as possible.

It has been argued that spending money on prevention is morally indefensible as long as there are people suffering and dying.[22] By that argument, carried to its extreme, we would never spend *any* money on prevention. This is a classic example of what Garrett Hardin called the "tragedy of the commons."[23] Hardin drew an analogy with a group of herdsmen sharing a cattle pasture. Up to a point, each herdsman could add more animals without harming the common grazing land, but as the number of cattle approached the land's capacity, adding further animals, although still in the interest of each individual herdsman, would finally destroy the commons. Hardin concluded that the clash between individual freedom and collective interests was insoluble. "Freedom in a commons," he wrote, "brings ruin to all." Similarly, with limited medical resources, giving moral priority to each desperately sick person's need for curative medical care denies the entire society, including that person, the chance of avoiding the illness altogether.

In an egalitarian health care system, medical and scientific research would also concentrate on disease prevention. The most common and disabling illnesses would receive the most attention from researchers, with the problems of the disadvantaged naturally having high priority in order to enhance total well-being. The products of publicly funded research, and ideally private research as well, would be made available to those for whom they could do the greatest good.

Skeptics may wonder what relevance these abstract goals and ideals have in the present context of hoped-for medical miracles and skyrocketing medical costs. It is at just such times, when pressures to compromise moral principles are greatest, that a little sober reflection is most important. In the case studies, ethical considerations were more often implicit than explicit. Most of the key figures in each case clearly believed they were "doing good," but this belief was seldom expressed in terms of justice per se or subjected to the critical scrutiny of those with divergent views. It was precisely these unexamined good intentions, in fact, that helped create many of the ethical dilemmas we now confront.

Access to scarce resources: Are some more equal than others?

Traditionally, medical resources have been distributed through impersonal mechanisms such as the market, or simply through local availability and impediments to access. What many new medical advances force us to do, by contrast, is to make distributive decisions knowingly and deliberately – to "ration" health care by design rather than default. Increasingly, we are faced with specific choices about who should receive sophisticated technologies like the artificial heart, given the extraordinary costs involved, or other heroic measures such as organ transplants, where access is severely limited by the supply of donor organs. The resulting allocation may be no worse in moral terms than that produced by impersonal mechanisms. However, merely having to make explicit choices – asking people to decide which of their fellow humans deserve to live and which will be condemned to die – creates an ethical dilemma we have historically been loath to confront.

Deciding who should have access to a scarce and costly new procedure is inevitably controversial, especially when the choices involve nonmedical criteria. Many artificial heart candidates have been rejected for psychological or social reasons, for example. One 43-year-old man was judged unsuitable for the artificial heart because, although then married, he had been divorced several times, was known to have had episodes of alcohol abuse, and was not himself sure of the appropriateness of his candidacy.[24] Another applicant was refused because his family, a wife and four daughters, did not much care whether he lived or died, according to the University of Utah's operating team. Although the man himself desperately wanted to live, the Utah team felt that his family's indifference could prove detrimental in the long and arduous recovery process following implantation.[25] Were such fears valid? "I have never seen an adult bull visit one of our calves," Dr. Robert Jarvik commented later, referring to the University of Utah's animal research program on the artificial heart, "and yet the calves do very well. They have no family support."[26] He argued that people should not be disqualified for implants simply because they have no families, when no one really knows just how important family support is in the recovery process.

A high-level "Working Group," appointed in 1985 to advise NIH on the artificial heart, endorsed the use of various nonmedical criteria in selecting candidates, including psychological and social factors such as the candidate's intelligence, "ego-strength," and familial and personal relationships. The group noted that these judgments might weigh against candidates with life-styles different from those of the people making the decisions, but concluded, somewhat lamely, that such "suitability criteria" were reasonable "as long as the possible sources of bias are recognized and guarded against."[27] Choices that seem reasonable to decisionmakers, of course, may not seem so to others. A member of Stanford Hospital's heart transplant team acknowledged that patients rejected because of psychosocial factors, say, a young man estranged from family and friends with a poor history of medical compliance, sometimes arouse indignation: "You mean to tell me you don't have families that can adopt and take care of someone like this poor young man? He's had a terrible life, and now you are going to let him die because of it."[28]

Such accusations are reminiscent of the early days of kidney dialysis, when life and death were judged in part by social worth. The technology for dialysis was perfected in a Seattle, Washington, hospital in 1960, and brought new hope to patients whose kidneys had failed. But there were only a few machines initially, and the hospital had to decide which patients would be allowed to use them – that is, quite literally, who would live and who would die.[29] One of the simplest ways would have been to allow the market to determine the price of dialysis, by the laws of supply and demand. This would have favored the wealthy, and the hospital found it unacceptable.[30] Several other more egalitarian approaches were tried, including first-come, first-served, but these too proved unsatisfactory. Finally, the hospital decided to set up a system by which society, not just medical professionals, would share in the burden of choice. The final selection of patients to receive dialysis, from all those judged medically suitable by physicians, was the responsibility of what came to be called the "Seattle God Committee."

Members of this committee were appointed by the County Medical Society, and served anonymously, without pay. The committee tried to balance many factors in selecting candidates for dialysis, including the patient's social and economic circumstances, psychological stability, and future potential.[31] Social contributions were also valued. One observer commented: "To have a record of public service is a help – scout leader, Sunday School teacher, Red Cross volunteer."[32] This meant that people whose interests fit the normal social stereotypes were given priority. "This rules out creative nonconformists," critics complained, ". . . the Pacific Northwest is no place for a Henry David Thoreau with bad kidneys."[33] Amid mounting criticism and fraught with increasing guilt, the committee disbanded with relief in 1967 when treatment facilities had expanded enough to accommodate most of the patients judged medically suitable.

The "Seattle God Committee" was a courageous attempt to get a group of lay citizens to resolve the dilemma of access to scarce medical resources. It drew fire from many physicians for exactly that reason: that ordinary citizens rather than medical experts were deciding who would get dialysis.[34] But most of the controversy centered on the committee's attempts to compare the value of human beings, reflecting, inevitably, its own predominantly middle-class values and prejudices – "the bourgeoisie sparing the bourgeoisie," some critics called it.[35] Committee members themselves were extremely uncomfortable with the choices they had to make. Judgments about people's "social worth" run contrary to our democratic ideals as well as our legal tradition. In the law, as Harvard law professor Paul Freund has pointed out, all human life is equally sacred: "The governing standard [of legal justice] is not the merit or need or value of the victim but equality of worth as a human being."[36] It has been suggested, in fact, that using social value criteria may violate the equal protection clause of the Constitution.[37]

So far, such charges have not arisen publicly in connection with the artificial heart, although it is not hard to imagine that they might. On the other hand, government advisory groups have officially sanctioned the use of nonmedical criteria in selecting candidates for the artificial heart, a policy that proved unacceptable with kidney dialysis. Are we gradually becoming inured to practices that were once considered morally intolerable?

The Seattle experience also cast a spotlight on the shortage of dialysis facilities. Around the country, pressure began to mount to expand funding and resources. Kidney patients lobbied Congress for federal assistance, giving live demonstrations of dialysis in front of several key House committees.[38] Their efforts succeeded; in 1972, Congress passed legislation covering kidney transplants and dialysis under Medicare. Finding no comfortable way of rationing dialysis, the government hoped to sidestep the problem entirely by providing access to all those in need.

This response was not unusual. Most contemporary societies have been extremely reluctant to confront what legal scholars Guido Calabresi and Philip Bobbitt aptly termed the "tragic choices" involved in rationing access to lifesaving technology. Resistance has been particularly strong in the United States, where it has often been assumed that we would be able to afford whatever new techniques medicine could develop. By expanding the supply of kidney dialysis facilities, however, Congress simply moved the problem to another level. As Calabresi and Bobbitt noted, "A system which offers 'kidneys for everyone' distinguishes those dying from renal failure – and prices their lives exceedingly high – from those dying from other diseases, who for similar expenditures could also have been saved."[39] The problem of access to scarce resources within a single disease category was submerged in the larger problem of allocating resources across diseases.

Access to the artificial heart

The question of who should have access to the artificial heart was first brought to public attention in 1973 by the Artificial Heart Assessment Panel, appointed to review the program's "legal, social, ethical, medical, economic, and psychological implications." Remembering the experience with kidney dialysis, the 1973 panel strongly opposed the use of any measure of social worth in selecting candidates. "The very essence of making decisions based on 'social worth' runs counter to basic principles of equality in our society," stated the panel's report.[40] Instead, the panel recommended allocation based on medical need, and proposed coverage by private insurance or a public program, such as Medicare, to eliminate problems of financial access. "Particularly in view of the substantial commitment of public funds for development of the artificial heart," the panel wrote, "implantation should be broadly available, and availability should not be limited only to those able to pay."[41] If medical criteria were not sufficient, random selection among equally suitable patients was proposed. The panel's ethical framework was clearly an egalitarian one.

This advice was consonant with the influential report in 1978 of the National Commission for the Protection of Human Subjects. The 1978 commission concluded that justice requires that people most in need receive the greatest benefits from health care, stating specifically: ". . . whenever research supported by public funds leads to the development of therapeutic devices and procedures, justice demands . . . that these not provide advantages only to those who can afford them."[42] Although the National Commission's report did not refer directly to the artificial heart program, the implications were obvious.

An advisory group appointed by the NIH in 1981 to review the artificial heart program also recommended, at least implicitly, public coverage of implants. It did so by analogy to kidney dialysis.

The group concluded that since the total cost of artificial hearts, which it estimated at around $600 million annually, would be well below the $1 billion the government was then spending annually for kidney dialysis, government coverage of artificial hearts was "not outside the bounds of social acceptance."[43] Both of these projections proved hopelessly unrealistic. By 1983, the costs of kidney dialysis had doubled, with annual program expenditures approaching $2 billion. At that point, even kidney dialysis costs had apparently exceeded the "bounds of social acceptance," and for the first time government funding was reduced.[44] Cost projections for the artificial heart were even farther off. The 1985 Working Group on the artificial heart estimated that the cost per implant would be about $160,000, resulting in total costs of around $2.5–$5 billion per year – *four* to *eight* times the 1981 group's estimate.[45] Furthermore, based on the steep rise in the number of patients on kidney dialysis during the past decade (from 11,000 in 1974 to 73,000 in 1983),[46] the pool of eligible candidates for artificial hearts will undoubtedly expand with time, leading to further cost escalation.

The 1985 Working Group's cost projections for the artificial heart were sobering, especially in light of the dismal record of the implants to date and increasing national concern about escalating medical costs. Members of the group worried, legitimately, that coverage of artificial hearts might divert resources away from more important – and more cost-effective – services. "Even without the availability of artificial hearts," the 1985 group observed, "our mechanisms for sharing health care costs across society are being stressed by the growing number and cost of available therapies."[47] On these grounds, the 1985 Working Group decided, in a sharp departure from its predecessors, to recommend *against* government coverage of artificial hearts and assist devices "until their clinical effectiveness and reasonable cost-effectiveness are demonstrated."[48]

However, the group's report suggested some understandable discomfiture with this recommendation. Without government coverage, artificial hearts will be available mainly to wealthy patients, especially if private insurers continue to exclude "experimental" procedures.[49] Given the higher prevalence and severity of heart disease among the poor, such a "reverse distribution" seems particularly unjust. The 1985 Working Group warned:

> The older segment of our population, the primary beneficiary [of the artificial heart], is unlikely to let pass unprotested a policy of distribution solely according to individual means.

> Perhaps a benefit of the artificial heart will be to stimulate public discussion of the limits on society's responsibility for the health care of its members and whether some lifesaving therapies can reasonably be considered luxuries rather than components of adequate care.[50]

Meanwhile, however, the artificial heart appears to be headed for the private market, where it, like so much of medicine, will be out of reach of many of the people who need it the most.

Such a prospect raises troubling questions about our commitment to justice and egalitarian ideals. Here is a technology, developed almost entirely at taxpayer expense, which earlier government advisory groups all said should be made equally available to those in need. Are we really prepared now to treat it just like any other "luxury" product – those who can afford it get it, and those who can't don't? Artificial hearts are *not*, of course, like other luxury goods in one crucial respect: For some people, they could mean the difference between life and death. Is our society now willing to make the "tragic choices" we tried so hard to avoid with kidney dialysis, whereby we knowingly and voluntarily ration life itself by ability to pay?

Such questions are "honorable to be raised," say those involved in the artificial heart program, but they caution that it would be "disastrous" to set any limits on continuing development.[51] At the time of Barney Clark's implant, a University of Utah official defended work on the artificial heart as a matter of freedom of scientific inquiry. To let such concerns stand in the way, he said, would be like "telling Galileo, don't look at the heavens because you may see something that will give us a problem that doesn't yet have an answer."[52]

Freedom of inquiry is indeed important for *basic* research, which generally has little direct social impact. But it is not a license for unfettered development of technological applications of research, like the artificial heart. The desires of scientists to pursue work on the artificial heart and other high-cost technologies cannot be allowed to override the right of society to determine whether it can accept the moral ramifications of these developments.

Resource allocation in an era of medical limits

Artificial hearts are only the beginning of the problem of resource allocation. Medical research and development are prolific; there is no end in sight to the generation of new procedures and equipment, many of them increasingly expensive. The problem is paying for them. Within what is technologically possible, we must decide what is medically and socially desirable.

An egalitarian allocation of health resources would correspond, in some fashion, to the distribution of preventable illness in the population. Such a policy raises countless questions. How should illness be measured? Should it reflect such things as severity, extent and length of disability, mortality, and burden on the health care system? How should cost-effectiveness (the relation between treatment benefits and costs) be measured? Although available data permit only the crudest comparisons of health needs across different social groups and disease categories, most of the evidence suggests that the present allocation of money and resources bears little relation to either illness or potential medical benefit.

A few examples: In 1981, $2 billion – about 1 percent of the nation's total health bill – was spent on coronary artery bypass grafts, a $25,000 surgical procedure received by less than 0.04 percent of the population.[53] Kidney disease patients receive about 10 percent of total Medicare expenditures on physician services, yet comprise less than 0.25 percent of all Medicare beneficiaries. In one state, Medicare payments in 1978 averaged $44,000 for kidney dialysis patients in the last year of life, compared to roughly $6,000 for other terminal-year beneficiaries.[54] Looking at rates of selected procedures in different communities, researchers have found wide geographic variations in medical practice which are apparently unrelated to the health needs of patients. Rates of surgery, including major operations such as heart surgery, vary as much as fifteenfold, with no evident differences in patients' health outcomes.[55]

For many of the nation's poor, the relation between health needs and medical resources is not just random but actually inverted. Hospitals and doctors are scarce in low-income communities and rural areas, while affluent suburban areas are overrun with physicians. Despite public programs such as Medicaid and Medicare, poor and minority groups, although suffering the highest rates of disease and disability, have less preventive care than the rest of the population, receive fewer physician services relative to health needs, face greater organizational and financial barriers to access, are less likely to have a usual source of care, and tend to be sicker when admitted to hospitals.[56] In 1973, four times as many low-income children as high-income children had never had a physical examination. Indeed, poor and minority groups have lower rates of virtually every preventive service, including dental care, breast examinations, immunizations – even vitamins![57] Such patterns led one British analyst to conclude that health care generally follows what he called the "inverse care law": The greater the medical need, the less care available.[58] The current system falls far short of achieving even equality of access to care, let alone equality of health.

The "inverse care law" was evident in the swine flu program, despite the federal government's efforts to counteract it. Influenza is almost twice as common among poor adults as the affluent, and age-adjusted death rates for influenza and pneumonia are about 50 percent higher among nonwhites than whites.[59] Accordingly, the government organized special media appeals to boost participation in ghettos and other poverty areas. Nevertheless, immunization rates in these areas remained well below the national average. Had the epidemic come, the poor and minorities would have been hardest hit.

The mass immunization campaign also diverted money and resources from other areas of potentially greater need, both medical and nonmedical. Although President Ford easily secured $135 million for swine flu shots, his administration refused to budget more than $5 million a year in federal funds to aid state programs aimed at immunizing children against much more serious diseases than flu, diseases such as diphtheria and rubella, which cluster in poverty areas.[60] The swine flu program cost almost as much as the entire budget that year for occupational health ($170 million), and represented about 13 percent of the total federal outlay for prevention and control of all health problems.[61]

The artificial heart is likely to create much greater disparities in resource allocation. If the government does not pay for implants, as discussed previously, the artificial heart will be financially unavailable to most low-income people, among whom heart disease is most common and most severe. Even if financial coverage were eventually provided, current projections suggest that the device will be suitable for only a tiny fraction of heart disease patients – probably less than 1 percent of the almost 5 million people with heart conditions serious enough to restrict activity, and about 0.2 percent of the 16 million persons with any type of heart condition.[62] Yet the artificial heart program accounts for more than 3 percent of all government funding for heart disease research, and has received roughly ten times as much government support as research on heart transplants.[63] Furthermore, the artificial heart offers limited medical benefits, since it cannot restore natural function but can only prolong the lives of people who are seriously or terminally ill.

Questions about allocation of resources were raised at the very outset of the artificial heart program. In 1966, the original Hittman report asked, "Why incur large financial obligations for a small percentage of the population?"[64] The report's answer was vague and facile, but reassuring: "It is generally agreed by the contractors that the questions raised will be satisfactorily answered once the program is underway and the public is acquainted with the total planning involved."

Two decades later, after a number of "permanent" artificial heart implants and many more temporary implants, there are still no satisfactory answers, and the questions are more urgent than ever. The projected costs of the program continue to climb, and implants still provide neither medical benefits nor cost savings for end-stage cardiac disease.[65] Indeed because artificial hearts and assist devices extend the life of patients who would otherwise die fairly quickly, they often double or even triple the costs involved in caring for people with fatal heart disease. And thus far, given persisting technical difficulties, the implants have simply substituted a new problem, "mechanical heart disease," for natural disease. "We hope that eventually, mechanical heart disease will be much less severe than human heart disease," a doctor at Humana said in 1985. "But at this point," he acknowledged, "it has certainly not been that successful."[66]

Use of the artificial heart as a temporary "bridge" to human heart transplants has also been sharply criticized. As long as there is a shortage of transplantable human hearts, critics argue, temporary devices serve mainly to redistribute human hearts among potential recipients, giving higher priority to those who receive temporary implants because their need is then considered more "urgent." If temporary devices are given to sicker patients, those who might be more likely to benefit from a human heart transplant may be deprived. Moreover, the double surgical procedure leaves the patient in weakened condition. In short, as long as there is a shortage of human hearts for transplants, temporary implantation of artificial hearts may do more harm than good.[67]

In view of such problems and limitations, serious doubts have been raised about the cost-effectiveness of the artificial heart program relative to alternative uses of the funds. Dr. David Eddy, a member of the 1985 Working Group and director of Duke University's Center for Health Policy Research and Education, suggested, in an appendix to the group's report, that development of the artificial heart was not compatible with the goal of maximizing the health of the population, since other uses of the money would almost certainly yield greater health benefits. Noting that many studies have linked cigarette smoking with heart disease, Dr. Eddy concluded that if the projected costs of artificial hearts – $2.5–$5 billion annually – were spent instead on antitobacco education or treatment and achieved even a 1 percent reduction in tobacco usage, the cost-effectiveness of tobacco control would substantially exceed that of the artificial heart.[68]

Even with major technical improvements in artificial hearts and assist devices, the medical costs of implantation and follow-up will still be orders of magnitude greater than the expenditures for many other heart disease patients who are comparably ill but who choose to forego, or cannot afford, heroic medical intervention. Whether we intend to or not, we will be spending much more per person to prolong a small number of lives of dubious quality than to relieve the suffering of those with manageable disease, or to spare others the burden of illness altogether.

Questions about cost-effectiveness and resource allocation become more pressing as high-cost procedures put increasing strain on public and private budgets. Such concerns led the 1985 artificial heart Working Group to recommend against government coverage of the cost of implants, as already noted. On the other hand, the 1985 Working Group supported continued federal funding of artificial heart research, noting that the projected cost of an implant is "within the same range as the costs of a number of accepted expensive procedures," such as kidney dialysis ($30,000 per year), heart transplants ($170,000–$200,000 for surgery plus a year of follow-up care), bone marrow transplants (over $100,000), and liver transplants ($230,000–$340,000).[69] Yet these procedures themselves involve difficult questions about cost-effectiveness and resource allocation. In 1983, the cost of a single liver transplant plus a year of follow-up care was estimated at $34,000 *more* than the entire budget of one inner-city health clinic in San Francisco, which provided 29,000 office visits to the area's poor.[70] Moreover, liver transplants are believed to have clear value for only 4,000–5,000 people in the United States. They keep most adult recipients alive for less than a year and have a postoperative mortality rate of 20–40 percent.[71] A bone marrow transplant for one child costs more than the entire budget for all crippled children's services in many counties; and the treatment often fails. In 1985, at least $800 million was spent in the United States for transplants – primarily heart, liver, kidney, and pancreas – and the figure could easily exceed $20 *billion* if the supply of donor organs were expanded to meet present demand.[72]

Efforts to restrain the dissemination of high-cost, new medical procedures and technologies, even those whose benefits are debated, have gotten essentially nowhere. The shifting status of heart transplants is illustrative. In 1980, the federal government withdrew Medicare coverage of heart transplants pending results from a four-year study of the procedure's benefits, announcing that some medical procedures might be simply too expensive to cover under government programs, especially if the number of potential patients was small and the benefits were uncertain.[73] This was a landmark step: For the first time, the government was attempting to assess the consequences of a significant new technology prior to widespread distribution. That same year, the Massachusetts General Hospital, a Harvard teaching facility and one of the world's foremost research institutions, took a similar step locally, deciding not to go ahead with a program of heart transplants because of the resources required. The hospital concluded that other patients had a greater need for hospital resources than did the relatively small number of heart transplant candidates. "We are now in an era when the decision to act in one way reduces or forecloses our options to do other things that we may want to do or have to do," explained the hospital's Chief of Medicine.[74]

That was in 1980. The picture has changed dramatically since then. With accumulating experience, previously experimental procedures such as heart and liver transplants are rapidly becoming "established" medical therapies. In 1983, Massachusetts General quietly resurrected its previously foresworn heart transplant program. Blue Cross–Blue Shield now provides private insurance coverage for heart transplants in Massachusetts and some other states, and by 1985 more than 80 U.S. hospitals were performing them.[75] A number of state Medicaid programs now cover heart transplants, and in 1986 Medicare reinstated coverage of heart transplants on a limited basis. The federal government designated liver transplants as a "nonexperimental" treatment for certain conditions in order to facilitate reimbursement under public programs.[76] Bone marrow transplants are also becoming increasingly accepted. In 1984, the State of California paid for several bone marrow transplants, and state legislation was proposed to cover the procedure under existing public programs.[77] As with medical care generally, the Reagan administration contends that paying for organ transplants and other high-cost procedures is not the federal government's responsibility. Yet while federal contributions to state Medicaid programs are being reduced, the Reagan administration has pressured at least twenty state Medicaid programs to pay for liver transplants, forcing them to divert funds away from broader public needs.[78]

Covert rationing, overt injustice

The growing acceptance of these sophisticated and extraordinarily expensive new therapies reveals the contradictory forces at work in medical innovation. While new technologies continue to push costs up, both the federal government and private employers are searching for ways to limit expenditures. Public programs as well as major insurers are requiring patients to bear a greater portion of their medical costs and encouraging them to enroll in cost-reducing health maintenance organizations. The federal Medicare program, which covers health care costs among the elderly, has introduced fixed reimbursement for specific diagnoses ("DRGs"), and the government has significantly cut both budgets and eligibility in the Medicaid program, which provides medical coverage for the disadvantaged. Lower federal contributions to Medicare and Medicaid have forced most states to restrict coverage, reduce provider reimbursement rates, or increase patients' charges. Even services proven to be effective have been cut, including nutritional prenatal, and delivery services for needy pregnant women,[79] preventive health examinations for low-income children,[80] immunization programs against infectious diseases,[81] genetic screening of infants for phenylketonuria,[82] teenage family planning services,[83] comprehensive community health centers in poverty areas,[84] community-based heart disease prevention programs,[85] and environmental measures like improved sanitation for rural areas.[86] Public support has also been reduced in equally essential nonmedical areas such as housing, jobs, and education, which also relate strongly to health. These various austerity measures cannot help but exacerbate existing inequalities in health status and service utilization. Many poor and handicapped people depend on public programs to meet their basic medical needs, and have gone without care or suffered financial hardship because of federal cutbacks; some have died.[87]

Even in strictly economic terms, many of these cutbacks may end up costing more than they save. Restrictions on prenatal care are a vivid example. In the early 1980s, federal funding for maternal and child health was reduced by some 20 percent, leading to an increased proportion of women receiving inadequate or no pre-natal care, and, in turn, to higher rates of prematurity and infant mortality.[88] Savings in prenatal visits are trivial compared with the costs of prematurity and related developmental disorders such as mental retardation and cerebral palsy. Sustaining a single pre-mature baby can cost upwards of $100,000. The Department of Consumer Affairs in California estimated that the state could save approximately $66 million annually in neonatal intensive care costs if all women received adequate prenatal care – not to mention the tragic waste of human resources that would be prevented.[89]

Reductions in public aid for the disadvantaged, coupled with continuing technological development, are inexorably shifting medical resources away from primary care and toward high-cost, high-technology services. Public programs, under pressure to cover ever more expensive new procedures, take increasingly dras-tic steps to control expenditures. Meanwhile, clinical research and development continue to generate more new procedures, for which reimbursement is eventually sought. Public programs can-not accommodate escalating demands on shrinking budgets. Something has to give; so far, it has been mainly basic medical assistance for the disadvantaged. The result is an allocation of resources that is becoming steadily not only less equitable but also less cost-efficient.

It is sometimes argued that research and development represent an investment in the future, and that reductions in public medical assistance are necessary to finance this investment. But if sacrifices are to be made on behalf of future generations, justice demands that the burden of these sacrifices be distributed fairly within the present generation, with special regard for those who are least well off.[90] To increase government support for future-oriented research and development while contracting current public programs for the poor is in direct conflict with egalitarian ideals. Future prom-ises, however alluring, do not justify present losses to the disadvantaged.

These trends are also shortsighted from the standpoint of human development. Reductions in public medical programs, many of

which provide services to children, together with basic demographic shifts in the population, have led to growing intergenerational inequities. During the past decade, government spending
on the elderly has been rising while government spending on
children has been falling; as a result, the proportion of the elderly
living in poverty has been declining while the proportion of children in poverty has been increasing. Together, federal and state
programs lifted almost half of the elderly out of poverty in 1981,
compared with less than 5 percent of children.[91] In 1984, estimated
federal expenditures per child through the major child-oriented
programs were less than a tenth of total federal expenditures per
person aged sixty-five years or more.[92]

Differential medical payments contribute to these intergenerational disparities. The elderly are frequent recipients of dramatic,
"last-ditch" medical interventions, which often carry an astronomical price tag. Medicare spends roughly three to six times as
much on people during their last year of life as during their non-
terminal years.[93] Of Medicare's $70 billion expenditures in 1985,
28 percent went to maintaining people in their last year of life –
30 percent of that amount in the last month alone. Each year, at
least 5,000 permanently comatose patients in the United States –
people like Karen Ann Quinlan and others – are kept alive at a
cost of about $100 million.[94] Indeed, provision of intensive treat-
ments to people who are critically ill appears to be a primary cause
of rising hospital costs.[95]

Allocation of medical resources to people who are hopelessly ill
may be fully justified morally, but not if it means depriving others
of basic health and social needs, particularly the young, who have
a full lifetime ahead. The most important investment a country
can make is in the health and well-being of its children. In 1987,
a committee of private business leaders recommended increased
public investment in the health and educational needs of young
children as a cost-saving measure over the long run and the best
means the group could identify of enhancing the future prosperity
of the American economy.[96] Moreover, there is no need to deprive
the young of needed health and social services. As discussed in
the concluding chapter, with different and more humane medical
priorities, a great deal more could be done for all age groups even
within the present health care budget.

Living with limits cannot be avoided. The only choice is whether
we make allocation decisions deliberately, through a process of
public discussion and coordinated planning, or by default, under

the relentless pressures of proliferating medical technologies and spiraling medical costs. Calabresi and Bobbit argue that the only way to avoid further "tragic choices" in rationing access to life-saving technology is by limiting the initial development of additional technologies: It is easier not to develop a lifesaving device in the first place, they maintain, than to deny it to those in need once it is developed. This requires deciding what *not* to do as well as what to do, something as alien to the medical ethic as it is to the American dream of boundless opportunity.

As long as medical resources seemed plentiful, we could avoid this unpleasant reality by simply devoting more resources to each new demand that arose. However, resources no longer seem infinite. If we refuse to make explicit decisions about their use and allocation, these decisions will be made by default – as indeed they already are – in ways not always to our liking. Harry Schwartz, medical writer for the *New York Times*, put it well:

> If we refuse to adopt formal rationing based on a consensus developed after proper debate, we shall continue doing what is now increasingly the case: irrational and cruel rationing of medical care imposed by administrative fiat . . . and more draconian measures to halt the flood of medical bills before they drown our society. There is no ideal alternative because our resources are finite, while our medical-care ingenuity has no limits.[97]

Private profit at public expense

Another important distributive question raised by the cases involves the relation between public support of medical innovation and private benefit from the results. Private industry, as we have seen in the case studies, has largely taken over both the artificial heart and genetic engineering, even though both technologies were developed primarily at public expense. Like many medical advances, both technologies resulted from decades of NIH-funded research conducted mainly in university settings; commercial production and marketing were merely the final stage in the long process of discovery and development.

The successful corporate ventures of molecular biologists were apparently an inspiration to Dr. Robert Jarvik, inventor of the Jarvik-7 heart and cofounder of Symbion, Inc., a company he formed with Dr. Willem Kolff in 1976 to produce artificial hearts and other organs. "What Genentech is doing with gene-splicing products," Jarvik said, "we want to do with artificial organs over the next decade."[98] One thing the two companies clearly had in common was that both were spawned by federally funded research. Altogether, the NIH has invested several billion dollars in the development of gene-splicing and related techniques, and at least $200 million in the artificial heart program. The University of Utah alone received over $8 million of NIH funding for artificial heart research, plus another $7 million from other sources.[99] Commented one of Symbion's top officials: "That was the company's biggest advantage – it had the university as its research-and-development arm subsidized by the government."[100]

Symbion also took full advantage of the publicity surrounding Barney Clark's implant at the University of Utah, as have biotechnology companies from strategically timed press conferences and scientific announcements. Symbion, in shaky financial condition before Clark's implant, subsequently raised some $19 million from stock offerings and another $5 million in seed capital from large corporate investors, one of which was Humana.[101] Jarvik and others at Symbion, like their counterparts at Genentech, were suddenly worth millions. In return, the University of Utah received a 5 percent share in Symbion's net revenues, some stock holdings, and occasional research contracts from Symbion.

It would be difficult to argue that this was a fair exchange. Symbion's financial contributions to the University of Utah are likely to total a few million dollars at most, whereas Symbion has had access to at least two *hundred* million dollars of government-funded research, use of university personnel and facilities, and the public relations advantages of university connections. More importantly, it is not obvious that the benefits Symbion received were really the university's to sell, no matter how good the price. It was taxpayers that had funded the research, and the public purse that had nurtured the university's resources and reputation.

Ties between academic scientists and private industry are relatively new to biomedicine, although they have long been commonplace in engineering. With the federal administration's active encouragement, new types of cooperative arrangements have emerged in both biological research and clinical medicine, along with record levels of direct corporate sponsorship of university research. Such relationships provide a welcome new source of capital for universities, whose research costs are outstripping federal funding, but at a price too little appreciated. Loss of public financial profits from public investments is only a small part of that price. Loss of public control over research and development is far more significant, for it means that some applications, however socially worthwhile, will not be pursued because they are not profitable to the private sector. An obvious example is the low priority given by biotechnology companies to research aimed at producing vaccines for devastating Third World diseases; the people who would benefit most from such products cannot afford them.[102] Even within industrial countries, as we have seen, most biotechnology companies emphasize products that are likely to be profitable rather than those of greatest social need.

There are costs to medical science itself in allying with commercial enterprise. Firms involved in artificial heart research have jealously guarded their research designs and protocols as "trade secrets," impeding improvements and potentially exposing patients to unnecessary risks. Rivalry among competing biotechnology firms has had a distinctly chilling effect on communication at scientific meetings and among colleagues, according to many insiders. Such trends threaten to distort or inhibit the free flow of ideas that is vital to scientific progress. Many genetic researchers are no longer viewed as "objective" participants in scientific controversies because of their ties with private industry. Public confidence in science, already faltering, is likely to be further eroded as the search for truth becomes increasingly indistinguishable from the pursuit of profits.

Corporate involvement in health care delivery has also expanded dramatically in recent years, bringing with it a similar set of problems. Large profit-making companies now own or manage more than 20 percent of all U.S. hospitals – double the share they controlled five years ago – and are making corresponding gains in affiliated areas such as health maintenance organizations and walk-in clinics. Fierce competition in the health care market has forced even nonprofit agencies to become more cost-conscious and "businesslike," as profitability becomes the condition for survival. This usually means ignoring health-promoting social and behavioral changes unless they are also profitable, and serving only people who can afford to pay privately or are covered by insurance. Growing numbers of poor people are being denied care for financial reasons, and experts predict that the next decade will see "a shift away from the traditional view of health care as a social good that is exempt from market forces, and toward a view that it is an economic good subject to the influence of supply, demand and price."[103] In short, a new ethos is emerging in medicine, in which humanitarian considerations are overshadowed by financial concerns.

The new business-oriented approach to medicine is exemplified by Humana, Inc., the large health-care chain based in Louisville, Kentucky, which lured Dr. William DeVries away from the University of Utah by promising to pay for 100 implants at a cost of up to $35 million. Whatever other motives Humana may have had, this move was carefully calculated to promote corporate interests. Bringing in DeVries, according to a stock market analyst who follows Humana, "will provide Humana with incredible opportunities because so too will come more cardiovascular operations, more patients and more hospitals."[104] "The only thing Humana can get out of it is an enhanced reputation," said David Jones, Humana's chairman and chief executive. "But because our business is health, it is a wise investment."[105] "Humana will become a household word in health care," one Humana official predicted.[106]

He was right. Artificial heart implants kept Humana in the public spotlight throughout 1984 and much of 1985. Earnings were up 20 percent in 1984, making Humana by far the most profitable company in the industry.[107] Humana's high visibility proved invaluable when cost-cutting efforts by the federal government and employers began to lower hospital admission and occupancy rates around the country, and Humana decided to expand into other aspects of health care, including a prepaid health plan. In the six months following its first implant, in November 1984, the number of enrollees in the prepaid plan more than doubled – "tangible evidence that the brand-name strategy works," in the words of Humana's senior vice-president.[108] Another Humana official summed up the situation: "The rising tide lifts all boats," he said. "As the process receives attention and the tide rises, Humana rises with it."[109]

Again, however, the public pays a price for this private company's gains. Humana hospitals have relatively little experience with clinical investigation, and are probably not the best site for what is, or should be, an ongoing medical experiment. In the view of a number of medical leaders, Humana's "excessive interest in publicity" is both harmful to patients and at odds with the need for careful and objective evaluation and reporting.[110] Other critics fear that the ethics committee at the Louisville hospital will not "remain independent and objective in an institution with a heavy financial stake in the rapid success of the device," and thus may not adequately protect patients' interests.[111] DeVries himself, admittedly "frustrated" by the delays caused by the Utah ethics committee's concerns about risks and informed consent, observed that at Humana "the opportunities are better for less red tape."[112] There is also a direct financial cost to the public in turning the artificial heart over to private industry for production and marketing. When the device is ultimately perfected, taxpayers will be expected to pay again – billions of dollars – to Humana and other companies for the privilege of purchasing a device that they have already paid millions in taxes to develop.

In the end, the greatest price paid by the public is probably again loss of control over the artificial heart's future use and distribution. Dr. Jarvik has talked about developing regional centers that could implant "emergency" artificial hearts in tens of thousands of Americans suffering potentially fatal heart attacks each year.[113] Questions about eligibility, cost, availability of human hearts if the implants are intended as temporary, and, most importantly, whether this is the best use of the country's medical resources, all remain unanswered. Although such questions have significant ethical and economic implications for the entire society, they are now in the hands of private companies like Humana and Symbion. Inevitably, the answers will be based more on corporate priorities than the public interest.

It is not the responsibility of private businesses to protect the interests of society as a whole, and it would be naïve to expect them to do so. For companies, medicine and clinical research are simply good investments. Yet in medicine, good investments may result in bad health policy. Analyzing the medical technology industry, a Congressional Office of Technology Assessment study concluded that "the market has generally rewarded attention to technological sophistication but not to price or cost-consciousness and has fostered the development of devices used in acute care rather than in prevention and rehabilitation."[114] Nor is it safe to assume that private companies will at least deliver health care more efficiently than nonprofit agencies do. A major study by the National Academy of Sciences concluded that proprietary hospitals' claims of greater "efficiency" were unfounded; that in fact for-profit hospitals generally charge higher prices than nonprofit hospitals while serving proportionally fewer uninsured patients and providing less charity care.[115]

An even more fundamental problem with the growing commercialism in science and medicine is its tendency to distribute resources according to income. Over time, the more effectively commercial incentives operate, the more resource allocation will come to correspond to the distribution of income – and the less to population health needs. The richest 20 percent of all U.S. households have eleven times as much income as the poorest 20 percent. Any efficient market mechanism, if allowed to operate unhindered, will thus end up giving eleven times as much medical care to the wealthiest and healthiest segment of the population as to the poorest and sickest segment.[116] Commercial incentives, in short, will tend to produce an even more extreme version of the "inverse care law" in medicine than presently exists.

Roots of the paradox

The government responds to soaring health care costs by cutting back on preventive and primary care services known to be highly cost-efficient; at the same time, private industry's expanding role in both basic research and health care delivery is taking a new slice of profits from the shrinking biomedical pie. Unchecked, these trends will squander resources on the rich, while constricting basic services for the poor. Why, in medicine, long considered the most humane of professions, is this happening?

Many factors are surely involved. At the most obvious level, our society often looks upon technology as a "quick fix" for social problems. Medical technologies are particularly seductive, with their heart-stirring tales of medical heroism and human triumph. News coverage of the first permanent artificial heart implant painted the event in dramatic terms. "This man is no different [from] Columbus," a University of Utah official told the press. "This man and these people are on the threshold of something as exciting and thrilling as has ever been accomplished in medicine. It is a moment of great human hope."[117] United Press International reported that the event was "on a scale of greatness . . . right up there with man's first step on the moon."[118]

No such accolades have greeted heart disease prevention, even though prevention has unquestionably saved many, many more lives. The 1964 surgeon general's report on the dangers of cigarette smoking led to new concern about the importance of a healthy life-style. People now smoke less, exercise more, and eat less fat – many for health reasons.[119] Heart disease death rates have fallen by roughly 30 percent over the past two decades, a decline experts attribute in significant part to life-style changes along with medical and surgical advances.[120]

One explanation for medicine's emphasis on dramatic "high-tech" interventions rather than low-cost, low-profile forms of care and prevention is that the latter deal with "statistical" lives rather than with the identified lives of known people. As Calabresi and others have pointed out, society is typically willing to spend much more money to save the life of a known person in peril than to reduce the future level of fatal accidents.[121] This discrepancy has been ascribed to a "time bias" in favor of present lives over future ones and to our emotional response to a known person and the person's family. Although these factors may help explain the downgrading of "statistical" lives, they do not, of course, make it morally right.[122]

There are also powerful vested interests opposing many forms of disease prevention. According to the World Health Organization, control of cigarette smoking "could do more to improve health and prolong life in (developed) countries than any other single action in the whole field of preventive medicine."[123] The number of people smoking in the United States has declined substantially since 1964, yet per capita cigarette consumption was still the highest in the world as of 1978.[124] There is no organized, well-funded antismoking campaign, and the government continues to subsidize tobacco farmers. The tobacco industry spends $2 billion a year on cigarette advertising, more than the amount spent on any other product the American consumer buys, and far more than that spent on smoking prevention.[125]

Broader social changes could have an even greater impact on health. Eliminating the illness-breeding conditions associated with poverty, long known as the "mother of disease," would dramatically reduce both morbidity and mortality. Poor nutrition is a risk factor for many chronic and acute conditions, and is widespread even in the United States. In 1984, one out of every five Americans reported not always being able to afford food for their families.[126] Unhealthy and dangerous workplace conditions are directly responsible for a variety of unnecessary and disabling conditions. More than a century ago, the eminent pathologist Rudolph Virchow said, "The improvement of medicine would eventually prolong human life but improvement of social conditions could achieve this result more rapidly and successfully." Time has revealed the accuracy of this prediction.[127]

Improvement of social conditions is rarely an explicit medical objctive, even in preventive medicine, regardless of the potential impact on health. "Prevention," in the medical model, is generally confined to therapeutic interventions in the disease process, while the precipitating social circumstances remain unaddressed. Such "preventive medical interventions" are usually least available to poor and minority groups, as noted earlier, and the benefits of such interventions are often marginal compared to the damage done by social and economic deprivation.

The swine flu program, critics charged, illustrated the limits of the medical approach to prevention. No one knows exactly why flu strikes the poor and minorities so much harder than others, but it may be because their resistance has been lowered by nutritional deficiencies, stress, and other hardships. With these factors unchanged, critics contended, purely medical interventions like flu shots would inevitably be inadequate. They argued that mass immunization should be coupled with efforts at "augmenting the immunological defenses of the body through better nutrition, reduced physical and mental stress, secure living conditions and adequate income and housing." A "vaccine alone," they warned, "however effective, is a palliative, not a cure."[128]

They had a good case. A subsequent study of children's hospitalization for measles and whooping cough found that admission rates were more closely related to family economic circumstances than to whether the child had been vaccinated.[129] Indeed, the close relationship between poverty and most measures of morbidity and mortality implies that fully effective disease prevention is impossible without major changes in the structure of society itself. It is far easier – and considerably more acceptable politically – to focus on the consequences of that structure. As one medical observer lamented: "We have conquered polio but not poverty, tuberculosis but not truancy, syphilis but not slums. Somehow, we seem condemned to triumphs of biological wizardry and failures of social management."[130]

Although medicine alone cannot be expected to remedy all of society's ills, medical professionals and institutions could surely do more to call attention to the impact of adverse social conditions on health, to help reallocate medical resources, and to support efforts to solve the underlying social and economic problems. Instead, new medical technologies too often become a substitute for broader social reforms. Emerging genetic technologies have aroused this concern. Some labor unions fear, for instance, that the possibility of a "genetic fix" will lead to slackened efforts to eliminate occupational and environmental causes of genetic damage. They cite the apparently growing use of genetic testing by major corporations to identify employees who are particularly susceptible to toxic substances in the workplace. Although such screening may offer valuable protection for workers, critics worry that it will be used by industry to avoid cleaning up the workplace, creating a group of "genetically unemployed" workers and allowing potentially dangerous levels of risk to persist.[131]

The prevailing orientation toward high-cost, high-technology medicine is both reflected in and reinforced by biomedical research. Many scientists believe that biomedical research will ultimately unlock the mysteries of disease causation, permitting disease prevention through direct biological intervention in the disease process.[132] To some, a dramatic "rescue" technology like the artificial heart only highlights our failure to prevent or cure heart disease, underscoring the need for more biomedical research. "The greatest potential value of the successful artificial heart," wrote Lewis Thomas, the renowned medical scientist, shortly after Barney Clark's implant, "is, or ought to be, its power to convince the government, as well as the citizenry at large, that the nation simply must invest more money in basic biomedical research."[133]

Unfortunately, the promise of biomedical research remains largely unfulfilled; Thomas himself notes that "there are nothing like as many [truly decisive therapeutic discoveries] as the public has been led to believe."[134] What biomedical research clearly has produced is a growing supply of what Thomas calls "half-way technology": techniques which, like the artificial heart, are both highly sophisticated and — since they rarely prevent or cure — profoundly primitive. Less than a third of NIH funding is for research officially classified as related to prevention (and many experts consider this to be a considerable overestimate).[135] For the government as a whole, the figure is far lower — an estimated 4 percent of the federal health budget is spent on "prevention-related activities," including both research and medical services.[136] Increasing corporate involvement in biomedical research is likely to mean an even greater emphasis on new medical technologies, as industry sponsors seek marketable clinical interventions rather than nontechnological modes of prevention.

The federal health bureaucracy is complex and uncoordinated, and has neither the institutional mechanisms nor adequate data to allocate resources systematically according to measures of population health need.[137] As the four case studies illustrate, the research agenda emerges from a complex mix of political, economic, and scientific interests, more by default than design. And yet, from the seeming helter-skelter of decisionmaking, a design does emerge: Research priorities in both the public and private sectors tend to emphasize medical "commodities" which can be bought and sold – biological, clinical, and mechanical interventions in the diseases of individuals – rather than seeking ways to change the health-damaging conditions of the home, workplace, and environment. Indeed, the medical model of disease treats individual medical interventions as essentially the only meaningful weapons against disease, relegating social and behavioral changes to the domain of public health or private action. Medicine's emphasis on the provision of individual clinical "commodities" mirrors the economic structure of the larger society; the buying and selling of commodities is the lifeblood of capitalism. Medical commodities are the natural consequence of health care delivery and private enterprise; they address individual medical problems and they generate jobs and income for the professionals and companies that sell them. While promising hope to suffering individuals, they leave the social and economic organization of society intact.

The federal government does not hide its interest in using biomedical research funding to foster economic productivity. During the formative years of the artificial heart program, Dr. Frank Hastings, director of the National Heart Institute, stated candidly that "the most important part of the Artificial Heart Program is not to develop an artificial heart but to develop a biomedical capability in industry."[138] The Reagan administration has repeatedly stressed its intention of harnessing science, particularly genetic engineering, in the service of the economy. Federal research funding has been tied increasingly to what the director of the National Science Foundation called "this number one policy topic, economic competitiveness," and legislation has been proposed to expand the role of the Commerce Department in the development of long-term, high-risk technologies.[139] Not surprisingly, areas such as biotechnology, which promise commercial rewards as rich as their intellectual ones, rank high in federal funding priorities.

Scientists with direct ties to industry may consciously steer their research toward areas likely to yield commercial products, as appears to be happening in biotechnology. But all researchers are constrained by the funding available, and funding patterns – whether federal or private – gradually shape intellectual priorities.

Fields with money attract people, and these people naturally believe, or come to believe, that such fields are important and others less so. As investigators gravitate to funded areas, they imbue them with prestige and excitement, and at the same time their work confirms and enchances the prominence of the field. In the process, an "ideology" is created, a set of values that permeate not only medicine but the entire society, casting biomedical and clinical innovation in their mold far more effectively than explicit controls ever could.

That ideology masks the underlying contradiction between health and economic goals. Continued technological innovation to fuel the economy is in direct conflict with the social goal of maximizing the public's health within a fixed budget. The price we pay for allowing the needs of the economy to shape biomedicine is the curious paradox we now witness: excessive services for the affluent and inadequate care for the poor, proliferation of high-cost technology and indifference to prevention, and the gradual erosion of humanitarian values and their replacement by the ethics of business and profits. The logical consequence of the private sector's expanding role in both science and medicine is a steadily growing gulf between medical priorities and the population's health care needs.

Conclusion

The present allocation of medical resources has little to do with justice, no matter which ethical theory one prefers: It does not provide the greatest good for the greatest number, the classical utilitarian goal; it does not favor the least advantaged groups in society; and it certainly does not promote equal health. Nor is it compatible with efforts to contain costs. Medical priorities are, rather, the result of the structure of modern medicine, and of the larger society, in which power resides with medical and corporate interests and not with the general public, and in which political and economic incentives favor technological innovation rather than social and behavioral change.

Gradually, these priorities are transforming not only the face of medicine but its underlying values as well. The changes are subtle, and enthusiasm for the new may blind us to the loss of old standards and ideals. Let us look briefly at what the future portends while we can still remember what is good about the past.

Events once in the realm of macabre fantasy are now close at hand. In 1974, Willard Gaylin, a prominent medical ethicist, described how it might some day be possible to "harvest" human body parts from "living cadavers" housed in special "bioemporiums" – "warm, respiring, pulsating, evacuating, and excreting bodies," without brain function but attached to respirators – and, he might have added, artificial hearts.[140] Gaylin's vision now approaches reality. We do not yet have bioemporiums filled with "neomorts," as he called them, but artificial hearts have been implanted in brain-dead bodies, and researchers project that the devices may eventually be used to preserve "living cadavers" until their organs can be transplanted.[141] Pressures to promote organ donation are mounting; there are several privately run regional and national organ procurement programs, and a federal clearinghouse to coordinate the distribution of body organs.

What Gaylin may not have foreseen was that even body organs would not escape the omnipresent influence of commercialization. In 1983, reports began to surface about companies being formed to serve as brokers for donors who would be recruited to sell their organs for profit. Many physicians were outraged, charging it was "morally repugnant and reprehensible," an "obscene" violation of human ethics that could only victimize the poor.[142] Legally, the sale of nonvital organs is considered an act of such desperation that voluntary consent is impossible.[143] The federal clearinghouse prohibits organ purchases. Both the Reagan Administration and the American Medical Association have opposed federal involvement, preferring to leave distribution in the hands of the private sector. As the initial revulsion with the idea of buying and selling body organs has faded, articles have begun to appear extolling the virtues of this practice for individuals as well as society.[144]

Genetic engineering is another scientific advance that, for many people, falls in a moral twilight zone. In 1980, leaders of the country's major religions signed an open letter warning of the "fundamental danger" triggered by the rapid growth of new genetic technologies, especially their envisioned use on humans.[145] "History has shown us," they wrote, "that there will always be those who believe it appropriate to 'correct' our mental and social structures by genetic means, so as to fit their vision of humanity. This becomes more dangerous when the basic tools to do so are finally at hand. Those who would play God will be tempted as never before."

Scientists now stand at the threshold of human genetic engineering. Many awesome questions loom, although the President's Commission for the Study of Ethical Problems was optimistic that the benefits of human gene therapy would outweigh the harms. Not everyone agreed. At a hearing on the commission's report in 1982, science reporter and author Nicholas Wade accused the group of ducking its responsibility for considering potential future abuses and recommending appropriate safeguards.[146] Another witness, a molecular biologist, wondered uneasily whether "there is anything unique about humanness." A year later, still alarmed, leaders of virtually all of the country's major religions issued another statement of concern, this time urging an outright ban on genetic engineering of human reproductive cells.[147]

Such protests have not visibly slowed the genetic revolution. Research is proceeding rapidly, and scientists have predicted that gene therapy techniques for certain inherited genetic disorders will soon be developed.[148] In the meantime, diagnostic capabilities are outpacing the capacity to treat most genetic disorders, causing some scientists to wonder whether it may be more harmful than helpful to label people "genetically deficient," when there is little possibility of prevention or cure.[149] Also troubling are the ethical implications of using genetic fitness, over which an individual has utterly no control, as a means of allocating jobs and other rewards. "Unless applied with great care and sensitivity," warned one ethicist, "genetic screening could be perceived as a threat to the egalitarian ideals we cherish in a democracy."[150] The threat, bluntly put, is the prospect of a medical eugenics.

These developments reflect the extent to which modern scientific medicine, with its reliance on techniques and technologies, is gradually transforming the "art of healing" into the "science of body repair." That this transformation no longer evokes the kind of moral revulsion it did a decade ago, when Gaylin wrote his prophetic description of harvesting the living dead, is a measure of the degree to which our own attitudes, including notions of what is right and wrong, have also changed.[151] During the past decade, aspects of human existence which were once sacred or spiritual have become secular, even mundane. Traditional moral values have been replaced by the "scientific" objectives of medicine's significantly self-defined, and increasingly commercially oriented, mission. The loss of these values, Gaylin warned, "would diminish us all, and extract a price we cannot anticipate in ways yet unknown and times not yet determined."[152] Ultimately, that price involves the erosion of human dignity and personal autonomy.

It is perhaps inevitable that, as science pushes back the frontiers of medicine, it will strain the limits of traditional morality. These strains reflect contemporary society's struggle to be master of modern technology rather than slave, to reap the fruits of scientific progress while avoiding its social costs. The conflicts provoked by new medical developments are often highlighted by the stark contrast between secular and spiritual values, as in the divergent reactions to human genetic engineering. In such controversies, the greatest challenge is not how to achieve the best, or most "utilitarian," balance of harms versus benefits, but rather how equitably to reconcile opposing views and values.

There are no simple answers, no painless choices. Because people do have conflicting interests, so they will have different ideas about their health needs and about what constitutes a just distribution of medical resources. For a system of formal rationing to be effective, it must be able to balance these competing claims in an equitable manner. This is the central dilemma of justice in medicine.

Questions of justice thus lead inexorably to questions of power. We cannot ask how to distribute justly without also addressing the question: Who shall decide how to distribute and what is just? In the Kantian tradition, in fact, it is the very process of recognizing and accommodating the relative nature of ethical standards that preserves individual autonomy and, therefore, social justice. Justice, in this view, is whatever emerges from a process of informed public deliberation and political action, provided that this process allows citizens to realize their full dignity and powers as responsible agents and judges.[153]

The alarms have sounded from the headlands of medical innovation, yet they signal broader dangers. At risk are basic notions of morality and humanity, values that reflect the ideals that society holds and wants to perpetuate. They are more precious than any social mission, even one as noble as medicine's mandate to sustain life. For these are the values that tell us who we are as a society, and who we shall become.

10

The role of the public

The scepticism that I advocate amounts only to this: that when the experts are agreed, the opposite opinion cannot be held to be certain.
   –Bertrand Russell, *Sceptical Essays*, 1928

When the Cambridge (Massachusetts) City Council met on January 5, 1977, there were klieg lights, TV cameras, and scores of newspaper reporters. This was no ordinary meeting; it was when Dan Hayes would present the report of the Cambridge Experimentation Review Board – a citizens' panel charged with reviewing the risks of recombinant DNA research, of which he was chairman. Not only would the panel's report determine the future of genetic engineering at two of the world's leading universities, it would also reveal how well a group composed entirely of nonscientists had been able to deal with a complex and highly charged scientific controversy. Such a twist in science policymaking was as unprecedented as the research itself.[1]

The citizens' panel had been appointed by the Cambridge city manager the previous summer, at the height of public alarm about potential hazards of the research. Its seven members were carefully chosen to be broadly representative but, in order to avoid "scientific elitism," included no practicing scientists. Chairman Hayes ran an oil distribution company in North Cambridge, and the other members included a nurse, two physicians, a nun, a community activist, an engineer, and a professor of urban and environmental policy. Most of the members knew nothing at all about genetic engineering when they began meeting in August, so their sessions were a combination of inquiry and crash course in molecular biology. In all, they received a stack of background materials some three-feet high to read, heard testimony from more than thirty-five experts, had countless hours of discussion, and even conducted a mock "trial" with eminent witnesses presenting different views.

Their work was not in vain. The group produced a document that even so staunch a critic of the research as Harvard Professor George Wald, a Nobel Prize winner and outspoken opponent of genetic engineering, called "a very thoughtful, sober, conscientious report," even though it basically allowed the research to proceed, adding only minor restrictions to existing federal requirements.

The panel itself was very proud of the report, according to Chairman Hayes. He pointed out that all of the recommendations, including some sophisticated measures overlooked or avoided by NIH officials and experts, came from members of the citizens' group itself, not from its scientific advisors. Over the course of its work, the group had gained both technical competence and self-confidence. Some members who "couldn't even formulate a question" in the beginning learned not only to ask cogent questions but to pursue unsatisfying responses with a series of follow-up inquiries. A few could sometimes even spot instances where a witness was quoting someone out of context, Hayes said.

Whatever one thinks of the panel's final verdict or the process by which it was reached, one thing seems clear: The Cambridge Experimental Review Board showed that a broadly representative group of nonscientists could tackle a monumentally difficult science policy issue and arrive at a solution widely regarded as intelligent and responsible.[2]

Four years later, another controversy over genetic engineering erupted in and around Cambridge. With fledgling biotechnology companies sprouting up rapidly, city councils again had to decide for themselves (since mandatory federal guidelines did not cover private companies) what, if any, conditions to place on the research. Cambridge and some other neighboring cities enacted regulatory legislation extending the NIH guidelines to private industry and adding other requirements such as formal medical surveillance.[3] At stake this time were public and private wealth as well as public health. Communities had to weigh the medical and environmental risks of inadequate safety precautions against the economic and social risks of overly strict provisions which could drive companies away and lose needed tax revenue. As it turned out, the laws that were passed served to stabilize the regulatory environment without unduly impeding the research. After the initial flurry of anxiety had subsided, some companies came to favor them as much as local governments did.

However, the controversy was not over. A few years later, in the mid-1980s, a new struggle emerged in scattered locations around the country, as researchers sought permission to conduct open-air field tests of gene-altered organisms. Not surprisingly, the regulatory structure that had been so carefully designed to ensure that organisms would remain safely contained in laboratory facilities proved largely irrelevant to experiments in which organisms would be deliberately released into the open environment. In the debates that ensued over this new phase of genetic engineering, there was once again a sharp split between scientific experts and the biotechnology industry on the one hand, and lay citizens and public interest groups on the other.

The evolving controversy over genetic engineering exemplifies a fundamental dilemma facing society: how to assure the general public a meaningful voice in directing and controlling developments in modern science and medicine, some of which may irrevocably alter the nature of life as we know it. Democratic principles promise citizens a voice in such matters, but existing governmental institutions fall short of providing it. Polls indicate that the public wants more control, especially over technology.[4] For several decades, direct citizen influence has been a persistent theme in various social arenas, from the civil rights, antiwar, and women's movements of the 1960s, to the consumer and antinuclear movements of the 1970s, to the political promises of the 1980s about the "New Federalism." Increasingly, many political theorists have looked to direct participation as a way to combat the apathy and alienation characteristic of postindustrial society.[5] Yet serious questions remain. Some observers doubt that true "participatory democracy" is possible in the modern context of mass organization and technological interdependence. Others maintain that it is unrealistic to expect people to participate in long-range decisions about biomedical policy when personal matters are more pressing. Perhaps the most important question of all from a practical standpoint is whether, in the present society, ordinary people – people from all walks of life – have anything useful to say about technical issues whose intricacies defy even the experts.

The controversies played out in the four case studies provide a rich store of information about the kinds of influence that lay citizens had, or tried to have, in a range of biomedical policy issues. We examine here mainly the different forms of direct intervention in the policy process attempted by citizens or outside experts rather than the conventional functions of representative government. Although the latter are undeniably important, it is the less traditional forms of direct involvement that are most likely to offer citizens a greater sense of control over policy decisions. Direct involvement can occur both within the formal political system – in public hearings, for example – and outside it, through the creation of grass-roots citizen groups. Specific forms include participation on advisory panels of task forces, organized public interest groups, individual "whistleblowers" (employees who "go public" with grievances), and citizen review boards with delegated responsibility for particular issues.

By the "public," we mean lay citizens, professionals in fields unrelated to the innovation in question, and representatives of public interest organizations: anyone, essentially, who does not have a professional, political, or pecuniary interest in the issues in question. This excludes employees of industries that would benefit financially from the innovation's development, as well as most scientists and government employees who would benefit professionally, unless these individuals were clearly not acting in their own immediate organizational or career interests, as in the case of "whistleblowers" and "dissident" scientists who formed symbiotic alliances with lay groups.

In assessing public efforts to intervene in policymaking, it is important to consider the nature and goals of those efforts independently of their ultimate success in bringing about change. To judge public contributions primarily by their degree of impact would be to ignore the significant disadvantages that most lay groups face with regard to resources, information, experience, and contacts. We therefore focus primarily on the *kinds* of public interventions attempted in the four cases, returning briefly at the end of the chapter to the question of why some efforts failed whereas others were more successful.

The nature of public concerns

Medical innovations usually come to public attention only after an important breakthrough has occurred. Prior work and decisionmaking are confined almost exclusively to medical and scientific experts, people whose professional lives are devoted to a given area of investigation. It was heart surgeons and cardiac researchers who proposed the artificial heart program, molecular biologists who developed the recombinant DNA technique, biochemists who first synthesized DES, clinical researchers who initiated the experimental use of DES as a morning-after pill, and virologists and epidemiologists who conceived and promoted the swine flu immunization program.

The very beginning of an innovation is a particularly critical period, for it tends to shape the agenda for later deliberations. Because of the central role of experts initially, issues are often interpreted and boundaries of discussion set in a way that renders public concerns irrelevant and builds expert dominance into the very framework of debate. Once an initial commitment is made to an innovation, jobs and professional prestige come to depend on continuing momentum, generating further resistance to concerns that might threaten development. Such impediments were apparent during the early years of the artificial heart program. Although it rapidly became clear that the initial plans contained major miscalculations, National Heart Institute officials refused to allow the 1973 Artificial Heart Assessment Panel, an appointed group of largely nonmedical professionals, to include in its evaluation such basic issues as the need for the device and its technical feasibility.[6] The program never received more than ritual attention from Congress (probably less than three hours a year),[7] nor did many deeply probing questions come from the Heart Institute's Advisory Council, ostensibly a vehicle for public review but, in fact, heavily dominated, especially during the early years, by cardiac researchers and others in the medical community.

Despite such initial impediments public groups and their allies found ways to participate in many subsequent policy choices. Efforts by outsiders to influence decisions on genetic engineering and DES were relatively broad and publicly visible, whereas in the swine flu and artificial heart programs they were much more limited and restricted largely to professionals from other fields. Three themes recur repeatedly in those efforts: reassessing risks and benefits; broadening and humanizing the perspective taken on issues; and trying to ensure sound and legitimate decisions through a more democratic decision process.

Reassessing risks and benefits

With new technologies typically being developed and promoted by powerful professional and business interests, the aim of many attempted public interventions has often been to highlight possible risks. In all four cases, consumer advocates, outside experts, and a handful of dissenting scientists pressed for more attention to known or potential dangers, while questioning optimistic assessments of benefits. In the swine flu program, many Americans expressed their own personal concern about risks by deciding not to roll up their sleeves. The danger of radiation hazards led the 1973 Artificial Heart Assessment Panel to oppose further development of the nuclear-powered heart. Risks have also been the continuing focus of public concerns over both genetic engineering and DES.

When Dr. Arthur Herbst and his colleagues discovered the association between DES and clear-cell vaginal cancer in daughters of women who had taken the drug during pregnancy, the first priority for most women's organizations was to identify and aid DES daughters, for whom risk had become tragic reality. This discovery reinforced the long-standing concerns of environmentalists and consumer groups about the use of DES as a growth stimulant in livestock.[8] And it provided grim confirmation of the fears of feminist health critics, who had denounced the use of synthetic hormones for birth control and for menopause as a classic example of exploitation by the male-dominated medical care establishment.[9]

Women's groups and health activists also became alarmed about the growing use of DES as a postcoital contraceptive, a medical "experiment" that quickly became routine gynecologic practice during the early 1970s, even though it was supposedly restricted to "emergency" situations and was never marketed under the conditions approved by the FDA. By 1973, according to data cited by Nader's Health Research Group, about two million women had received DES or another estrogen as a morning-after pill.[10] At the University of Michigan, research by health activists, notably Kathleen Weiss and Belita Cowan, revealed other disturbing findings: that few women who had been given postcoital DES reported being asked about factors (such as prior DES exposure) that might contraindicate further exposure to DES, or being told of its experimental status and carcinogenicity; that debilitating side effects, mainly nausea, were common and many women failed to complete the full regimen of pills; and that medical follow-up to verify the treatment's success was apparently rare.[11] Weiss released these results at a press conference in December 1972, and they were the subject of an "ABC News" documentary.

DES activists objected that, since all estrogens were known to be carcinogenic, the morning-after pill exposed women to unnecessary health risks. As Belita Cowan, representing a network of women's health groups, pointed out in 1975 congressional hearings on DES, very few women (roughly 5 percent) become pregnant from a single unprotected sexual encounter. There would be less risk, she argued, if, rather than all women taking the morning-after pill, they waited for the results of pregnancy tests; then, only the 5 percent who turned out to be pregnant would, if they chose, undergo the risks of therapeutic abortion.[12] Critics also challenged the actual, as opposed to theoretical, effectiveness of the morning-after pill. At the 1975 congressional hearings, Cowan debunked the much-heralded clinical study by Kuchera purporting to show 100 percent efficacy, citing major methodological problems, including the lack of a control group and inadequate follow-up.[13] (Kuchera herself later admitted that some pregnancies had occurred among subjects excluded from the research.)[14] A study of the routine prescription of DES to rape victims in an Atlanta municipal hospital showed that the failure rate in that setting was at least 6 percent, and that many of the women who received DES were not even at risk of pregnancy.[15]

By investigating real-world settings, DES activists found risks that were not revealed in research situations. To ignore such problems, they argued, could be not only misleading but dangerous. For example, the assumption that DES was 100 percent effective and that patients would always complete the full regimen could lead doctors to pay less attention to informing patients of the dangers of DES or to verifying the treatment's success, thereby increasing the chance that some women might unwittingly give birth to DES-exposed babies.

Another risk that these critics helped to publicize was the higher breast cancer rate among DES mothers. In 1977, Nader's Health Research Group obtained, through the Freedom of Information Act, a copy of preliminary data from a study of cancer in DES mothers led by Dr. Arthur Herbst. According to Herbst, the data did "not show increased risk of breast cancer among DES-exposed women in comparison to those unexposed."[16] Dr. Sidney Wolfe, the Health Research Group's director, disagreed, claiming that the data showed "a substantial increase in breast cancer and other hormone-related cancer" among the DES mothers.[17]

The disagreement was not over the incidence of breast cancer, which was clearly higher among the DES-exposed women (4.6% versus 3.1% among the unexposed), but over whether this difference was statistically significant. Wolfe argued that it was, because, given animal studies linking DES to breast cancer and suggestive evidence in humans, "significance" should be judged by a less restrictive standard. Choosing the proper standard for judging statistical significance is a highly technical, and ultimately subjective, matter about which even statisticians often disagree. But Wolfe found strong support for his position from Dr. Adrian Gross of the FDA's Bureau of Drugs, who called the Herbst team's conclusion "nothing short of absolute nonsense."[18] Wolfe reported his findings to HEW Secretary Califano in a letter he also released to the press, and presented them at the January 1978 meeting of the FDA's Fertility Drugs and Maternal Health Advisory Committee. After much wrangling, this committee finally agreed that the data did provide "evidence for concern" and urged the FDA to inform physicians and the public of the findings.

The data caused quite a stir. Lobbying by DES activists had put mounting pressure on the government to "do something" about DES, and when the breast cancer controversy broke, Secretary Califano appointed a National Task Force to study all aspects of DES exposure and advise HEW. Different attitudes toward risks again emerged between many of the experts on the task force and the four "public" members who had been appointed as consultants at the insistence of consumers' and women's groups. For example, despite the lack of conclusive scientific evidence and opposition from some of the medical experts, the public members were able to convince the task force to recommend that oral contraceptives and other estrogens should be avoided by DES-exposed women.[19]

The majority of subsequent studies bore out Wolfe's contention that DES mothers were at substantially higher risk for breast cancer.[20] Prodded by Wolfe and others, the government convened another DES Task Force in 1985 to review recent research results and offer recommendations. The 1985 task force, again consisting of government scientists with consumer representatives and others as consultants, acknowledged that the new evidence gave "greater cause for concern than there was in 1978," but still concluded that in view of possible methodological flaws in these studies, "a causal relationship has not been established."[21]

A similar split between experts and nonexperts developed early on in the controversy over genetic engineering. Scientists, focusing primarily on specific biological hazards, devised elaborate laboratory "containment procedures" for protection against the unintended escape of organisms, and a series of "risk assessment" experiments to clarify the toxicity and survival capacity of different strains of bacteria.[22] What worried many nonscientists, in addition, were various social, ethical, and ecological risks that might not be immediately manifest and that most laboratory experiments would never address. These ranged from the moral dilemmas of human genetic engineering, to the danger of disturbing "the infinitely complex and delicate balance among living things";[23] and, as commercial and military involvement in genetic engineering increased, from the unknown hazards of mass production and deliberate release, to the specter of new and more dangerous forms of biological warfare.

A key issue underlying such concerns was the unique power of genetic manipulation to create new life forms with potentially irreversible consequences. Given these stakes and the admitted scientific uncertainties, public interest organizations argued for caution. In 1977 congressional hearings on proposed federal regulatory legislation, several environmental and labor groups called for an immediate moratorium on the research pending further investigation. Representing the Oil, Chemical, and Atomic Workers Union, Anthony Mazzocchi charged that continuing the research would amount to allowing "industry and academic institutions to play roulette with the lives of this particular generation and those of generations to come."[24] A few well-known intellectuals, such as anthropologist Margaret Mead, also pleaded for restraint while the uncertainties were being explored.[25]

Most of the research community took a very different position. Dismayed by the public apprehension and criticism that had developed, many scientists found considerable comfort in the results of early risk assessment experiments. An effort began to convince the public that the risks, as scientists defined them, were far less than originally feared. By 1977, leading investigators were backpeddling hard. Speaking at an NIH meeting, James Watson, a Nobel Laureate biologist, compared the risks of recombinant DNA research to those of "being licked by a dog." "Science," he stated flatly, "is good for society."[26] Others questioned the intelligence and training of anyone concerned about risks. The president of the American Society for Microbiology called critics "uninformed laymen and scientists who . . . have little understanding of the mechanisms of spread of infectious agents." "Kooks, shits and incompetents" were Watson's terms for them.[27] Such attacks, in addition to more subtle forms of censure, took a heavy toll on many scientist-critics.

Mainstream scientists testifying against regulation spoke little of the broader risks that concerned many public groups, and when they did it was mainly to discount them. For example, Dr. Oliver Smithies, a noted molecular geneticist, predicted in 1977 congressional hearings that human applications of genetic engineering were scientifically remote. (Three years later, an unsuccessful attempt at human gene therapy occurred.) "But," Smithies added with supreme confidence, "even if the chances were good, I would not be concerned. Why would rational individuals want to create monsters? They would not."[28]

The deliberate release of genetically engineered organisms also evoked radically different responses from scientists and citizens. Local opposition to the first approved outdoor tests was so strong in Monterey County, California, that the board of supervisors, meeting before a capacity crowd in early 1986, voted to place a moratorium on the experiment. In the field test, proposd by a small California biotechnology company called Advanced Genetic Sciences (AGS), microbes that had been genetically altered to prevent frost damage to crops were to be sprayed on 2,400 strawberry plants. The board of supervisors and county residents listened to a series of scientific experts, AGS officials, and federal regulators testify that the experiment posed few if any hazards, but were more convinced by testimony stressing the inadequacy of current scientific information regarding risks. The board was hardly reassured that the bacteria were harmless when it learned that the company employees who were to spray them on the plants would wear protective gear, and that the test plot was to be incinerated afterward. "We are playing Russian roulette," said one board member, prompting a round of applause.[29] Alarmed citizens formed a group called ALERT (Action League for Ecologically Responsible Technology), whose president, a tree farmer in Monterey County, chastised federal and company officials for their failure to inform the community of the intended tests, warning that such disregard had fueled fears that the whole biotechnology industry was "a disaster waiting to happen."[30] A few months later, community opposition brought a similar field test in northern California to a halt.[31]

What accounts for the divergent reactions of experts and nonexperts to risks? First, as in the case of DES, the range of concerns often varied markedly. Experts considered primarily technical and scientific matters, whereas lay fears encompassed a broader array of issues. The context within which risks were interpreted was also different, with nonexperts tending to judge risks on the basis of real-world experiences. OCAW's Mazzochi, for example, viewed the containment of recombinant organisms against a long history of occupational hazards, and concluded: "We know of no situation where any toxic substance has been successfully contained in any industrial establishment . . . What is in will out."[32] Psychological research comparing expert and lay perceptions of risk confirms that although lay people may lack certain factual information about hazards, "their basic conceptualization of risk is much richer than that of the experts and reflects legitimate concerns that are typically omitted from expert risk assessments."[33]

Experts and nonexperts also seem to have different attitudes toward uncertainty. To many nonexperts, the possibility of unknown dangers justified considerable caution. This response is consistent with another body of psychological research showing that most people prefer to forego benefits rather than to risk harm in the hope of future gain.[34] Experts, in contrast, tended to be more optimistic about benefits, as has been discussed, and less worried about unknown risks, especially human errors. In the recombinant DNA debate, human fallibility was one of the crucial points stressed by those who advocated caution. "We know so much about genes that we forget how much more we don't know," warned Ethan Signer, a prominent critic, in 1977. "We will slowly move from high level containment to low level containment to large scale production to buying the hybrids at the local drugstore. At each stage we will get a little cockier, a little surer we know all the possibilities. And then one day we will learn that the hybrid can also do something we hadn't thought to look for . . ."[35]

Finally, divergent views of risks may stem from underlying value differences. Perceptions of risks reflect more than just the perceived likelihood of adverse events. They also incorporate the values people attach to those events. Viewed in this light, public attitudes toward risk represent not "irrational" perceptions of present reality so much as people's hopes and preferences for the future.

Soft issues and hard choices: broadening and humanizing
biomedical policy

Costs, risks, and benefits are major elements of today's "rational" policy discussions. But not all issues fit neatly into this framework. The emotional anguish of a DES mother cannot readily be translated into dollar costs. Nor is it easy to balance the possible environmental and ethical risks posed by genetic engineering against its considerable commercial promise. Choices about such issues are fraught with uncertainty and depend on widely varying personal and social values. They are, inevitably, "soft." Yet in many ways, these choices are the most important of all, for they reflect basic notions of justice and humanity. Calling attention to these choices – the "soft" side of policy decisions – has been another valuable function of public participation.

For example, at congressional hearings in 1975, two DES mothers and a DES daughter recounted their personal tragedies, pleading for a total ban on DES. Their testimony was dramatic and moving; it reached millions of Americans on national news that evening. Consumer representatives on the 1978 DES Task Force, including DES mothers and daughters, likewise drew attention to the emotional and psychological problems of the DES-exposed, as well as to practical issues such as appropriate forms of contraception for DES daughters and their fertility problems. According to one physician member of the task force, the presence of individuals who had been personally affected by the drug helped to remind the group that it was "discussing a *human*, not an academic problem."[36] Although there are no handy formulas for factoring the emotional dimensions of problems into decisions, simply making them visible, and evocative, helps counteract the tendency for policy discussions to become so abstract that, as one authority put it, they "anesthetize moral feelings" of decisionmakers.[37]

Wider involvement of professionals in humanistic or ethically oriented disciplines can also help to promote awareness of the "soft" side of policy issues. In the artificial heart program, it was not until the largely nonmedical 1973 Artificial Heart Assessment Panel issued its report that underlying moral and social problems received serious consideration. The panel stressed the need for ongoing public monitoring of the program, and raised ethical concerns about human experimentation, informed consent, criteria for patient selection, and the quality of life of device recipients. It urged that access to the artificial heart be based on medical criteria alone, excluding any consideration of patients' "social worth" or ability to pay. Looking back on the work of the panel, Harold Green, its chairman, concluded that its most important contributions had been to demonstrate "that a broadly interdisciplinary group of nonexperts could comprehend and effectively cope with the potential impact of a relatively narrow piece of technology," and that "technology assessment and cost–benefit analysis could be performed meaningfully in qualitative rather than quantitative terms."[38]

Social and ethical issues have figured prominently in the controversy over genetic engineering, due largely to the efforts of religious groups, public interest organizations, and a few outspoken scientists. Many warned of the consequences of scientists "playing God" and the perils of using genetic engineering methods on humans. Others foresaw a whole new set of moral dilemmas for which society was unprepared, choices which would ultimately affect "our understanding of ourselves and of our relation to nature."[39] Organized religion has played an active role in the debate. The National Council of Churches, representing the major Protestant denominations, sponsored a number of meetings and symposia, and appointed a special panel to review the emerging theological, ethical, and policy implications of the new technology. In 1980, after the Supreme Court's decision allowing patents on new life forms, the general secretaries of the Protestant, Jewish, and Catholic national organizations called for the government to launch a thorough investigation of the issues raised by genetic engineering and to "provide a way for representatives of a broad spectrum of our society to consider these matters and advise the government on its necessary role."[40] Seven years later, religious groups joined members of Congress and a broad coalition of public interest organizations in an effort to prevent the U.S. Patent Office from issuing patents on genetically engineered animals until Congress had explored the moral, ethical, and economic implications of this step.[41]

Although changing scientific attitudes toward the risks of genetic manipulation have partly allayed some of the initial fears, the technique's rapid proliferation and future application to humans remain a source of continuing concern. A 1986 poll showed that over half of the general public believed that genetically engineered products were at least somewhat likely to represent "a serious danger to people or the environment," and that 42 percent thought that altering the genetic makeup of human cells was morally wrong, regardless of the purpose.[42]

To some observers, the most fundamental ethical issue was how the new technology would be used. Critics argued that "genetic fixes," which themselves posed potential risks, should not be used as "Band-Aid" solutions for problems whose roots were economic and political. Sophisticated new technologies, they warned, unless accompanied by more basic forms of social change, would do little to eliminate the underlying conditions of hunger and poverty. Ethan Signer pointed to the failure of the Green Revolution, the last "technological miracle," in relieving world hunger, observing that it had "only made the rich richer and the poor even poorer." "The problem is political, not technological," he concluded, "and it's going to be with us until there's a political solution. People are hungry, but it's not for lack of recombinant DNA research."[43] The National Conference of Catholic Bishops, although more circumspect, also implied that other forms of "progress" might sometimes be preferable to genetic engineering.[44]

Such arguments took genetic engineering out of the narrow framework of health and safety and placed it in the broader context of appropriate technology. The new genetic techniques might be positively harmful, warned groups such as Science for the People, if they were substituted for preventive or nontechnological measures: A handy "bug" that can clean up oil spills would do little to promote more effective efforts to prevent the spills from occurring; and ready supplies of insulin and other genetically engineered drugs might discourage the use of nonpharmaceutical methods of disease prevention and treatment.[45] Such fears were not long in being fulfilled. A bias toward chemical technology over natural prevention was evident in a 1981 article in the *New England Journal of Medicine* reporting the successful treatment of short children with genetically produced human growth hormone. An accompanying editorial noted that although "commercial DNA technologists will applaud" the findings, the study had virtually ignored the role of nutrition, the major influence on growth, as either cause or possible cure.[46]

The controversy over genetic engineering also raised basic questions about the nature of contemporary science and medicine. Critics claimed that a more "popular" approach to scientific investigation, one involving lay people directly, might yield more socially useful knowledge. Although this vision remained rather abstract, it enjoyed passing popularity, even among the scientific community. Far more contentious was the question of whether some forms of investigations might ultimately be too dangerous to pursue. This was a troubling reminder of the role of American research in developing the atomic bomb, and a number of scientists held that perhaps certain types of knowledge should not be sought. This argument shook the very foundations of the scientific establishment, and it reacted accordingly. The rallying cry was freedom of scientific inquiry, and stories of Galileo and Lysenko were resurrected.

In all four cases, the concerns of religious organizations, public interest groups, and dissident scientists expanded the boundaries of policy discussions, amplified the human and spiritual side of policy problems, and, by focusing on "the big picture," linked technical issues to the larger social, political, and economic context. Yet, policymakers frequently ignored the concerns and options raised, especially those that ran counter to mainstream professional and economic interests. Accordingly, the third persistent objective of public groups in their efforts to affect policy was to change the process of decisionmaking itself.

Democratizing decisions

On one level, efforts to make the policy process more open, accountable, and democratic may be viewed as purely procedural. They offered public groups the hope of greater access to policy decisions and were typically justified in terms of public rights. Yet such changes also had important substantive implications, since they often essentially ensured different outcomes. Furthermore, many choices simply cannot be evaluated in terms of results. The issues involved are so complex and ambiguous that the only standard possible is the soundness of the decision process. A persuasive proponent of this view has been U.S. Appeals Court Judge David Bazelon, drawing on extensive experience with federal regulatory decisions. Bazelon advocates public scrutiny, discussion, and criticism to encourage the fullest possible consideration of relevant data and viewpoints, which, he argues, provides the essential basis for good decisions.[47] On this more substantive level, then, open and democratic policymaking can also be justified in terms of improved decisions.

In the four cases, proposed changes were most often based on an appeal to public rights. For example, the Artificial Heart Assessment Panel, citing the notion of "societal informed consent," maintained that, at a minimum:

1. The public has a right to reject a biomedical innovation if it conflicts with important social values.
2. Researchers have an obligation to disclose the risks involved in any development.

3. It is desirable for society to evaluate new technologies carefully before they are put into use.

To implement these goals, the Panel recommended the creation of a "permanent, broadly interdisciplinary, and representative group of public members" that would monitor and help direct the program.[48] This recommendation was never acted on, however, and following DeVries's move to Humana, control of the artificial heart program passed largely from the public sector into the hands of private corporations.

To some participants in the recombinant DNA debate, the public's right to participate in policy decisions was the paramount issue, transcending concerns about risks and benefits. As David Clem, a member of the Cambridge City Council, put it at a 1977 Senate hearing: "The real issue before us is not recombinant DNA research. The basic issue is the right of the public to know and the right of the public to decide."[49]

That right was taken quite literally in Cambridge and in the handful of other towns and states that also passed ordinances regulating the research. For a time, it also had strong support in Congress, most notably from Senator Edward Kennedy. Kennedy told a 1977 Senate hearing: "The assessment of risk and the judgment of how to balance risk against benefit of recombinant DNA research are responsibilities that clearly rest in the public domain. And it follows that these issues must be decided through public processes."[50] The *New York Times* declared bluntly: "Regulation is too important to be left to the scientists."[51] Various forms of regulatory legislation were submitted to Congress during the winter and spring of 1977. The bill proposed by Kennedy had the most far-reaching role for the public: a freestanding commission, including a majority of nonscientists, which was to make regulatory decisions and also to consider the broader, long-term policy implications of the research. Through this commission, the public would participate in the crucial initial step of delineating the issues to be decided as well as in the decisions themselves.

In the political arena, the struggle over public rights quickly became a struggle over power. The scientific community viewed all of these developments with growing alarm. The prospect of national legislation was distasteful, but even more threatening was the specter of Cambridge and the possibility of a hodgepodge of regulations varying from one town to the next. To head off this possibility, scientists organized a lobbying effort in Washington representing a broad coalition of scientific organizations and professional societies. After consulting with professional lobbyists for advice on strategy and tactics, coalition leaders agreed on nine target principles. High among these was preventing local communities from setting their own safety standards for the research, a goal that was also strongly endorsed by pharmaceutical companies and others with an interest in the nascent biotechnology industry. In explaining the involvement of the once pristine professional societies in the dirty world of politics, the president of the American Society for Microbiology cited the need for "scientific expertise" in evaluating the issues.[52] Critics charged it was a self-serving attempt to protect professional autonomy.

Whatever the motives, the science lobby had a powerful impact on Congress. Legislators were deeply disturbed by the ominous warnings they heard from scientists forecasting the loss of valuable health, medical, and economic benefits if restrictive regulations were adopted. In vain, public interest groups defended the need for legislation, arguing that local initiative was vital in promoting wider understanding and acceptance of the new technology, and that the Cambridge experience had shown that public decision-making could be intelligent and responsible. Support for federal legislation of any sort eroded rapidly under the influence of the science lobby combined with the impact of new data suggesting fewer research risks. Finally, Kennedy withdrew his support for legislation, citing scientists' fears of overly restrictive regulation.[53]

With legislation doomed, public interest groups redirected their energies toward the NIH guidelines governing the research, which were undergoing major revision. Open and democratic decision-making was again a principal concern. Environmentalists, labor representatives, and dissident scientists all criticized the draft revisions on numerous counts, including their vague provisions for public participation. Several critics also reiterated their by-then familiar objection to the potential conflict of interest in having NIH, the major sponsor of the research, draft guidelines intended to regulate it. Revised guidelines, issued in 1978, responded to many of these criticisms. Although containment levels were lowered, the new guidelines went further than any previous draft in providing for more broadly based policymaking. They mandated wider representation in both federal and local policymaking, requiring 20 percent of the NIH's Recombinant DNA Advisory Committee ("RAC") members to have legal, social, environmental, or occupational health backgrounds, and 20 percent of each local institutional biosafety committee to be drawn from outside the institution. Moreover, to the surprise of some, a few known critics were also appointed to the RAC.

This was a gratifying, if ultimately somewhat hollow, victory for public interest advocates, who had long complained about RAC's narrow composition. For example, MIT professor Jonathan King, a vocal critic of the research, had charged back in 1975 that the committee was designed to "protect geneticists, not the public," and that having the committee chaired by a recombinant DNA scientist was akin to "having the chairman of General Motors write the specifications for safety belts."[54] Faced with such criticism, RAC added a few nonscientists in 1976, but with little enthusiasm. The committee's chairperson wrote candidly to NIH: "Like many other present members of the committee, I'm not sure [a nonscientist] could contribute to the deliberations, but I *am* sure that we need one for the purpose of being able to say we have one when there are complaints."[55] Even after 1978, nonscientists remained in the minority on RAC and had little impact on key policy decisions. Containment levels were steadily lowered, and public accountability remained largely symbolic. Community involvement in local biosafety committees likewise had no dramatic effect on policy trends. Yet it did show that lay participation could be both feasible and constructive; the presence of nonscientists seemed to encourage more comprehensive discussion and more critical scrutiny of research proposals. According to one committee chairman, "the public members proceeded to sensitize the entire committee to issues that they would otherwise not have considered."[56]

Sometimes demands for broader decisionmaking couched in terms of public rights converged with arguments based on better results. One example was the involvement of laboratory workers in safety decisions. Critics argued that workers had a special right to participate in decisions about safeguards because they would suffer greatest harm if health hazards developed; and, furthermore, their familiarity with day-to-day laboratory practices would improve the final decisions. Broader public involvement was defended on similar grounds. To strike the proper balance between costs and benefits in research policies, some observers proposed, decisionmaking should involve both the larger public and laboratory workers, who were potentially at greatest risk from the research, as well as researchers and corporate officials, who were likely to receive the most direct benefits.[57]

That these various demands, which clearly implied a shift in existing patterns of authority, provoked a struggle for power was not in itself surprising. But the intensity of this struggle suggested a deeper clash of values. When Ford administration officials refused to debate Sidney Wolfe about the swine flu immunization program, for instance, Wolfe protested to the Secretary of HEW: "You want to stifle public debate on these serious problems?... If HEW can't live with such debate, perhaps its leaders and Mr. Ford should consider whether they are comfortable living in a democratic society?"[58] Critics in the recombinant DNA controversy claimed that attempts to suppress dissent reflected the subversion of democratic principles by narrowly defined expertise. As a widely cited article appearing in 1977 in the *Atlantic* put it, "the ultimate question is not whether bacteria can be contained in special laboratories, but whether scientists can be contained in an ordinary society."[59]

Even when the controversy shifted from containment to deliberate release and biotechnology company entrepreneurs joined scientists as central decisionmakers, the issue of public accountability remained central. In Monterey County, for example, citizens were outraged when they learned that the world's first "authorized release" of gene-altered organisms was to take place in their own back yard without anyone – the EPA, the company (AGS), or the state – so much as even informing local governments or community residents. If the experiment was not dangerous, people said, then they would not have been kept in the dark, and the company would not have insisted on keeping the test site secret. Even more shocking to many was the revelation that neither EPA nor the state had actually visited the company's proposed "remote" test site prior to approving the experiment, so were unaware that it in fact bordered a residential area. "The company has no credibility now with local citizens," said Glenn Church, president of the citizen's group ALERT, "and it was clear that we couldn't rely on the regulatory agencies to do their jobs on their own."[60] To AGS, contrite about its failure to communicate with the community, it was a pointed lesson in the animosity and suspicion that arise when people feel they have been excluded from regulatory decisions. "What we saw at the hearing was wrath because they were not part of the process," said the AGS director of marketing. "We didn't anticipate this."[61]

It is perhaps peculiarly American to put so much emphasis on due process and democratic decisionmaking. Yet, with fundamental disagreement about the "right" decisions on complex scientific and technological issues, a fair and open process may be the only way to arrive at results that are accepted as sound and legitimate. Whatever the reasons, opinion polls reveal steadily growing support for increased control by society over science and technology.[62]

The impact of public efforts

Although public involvement in the cases generally led to more discussion and a broader view of issues and options, it often had less effect on policy outcomes than proponents would have liked. Yet the impact should not be underestimated. Although the swine flu program went forward, adverse publicity generated by public interest critics and dissident experts undoubtedly lowered immunization rates. Research on the artificial heart continues; surgical heroics are still glamorized in the popular press; and there is no mechanism for public involvement in the program's direction. On the other hand, federal funding for research on the nuclear-powered heart was discontinued following strong opposition from the 1973 Artificial Heart Assessment Panel.

In the case of genetic engineering, although federal regulatory legislation was never passed and commercial applications of biotechnology have mushroomed, some cities and towns passed local ordinances to regulate the growing commercial market. In some communities, grass-roots protests succeeded in delaying proposed field tests of gene-altered organisms while more information was obtained, or in getting the test sites moved to more remote areas. And, while broader representation in federal policy on genetic engineering did not appreciably alter previous policy trends, community participation in local biosafety committees was shown to be both feasible and worthwhile.

For DES, the picture is likewise mixed. The use of DES and other estrogens as postcoital contraceptives continues to elude regulatory controls, and most DES victims receive no special medical assistance or financial compensation for their injuries. But activities in many states and communities have done a great deal to inform and help DES victims. Most of the credit for these activities belongs to DES Action, an organization based almost entirely on local grass-roots efforts. Its accomplishments are worth recounting briefly.

The origins of DES Action go back to 1974, when Pat Cody and a few other women in Berkeley, California, dismayed at the continuing lack of information and resources for DES victims, decided that something had to be done.[63] They formed a group to work on the problem and developed an informational pamphlet. This pamphlet circulated widely through informal networks and, as the information spread, so did concern and commitment. By 1978, at least five other DES groups had formed around the country. They agreed to establish a national network, with a common name and common objectives. Their motto: "Don't Mourn, Organize!!!"[64]

This motto could hardly have been more apt. By 1987, there were over sixty DES Action groups around the country and throughout the world. Their activities have been manifold, ranging from public and professional education, to the development of technical resources such as audiovisual materials and physician referral lists. Two national DES newsletters provide current information on medical, legal, and legislative developments.[65] Largely as a result of lobbying by DES groups, more than fourteen states have considered or passed bills or resolutions dealing with DES, and at least three (New York, California, Illinois) also appropriated funds. One of the most effective laws was New York's, passed in 1978, which set up special DES-screening centers around the state and a DES registry for research and follow-up. An intensive media campaign mandated by the N.Y. legislation almost doubled public awareness of DES in the state.[66]

DES advocates have been active nationally as well, testifying at various congressional hearings. DES Action lobbyists were instrumental in getting HEW to form federal task forces on DES in 1978, and again in 1985. Letters from local DES action groups helped convince Congress to declare a "National DES Awareness Week" in 1985.[67] The efforts of DES Action may also have helped sensitize judges to the special legal problems of DES victims. Starting with the *Sindell* decision in California, as discussed in Chapter 8, a number of courts have allowed DES daughters to sue major DES manufacturers without having to identify the specific brand that caused the injury – in effect, shifting the burden of proof from the victim to the manufacturer. In California, several industry-sponsored attempts to overturn the *Sindell* ruling were defeated, and press coverage gave credit for these defeats to "DES victims," in particular the opposition of DES Action.[68] In New York, six years of lobbying by DES Action finally led, in 1986, to the passage of a law adopting a three year "discovery rule" for filing DES lawsuits, thereby removing an important obstacle to recovery by DES victims.[69]

A crucial factor in DES Action's success has been the intense commitment of the women involved, many of whom have been personally affected by DES. The consequences of DES exposure are frightening and potentially lethal. Ties to the women's health movement helped transform these personal fears into political action. Women's health activists, all too familiar with being marginal to mainstream medicine, financially threatening to the drug industry, and low priority for the federal government, have schooled DES advocates in the tactics of self-help through collective action, sharing resources and contacts. DES Action has also benefited greatly from the involvement of compassionate physicians.[70] But it has been DES Action's ties to its local constituents that are its greatest source of strength. This community base has kept the organization publicly accountable while allowing it to flourish entirely outside any formal structure of government, academia, or industry. With this base, DES Action has shown growing numbers of women how to take political as well as personal control of their medical destinies.

Keys to public involvement and influence

The public interventions attempted in the four cases varied widely in their impact on policy outcomes. From these experiences, we can identify certain factors that tended to promote broader and more effective lay involvement.

Both DES and genetic engineering posed real or potential health risks that first sparked public concern and activity. As these issues came to reflect lay rather than scientific priorities, the focus on risks typically expanded to encompass a wider spectrum of concerns: the long-run social, ethical, and political implications of genetic engineering, the real-world problems of the DES morning-after pill, and various medical, financial, and emotional needs of the DES-exposed.

Along with this natural broadening of focus, interested groups also made a deliberate attempt to relate specific concerns to larger issues, in order to create working coalitions. Genetic engineering was linked with environmental and occupational health problems, and DES was linked with other women's health issues. Although some new ties were temporarily forged, however, most existing organizations retained a predominantly single-issue focus.

Adequate access to technical resources and expertise was critical in allowing nonexperts to take part in disputes that were often highly technical. After scientists and medical experts had first called attention to risks, only a few "dissident" experts were willing (sometimes at considerable personal sacrifice) to work with lay groups as the controversies evolved. Women's groups working on DES allied with sympathetic physicians, and public interest advocates in the recombinant DNA debate depended heavily on a small number of scientist-critics. This expertise was indispensable in interpreting technical developments, and it lent credibility to the arguments of lay groups.[71]

Also important were the organized constituencies and political contacts of the public groups concerned about genetic engineering and DES. Scientist-critics, some associated with Science for the People, activists in the women's health movement, and environmental, labor, and consumer groups had all dealt with political controversies before, and made use of the skills and networks they had developed. Their larger constituencies helped supply added political pressure.

Barriers and boundaries

A number of significant barriers to public participation are also apparent in the case studies. One obvious problem was the extremely limited opportunities for formal citizen involvement in decisionmaking about the four cases and the constraints placed on those opportunities that did exist. Official public representatives on government committees and review boards were usually in the small minority, often could not vote, and typically had the least financial and technical resources.[72] Local citizen participation was more feasible, but less likely to influence national policy.

Much formal public involvement thus turned out to be largely symbolic, allowing token contributions without any real control over decisions. The nonscientists and critics added to the RAC in 1978 comprised a small fraction of the total membership, and could neither slow the steady dismantling of the NIH guidelines nor get the RAC to consider ethical issues or troubling questions about the technology's commercial or military exploitation. Moreover, while these members privately questioned their own effectiveness on the RAC, their presence created the image of balanced, broadly based policymaking. By silencing complaints about exclusion of dissent in this way, symbolic participation may end up legitimizing the existing distribution of power.

Informal ad hoc public intervention was generally equally difficult. Outsiders, especially those with controversial views, often had trouble gaining effective access to the massive and byzantine federal policymaking bureaucracy. Even the boldest and most persistent efforts by recombinant DNA and DES activists had relatively little impact on national decisions. In turn, bureaucratic and ideological obstacles were compounded by the apathy and cynicism of the large majority of the public, accustomed to powerlessness and passivity vis-à-vis science and technology. If "let the experts decide" was no longer the comfortable solution it once seemed, ready alternatives were not, for the most part, apparent.

Another barrier to public participation was the inadequate and uncertain data available on risks, especially the human, social, and ethical consequences of scientific and medical innovations. Because these concerns were also difficult to integrate with standard quantitative criteria for decisionmaking, a common, if hardly optimal, solution was simply to exclude them altogether from official policy deliberations.[73] And, of course, it was usually scientific or medical experts who decided what information on risks should be collected, and how it should be interpreted and presented for public consumption. Although dissenting experts could offer alternative views, the positions taken by mainstream scientists and physicians nevertheless had a powerful impact on public opinion.

The diffuse and uncertain character of many of the issues meant that, like most areas of health policy, they were not high priority for most people – until problems arose. DES did not lack for a mobilized public once the link with adenocarcinoma was discovered, but by then much of the damage had been done. Public apprehension over the risks of genetic engineering, by contrast, as with nuclear power and weaponry, suggests an increasing desire to prevent future tragedies rather than merely mopping up their aftermath. But concern with prevention still does not match the catalyzing effects of a disaster that has already happened.

Public interventions that were effective invariably produced a powerful backlash from established interests. In California and other states, medical and industry groups tried to pass legislation that would limit consumer protection and block many DES lawsuits. Similarly, faced with the prospect of federal regulation of recombinant DNA research, scientists mounted a well-orchestrated lobbying campaign which succeeded in derailing proposed national legislation. Many "dissident" scientists who sided with public groups in the struggle over legislation were ostracized by the scientific establishment.[74]

The private sector, a major source of biomedical innovation, has always been essentially off-limits to direct public intervention. The Artificial Heart Assessment Panel could recommend discontinuation of federally funded research on nuclear-powered hearts, but was powerless to stop corporate research. Private industry is well represented on most government advisory committees, and has considerable influence in government decisions as well as enormous informal sway. In 1977, it is estimated that corporations spent 50–100 times as much as consumer groups lobbying federal agencies.[75] The promised commercial benefits of biotechnology reinforced the push for rapid and unfettered development of genetic engineering. Pharmaceutical and bioengineering firms joined with scientists in defeating federal regulation in 1978, and lobbied heavily against state regulation. These events point to definite boundaries on what citizen groups can accomplish when private sector interests are threatened.

Corporate ties with academic institutions gave a veneer of intellectual respectability and scientific credence to private industry's pursuit of profits. These ties tended to inhibit independent criticism of corporate practices by academic experts, effectively silencing much potential opposition to private interests in matters of social or political controversy. Several members of the 1973 Artificial Heart Assessment Panel reported, for instance, that people with dissenting views were sometimes afraid to speak up "on the record" for fear of losing favor with federal or industrial sponsors.[76] The involvement of academic scientists in biotechnology companies has clearly had a chilling effect on the debate over genetic engineering. The more extensive the ties with private industry become, the more the norms of science and democracy are subordinated to the demands of power and privilege. The result, in the words of two astute observers of recent trends in science and technology policy, is "not merely a technically more complex political universe, but a quieter one, in which the voices of popular resistance and control are no longer heard."[77]

The most formidable obstacle to broader representation of public interests in biomedical policy is the present structure of economic and political power. Scientific, professional, and corporate groups have a strong vested interest in seeing that innovations go foreward, and they typically have substantial financial, organizational, and technical resources to invest in pursuing those interests. Most public groups, by contrast, as well as society as a whole, have a much less direct stake in the outcome of given policy choices, and usually can draw on only limited economic and institutional resources.

This imbalance in stakes and resources is structural; it is not alleviated by procedural reforms alone. Requirements for open meetings enacted during the 1970s, for example, ended up benefiting special interests more than the public. "When we opened the doors," remarked one Senator, "it wasn't the American people who came in, it was all those interest groups. The American people aren't here, but the National Rifle Association is."[78] For participation to counteract, rather than magnify, the influence of vested interests, special provisions are necessary to compensate for the relative disadvantages of groups with less power, fewer resources, and hard-to-organize stakes. Without substantial redistribution of decisionmaking power, any form of public participation will be an empty and frustrating process for the powerless, and can actually serve to reinforce the status quo.[79]

Questions and arguments about public participation

In the case studies, the most common justification for public participation was the right of citizens in a democracy to a voice in major social issues. Many biomedical policy developments have important and far-reaching consequences for society as a whole and involve large expenditures of public funds. The case studies also suggest that broader public involvement can benefit medical science, by introducing a wider range of concerns, perspectives, and relevant information, and by preventing narrowly defined special interests from having inordinate influence. The resulting decisions should thus be more fully informed, more thoroughly scrutinized, and hence more socially acceptable. In this sense, public participation can be said to produce "better" decisions.

What are the principal arguments against public participation? A number revolve around questions of feasibility. It is obviously not possible to involve every citizen in every policy decision. That is the extreme case, however, and not a convincing argument against involving *some* lay people in *some* aspects of biomedical policy. Yet it does raise the valid question of which people should be involved in which policy issues. It is no simple matter to decide how "the public," in all its rich and subtle diversity, should be represented.

Political theorists have devised a multitude of representational schemes, but the most compelling ones, for matters of health policy, are based on the concept of interests. Political scientists Theodore Marmor and James Morone have argued that all of the principal interest groups affected by policy decisions, either positively or negatively, have a right, through their surrogates, to a voice in those decisions.[80] If the choices potentially affect the whole society, as do most of those concerning medical innovation, the interests of all major population groups should be represented in decisionmaking, although participation might vary depending on the type of innovation. It is not difficult to specify a set of categories that, although necessarily arbitrary, would significantly broaden the range of interests that presently dominate biomedical policymaking. As examples, Marmor and Morone list the poor, racial minorities, the elderly, women, and third-party payers. Representation should be designed to redress basic structural inequalities in stakes, power, and resources. Various forms of support, including educational and organizational assistance, funding for expert consultants, and financial reimbursement for time and expenses, could all help to compensate for the unequal skills and resources of groups that have traditionally been excluded.

In designating interests to be represented, special consideration should be given to those likely to be harmed by biomedical innovations, such as communities exposed to particular risks, or population groups likely to lose services if funds are diverted into new areas. At present, as noted, these groups generally have less access to biomedical policymaking than those likely to benefit from technological innovation – well-endowed corporate, scientific, and professional interests. To correct this imbalance, Sheldon Krimsky, a former member of the Cambridge Experimentation Review Board and now an authority on the genetic engineering controversy, has proposed a "weighted input principle" of decisionmaking in which the groups or individuals asked to bear the greatest risks from policy decisions would have the largest role in the choices made.[81] The rationale for such a principle is similar to that proposed by John Rawls in *A Theory of Justice*, which argues, in essence, that if society chooses to distribute resources unequally, further inequalities should not make the least advantaged worse off.[82]

Another structurally underrepresented interest that deserves special consideration is the collective "public interest" – concerns we all share but which no one individual or organization has a particularly strong incentive to pursue on its own, especially when they conflict with private aims. In medicine, as we have discussed, the lack of effective representation of the common good results in what has been called the "tragedy of the medical commons." The pursuit of individual interests, regardless of how broadly representative, does not always produce outcomes that meet collective needs. There should be explicit provisions for representation of society's collective interests in biomedical policy decisions that cut across, but may be neglected by, individual interests.

It serves little purpose to represent these various interests, of course, without some way of holding the representatives accountable to their constituencies. Accountability is a special problem in health matters because of their low salience for most people most of the time. The traditional accountability device of democracy, selecting representatives by majority vote, does not work very well, as the low voting rates in most community-based health organizations attest. Again, there are no perfect solutions to this problem, but a variety of less-than-perfect approaches that would still make for a considerable improvement over present practices. For example, simply allowing designated constituencies to choose their own representatives, by whatever means they want, should lead to greater accountability than when, as is now generally the case, they are chosen by government agencies.[83] Good old-fashioned political debate also serves as an important, albeit underappreciated, mechanism of accountability, by requiring participants to air their views about the impact of different policy measures on their constituencies. If it is understood that these views will be widely known, moreover, participants are more likely to attend not only to their constituencies' interests but also to cross-cutting common interests.[84]

Some opponents of broader participation in biomedical policy issues argue that it is unnecessary, given existing representative institutions of government, and that adding yet another layer of bureaucracy to that system will paralyze an already tortuous policymaking process. Moreover, broader public participation introduces new views and priorities that are bound to conflict, as previously excluded groups fight for their share of the shrinking biomedical pie. Democracy is rarely defended on the grounds of efficiency (although how one measures "efficiency" in the absence of agreed-on common goals remains unclear). While wider involvement is indeed more cumbersome administratively, the extra administrative effort required might well be fully justified if it helped to stem faltering public confidence in government. Furthermore, in a democracy, conflict over divergent priorities, at least within limits, is preferable to consensus based on repression of interests and exclusion of dissent – even if it does slow things down.

The most basic argument against public participation is that even if it is feasible, it is not desirable. Many scientists, and many nonscientists as well, believe that science and medicine are so highly specialized that only experts are qualified to make competent judgments. The advancement of knowledge, it is held, is best served by leaving decisionmaking to scientists, unburdened by public demands. Indeed, many scientists view public intervention as a dangerous intrusion of inexpert and alarmist ideas. For instance, Philip Handler, former president of the National Academy of Sciences, has stated that "most members of the public usually don't know enough about any given complicated technical matter to make meaningful informal judgments. And that includes scientists and engineers who work in unrelated areas."[85] He argues that such judgments should be left to "knowledgeable wise men" serving as trusted representatives of the public; citizens should assess the options determined by established government institutions and communicate their opinions to elected decisionmakers.

As this argument implies, the debate about how science should be governed is closely related to the larger question of how society should be governed. At the root of this question is the role of citizens in a democracy. According to the so-called classical theory of "participatory democracy," often identified with such philosophers as Mill and Rousseau, direct citizen involvement in political decisions is critical in ensuring wise and legitimate decisions and, equally important, in allowing the continuous education of citizens about important issues and in sustaining their sense of political "efficacy." Indeed, this theory holds that the human development that results from the participatory process is essential for the survival of democracy. By contrast, many contemporary political theorists contend that democracy is quite compatible with a much more limited role for citizens, confined essentially to selecting and rejecting leaders through the vote. In their view, leaders are the critical element of the system and the competition for leadership by elites is the defining characteristic of democracy.[86]

This contemporary view of democracy was widely embraced after World War II by political theorists seeking to explain and justify declining trends in voter participation. Schumpeter, Huntington, and others argued that direct participation in government was neither feasible nor desirable in the modern age – that inactivity on the part of a majority increases political stability. "The effective operation of a democratic political system," wrote Huntington, "usually requires some measure of apathy and noninvolvement on the part of some individuals and groups."[87] Others noted that the noninvolvement of lower socioeconomic groups in particular, characterized as having "authoritarian" personality traits, was especially functional in maintaining stability.[88]

The recent forays of lay groups into science policy have extended this larger debate into a new arena. Many of the issues and problems are the same, despite the esoteric nature of modern research, and the arguments invoked against broader participation have a familiar ring. Since the late 1700s, each successive extension of voting rights – first to all white males and eventually to women, blacks, and youths – was opposed on virtually the same grounds as many now give for opposing broader public involvement in science policy: that the unenfranchised group was too ignorant, uninterested, and unreliable to be allowed to participate, or that the group's participation would be too costly or disruptive to the system's efficiency and stability.[89] Today, universal suffrage is essentially unquestioned as a political right, even though never fully realized as political practice.

Evidence from the four cases casts doubt on dire predictions about the consequences of broader public participation in science policy. Many citizens and public groups clearly made useful and responsible contributions to policy deliberations, providing information and perspectives that often complemented and enriched those of scientific and medical experts. In general, wider involvement resulted in more comprehensive consideration of issues, greater attention to potential risks and possible alternatives, and a more realistic assessment of likely benefits. Although outsiders were often critical of existing policy, they were for the most part not trying to block progress. Rather, they wanted decisionmakers to incorporate a wider range of social values and priorities in policy choices. These and other experiences with lay involvement in policymaking do not suggest that the public is "antiscience" and wants to stop scientific and technological development. To the contrary, polls show that although most people recognize the harmful effects of science and technology, they nevertheless view science as instrumental in achieving important social goals.[90]

Opponents of participatory democracy down through the ages have warned of the dangerous and irresponsible passions of particular groups. This argument, too, arises in the ongoing debate over the control and direction of medical science. Even if the majority of the public is trustworthy, skeptics ask, should decisions about medical innovation be vulnerable to those with views that many consider to be extreme or wrong-headed? Populist movements in the past have been susceptible to the influence of extremists, and some have been intolerant of minority views. In the final analysis, the only cure for such problems is the education and understanding that come from political involvement. To the advocate of public participation, there is no better response than Thomas Jefferson's reply in 1820 to similar concerns:

> I know no safe depository of the ultimate powers of the society but the people themselves; and if we think them not enlightened enough to exercise their control with a wholesome discretion, the remedy is not to take it from them, but to inform their discretion by education.[91]

With the "ultimate power" that science now has in society, this response is as valid today as it was in Jefferson's time.

Any significant effort to encourage broader citizen involvement in biomedical policy will certainly lead to new questions and problems; and it will undoubtedly involve added time and expense. Yet, in the end, these costs may be a small price to pay if they help to build a more respectful and responsive relationship between science and society.

11

What is possible? Toward medical progress in the public interest

Medical innovation directly and intimately affects whether and how we live, and when and how we die. Research has unlocked technical capabilities that defy the imagination, and will continue to do so. With genetic engineering, scientists may leave their mark on the genetic heritage of the entire planet. Yet, as the case studies in this book document, we who use medicine's innovations, who bear their risks, and who, both as taxpayers and as consumers, pay for the products and the research that made them possible, have little if anything to say about the nature and pace of medical progress.

This book argues for an end to that exclusion; this chapter suggests the kinds of changes that might bring it about. It draws together many of the themes and issues that have been discussed throughout the book, looking at the lessons that can be learned and how they apply to present and future policies in the United States. These lessons become all the more urgent in light of medicine's deepening dilemmas and the failure of present policies to resolve them.

It is perhaps useful to stress again, as we did at the book's outset, that the four case studies were chosen to illuminate some of the major dilemmas posed by contemporary medicine. Although questions of costs, risks, efficacy, and equity are clearly of widespread concern, we do not intend to suggest that all medical innovations will encounter exactly the same types of problems as occurred in the four cases. Nor do we mean to deny the real and important benefits provided by many medical advances, or to ignore technology's undeniable potential for bettering the human condition. But if the case studies teach anything, it is that medical and scientific "advances" can do harm as well as good, and that judgments about both risks and benefits are not automatic or absolute but are made by individuals and agencies acting in the context of particular social, political, and economic pressures. For modern medicine to realize its full capacity for doing good, these pressures must be realigned to enhance the health of society as a whole. This is possible, but not without the active effort and genuine commitment of an informed citizenry.

Common pitfalls in decisionmaking

Previous chapters have discussed three critical policy issues illustrated by the cases – communication and control of medical risks, compensation of medically related injuries, and the allocation of health resources – and some of the underlying "macro-level" political and economic forces that contributed to them. Here we look at the impact of these same forces at a more "micro-level": how they shaped and conditioned day-do-day decisionmaking in each of these policy areas.

Although the four cases involve separate medical and scientific arenas, diverse actors, and varying time periods, certain characteristic flaws in decisionmaking can be identified in all four cases which, although not necessarily relevant to every situation, were typical of the general mindset and inclinations of key figures:

> *Technological optimism*: Experts tended to be overly optimistic about their ability to overcome scientific or technical obstacles, assuming that more research would permit mastery of scientific unknowns. They often saw benefits as nearer, and greater, than they were, while similarly inflating the dangers of *not* developing or applying the new technology.
>
> *Underestimation of risks:* Authorities tended to underestimate and undervalue the likelihood and seriousness of possible problems, especially those that were difficult to quantify. The need for more effective forms of risk assessment and detection was sometimes denied using the (self-fulfilling) argument that no risks had yet been demonstrated. No evidence of risks was taken as evidence of no risks.
>
> *Suppression of doubt and dissent*: Decisionmakers frequently downplayed the existence of substantial uncertainty surrounding benefits as well as risks, while minimizing or ignoring controversies about the inter-

pretation of scientific data and disagreements about the assumptions underlying such evidence.

"Hard" data dwarf "soft" concerns: Technical aspects of policy decisions tended to get most of the attention whereas less tangible concerns, such as the innovation's impact on social, ethical, and political values, were lucky even to get lip service. Secondary effects of new developments were often ignored, along with the likelihood of practical problems and human error.

Fragmentary and myopic view of problems: To cope with the complexity of most biomedical policy issues, decisons were piecemeal, reductionist, and incremental. Narrow definition of issues and options, based on existing professional or bureaucratic interests, made it difficult to understand the broader context and dimensions of problems, allowing only marginal changes in attempts to solve them.

Assumption of unlimited resources: Budgetary decisions were similarly piecemeal, with little if any explicit coordination of priorities across projects and programmatic areas. Especially during the 1960s, when NIH funding levels were rising rapidly, government officials assumed that additional resources could always be found and that explicit trade-offs were therefore unnecessary.

Inflexibility of decisions once made: Once having begun an innovation, those responsible were often resistant to new information that might threaten further development or alter its course. The initiation of a new program or research effort created a momentum of its own which could often override signs of impending trouble.

These problems did not go entirely unnoticed, even at the time. In all four cases, critics, including not only dissident medical and scientific experts but also lay people and public groups, raised one or more of these concerns. In the artificial heart program, for instance, outsiders criticized the overoptimistic assumptions implicit in early planning and the undervaluation of the radiation risks of the then-intended nuclear-powered heart. In the DES case, women's groups and health activists were considerably more alarmed than most experts about the risks of DES exposure, including the likelihood of an association between DES and breast cancer and the potential hazards of the morning-after pill. Throughout the history of genetic engineering, environmental, religious, and public interest groups have consistently tried to broaden policy debates to include questions about the possible misuse of the technology and the danger of adverse effects on the ecosystem and on human moral and spiritual values. Likewise, in the swine flu immunization program, lay critics ranging from the *New York Times* to public interest groups maintained almost from the outset that mass vaccination was unneeded and risky, and repeatedly urged program officials to alter plans in the face of mounting evidence that there would be no pandemic.

Yet, for the most part, such warnings, and criticisms appeared to have little impact on official policy decisions. In each case, a relatively small group of medical and scientific experts, sometimes only a handful of key people, easily swept aside most doubts and objections, launching medical innovations of unproven merit and unprecedented consequence for society, with only minimal outside scrutiny. By virtue of their central role in the early, formative stages, scientific and medical experts set the terms for much of the subsequent policy debate, from the range of options considered to the types of issues and expertise deemed relevant. Dissident scientists and medical professionals were generally discounted or demeaned, while lay critics were ridiculed or simply ignored.

A number of circumstances contributed to the silencing of dissent. Faith in the unalloyed benefits of science and medicine was much greater prior to the 1970s, and probing questions more likely to be brushed aside. Key members of Congress, sharing this optimistic outlook, supported most of the proposed innovations and generally took the experts' promises and predictions quite seriously. Many of the top officials in the NIH, FDA, CDC, and other government funding and regulatory agencies were themselves scientific or medical experts, whose backgrounds, biases, and outlooks were similar to those of the investigators proposing the innovation. Proponents of medical innovations had relatively easy access to administrative and congressional deliberations in comparison to critics, especially during the innovation's startup phase, and also more informal access to decisionmakers through overlapping professional and social relationships. Furthermore, as discussed in Chapter 10, proponents generally had a greater stake in the innovation than lay groups and critics, both professionally and economically, as well as greater financial and institutional resources and personal influence to invest in lobbying efforts and political maneuvering.

Expert-based, bureaucratic decisionmaking, with its characteristic weaknesses, is hardly unique to biomedical innovation. Analogous problems have been identified in policy arenas as diverse as nuclear power and foreign policy.[1] Commonly accepted precepts for effective decisionmaking can help to avoid many of these deficiencies. Most analysts agree that sound policymaking should accomplish five basic tasks:

1.   ensure that sufficient information is obtained and is analyzed adequately;
2.   consider all the major values and interests affected;
3.   ensure search for a wide range of alternative actions and expose uncertainties affecting each option;
4.   consider carefully the problems that may arise under each option; and
5.   maintain receptivity to indications that current policies are not working out well, and an ability to learn from and respond to experience.[2]

Evidence from the case studies suggests that broader public involvement in biomedical policy tends to promote each of these goals, as discussed in Chapter 10. By helping to protect the policy process against subversion by narrowly defined private interests, broader participation should increase medicine's responsiveness to public needs and wishes. We shall come back to possible mechanisms for expanding public representation in the biomedical arena later. Let us first look briefly at some of the consequences of medicine's lack of public accountability, as reflected in present policies toward medical risks, compensation of injuries, and allocation of health resources.

Present realities, future visions

The following sections contrast current trends in each of these three policy areas with expressed public interests and concerns. Alternative approaches that would be more consistent with public wishes are suggested, many of which have been shown to be practically feasible, if not always politically palatable to powerful interest groups. These suggestions do not represent a comprehensive solution to policy problems but rather illustrate general directions and strategies. They are intentionally modest, drawing for the most part upon measures with which there has been at least some actual experience.

Individual and collective risks

Risks to individuals and society as a whole are one of the prices we pay for medical progress. Some hazards are known, or can be discovered with more extensive testing, but new and untried therapies will always entail unpredictable risks. Only the people who bear those risks can legitimately decide whether they are worth taking. Patients have the right to be informed of all known major adverse effects of treatment before giving their informed consent. Informed consent seeks at a minimum to protect patients against unwitting exposure to unacceptable risks, and, ideally, to enhance patients' autonomy and permit their enlightened participation in medical decisions. It can be argued, likewise, that society as a whole has an analogous right to a voice in decisions about the development of major therapeutic innovations.

Unfortunately, even the minimal goal of informing individual patients of treatment risks is often unfulfilled, as the case studies demonstrate. Pregnant women given DES in a formal clinical trial reported being unaware of taking an experimental drug, thinking they were simply taking vitamin pills. The initial swine flu informed consent form made no mention of the known risk of serious neurological illness. Lack of informed consent was the legal charge brought against Dr. Denton Cooley by Mrs. Karp, wife of the first artificial heart recipient, and it was a charge made against Dr. Martin Cline by some of his peers for his unauthorized use of genetic engineering on humans.

Such failures of communication are surely not uncommon. Although the patient's right to informed consent has gained increasing acceptance in the medical community, many physicians still express grave doubts about whether patients can, or even should, be fully informed about treatment choices. Implicit in these objections is the essentially paternalistic assumption that patients are incapable of weighing risks and benefits intelligently and that physicians can and should make these judgments instead. When information is provided to patients, there is often little effort to make it comprehensible. Informed consent documents typically require college-level reading skills, and those that are longer and more detailed tend to be even more difficult (by 1985 the informed

consent form for the artificial heart was seventeen pages long).[3]
With the growing concern about liability, moreover, the underlying purpose of informed consent has been shifting from protection of patients against risks to protection of physicians and
institutions against lawsuits. One national survey revealed that 79
percent of those interviewed believed that the informed consent
document primarily protects the doctor – and 55 percent of doctors
agreed![4]

Opposition from the pharmaceutical industry and most of the
medical profession has also prevented development of another
promising source of information for patients: "patient package
inserts," leaflets describing in simple language a prescription drug's
proper use, risks, and side effects. Most people, including those
with less education, read such leaflets and learn from them, according to government-sponsored evaluations. Consumer organizations and women's health groups have expressed strong
support for them. Nevertheless, previous plans by the FDA to
develop and distribute patient package inserts on all prescription
drugs were shelved in 1981 by the Reagan Administration in response to fierce pressure from industry and professional groups.[5]

The issue, then, is not merely the failure of physicians to disclose
therapeutic risks and uncertainties but the abuse of professional
power and privilege. This larger problem can be solved only by
creating a new view of the doctor–patient relationship, one in
which both patients and physicians acknowledge their interdependent roles, share information fully, and reach decisions jointly.
Patients must be encouraged to think independently and critically
rather than to accept passively the paternalistic authority of physicians, while physicians must recognize the patient's authority as
an integral component of medical decisionmaking. In this context,
informed consent would not be an empty gesture toward liability
reduction but a means of transforming the risks of medical treatment from a threat to the doctor–patient relationship into the very
basis on which an alliance can be formed.[6] The chairman of the
President's Commission for the Study of Ethical Problems in Medicine called the formation of such a therapeutic alliance "the best
prescription for improving medical care and quickly reducing
medical malpractice claims and expensive verdicts."[7]

Many factors contribute to physician dominance in the doctor–
patient relationship, and actions at various levels will be necessary
to alter this imbalance. Leaders of academic medicine have recommended reforms that would give greater emphasis in medical

school selection and training to humanitarian and ethical concerns and to physicians' skills in communicating with patients.[8] The legal system could also encourge better dialogue between doctors and patients by shifting its attention from the pro forma disclosure of particular risks to the adequacy of the entire process of continuing communication between doctor and patient. The President's Commission suggested that courts might consider such questions as whether the doctor made reasonable efforts to impart information, to determine whether the patient understood it, to elicit the patient's values and preferences, and to encourage the patient to participate in therapeutic decisions.[9] Informed consent in clinical experiments might be improved by adding more lay members and professionals trained in communications to institutional ethics committees to complement their present predominantly scientific composition, and by giving these members special responsibility for developing and monitoring consent procedures. Finally, patient package inserts could be included with all prescription drugs.

Through these or other steps, the medical profession must be encouraged to adopt a somewhat different philosophy regarding the practice of medicine. Like other experts, physicians are generally optimists about what their field can do. Accordingly, they tend to employ all the treatments that might possibly help and avoid only those certain to cause harm. Greater caution (realism) is in order. Physicians ought to be more skeptical about new or untried therapies until they have been carefully studied and shown to be safe and effective; conversely, physicians should pay more attention to the various possible adverse effects of the care they do provide.[10] Listening more closely to patients' fears and concerns might help to counterbalance doctors' natural optimism.

Just as individual informed consent seeks to protect the individual's right to self-determination, so society as a whole is entitled to give its "informed consent" to the development of major therapeutic innovations. Such consent is rarely obtained, even indirectly. All four of the medical innovations examined here were initiated with scarcely any input from the public at large, or even from anyone in government outside of the immediate funding agencies involved. Although the 1973 Artificial Heart Assessment Panel saw itself as a vehicle for providing a belated form of "societal informed consent" for the artificial heart, at least one member had serious reservations about whether the panel could claim to speak for society as a whole, and of course by then the program's continuation was no longer in any real doubt. More recently, federal officials authorized two momentous steps in genetic engineering – the deliberate release of genetically engineered organisms into the environment, and the application of "gene therapy" to humans – despite many unanswered scientific and social questions and evidence of significant public apprehension. Nor have such doubts and uncertainties visibly slowed the rush to apply this new technology to a rapidly expanding array of human and agricultural uses.

More than three of four Americans believe life is riskier today than twenty years ago.[11] Public perceptions of risk reflect a mixture of facts and values, including not only the likelihood that something bad will happen but many other considerations such as whether the risks are under one's control or not, their catastrophic potential, the degree of uncertainty involved, and issues of equity and threat to future generations.[12] Risk aversion was a consistent theme running through almost all efforts by lay groups to influence policy in the four cases, in striking contrast to the confidence expressed by scientific and medical experts. Yet most experts appear to be unaware that their views differ from those of the public. In one survey comparing people's opinions about whether science and technology would do more good than harm over the next twenty years, corporate executives and members of Congress were not only substantially more optimistic than the general population, but both groups overestimated public optimism by significant margins when asked what they thought the public's expectations were.[13] Many things contribute to divergent attitudes among experts and nonexperts, as discussed in previous chapters. Whatever the reasons, the result is a more aggressive approach to scientific and technological innovation than many Americans may be comfortable with, especially those sectors of society that stand to receive fewer of the benefits.

Regulatory safeguards reflect society's basic aversion to risks. However, rather than responding to public concerns about medical and technological risks by strengthening safeguards, the present administration has done just the opposite. Federal agencies have steadily relaxed regulations governing genetic engineering, and gene-altered products are being handled by existing regulatory agencies following procedures whose adequacy has been questioned. Moreover, relatively little research on the risks of biotechnology is being conducted compared to the vast amount directed toward its applications.[14]

The regulation of drugs and medical devices has also been scaled back, exposing the public to greater risks, critics say. FDA requirements for approval of new drugs prior to marketing have been streamlined to reduce "administrative delays," and a "fast-track" approval process has been established for drugs deemed therapeutically valuable. A 1984 government study concluded that "major portions" of a 1976 law expanding FDA regulation of medical devices had never been implemented and that "systematic information on the hazards associated with device use" was lacking.[15] And in 1982, objections from medical device manufacturers and the American Medical Association brought about the demise of the National Center for Health Care Technology (NCHCT), an agency that had been set up during the Carter administration to monitor and evelute emerging medical technologies. Occupational safety regulations have been sharply curtailed, even though one national poll showed that 75 percent of the public favored keeping, without weakening, existing government regulations aimed at protecting the health and safety of workers. The Reagan administration has also cut support for environmental protection, which is favored by even larger majorities of the public.[16]

A societal version of informed consent would involve a partnership between science and society characterized by mutual respect, full disclosure of risks, and shared decisionmaking. Much of the framework for that partnership already exists. Federal regulatory agencies have reasonably broad legislative authority; what they need is the political mandate and independence to use the powers they have in the public interest. Two new regulatory mechanisms are also needed for medical innovation: a systematic surveillance system to detect rare or long-term adverse drug reactions after drugs are on the market, and an independent agency, similar to the now–defunct NCHCT, to assess the likely costs, risks, and benefits of emerging technologies.[17] Because proponents of new technologies naturally emphasize the benefits of innovations, the assessment agency should give special attention to possible negative consequences. Legal scholar Harold Green suggested, more than a decade ago, that technology assessment include some means whereby "the negative factors, particularly the risks, will be vigorously, effectively, and responsibly pressed upon the decision-makers in a manner that will permit the Congress to make its own judgments and that will permit the public to make its own judgments, so that its views will become known to the Congress."[18] However, any new agency, if it is to avoid NCHCT's fate, will have to be adequately insulated from the enormous economic and political pressures likely to be applied by advocates of new technologies.

One technology assessment strategy that builds in a critical component is called "critical review and public assessment," a process that has been used in some European countries to deal with controversial environmental issues such as nuclear waste disposal. In this process, the government solicits multiple analyses by competing groups of experts and critics, and submits them to public hearing for review and comment. It differs from the medical "consensus conferences" held by the U.S. Department of Health and Human Services in its deliberate emphasis on contending views as well as its reliance on public hearings to explore areas of disagreement.[19]

"Informed consent" is really a misnomer for what is needed, whether for individual patients or society as a whole. Patients must transcend the passivity implied by the term "consent" and become full-fledged partners in the therapeutic process. Likewise, members of the public deserve full information about the risks of new medical and scientific technologies and a meaningful voice in their development, deployment, and regulation. Had such participation occurred in the four cases, some of the problems encountered might have been avoided. In the final analysis, the only valid measure of "good" decisionmaking is the degree to which the resulting choices reflect the views and values of the people affected.

Compensation of injured victims

Social policies on compensation should be inversely related to those on risks: The less protection society affords its members against individual and collective risks, the more generous it ought to be in compensating injuries. In the United States, by contrast, reductions in protective regulations during the early 1980s were accompanied by a *tightening* of federal disability requirements and a massive, industry-backed campaign to limit recovery in personal injury lawsuits. Individual citizens have not only had to accept greater risks than they might wish, but are also being forced to shoulder a growing share of the financial and medical burden of the resulting injuries.

Lawsuits over DES and swine flu injuries reveal the American legal system's characteristic problems: protracted trials, delayed awards, widely varying damages, and huge legal costs. Courts are poorly equipped to handle these and other cases typical of modern mass production, involving widespread exposure to risk and an uncertain or delayed link between exposure and harm. Tort law has been evolving in an effort to meet such needs and to provide what social policy does not: no-fault compensation for victims of large-scale injury, even when the harms are arguably unforeseeble. DES suits and so-called toxic tort cases have helped to fashion more equitable and efficient ways of handling mass injury cases and new ways of allocating damages when the manufacturer of the product involved is unknown.

While the courts have been gradually expanding manufacturers' liability, manufacturers and their insurers have been lobbying for product liability "reforms" that would limit recovery in damage suits. Liability insurance premiums have soared, and in the mid-1980s some towns and industries were left without any coverage at all. Many states passed legislation restricting manufacturers' liability in various ways, and the Reagan administration proposed federal legislation that would "preempt and supersede" state product liability laws. Federal legislation was required, the administration said, because the "recent explosive growth" in product liability lawsuits had jeopardized the health of many industries, impeding the "free flow of products in interstate commerce."[20]

Consumer groups, including DES Action, opposed such legislation. They argued that it would erode corporate accountability for unsafe or harmful products, allowing liability insurers to increase already inflated profits, and shifting much of the burden of injuries to the victims themselves. Moreover, they questioned the assumptions underlying it. The so-called litigation explosion was largely a myth, consumer advocates contended, citing evidence that tort litigation in state courts had merely kept pace with population growth, and that almost a third of the increase in federal product liability cases was due to asbestos suits.[21] They also challenged the insurance industry's claims that the sharp escalation in premium prices was necessary to cover costs, noting that in 1985, insurers made $2 billion in profits after tax breaks along with a record-setting $77 billion of surplus capital.[22] In any event, the evidence on how much tort reforms held down insurance rates or increased the availability of coverage remained unclear. Nevertheless, most of the proposed "reforms" rested on the assumption that insurance rate increases were warranted, and that the liability crisis should be resolved by curbing the recoveries of injured victims rather than the business practices of insurers. As in the swine flu program, the insurance industry was again more or less naming its price.

In the long run, restrictions on tort law recovery will only exacerbate the underlying problem reflected in rising liability costs: the effort by the courts to produce what the political process has withheld, and the consequent transformation of tort law into an instrument of social policy for achieving such goals as loss spreading, injury deterrence, and redistribution of wealth from the deeper pocket.[23] The *New York Times* called tort reform "half a response" to the liability crisis, and urged the federal government to follow the recommendations of a New York study commission by limiting the extent to which insurers could change premium prices in a given year.[24] Another option would be to create public alternatives to private insurers, which would be managed by cities and states and therefore publicly accountable. The State of Wisconsin has been offering life insurance for decades, at rates that are now well below those of many private insurers.

Although the insurance industry and manufacturers have portrayed personal injury damage awards as outlandish and irresponsible, in fact they are part of a worldwide trend toward compensating the victims of medical and technological injury, regardless of fault. The sense of social responsibility is particularly strong in cases where individuals have taken risks that benefit society as a whole, for example, by participating in mass immunization programs or in clinical research. American courts have been gradually recasting traditional legal doctrine to cover such situations; many other industrial countries provide no-fault compensation through publicly funded social insurance as well as private sector programs. While most no-fault systems are less generous than American juries tend to be in personal injury suits, they generally compensate a much higher proportion of all covered injuries. However, this country's closest analogues, the workers' compensation system and the federal Social Security Disability program, have been widely criticized, underscoring the political and economic vulnerability of nonjudicial systems in the United States.

With that single important qualification, no-fault compensation run by the government appears to have a number of advantages over the present fault-based adversarial legal system: Compensation should be easier to obtain, awards would be more predictable, and the enormous expense of trials and lawyers would be avoided. There is considerable support among both consumer and industry groups for a no-fault compensation program for vaccine-related injuries, although the details of different proposals vary. Another approach worth exploring for mass injury cases is separating compensation awards from judgments of responsibility, as in Japan's system of compensating pollution-related disease. Determining which injuries to reimburse would undoubtedly be difficult and to some degree arbitrary, as discussed in Chapter 8, but for large-scale exposures, harms could often be identified more readily in aggregate statistical data than in individual cases. It would be essential in such a system to ensure that damage awards were not compromised by pressure from manufacturers seeking to avoid liability. The equity and performance of any administrative system depends crucially on its independence and integrity. With appropriate political protection, no-fault compensation run by the government and funded at least in part by drug and medical device manufacturers would, ideally, allow more equitable reparations for victims, while separate aggregate decisions about cost allocation would create economic incentives for manufacturers to reduce future injuries.

Allocation of health resources

The unending proliferation of costly new therapies confronts us with painful choices about the allocation of medical resources. By default as much as by design, federal cutbacks and private cost-control efforts are inexorably shifting care away from basic services for the poor and elderly, while expensive new procedures claim an ever increasing share of public reimbursement programs. The resulting distribution of resources is neither equitable nor, in the long run, efficient.

No medical advance highlights these issues more starkly than the artificial heart. Ethical dilemmas haunt almost every aspect of this dramatic new technology, from decisions about how individual patients will be selected to larger questions about the wisdom of investing finite resources in a technique that may provide dubious benefits to a tiny fraction of people with heart disease. If government programs do not cover implants, likely to cost several hundred thousand dollars apiece, this taxpayer-developed technology will be beyond the reach of most of the nation's citizens, especially the poor, among whom heart disease is most common and most severe. On the other hand, if the government does pay for implants, this would further strain already overburdened public programs, increasing the pressure to divert funds away from proven forms of prevention and primary care.

This diversion is already occurring; state Medicaid programs have been forced to extend coverage to a widening range of extraordinarily expensive new procedures. A 1986 federal task force recommended tht Medicare and Medicaid pay for kidney, heart, and liver transplants for everyone in need, at a projected additional annual cost of $42–$70 million (other estimates are higher).[25] Yet at the same time, the budgets of publicly funded health and social programs have been slashed. Cutbacks under the Reagan administration brought an end to the trend toward expanded access to care for poor, minority, and elderly groups that had prevailed since the mid-1960s, undermining previous gains in access and health status.

Continued development of new high-cost procedures, creating further demands for expanded public coverage, will only exacerbate the problems created by federal cutbacks in the early 1980s. Consider some of the following facts:

- Over 600,000 people lost Medicaid coverge between 1981 and 1983 because of funding cuts. In some states, Medicaid covered fewer than 20 percent of the poor.[26] A congressional study found that in 1983 the poorest fifth of all families had, on average, roughly 30 percent less income to meet their basic needs than in 1968.[27]
- Federal cutbacks forced over 250 community health centers to close. A million poor children were excluded from child nutrition programs, and a million people declared ineligible for food stamps. Nutrition programs for poor pregnant women could serve only one-third of those eligible, despite the proven cost-effectiveness of improved nutrition and prenatal care (each $1 spent yields between $2 to $11 in savings on other services).[28]
- The Social Security disability rolls were pruned, denying benefits to thousands of the disabled. The *New York Times* called this action "pettifogging cruelty to the disabled," and the Reagan administration lost most of the many court challenges, including a Supreme Court ruling vindicating the handicapped.[29]
- The average hospital deductible paid by Medicare be-

neficiaries doubled between 1981 and 1985. The elderly now spend a larger fraction of their income on out-of-pocket medical expenses than before Medicare was enacted.[30]

- An estimated 200,000 Americans were denied emergency hospital care in 1983, and 800,000 were denied routine care for lack of money. Nearly half of all people with incomes under $10,000 reported that there were times when they did not have enough money to pay for medical care.[31]
- "Dumping" of indigents from private hospitals to public hospitals has become increasingly common, jeopardizing patients' health. So serious was this problem in some places that both state and federal legislation were passed to try to prevent further abuses.[32]
- Preventable childhood diseases began to increase in some disadvantaged populations, as did rates of infant mortality and low birth weight. One study of patients excluded from Medicaid found that health status deteriorated and some patients died because of inadequate care.[33]

For many of the nation's poor, in short, the "safety net" clearly failed. Still the Reagan administration pressed on, proposing to cut Medicaid and Medicare by an additional $70 billion between 1986 and 1991, while continuing to encourage the development and dissemination of high-cost medical technologies.

These priorities run directly contrary to public wishes. Poll after poll has shown that most Americans, while supporting efforts to balance the federal budget, think our society is spending too little, not too much, on health care. Nearly two out of three people favor national health insurance, even if it means increased taxes. Most people want continued development of high-cost medical procedures such as organ transplants, but believe that everyone should have access to them. Although people tend to react negatively to the term "welfare," with all its invidious connotations, more than three-quarters of the respondents in a 1977 national survey approved of "taxes being used to pay for health care for poor people."[34] In a 1984 "ABC News" poll, 75 percent of the respondents indicated that "the government should institute and operate a national health program."[35]

Even more out of line with public priorities are federal cutbacks on health coupled with increased military spending. While seeking reductions of $70 billion in Medicaid and Medicare between 1986 and 1991, the Reagan administration proposed to increase the military budget by $120 billion.[36] Similarly, the administration proposed to reduce NIH funding for medical research by $290 million in 1987 while increasing support for military research and development by $9 billion, but was thwarted by Congress, which instead expanded the NIH's budget by $800 million over 1986. If implemented, these proposed changes would have put nearly three-quarters of all federal R&D dollars into military programs, whereas in 1980, when Reagan took office, federal spending on military and nonmilitary R&D was about equal.[37] Yet opinion polls have repeatedly shown that most Americans prefer cuts in military spending over cuts in health and social programs by large margins.[38]

Signs of public dissatisfaction with medicine's structure and priorities are not hard to find. Faith in modern medicine's ability to address important human and social needs has been severely eroded. More than half of the respondents in a 1984 Harris poll believed medicine should focus less on curing existing illness and more on prevention of disease.[39] Three-quarters of the general public favor "major" changes or a complete restructuring of the health care system.[40] (Interestingly, only one-third of the physicians polled thought "major" changes were necessary and none favored total restructuring.) There is mounting evidence that advanced technology does not always mean better care, and in some cases may be at the root of public discontent. In the burgeoning "right to die" movement, for example, patients are rebelling against the use of heroic medical technologies against their wishes. Support for this movement has increased dramatically during the past decade.

Growing corporate involvement in health care and medical research is likely to intensify problems of equitable and efficient resource allocation. Corporate priorities, as discussed in Chapter 9, necessarily emphasize salable medical "commodities" – chemical, biological, or surgical interventions – rather than health-promoting social or behavioral changes. To date, market forces have favored the development of high-cost, technologically sophisticated acute-care interventions.[41] Commercial incentives respond to purchasing power, not health needs, and the two are often inversely related.

It is the government's duty to meet social needs that the private sector fails to fulfill. Medical care and research are, more than ever before, one such need. For resources to be allocated justly and efficiently, the government must play a central role in the planning, organization, and financing of both medical care and clinical research.

Government coverage of basic services is essential in getting health care to those suffering the greatest burden of disease and disability. Almost every industrialized country except the United States has some form of national health insurance that removes major financial obstacles. Universal access appears to improve health, and government regulation or management permits rational allocation of resources as well as more effective cost control. In Canada and Great Britain, for instance, which have national insurance and health service programs respectively, per capita health expenditures average 20–70 percent less than in the United States. Age-adjusted mortality rates have declined sharply in both countries and are now slightly lower than in the United States, whereas before the introduction of universal free access to care they had been higher.[42] Furthermore, antigovernment rhetoric notwithstanding, these government-run systems are considerably more efficient administratively than this country's private system. It is estimated that we waste anywhere from $30 to $40 billion a year – 8–10 percent of total U.S. health spending – on unnecessary administrative overhead, costs that would be avoided if we had a national health insurance system as in Canada, or a government-run national health service as in Great Britain.[43] These savings are greater than those envisaged by any cost-control program that has received serious consideration in Washington, and they would not compromise health care equity.

Medicaid and other public programs have shown that the "inverse care law" of medical utilization can be changed if financial and organizational barriers are reduced. Improved access to care does not guarantee equal health, of course, for medical care is a relatively minor influence on health compared to social and economic circumstances. As long as there are significant inequalities in people's home and work environments, disparities in health will persist. In 1986, America's Catholic bishops released a pastoral letter on the U.S. economy, which stated: "The obligation to provide justice for all means that the poor have the single most urgent claim on the conscience of the nation."[44] Public opinion supports that view. In one national poll, nearly two-thirds of those interviewed said that "the government should work to substantially reduce the after-tax income gap between rich and poor."[45] (In 1986, the top and bottom fifths of the U.S. income distribution received 46 and 4 percent of total income, respectively.) Again, present federal policies have moved in the opposite direction.[46]

In a coordinated national system, general priorities for allocating resources among regions and communities could be based on routinely collected epidemiologic and mortality data reflecting population health needs. The United States, for all its free-market rhetoric, has had a fair amount of experience with health planning. In the mid-1970s, regional Health Systems Agencies were established around the country with a majority of consumers required on their governing boards, to try to plan for more rational and equitable health service delivery. Although hampered by various political and financial constraints, these agencies broadened participation in local health policy, and some had considerable impact.[47] Around the same time, the federal government also launched a comprehensive, five-year planning effort for medical research, beginning with the development of explicit health priorities. A blue-ribbon committee of the Institute of Medicine "strongly endorsed" long-range planning for medical research, but called for more public involvement in the process.[48] This and most other forms of health planning were abandoned by the Reagan Administration, which has stressed market forces rather than governmental control, and domestic budget-cutting rather than research priority-setting. There is still substantial support in the scientific community for a more "rational" system of directing scientific and technological innovation, however, stemming in part from dissatisfaction with the influence of special interests in Congress and administrative agencies.[49] Many European counties as well as Canada also have a strong tradition of health and technology planning, especially for important and controversial new technologies.[50]

In many cases, public priorities for medical innovation will involve programs or activities that are not profitable for the private sector. The government has a special obligation to support research on areas likely to be neglected by private interests, such as the reasons for higher rates of disease and disability among the disadvantaged, or ways of making existing service programs more effective. This could be accomplished through present funding mechanisms or through the use of special tax breaks, low-interest loans, or other economic incentives to promote specified types of research or service programs. These subsidies could be financed from general tax revenues, or, alternatively, from a set levy on the profits from patents on products of publicly funded research. Another option that has been suggested is to divert a fraction of the taxes on biotechnology companies and other high technology industries into a federal fund dedicated to basic research.[51]

With finite medical resources, new techniques should be not only equitably distributed but also reasonably cost-efficient. The artificial heart, DES, and the swine flu program illustrate how easy it is for unproven therapies to be adopted without adequate assessment. Cochrane's classic study, *Effectiveness and Efficiency*, provides other examples of the haphazard introduction of new techniques.[52] Cost-control efforts should begin with the identification and elimination of marginal medical procedures, including some that may even be harmful to patients. Great Britain spends about 70 percent less than the United States, per capita, on chemotherapeutic agents for incurable metastatic cancer. British oncologists treat curable cancers as readily as their American counterparts, but they see no reason to subject terminally ill patients to what one doctor called "treatment which brings them nothing but unpleasant side effects and is of no benefit."[53] The clinical, social, and economic consequences of all major new procedures should be rigorously evaluated prior to dissemination, and the results implemented through planning and regulatory mechanisms. For scarce and expensive technologies like the artificial heart and organ transplants, national guidelines are needed to determine how candidates will be selected; what nonmedical criteria, if any, are allowable; how selection decisions will be made; and how financing will be provided.[54] If public coverage is offered, it should be a separate budget item to guard against diversion of funds from programs that are more widely beneficial, albeit less glamorous.

A final element in making medical innovation both more equitable and more efficient is to reverse the current trend toward expensive, high-technology procedures and give greater emphasis to lower-cost forms of prevention and primary care. Cost-cutting, depending on how it is done, need not mean poorer quality care; indeed, some lower-cost approaches may actually enhance patients' well-being. Care of the elderly is a good example. In this country, aggressive medical intervention and frequent use of hospitals and nursing homes can bankrupt families and leave them anguished and dissatisfied as well. The approach taken by some European countries is both more humane and less expensive. In

Denmark, for instance, the elderly are encourged to live at home as long as possible and, in many cases, to die there. The government will renovate people's homes to accommodate their physical disabilities and provide necessary support and medical services at home. This arrangement is vastly preferable to most people, and it is a good deal cheaper for the government to provide all these services than to institutionalize elders in hospitals or nursing homes. So strongly committed is the Danish government to the principles of independence and self-reliance for the aged that it has said that no new nursing homes can be built after January 1987.[55]

There are also examples of creative, socially responsive medical frugality in the United States. One is an experimental, fee-for service health plan in northern California, the Mid-Peninsula Health Service (MHS). MHS is governed by an elected board, and seeks to provide as much care as possible in the home and community, to eliminate redundant or indecisive therapeutic practices, and to establish a partnership between health professionals and patients, using supervised self-care whenever possible. Studies comparing MHS members with similar patients in the community show that MHS has succeeded admirably in its goals: Patients have lower costs, less frequent hospitalization, and fewer diagnostic tests and prescription drugs than people using other providers, with no apparent adverse effect on health status; furthermore, MHS patients and doctors both report greater satisfaction with this style of care.[56]

Low-cost, low-technology approaches may be even more effective in meeting the needs of disadvantaged populations. Nowhere in this country has that been more apparent than in the Community Health Center program, launched in the mid-1960s. The most ambitious centers, like one in Mound Bayou, an extremely poor region in the Mississippi delta, sought to improve health not only through traditional medical care but also through efforts to alleviate the basic social and economic conditions that cause or perpetuate illness.[57] The health center dug wells, installed sewage lines and sanitary outhouses, provided screens and other home improvements, and offered social services ranging from transportation to education. For nutritional problems, center doctors wrote prescriptions for food, while other staff helped organize a cooperative vegetable farm. The hope was to create self-sustaining initiatives that would enable the community to lift itself out of impoverishment and dependency. Although not all of Mound Bayou's pioneering programs survived, its achievements have been impressive. The infant mortality rate dropped 40 percent during the center's first four years, hundreds of townsfolk have better housing and sanitary water, and over 100 people earned post–secondary-school degrees, including thirteen M.D.s, as a result of health center-sponsored programs – in a community where the average education had been four years of school.

Many more examples of highly effective and less costly "low-technology" medicine could be listed. They all go to show that much can be done to improve health even amid the most adverse social conditions, that it does not take large amounts of money, and that standard medical practices, with their emphasis on technology and medical intervention, may often be irrelevant, and sometimes downright harmful. In pledging to achieve "Health for all by the year 2000," the World Health Organization listed treatment of disease last among its goals, after provision of food, clean water, and a sanitary environment, recognizing that medical care has less to do with health than social and environmental circumstances, especially for impoverished populations.[58] Of course, isolated examples can only point the way toward a general redirection of medical priorities; they do not provide a complete blueprint. Nevertheless, these experiences illustrate the types of approaches that could achieve a more equitable allocation of medical resources at little if any additional expense.

Indeed, such a reallocation of resources might even end up saving money. In 1987, a committee of private business leaders in the United States concluded that increased public support for the health and educational needs of poor children would more than pay for itself over the long run in reduced tax costs for later medical care, special education, public assistance, and crime, while also adding needed workers to the country's labor supply. "This nation cannot continue to compete and prosper in the global arena when more than one-fifth of our children live in poverty and a third grow up in ignorance," these executives warned. "The nation can ill afford such an egregious waste of human resources."[59]

The central dilemma: lack of public accountability

With steadily expanding powers, staggering costs, and intensifying ethical dilemmas, medical research and practice are – and should be – matters of public concern. In the 1980s, the question is not whether science and medicine should be held accountable to society, but how.

The inadequacy of present mechanisms of accountability is evident from the foregoing discussion. This is not to denigrate the public oversight that does occur. More than once in the four cases, Congressional oversight goaded the FDA or the NIH into action, and criticism from consumer groups, lay members of federal advisory committees, or dissident experts added new items to the policy agenda. Yet the overriding influence of mainstream medical, scientific, and corporate interests in funding and regulatory decisions was also abundantly clear. The impact of direct participatory initiatives was even more limited. Even at the height of public controversy over recombinant DNA research, only a small fraction of the general public played any active role in efforts to influence policy. Absent controversy or catastrophe, most people rely on existing political institutions to represent their interests in medical innovation.

Therein lies much of the problem. For, as the case studies illustrate, the world of health and science policy does not comform to the idealized image of a pluralist society, in which relevant competing interest groups are all adequately represented. In one political scientist's classic characterization: "The flaw in the pluralist heaven is that the heavenly chorus sings with a strong upper-class accent. Probably about 90 percent of the people cannot get into the pressure system."[60] Just as large and powerful interests dominate economic markets, so they tend to dominate virtually every level of the political process. Professional and industry groups have a direct stake in governmental policies that affect them and have the resources to promote their interests, whereas consumer, labor, and community groups have broader, more diffuse interests and generally far fewer resources. Political scientists have called this the problem of "imbalanced political markets."[61] By whatever name, it is democracy's central dilemma.

In the United States, in recent years, this imbalance has been growing rather than shrinking. Major changes in the political and economic structure of American politics over the past several decades have seriously compromised the ability of Congress and the executive branch to represent the interests of all segments of society, while federal actions have erected new impediments to public participation. Among the changes impinging most directly on efforts to make medicine more responsive to broad public needs are the following:

- The influence of well-endowed private interests has expanded significantly. Between 1981 and 1986, the number of registered Senate lobbyists increased from 5,662 to 20,400. In 1982, corporate, trade, and professional groups spent more than twice as much as labor groups on political contributions – $85 million versus $35 million – whereas only a decade earlier they had been spending the same amount. In one disillusioned Senator's words, "The plain truth is that Washington has become a sinkhole of influence peddling."[62]
- The gap in voting rates between rich and poor has been steadily widening. In 1968, voter turnout was 55 percent among laborers and 83 percent among white-collar professionals; by 1980, the gap had increased to 42 percent for laborers versus 78 percent for professionals. Overall, only 53 percent of all eligible voters went to the polls in 1980 – 76.5 million people did not participate.[63] While voting certainly does not guarantee political accountability, it serves as an index of popular opinion and a check on the actions of elected officials.
- Programs to facilitate citizen participation at all levels of government have been scaled down or eliminated, including a comprehensive federal "consumer program" developed by the Carter Administration, financial and technical support for public interest "interveners," public representation on certain federal advisory groups, and local Health Systems Agencies with specified consumer membership. Administrative requirements intended to give the public a voice in federal rule-making have been sidestepped or ignored.[64]
- Public information of all types has been curtailed. While enacting some of the most far-reaching legislative changes in recent history, the Reagan administration has slashed the budgets of nearly all federal data-gathering programs that might monitor the impact of such changes, and has restricted access to government documents through the Freedom of Information Act. "When a vessel is in stormy seas, it is foolhardy to cut corners on radar, navigational equipment, good maps, and ample, well-trained crews," the University of Chicago dean complained to Congress, referring to across-

the-board cuts in federal statistics collection.[65] The head of the Library of Congress, where reading room hours were reduced by 30 percent because of limited funding, called the information cutbacks "antidemocratic and antiknowledge."[66]

* Despite talk of the "new federalism" and bringing government "closer to the people," the Reagan administration has in fact sought to impose more, not less, federal control over some areas of public policy and personal responsibility. A growing number of laws preempt state and local authority in various arenas, a trend that has been attributed to the increasingly national or international scope of corporate markets. Meanwhile, federal funding for state and local governments has been substantially reduced, leaving them with fewer federal directives but too little money to do much on their own.[67]

These changes obviously present formidable obstacles to the creation of a legislative and planning environment that is responsive to citizens' interests in science and medicine. The challenge is to invest the existing mechanisms with the power they would need in order to represent the full spectrum of societal concerns, and to design new vehicles for citizen participation that expand public input into biomedical policy and enlarge its domain of accountability. Such a fundamental shift in the balance of political power will not occur easily. Only through the large-scale mobilization of an informed and determined citizenry is it even imaginable. Yet, for the nation's health and well-being, the need has never been greater.

Making medicine more publicly accountable

A full discussion of the implementation of participatory politics in the United States today is beyond the scope of this book.[68] We focus here on participatory mechanisms at the federal, state, and local levels, which together would be mutually reinforcing. Separately, the most important changes are probably those at the local level, because they are most likely to allow citizens to learn enough about biomedical issues to form their own opinions, and to gain confidence in their ability to articulate those opinions. Contacts at the local level also provide the basis for building the organized popular constituencies necessary to keep higher levels of government accountable to public wishes.

With growing public concern over the future of health care, people are increasingly likely to respond to opportunities to express their views. In 1983–4, some 5,000 people attended a series of small group and "town hall" meetings held in Oregon to elicit public views about various ethical dilemmas in health care. The local meetings culminated in a statewide "Citizens Health Care Parliament," which passed thirty-five resolutions addressing the issues raised. The consensus was that "Society must decide what should be the adequate level of health care it will guarantee to all its members."[69] Among the group's recommendations were increased funding for research on disease prevention, better coverage for the medically needy, and allocation of medical resources according to a definition of "adequate health care" developed through widespread public participation. This effort attracted national attention, and in 1985 similar projects were launched in six other states. Leaders of these projects view them as part of a growing movement founded on grass-roots involvement in health policy, a movement they believe could be as important for the 1980s as civil rights was for the 1960s.[70]

Certainly the success of a Massachusetts ballot referendum on health care revealed widespread support for similar principles. In 1986, Massachusetts voters approved by a 2 to 1 margin a measure proposed by a grass-roots coalition urging the U.S. Congress to enact a national health program that provides universal access to comprehensive preventive, curative, and occupational health services, and is community controlled, rationally organized, equitably financed, and cost-efficient.[71] Statewide initiatives could set an important precedent for national action. The Canadian health insurance system, which provides universal coverage, started in a single province and later became a national program when the Parliament adopted principles that each province had to meet to receive federal funds.

Another way to encourage local participation in biomedical policy would be to hold hearings on major issues around the country. There are precedents for this in Canada and European countries as well as in the United States. In the late 1970s, the National Science Foundation (NSF) held a series of public hearings in different parts of the country to elicit suggestions for research priorities with the aim, in the NSF Director's words, of "reinvigorating the contract between science and society."[72] Well publicized local hearings would give citizens a chance to learn about alternative policy options, to ask for further investigation of matters that are unclear or controversial, and to express their wishes and concerns. These views could feed directly into state and local policy decisions, and could also lay the groundwork for holding national decisionmakers accountable to the wishes of local constituents.

Open public hearings are vulnerable to all the problems of imbalanced interests discussed in Chapter 10. One way to minimize these problems would be to model local hearings on the court jury system, in which the hearings are public but the "jurors" themselves are preselected to represent a cross section of the community or designated constituencies. An example of this approach, which was judged quite successful by most observers, was the Cambridge Experimentation Review Board, the citizen group appointed in 1976 to decide the fate of recombinant-DNA research at Harvard and MIT. While the representatives in such a lay jury system are not formally accountable to their constituencies, their discussions and conclusions would nevertheless help clarify public sentiments. The juries could be presented with specific choices, such as the safeguards necessary for applying genetic engineering to humans, or they could be asked to choose among various policy alternatives, such as increasing Medicare deductibles, closing community health centers, foregoing the artificial heart, or reducing military spending. Although disagreements would surely emerge, the discussions should help to identify the trade-offs involved in policy choices and general public priorities. Unlike proposals for a "science court," which envision distinguished scientists deciding the merits of opposing scientific arguments, the citizen jury approach does not assume that facts and values can always be cleanly separated in scientific controversies. Citizen juries are also more likely to promote public understanding of the issues and public confidence in the results.[73] We entrust juries of our peers with judgments about civil and criminal matters, no matter how technically complex the issues and how disparate the views of testifying experts. Why not use a similar system for forming at least preliminary judgments about matters of biomedical policy?

It might also be desirable to decentralize some aspects of research funding, by setting aside some support to be allocated by state and local authorities for work they consider useful. A possible model is the "science shops" in the Netherlands – cooperative ventures between universities and communities set up "to promote socially relevant research in the universities, to relate this work to the needs of society, and to provide client groups with technical information."[74] Another precedent closer to home is the California Policy Seminar, a joint undertaking between the University of California and the state legislature, which attempts to match academic research projects with the state's health problems.

Individual medical and research institutions could also encourage broader involvement in their own biomedical policy decisions. All federally funded institutions conducting research on humans have committees to review ethical issues, and those involved in genetic engineering have "biosafety committees" to oversee the implementation of federal safety guidelines. Limited outside participation is required in both types of committees, and appears to have been generally constructive, although constrained in various ways.[75] These committees provide another potential mechanism for community participation in biomedical policy decisions. Their purview could be enlarged to include research priorities as well as ethical and safety issues, and public members could be selected by and held accountable to local governments or other constituencies rather than, as at present, appointed by the institutions. Approaches like these, operating at the local level, offer a feasible way to involve the general public in aspects of medical innovation while encouraging researchers to respond to community needs.

Local opportunities for involvement are necessary, but not sufficient. The federal government presently makes most major decisions about biomedical research and regulation, and probably always will. Enlarging public influence in federal decisionmaking is also essential if medicine is to be held more broadly accountable.

One step that would require no new governmental structures would be to give Congress more of a voice in the affairs of NIH and health regulatory agencies. In late 1985, Congress finally passed, over a presidential veto, legislation asserting its right and responsibility to set general priorities for the nation's health research budget. This capped a five-year struggle against the Reagan administration, NIH officials, and members of the academic research establishment, who claimed that additional congressional influence would undermine scientific freedom and autonomy. Representative Henry Waxman, the driving force behind the legislation, was not persuaded: "We feel that if the taxpayers' dollars, $5 billion a year, are being used for biomedical research, we ought to spell out some of the priorities," he insisted. "We ought to tell

NIH what we think they ought to be looking at."[76] Among the things slated for greater attention are disease prevention, and ailments ranging from arthritis to Alzheimer's disease. Congressional oversight of NIH could be expanded further. A 1987 advisory committee appointed to assess the National Science Foundation recommended tht Congress create its own system for reviewing research proposals that would bypass the regular scientific peer review process.[77] To ensure that public rather than private interests are served, however, more effective reforms are needed to limit the financial pressures that private interests exert through campaign financing and special interest lobbying.

A meaningful public voice in federal biomedical policy requires more than just enlarging Congressional input into NIH affairs. Members of Congress have neither the time nor the knowledge to become intimately involved in the daily decisions of federal health agencies. Decisionmaking within these administrative agencies themselves needs to be both broadened and better coordinated. One way to accomplish this would be to establish within the Department of Health and Human Services an overall policymaking body with representation from designated consumer, community, environmental, and labor constituencies as well as scientists, planners, and government officials.[78] These constituencies should be chosen with the goal of counteracting the major structural inequalities in stakes, power and resources discussed in Chapter 10, and representatives should be given appropriate technical and financial assistance.[79] To ensure accountability, each constituency group should select its own delegates, with final approval resting with Congress. This policymaking body could be charged with developing explicit health goals and priorities for the nation, soliciting public views both nationally and locally on the most important health and environmental problems facing the country. It would need to have wide jurisdiction in order to coordinate funding priorities across areas ranging from basic science to technology development, primary medical care to nonmedical forms of disease prevention.

Had such an umbrella agency existed in the early days of the artificial heart program, it might have given NIH Director Shannon the support he sought for shifting more of the program's funding into heart disease prevention. Likewise, when dissension about the swine flu program began to emerge, this more ecumenical group might have convinced CDC Director Sencer not to go forward with plans for mass immunization. Certainly it would be valuable to have this sort of diverse representation in setting priorities for the development of biotechnology and other important new technologies.

The foregoing suggestions for broadening public influence in biomedical policy are rudimentary and provisional; more and better mechanisms could undoubtedly be proposed. Their main purpose is to illustrate the kinds of steps needed, and to demonstrate that most are well within the realm of practical possibility given the political will to take them.

Final words

The four case studies and the issues they raise reveal a widening gulf between the present course of medical innovation and public concerns and priorities. Even technologies like genetic engineering, which represent genuine scientific breakthroughs, are not apt to address some of our most pressing social problems if present trends continue. These trends underscore the urgency of finding more effective ways of holding medicine publicly accountable. The present, essentially laissez-faire approach to scientific and technological development in medicine is no longer viable. Rather, we must be prepared to specify the social ends for which innovations are to be used, and we must find some means of enforcing those ends. In the simplest possible terms, we must be willing to ask – and answer – the questions: Science for what? and Science for whom?

What would broader accountability mean for medical innovation? The case studies as well as public opinion data reveal substantial differences between the views and values of lay citizens compared with those typical of decisionmakers in science, government, and business. On the whole, lay citizens tend to see a wider range of risks in new developments and to be more skeptical of the likely benefits. This perspective would call for a slower and more cautious pace of medical innovation as well as somewhat different goals and priorities.

Only history can determine whether such changes are desirable. Nevertheless, we should bear in mind that technological "progress" has a paradoxical relationship to the human condition. Sociological studies suggest, ironically, that technological advances and a higher standard of living have tended to make people less, rather than more, happy personally.[80] In medicine today, it seems painfully obvious that "more of the same" no longer makes sense: that whatever the virtues of humankind's ancient quest for power over nature have been historically, we have reached a point where to continue this same course, unmodified by today's needs and problems, threatens to diminish the gains that have been made and to degrade the very humanity it was meant to serve. We cannot undo technology or close scientists' minds. But we can look more realistically at what high-technology medicine can do for us, and what it cannot. We can listen to what people – ordinary folk, not just experts – are saying about what they want for their medical future, and what they fear; and we can remember this country's collective, humanitarian traditions, which are as much a part of our cultural heritage as individual freedom and faith in technology. Finally, we can heed the counsel of enlightened business leaders and many others to invest in the health and welfare of needy children, not only because it is morally right but because it is vital to the country's future social and economic well-being.

Broader public participation in biomedical policy would force decisionmakers to reconsider the present course of medical innovation. It would introduce new views and values into the policy process and encourage more thorough scrutiny of proposed choices and possible alternatives. It would help prevent domination of the policy agenda by special interests. It would inform scientific and medical experts of the wishes of their public benefactors and at the same time give the public more reasonable expectations about what medical progress can accomplish. It is the prerequisite for addressing the growing popular dissatisfaction with medicine's structure and priorities, and might help to counteract feelings of impotence and malaise among the general public in the face of advancing technology. If taken seriously, it could help restore some of the sense of community and shared values we have lost in our increasingly alienated and footloose society, as people worked, talked, and struggled together to devise equitable and democratic solutions to collective problems.[81]

These goals are, admittedly, idealistic. Yet we must remember that democracy, whether in medicine or any other area, remains our best, indeed only, hope for achieving lasting social change. In the final analysis, all social institutions arise from and require the consent and support of the polity.

There are reasons for hope. The wave of populist sentiment that swept the nation in the 1960s left America with a skepticism toward professionalism and experts that never entirely disappeared.

Demands for grass-roots empowerment continue to emerge in areas ranging from neighborhood and patient self-help movements, to the rights of tenants, women, consumers, environmentalists, and gays, to the appeal of the Catholic bishops and the nuclear freeze movement. One of the greatest challenges facing such efforts is to avoid the tendencies of populist movements in the past toward parochialism, antiintellectualism, and manipulation by demagogues. The best safeguard, most advocates of neo-populist efforts believe, is a democratic political structure that allows grass-roots initiatives to be both competency-inducing and also sensitive to the needs of minority interests.[82] Contemporary populist movements seek to integrate local activities with the national political decisionmaking that is an immutable feature of modern industrial societies. Efforts to broaden public control over biomedical innovation converge with this larger wave of grass-roots empowerment.

Some will regard the steps that have been suggested here to enlarge popular influence in medical science as too radical, others as too timid. But something new must be tried if we are to stem the hemorrhage of public expenditures on high-cost technologies of questionable benefit, the denial of care to those in greatest need, the overreliance on drugs and techniques that endanger patients while leaving the underlying causes of disease untouched, the unabashed pursuit of private profits at public expense, and the indefensible discrepancies between stated public wishes and governmental actions. It is time for lay citizens to heal some of medicine's wounds – a bitter pill for many experts to swallow, but one that is essential to the health of medicine and society over the long term.

Notes

1. Introduction

1. Robert Pear, "Medical-Care Cost Rose 7.7% in '86, Counter to Trend," *New York Times*, February 9, 1987, pp. 1, 9. Total medical expenditures in 1986 were $458 billion, or 10.9 percent of the Gross National Product (Department of Health and Human Services, "HHS News," news release, June 23, 1987).

2. See, for example, Archibald L. Cochrane, *Effectiveness and Efficiency* (London: The Nuffield Provincial Hospitals Trust, 1972); also, John Powles, "On the Limitations of Modern Medicine," *Science, Medicine and Man 1* (1973), 1–30.

3. *Competitive Problems in the Drug Industry*, hearings before the Subcommittee on Monopoly of the Select Committee on Small Business, U.S. Senate, 93rd Congress, 1st Session, February 5–8 and March 14, 1973, p. 9373.

4. Frederick Mosteller, "Innovation and Evaluation," *Science 211* (February 27, 1981), 881–6.

5. Thomas McKeown, *The Role of Medicine: Dream, Mirage, or Nemesis?* (Princeton, N.J.: Princeton University Press, 1979).

6. Marc Lalonde, *A New Perspective on the Health of Canadians* (Ottawa: Information Canada, 1975), p. 18.

7. Champion International Corportion advertisement, "Planting Seeds for the Future," *Newsweek* (July 14, 1980), 45.

8. Robert Jarvik, quoted in Marie-Claude Wrenn, "The Heart Has Its Reasons," *Utah Holiday* (July 1982), 47.

9. David Starr Jordan, "The Relation of the University to Medicine," speech at the dedication of Lane Medical Library, Stanford University (copy at Lane Library), November 3, 1912. For examples of biomedical advances resulting from the study of pathology, see Judith P. Swazey and Karen M. Reeds, *Today's Medicine, Tomorrow's Science: Essays on Paths of Discovery in the Biomedical Sciences*, Department of Health, Education and Welfare, Publ. No. (NIH) 78–244 (Washington, D.C.: Government Printing Office, 1978). Also, Richard M. Krause, "The Beginning of Health Is to Know the Disease," *Public Health Reports 98* (November–December 1983), 531–5.

10. Charles Rosenberg, "Science in American Society," *ISIS 74* (1983), 356–67.
11. R. H. Tawney, quoted in Baruch Fischhoff, "For Those Condemned to Study the Past: Reflections on Historical Judgment," in Richard Shweder and D. W. Fiske, eds., *Fallible Judgment in Behavioral Research*, Vol. 4 of *New Directions for Methodology of Social and Behavioral Science* (San Francisco: Jossey-Bass, 1980), pp. 79–93.
12. The lack of research on policy decisions involved in medical technology development is discussed in H. David Banta, "Social Science Research on Medical Technology: Utility and Limitations," *Social Science and Medicine 17* (1983), 1363–9.

2. Where are we and how did we get there?

1. Figures on health research funding are from National Institutes of Health, *NIH Data Book 1986*, Department of Health and Human Services, NIH Publ. No. 87–1261 (December 1986), p. 2. Aggregate health expenditures include, besides research funding, public and private expenditures on hospital care, physician and dentist services, nursing home care, drugs, eyeglasses, government public health activities, and supplies. Figures are from Department of Health and Human Services, "HHS News," news release, June 23, 1987.
2. H. David Banta, Clyde J. Behney, and Jane Sisk Willems, *Toward Rational Technology in Medicine: Considerations for Health Policy* (New York: Springer, 1981), p. 125.
3. National Science Board, National Science Foundation, *Science Indicators 1980* (Washington, D.C.: Government Printing Office, 1981), p. 167.
4. Milton Silverman and Philip R. Lee, *Pills, Profits and Politics* (Berkeley: University of California Press, 1974), pp. 262–4.
5. Ernest M. Gruenberg, "The Failures of Success," *Milbank Memorial Fund Quarterly/Health and Society 55* (Winter 1977), 3–24.
6. Erwin Chargaff, "On the Dangers of Genetic Meddling" (letter to the editor), *Science 192* (June 4, 1976), 938, 940.
7. "Risky Cancer Research Gets Go-Ahead," *San Francisco Chronicle,* January 1, 1983, p. 5.
8. John D. Archer, "The FDA Does Not Approve Uses of Drugs," *Journal of the American Medical Association 252* (August 24/31, 1984), 1054–5.
9. *1968 Congressional Quarterly Almanac,* 90th Congress, 2nd Session, Vol. 24, Major Legislation: Health and Education (Washington, D.C.: Congressional Quarterly Service, 1968), p. 703.
10. President's Commission for the Study of Ethical Problems in Med-

icine and Biomedical and Behavioral Research, *Summing Up*, Final Report (Washington, D.C.: Government Printing Office, March 1983).

11. Donald S. Fredrickson, "Biomedical Research in the 1980's," *New England Journal of Medicine 304* (February 26, 1981), 509. Similar views were expressed by Alexander Morin, Director of the National Science Foundation's Office of Science and Society, in Eliot Marshall, "Public Attitudes to Technological Progress," *Science 205* (July 20, 1979), 283.

12. "Will Biocommerce Ravage Biomedicine?" *Science 210* (December 5, 1980), 1102–3.

13. The Working Group on Mechanical Circulatory Support, National Heart, Lung, and Blood Institute, *Artificial Heart and Assist Devices: Directions, Needs, Costs, and Societal and Ethical Issues* (Bethesda, Md.: National Heart, Lung, and Blood Institute, May 1985).

14. See, for example, the report of the National Commission for the Protection of Human Subjects, *Federal Register 44*, No. 76 (April 18, 1979), 23194.

15. Guido Calabresi and Philip Bobbitt, *Tragic Choices* (New York: Norton, 1978).

16. Anne S. Moffat, "Fish Incorporate Rat Growth Hormone Genes," *Genetic Engineering News 7* (September 1987), 1, 29. See also Jeffrey G. Williams, "Mouse and Supermouse," *Nature 300* (December 16, 1982), 575; Richard D. Palmiter, Ralph L. Brinster, Robert E. Hammer, et al., "Dramatic Growth of Mice That Develop from Eggs Microinjected with Metallothionein: Growth Hormone Fusion Genes," *Nature 300* (December 16, 1982), 611–15. The quote is from a talk given by James Watson at Stanford University, "Nobel Prize-winner James Watson Lambasts Federal Regulations on Genetic Engineering," *Stanford University Campus Report* (March 27, 1985), p. 9.

17. "The First Genetic Experiment on Humans," *San Francisco Chronicle,* October 8, 1980, pp. 1, 20. The experiment, by Dr. Martin Cline of the University of California at Los Angeles, was criticized by scientific colleagues as premature, and the federal government eventually suspended Cline's funding (Marjorie Sun, "Cline Loses Two NIH Grants: Tough Stance Meant as a Signal That Infractions Will Not Be Tolerated," *Science 214* [December 11, 1981], 1220).

18. Dr. Claire Randall, Rabbi Bernard Mandelbaum, and Bishop Thomas Kelly, "Message from Three General Secretaries," National Council of Churches of Christ (110 Maryland Avenue, N.E., Washington, D.C. 20002), July 1980.

19. Lane P. Lester, quoted in Charles Austin, "Ethics of Gene Splicing Troubling Theologians," *New York Times,* July 5, 1981, pp. 1, 12.

20. Robert Jarvik, quoted in Marie-Claude Wrenn, "The Heart Has Its Reasons," *Utah Holiday* (July 1982), 34.

21. Data on public confidence are from Louis Harris and Associates National Surveys, 1966–82. The "distrust" statistic is from Arthur Miller, "The Institutional Focus of Political Distrust," paper presented at the Annual Meeting of the American Political Association, Washington, D.C., August 31–September 3, 1979, typescript. See also, Arthur Miller, "Regaining Confidence: Challenge for Yet Another Administration," *National Forum* (Spring 1981), 30–2.

22. Louis Harris and Associates National Surveys, 1966–82.

23. National Science Board, National Science Foundation, *Science Indicators 1976* (Washington, D.C.: Government Printing Office, 1977), p. 171. See also Todd R. La Porte and Daniel Metlay, "Technology Observed: Attitudes of a Wary Public," *Science 188* (April 11, 1975), 121–7.

24. See National Science Board, National Science Foundation, *Science Indicators: The 1985 Report* (Washington, D.C.: Government Printing Office, 1985), p. 153; and National Science Board (see n3), *Science Indicators 1980*, p. 161. Similarly, a 1986 national survey revealed that 62 percent of the respondents thought that over the next twenty years, the benefits to society resulting from continued technological and scientific innovation would outweight the risks, while 32 percent thought they might or would not. See U.S. Congress, Office of Technology Assessment, *New Developments in Biotechnology – Background Paper: Public Perceptions of Biotechnology*, Publ. No. OTA-BP-BA-45 (Washington, D.C.: Government Printing Office, May 1987), p. 29.

25. National Commission for the Protection of Human Subjects, *Special Study*, Department of Health, Education and Welfare Publ. No. (OS)78-0015 (Washington, D.C.: Government Printing Office, 1978), p. 207.

26. Merrill Sheils, Gerald C. Lubenow, and Ronald Henkoff, "The Key Breakthroughs," *Newsweek* (June 4, 1979), 64.

27. Barry Checkoway, "Public Participation in Health Planning Agencies: Promise and Practice," *Journal of Health Politics, Policy and Law 7* (Fall 1982), 723–33.

28. John H. Glenn, "Long-Term Economic Rx: Research" (editorial), *Science 215* (March 26, 1982), 1569; Alan F. Hofmann, "Economic Recovery and Scientific Research," *Science 214* (December 11, 1981), 1194. More generally, see David Dickson, *The New Politics of Science* (New York: Pantheon, 1984).

29. "New Regulations to Speed Drug Approvals, Improve Safety Monitoring," *FDA Drug Bulletin* (April 1985), 2–3.

30. Donald Kennedy, "A Calm Look at 'Drug Lag,' " *Journal of the American Medical Association 239* (January 30, 1978), 423–6.

31. Aaron Wildavsky, "No Risk Is the Highest Risk of All," *American Scientist 67* (January/February 1979), 32–7.

32. Marshall S. Shapo, *A Nation of Guinea Pigs* (New York: Free Press, 1979).

33. James Fallows, "American Industry: What Ails It, How to Save It," *Atlantic* (September 1980), 35–50. See also Denis J. Prager and Gilbert S. Omenn, "Research, Innovation, and University–Industry Linkages," *Science 207* (January 25, 1980), 379–84.

34. Department of Health and Human Services, National Institutes of Health, Office for Medical Applications of Research, "Biomedical Discoveries Adopted by Industry for Purposes Other Than Health Services," draft report for the Director, National Institutes of Health, March 2, 1981, typescript. The pragmatic appeals of scientists for continued research funding are described in Barbara J. Culliton, "Frank Press Calls Budget Summit: Scientists Link Federal Support of Research to National Security and a Strong Economy," *Science 214* (November 6, 1981), 634–5.

35. See, for example, the comments of Paul Berg, quoted in Sharon Begley and Pamela Abramson, "The DNA Industry," *Newsweek* (August 20, 1979), 53.

36. William J. Broad, "Whistle Blower Reinstated at HEW," *Science 205* (August 3, 1979), 476.

37. *Technovation* is the title of a journal that Elsevier Science Publishing Company, Amsterdam, began publishing in 1980.

38. This is a well-documented problem in health policy as well as other arenas. See David Banta, "The Federal Legislative Process and Health Care," in Steven Jonas, *Health Care Delivery in the United States* (New York: Springer, 1977), pp. 329–45.

39. Marjorie Sun, "For NIH, Business as Usual," *Science 210* (December 19, 1980), 1332–3.

40. The threat of legislation can be an especially powerful form of congressional influence. The prospect of a Senate bill proposing a complete ban on DES finally moved the FDA to restrict the use of DES in animal feed, after more than a decade of delay (Nicholas Wade, "FDA Invents More Tales About DES," *Science 177* [August 11, 1972], 503).

41. Nicholas Wade, "Gene Splicing: Senate Bill Draws Charges of Lysenkoism," *Science 197* (July 22, 1977), 348–9. Lysenko, an agricultural biologist in the Soviet Union, falsified data and argued for the inheritance of acquired characteristics. His views came to dominate aspects of Soviet biology.

42. Harold P. Green, "The Recombinant DNA Controversy: A

Model of Public Influence," *Bulletin of the Atomic Scientists 34* (November 1978), 13. A similar argument is eloquently presented in David L. Bazelon, "Risk and Responsibility," *Science 205* (July 20, 1979), 277–80.

3. DES and the elusive goal of drug safety

1. E. C. Dodds, L. Goldberg, W. Lawson, et al., "Oestrogenic Activity of Certain Synthetic Compounds," *Nature 141* (February 5, 1938), 247–8.
2. John C. Burch, M.D., quoted in William L. Laurence, "New Remedy Used in Ills of Women," *New York Times*, October 21, 1939, p. 13.
3. M. Edward Davis, "A Clinical Study of Stilbestrol," *American Journal of Obstetrics and Gynecology 39* (1940), 952.
4. E. C. Dodds, Cameron Prize Lecture, October 15, 1940, published in "The New Oestrogens," *Edinburgh Medical Journal 48* (January 1941), 12.
5. The primary sources for the early history of DES were Susan E. Bell, *The Synthetic Compound Diethylstilbestrol (DES) 1938–1941: The Social Construction of a Medical Treatment*, dissertation, Brandeis University, May 1980; Richard Gillam, "Prenatal DES: The Vagaries of Public Policy," unpublished paper prepared for the Stanford EVIST Project (February 1983); Robert Meyers, *D.E.S. The Bitter Pill* (New York: Seaview/Putnam, 1983); Stephen Fenichell and Lawrence S. Charfoos, *Daughters at Risk: A Personal DES History* (Garden City, N.Y.: Doubleday, 1981); Barbara Seaman and Gideon Seaman, *Women and the Crisis in Sex Hormones* (New York: Rawson Associates, 1977); Roberta J. Apfel and Susan M. Fisher, *To Do No Harm: DES and the Dilemmas of Modern Medicine* (New York: Yale University Press, 1984); and Cynthia L. Orenberg, *DES: The Complete Story* (New York: St. Martin's Press, 1981). The history of the DES morning-after pill is based mainly on Richard Gillam, "DES as a 'Morning-After Pill': The Failure to Regulate a New Biomedical Technology and the Limits of Expertise, II," unpublished paper prepared for the Stanford EVIST Project (1985); Meyers; Seaman and Seaman; and Kay Weiss, "Afterthoughts on the Morning-After Pill," *Ms 2* (November 1973), 22–6. References used for livestock uses of DES include two unpublished papers prepared for the Stanford EVIST project: David Schnell, "DES in Cattlefeed: A Study in the Politics of Safety Regulation" (June 1981), and W. B. Smith, "DES and the Politics of Regulation: A Policy Analysis" (May 1980); also, Edward W. Lawless, *Technology and Social Shock* (New Brunswick, N.J.: Rutgers University Press, 1977), pp. 70–80; Nicholas Wade, "DES: A Case Study of Regulatory Abdication," *Science 177* (July 28,

1972), 335–7; Susan G. Hadden, "DES and the Assessment of Risk," in Dorothy Nelkin, ed., *Controversy: Politics of Technical Decisions* (Beverly Hills: Sage Publications, 1979), pp. 111–24; Fenichell and Charfoos; Seaman and Seaman; and Meyers.

6. Price quote is from Bell (see n5), *The Synthetic Compound DES*, p. 209.

7. For a comprehensive review of these animal studies, see Brian L. Strom, "Summary Exhibit: State of the Art of DES Adverse Reactions to 1952," Clinical Epidemiology Unit, University of Pennsylvania School of Medicine, Philadelphia, July 1982. Quote is from "Estrogen Therapy – A Warning" (editorial), *Journal of the American Medical Association 113* (December 23, 1939), 2324.

8. S. J. Folley and H. M. S. Watson, "Some Biological Properties of Diethylstilboestrol," *Lancet 2* (1938), 423, quoted in Bell (n5), *The Synthetic Compound DES*, p. 80.

9. See, for example, A. S. Parkes and C. W. Bellerby, "Studies on the Internal Secretions of the Ovary. II. The Effects of Injection of the Oestrus Producing Hormone During Pregnancy," *Journal of Physiology 62* (October 30, 1926), 145–55; Margaret G. Smith, "On the Interruption of Pregnancy in the Rat by the Injection of Ovarian Follicular Extract," *Bulletin of the Johns Hopkins Hospital 39* (1926), 203–14.

10. A. S. Parkes, E. C. Dodds, and R. L. Noble, "Interruption of Early Pregnancy by Means of Orally Active Oestrogens," *British Medical Journal 2* (September 10, 1938), 559. After noting the relevance of their findings to women, the authors cautioned, "It would be unwise at the present juncture to make a comparison of the relative suitability of the two substances examined [DES and ethinyl oestradiol], or of other orally active oestrogens, for the purpose under consideration."

11. James Reed, "Doctors, Birth Control, and Social Values: 1830–1970," in M. J. Vogel and C. E. Rosenberg, eds., *The Therapeutic Revolution* (Philadelphia: University of Pennsylvania Press, 1979), pp. 109–33.

12. H. G. Lazell, former chief executive officer of the Beecham Group, a British pharmaceutical company, quoted in Meyers (n5), *D.E.S. The Bitter Pill*, p. 44.

13. W. R. Winterton and T. N. MacGregor, "Clinical Observations with Stilboestrol (Diethylstilboestrol)," *British Medical Journal 1* (January 7, 1939), 10–12; P. M. F. Bishop, Muriel Boycott, and S. Zuckerman, "The Oestrogenic Properties of 'Stilboestrol' (Diethylstilboestrol)," *Lancet 1* (January 7, 1939), 5–11. However, not all of the results from early human trials of DES were equally favorable. One French researcher concluded that the adverse effects produced by DES, notably nausea and vomiting, were too severe to justify substituting it for natural estrogens; see J. Varangot,

"Clinical Use of Stilboestrol" (letter), *Lancet 1* (February 4, 1939), 296.

14. J. A. Morrell, "Stilbestrol," *Journal of Clinical Endocrinology 1* (May 1941), 419.

15. For a more complete discussion of the drug industry in the early twentieth century, see Peter Temin, *Taking Your Medicine: Drug Regulation in the United States* (Cambridge, Mass.: Harvard University Press, 1980).

16. Ibid., p. 38.

17. Both Walter Campbell, Commissioner of the FDA, and J. J. Durrett, then Chief of the Bureau of Drugs worked actively for drug reform legislation during the 1930s (see Charles O. Jackson, *Food and Drug Legislation in the New Deal* [Princeton, N.J.: Princeton University Press, 1970]). At least fourteen separate women's organizations ultimately joined in the struggle for new drug legislation, playing a decisive role in its eventual passage (ibid., pp. 46, 198, 213).

18. Jackson (n17), *Food and Drug Legislation in the New Deal*, provides a detailed account of the five-year struggle to pass the 1938 drug reform law. See also Paul J. Quirk, "Food and Drug Administration," in James Q. Wilson, ed., *The Politics of Regulation* (New York: Basic Books, 1980), pp. 191–235; and Temin (n15), *Taking Your Medicine*, pp. 38–57.

19. Deposition of Don Carlos Hines, M.D., March 1, 1976 [hereafter Hines dep.], in *Abel v. Eli Lilly and Co.*, Wayne County, Michigan, Civil Action No. 74–030–070-NP, p. 52.

20. See, for example, A. Lacassagne, "Apparition d'Adenocarcinomes Mammaires Chez des Souris Males Traitées par une Substance Oestrogene Synthétique," *Compte Rendus des Séances de la Societé de Biologie 129* (November 19, 1938), 641–3; and Charles F. Geschickter, "Mammary Carcinoma in the Rat with Metastasis Induced by Estrogen," *Science 89* (January 13, 1939), 35–7. For a comprehensive review of studies of the carcinogenic effects of both synthetic and natural estrogens, see Strom (n7), "Summary Exhibit," pp. 1–24.

21. See Strom (n7), "Summary Exhibit," pp. 25–32, for studies relating to the question of whether animal and human fetuses exposed to exogenous estrogens develop physical abnormalities.

22. H. O. Burdick and Helen Vedder, "The Effect of Stilbestrol in Early Pregnancy," *Endocrinology 28* (April 1941), 629–32.

23. "Synthetic Female Hormone Pills Considered Potential Danger," *Science News Letter* (January 13, 1940), p. 31.

24. The possibility of liver damage is raised in Ephraim Shorr, Frank H. Robinson, and George N. Papanicolaou, "A Clinical Study of the Synthetic Estrogen Stilbestrol," *Journal of the American Medical Association 113* (December 23, 1939), 2312–18. The animal study

showing that DES caused liver damage in mice is Hans Selye, "On the Toxicity of Oestrogens with Special Reference to Diethylstilboestrol," *Canadian Medical Association Journal 41* (July 1939), 48–9.

25. Sir Edward C. Dodds, quoted in Fenichell and Charfoos (n5), *Daughters at Risk*, p. 19.

26. Council on Pharmacy and Chemistry, "Stilbestrol: Preliminary Report of the Council," *Journal of the American Medical Association 113* (December 23, 1939), 2312.

27. "Estrogen Therapy – A Warning" (editorial) (n7).

28. "Contraindications to Estrogen Therapy" (editorial), *Journal of the American Medical Association 114* (April 20, 1940), 1560.

29. Bell (n5), *The Synthetic Compound DES* (p. 222) states that eleven applications were on file at the end of 1940, although the January 6, 1941 Notice of Hearing lists only ten companies (Defendants' Appendix [hereafter Dep. App.], pp. 429–30b, in *Abel* v. *Eli Lilly and Co.*, 289 NW2d 20 [1980]). Most of the initial NDAs were based on British data (Hines dep. [n19], p. 56), These were withdrawn when the FDA indicated it would not accept either "work done abroad" or "studies in the Southern Medical Journal" (Bell, p. 220). The latter was in reference to work by Karl J. Karnaky, who, as we shall see, was one of the early advocates of prenatal DES therapy.

30. Frailey to Carter, December 27, 1939, Def. App. (n29), p. 424b.

31. See deposition of Theodore G. Klumpp, M.D., October 20, 1976 [hereafter Klumpp dep.], in *Abel* v. *Eli Lilly and Co.*, Wayne County, Michigan, Civil Action No. 74–030–070–NP, p. 36.

32. 1939 FDA *Annual Report*, Food Law Institute Series, 1951, quoted in Bell (n5), *The Synthetic Compound DES*, pp. 248–9.

33. Although the review of new drugs was one of the most demanding of the many new duties assumed by the FDA under the 1938 Act, its budget in 1940 included only a paltry $103,000 for this task (David F. Cavers, "The Evolution of the Contemporary System of Drug Regulation under the 1938 Act," in John B. Blake, *Safeguarding the Public* [Baltimore: Johns Hopkins Press, 1970], p. 163). The Klumpp deposition (n31), p. 21, states that there were two medical officers in the FDA's New Drug Section, and about six medical officers in the Drug Division. Such staffing levels are indeed startling in view of the fact that, between 1938 and 1941, NDAs were coming in at a rate of well over 100 a month (Cavers, p. 167). The FDA's position in those years was, as Cavers put it, "well-nigh desperate" (p. 163).

34. Klumpp dep. (n31), pp. 74–5. See also Hines dep. (n19), p. 59, for evidence that the manufacturers were fully aware of this strategy.

35. Klumpp dep. (n31), p. 31.

36. See Bell (n5), *The Synthetic Compound DES*, pp. 205–6, 253–4.

One previous joint application was for sulfathiazole, a lifesaving drug. See Klumpp dep. (n31), pp. 21–3.

37. Klumpp dep. (n31), p. 49; see also p. 32.
38. Hines dep. (n19), pp. 213–14.
39. Klumpp dep. (n31), p. 33. For confirmation, see Hines dep. (n19), pp. 70, 196.
40. Fenichell and Charfoos (n5), *Daughters at Risk*, p. 33; Bell (n5), *The Synthetic Compound DES*, pp. 251–4. See also Klumpp dep. (n31), pp. 33–5. For Frailey's telegram to prospective DES manufacturers instructing them to telegraph Campbell to withdraw their applications, see Frailey to Anderson, December 30, 1940, Def. App. (n29), p. 427b. For manufacturers' telegrams, see Plaintiffs' Appendix [hereafter Pla. App.] 5, in *Abel* v. *Eli Lilly Co.*, 289 NW2d 20 (1980).
41. See "Report of a Meeting of Pharmaceutical Manufacturers Interested in Stilbestrol," Washington, D.C., January 28, 1941, Def. App. (n29), pp. 436–9b; also Hines's report on this meeting to W. J. Rice, January 30, 1941, Hines dep. (n19), Ex. 1.
42. This important point is made in both Bell (n5), *The Synthetic Compound DES*, and Gillam (n5), "Prenatal DES."
43. See Klumpp dep. (n31), pp. 38–9, 51–2; and Fenichell and Charfoos (n5), *Daughters at Risk*, pp. 34–5. The fears of the "New York Group" concerning the toxicity of DES had also been published in the medical literature. See U. J. Salmon, S. H. Geist, and R. I. Walter, "Evaluation of Stilbestrol as a Therapeutic Estrogen," *American Journal of Obstetrics and Gynecology 40* (August 1940), 243–50.
44. Hines quoted in "Memorandum of Meeting Between A.D.M.A. and DES Manufacturers," March 25, 1941, Pla. App. (n40) 16.
45. Hines to Upjohn, March 5, 1941, Hines dep. (n19), Ex. 32.
46. Klumpp dep. (n31), p. 93.
47. For a more extensive exposition of this point, see Bell (n5), *The Synthetic Compound DES*, pp. 116–17, 156. Quote is from an interview of Shorr by Durrett, September 19 and 20, 1940, quoted in Bell, p. 116. Sevringhaus's interpretation of the New York group's unfavorable findings was communicated to the FDA in a letter to Durrett dated January 4, 1941 (quoted in Bell, p. 116).
48. Hines's report to Rice, "Meetings Regarding Stilbestrol, March 24 and 25, 1941," Def. App. (n29), p. 445b.
49. See Bell (n5), *The Synthetic Compound DES*, p. 224. Some drug companies encouraged physicians to lobby the FDA on behalf of DES, and others manipulated them into doing so – as, for example, when Sharpe and Dohme informed clinicians the company would no longer supply them with DES because they had to withdraw their application from the FDA. Withdrawal in no way required them to stop supplying DES for research purposes, but

this message prompted a flood of letters to the FDA praising DES (see Bell, pp. 252–3).

50. Hines to Rice, "Meetings Regarding Stilbestrol," Def. App. (n29), pp. 445–6b.
51. Frailey to Hines, May 12, 1941, Hines dep. (n19), Ex. 10.
52. Klumpp dep. (n31), pp. 42, 45, 48.
53. Bell (n5), *The Synthetic Compound DES*, p. 271; also Hines to Herwick, April 23, 1941, Def. App. (n29), p. 449b.
54. Klumpp dep. (n31), p. 47.
55. Quoted in Bell (n5), *The Synthetic Compound DES*, p. 141; also, pp. 154–5.
56. See notes 20 and 21 for some of this evidence.
57. For date of submission of Master File, see Hines to Carter, June 4, 1941, Pla. App. (n40) 32. For approval, see Frailey's telegram to Volwiler, September 12, 1941, Hines dep. (n19), Ex. 49.
58. Hines to Upjohn, October 8, 1941, Def. App. (n29), p. 455b.
59. Council on Pharmacy and Chemistry, "Report of the Council: Diethylstilbestrol," *Journal of the American Medical Association* 119 (June 20, 1942), 632–6. Also, see Klumpp dep. (n31), pp. 48, 52–62; also Bell (n5), *The Synthetic Compound DES*, p. 206.
60. See Bell (n5), *The Synthetic Compound DES*, pp. 271–3. Indicating the FDA's awareness of the proliferating uses of DES, an FDA official told a representative from Merck: "We realized certain experts in the field of gynecology were using this drug for treatment of various menstrual disorders, and . . . we did not decry the use of the drug in these conditions by such experts; however, . . . it was our judgment that the general practitioners should be extremely conservative when using the drug." (memo of interview, FDA, October 23, 1941), quoted in Bell, p. 272.
61. See Hines dep. (n19), pp. 120–1; also Bell (n5), *The Synthetic Compound DES*, pp. 173, 210–11. The contraindications listed for DES were similar to those for natural estrogen products and included evidence of cancerous or precancerous lesions of the breast or cervix, and a family history of breast or genital cancer (letter, FDA to Lilly, September 12, 1941, cited in Bell, p. 273).
62. For example, FDA Commissioner Campbell wrote to Sharpe and Dohme that with regard to the FDA's labeling and literature requirements for DES, "the public interest may well be served by the exercise of conservatism with respect to the drug at this time" (letter, Campbell to Sharpe and Dohme, November 3, 1941, quoted in Bell [n5], *The Synthetic Compound DES*, p. 273).
63. Bell (n5), *The Synthetic Compound DES*, pp. 171, 250.
64. Address before the American Pharmaceutical Manufacturer's Association, June 23, 1941, reprinted in Theodore G. Klumpp, "The Influence of the Food, Drug and Cosmetic Act on the Marketing of Drugs," *Connecticut State Medical Journal 6* (January 1942), 4.

65. Klumpp dep. (n31), p. 48; also, pp. 36, 40.
66. See "Legal Status of Approved Labeling for Prescription Drugs; Prescribing for Uses Unapproved by the Food and Drug Administration," *Federal Register 37*, No. 158 (August 15, 1972), 16503. Then, as today, the FDA had little to say about how physicians chose to use drugs. The main restraint on physicians was the threat of malpractice, but this threat carried little weight unless there were really gross violations of conventional medical practice, and even then, as Chapter 8 discusses, malpractice claims for drug injuries were often difficult to prove.
67. Quote is from William L. Laurence, "Time Is Reversed by New Extract," *New York Times*, May 14, 1939, p. 39. See also "New Method Used in Rejuvenation," *New York Times*, August 24, 1939, p. 13; and Laurence (n2), "New Remedy Used in Ills of Women."
68. Helen Haberman, "Help for Women Over 40," *Hygeia 19* (November 1941), 898–9. This article was also condensed in the *Reader's Digest* (November 1941), 67–8.
69. Morrell (n14), "Stilbestrol."
70. See Hines dep. (n19), p. 37. The theory behind the use of DES to stunt growth, which was derived from animal research, was that DES closes the bones so that they are unable to grow any more. For other clinical uses, including arthritis, see A. H. Aaron, Frank Meyers, Morton H. Lipsitz, et al., "Toxicity Studies on Stilbestrol," *American Journal of Digestive Diseases 8* (November 1941), 437–9; also, "Prostate Cancer Yields to a Drug," *New York Times*, December 15, 1943, p. 29.
71. Excerpts from McNeil-O-Gram, reprinted in Pla. App. (n40) 43.
72. Dr. Edward D. Allen, discussant of Lewis C. Scheffey, David M. Farell, and George A. Hahn, "The Role of Injudicious Endocrine Therapy," *Journal of the American Medical Association 127* (January 13, 1945), 79.
73. S. Charles Freed, "Present Status of Commercial Endocrine Preparations," *Journal of the American Medical Association 117* (October 4, 1941), 1175.
74. Dr. Joe V. Meigs, discussant in Scheffey et al. (n72), "The Role of Injudicious Endocrine Therapy," 79. Preceding quotes are from F. Jackson Stoddard, "The Abuse of Endocrine Therapy in Gynecology," *Journal of the American Medical Association 129* (October 13, 1945), 509; Emil Novak, "Postmenopausal Bleeding as a Hazard of Diethylstilbestrol Therapy," *Journal of the American Medical Association 125* (May 13, 1944), 98; and Scheffey et al., 79.
75. Karnaky began using DES in the treatment of endometriosis and menstrual problems very soon after its synthesis in 1938. See Karl J. Karnaky, "Stilboestrol, the New Synthetic Estrogenic Hor-

mone," *Urologic and Cutaneous Review 43* (1939), 633–4; and Karl J. Karnaky, "Toxicity of Stilbestrol" (letter), *Southern Medical Journal 32* (December 1939), 1250. For a good summary of the different clinical approaches to threatened abortion at this time and the introduction of prenatal DES therapy, see Gordon Rosenblum and Eugene Melinkoff, "Preservation of the Threatened Pregnancy with Particular Reference to the Use of Diethylstilbestrol," *Western Journal of Surgery 55* (November 1947), 597–603.

76. Interview with Karnaky, quoted in Meyers (n5), *D.E.S. The Bitter Pill*, p. 53.

77. Ibid, p. 53.

78. Karl J. Karnaky, "The Use of Stilbestrol for the Treatment of Threatened and Habitual Abortion and Premature Labor: A Preliminary Report," *Southern Medical Journal 35* (September 1942), 838.

79. Ibid., 839.

80. Ibid., 843.

81. See O. Watkins Smith and George Van S. Smith, "Prolan and Estrin in the Serum and Urine of Diabetic and Nondiabetic Women During Pregnancy, with Especial Reference to Late Pregnancy Toxemia," *American Journal of Obstetrics and Gynecology 33* (March 1937), 365–79, and its accompanying references. This paper was first presented at a medical meeting in 1935. For later papers, see the references listed in O. Watkins Smith, George Van S. Smith, and David Hurwitz, "Increased Excretion of Pregnanediol in Pregnancy from Diethylstilbestrol with Special Reference to the Prevention of Late Pregnancy Accidents," *American Journal of Obstetrics and Gynecology 51* (1946), 411–15.

82. See Smith et al. (n81), "Increased Excretion of Pregnanediol." The Smiths cited evidence showing that DES was 100 times as active as estrone (a natural estrogen) in stimulating increased secretion of gonadotropic hormones in rats. They also measured the level of pregnanediol, a hormone thought to correlate closely with progesterone itself, in the urine of pregnant women and found that it increased sharply when DES was administered and decreased when DES was withdrawn.

83. Priscilla White, Raymond S. Titus, Elliott P. Joslin, et al., "Prediction and Prevention of Late Pregnancy Accidents in Diabetes," *American Journal of the Medical Sciences 198* (October 1939), 482–92; Priscilla White and Hazel Hunt, "Prediction and Prevention of Pregnancy Accidents in Diabetes," *Journal of the American Medical Association 115* (December 14, 1940), 2039–40; Priscilla White and Hazel Hunt, "Pregnancy Complicating Diabetes," *Journal of Clinical Endocrinology 3* (September 1943), 500–11; Priscilla White, "Pregnancy Complicating Diabetes," *Journal of the American Med-*

ical Association 128 (May 19, 1945), 181–2; Priscilla White, "Pregnancy Complicating Diabetes," *American Journal of Medicine 7* (November 1949), 609–16.

84.   White and Hunt (n83), "Prediction and Prevention of Pregnancy Accidents," 2040.

85.   O. Watkins Smith, "Diethylstilbestrol in the Prevention and Treatment of Complications of Pregnancy," *American Journal of Obstetrics and Gynecology 56* (November 1948), 821–34.

86.   See Smith et al. (n81), "Increased Excretion of Pregnanediol," 414.

87.   Ibid., 414.

88.   See material in Def. App. (n29), pp. 475–8b, including Lilly to FDA submitting NDA, April 15, 1947; Abbott to FDA submitting NDA, May 15, 1947; McNeil to FDA submitting NDA, September 10, 1948; White to FDA submitting NDA, January 11, 1950; also, Pla. App. (n40) 40A, "Full Reports of All Investigations Which Have Been Made to Show Whether or Not Stilbestrol Is Safe for Use;" and the list of articles submitted for Lilly's 1947 supplemental NDA in Hines dep. (n19), Ex. 16. For further evidence and discussion, see Gillam (n5), "Prenatal DES: The Vagaries of Public Policy," p. 20; and Apfel and Fisher (n5), *To Do No Harm*, pp. 19–20 and note 13.

89.   For quotes and criticisms, see: Dr. William Bickers commenting on Karnaky (n78), "The Use of Stilbestrol," 846; Dr. Henry Dolger and Dr. William J. Dieckmann commenting on White (n83), "Pregnancy Complicating Diabetes," 182; and Dr. Ernest W. Page, Dr. Willard M. Allen, Dr. Ralph Reis, and Dr. William J. Dieckmann, commenting on O. Watkins Smith and George Van S. Smith, "The Influence of Diethylstilbestrol on the Progress and Outcome of Pregnancy as Based on a Comparison of Treated with Untreated Primigravidas," *American Journal of Obstetrics and Gynecology 58* (November 1949), 1005–8. Commenting in 1956 on White's twenty years of research with prenatal hormone therapy, Dr. Reis observed that it was "based on the investigation of the Smiths, which work in its entirety is confusing to most of us;" see discussion following Priscilla White, Luke Gillespie, and Lloyd Sexton, "Use of Female Sex Hormone Therapy in Pregnant Diabetic Patients," *American Journal of Obstetrics and Gynecology 71* (January 1956), 69.

90.   Smith et al. (n81), "Increased Excretion of Pregnanediol," 414.

91.   These studies are listed in Plaintiffs–Appellees Appendix [hereafter Abel App.], Vol. II, pp. 320–63b, in *Abel* v. *Eli Lilly and Co.*, 343 NW2d 164 (1984).

92.   "Contraindications to Estrogen Therapy" (editorial) (n28), 1561.

93.   Edgar Allen, "Estrogenic Hormones in the Genesis of Tumors and Cancers," *Endocrinology 30* (June 1942), 950. A 1941 study of

cancer of the cervix in mice following prolonged doses of estrogens concluded similarly that "the high incidence of cervical cancer in these experimental groups emphasizes that estrogen is a very important factor, not merely an incidental one, in cervical carcinogenesis" (Edgar Allen and W. U. Gardner, "Cancer of the Cervix of the Uterus in Hybrid Mice Following Long-Continued Administration of Estrogen," *Cancer Research* [April 1941], 365). The authors also made an interesting observation about other animal studies: "One reason why the incidence of cancer of the cervix uteri has been so low in mice treated with estrogens is that mammary cancer appears at an earlier age than does cervical cancer, and consequently animals may die of the former; i.e., many of them do not live long enough for the cervical cancer to develop" (p. 359).

94. J. S. Henry, "The Avoidance of Untoward Effects of Oestrogenic Therapy in the Menopause," *Canadian Medical Association Journal* 53 (July 1945), 33. Not all of these warnings were based on hypothetical inferences from animal studies. Henry described two patients who developed malignancies following prolonged high dosages of DES and other estrogens which were attributed to the therapy.

95. Novak (n74), "Postmenopausal Bleeding," 98–9. A widely read text published by the AMA entitled *Glandular Physiology and Therapy* contained a similarly sharp admonition. Pointing to the "definite trend toward massive dosage" of estrogens, the author urged doctors not "to be too incautious. . . . The increased incidence of malignant change following administration of estrogens seems a definite warning. . . ." (Edward Doisy, quoted in Fenichell and Charfoos [n5], *Daughters at Risk*, pp. 43–4). See also, Joe V. Meigs, "Gynecology: Carcinoma of the Endometrium," *New England Journal of Medicine 233* (July 5, 1945), 11–17.

96. Karl J. Karnaky, "Estrogenic Tolerance in Pregnant Women," *American Journal of Obstetrics and Gynecology 53* (February 1947), 315.

97. *Dispensatory of the United States of America*, 24th ed. (Philadelphia: J. B. Lippincott, 1947), p. 363.

98. A 1945 text stated, for example: "Oestrogen can traverse the placenta and so enter the foetal circulation. . . . Philipp (1929) found oestrin in relatively large quantities in the urine of newborn babies. . . . In 1930, the observation that estrogen can pass from the maternal into the foetal circulation was confirmed by Courrier . . ."; see Harold Burrows, *Biological Actions of Sex Hormones* (London: Cambridge University Press, 1945), pp. 258, 263. Also, S. Zuckerman and G. van Wagenen, "The Sensitivity of the New-Born Monkey to Oestrin," *Journal of Anatomy 69* (1935), 497–501; and Joachim Sklow, "Is Human Placenta Permeable to Gonado-

tropic and Estrogenic Hormones?" *Proceedings of the Society for Experimental Biology and Medicine 49* (1942), 607–9.

99. Karl J. Karnaky, "Prolonged Administration of Diethylstilbestrol," *Journal of Clinical Endocrinology 5* (July–August 1945), 280.

100. See, for example: R. R. Greene, M. W. Burrill, and A. C. Ivy, "Experimental Intersexuality: The Paradoxical Effects of Estrogens on the Sexual Development of the Female Rat," *Anatomical Record 74* (1939), 429–38; R. R. Greene, M. W. Burrill, and A. C. Ivy, "Experimental Intersexuality: Modification of Sexual Development of the White Rat with a Synthetic Estrogen," *Proceedings of the Society for Experimental Biology and Medicine 41* (May 1939), 169–70; R. R. Greene, M. W. Burrill, and A. C. Ivy, "Experimental Intersexuality: The Effects of Estrogens on the Antenatal Sexual Development of the Rat," *American Journal of Anatomy 67* (1940), 305–45. In addition to these, Strom (n7), "Summary Exhibit," lists at least seventeen other papers published before 1947 all relating to the question of whether animal and human fetuses exposed to exogenous estrogens develop physical abnormalities.

101. A 1947 study found that the offspring of pregnant mice injected late in term with urethane had a high incidence and number of pulmonary tumors at six months of age. See C. D. Larsen, Lucille L. Weed, and Paul B. Rhoads, Jr., "Pulmonary-Tumor Induction by Transplacental Exposure to Urethane," *Journal of the National Cancer Institute 8* (October 1947), 63–70.

102. Richard Gillam, "DES and Drug Safety: The Limits of Expertise," unpublished paper prepared for the Stanford EVIST project (September 1984), p. 25. Gillam quotes the following excerpts from letters received by the FDA in April 1947: "I do not believe that the profession at large is prepared to use such massive doses of diethylstilbestrol." A second physician argued "that estrogen therapy with such tremendous strength as this . . . would be very likely to be greatly misused by the profession at large. . . . I think it might lead to more harm than good if it were advertised to and distributed widely to the profession at large." Another correspondent noted the "experimental" nature of prenatal DES, urging the FDA to "postpone any decision in regard to the packaging of this drug in concentrated form until such time as more evidence is available that the use of large amounts of diethylstilbestrol is desirable and free of any possible danger."

103. Rosenblum and Melinkoff (n75), "Preservation of the Threatened Pregnancy," 601.

104. Def. App. (n29), pp. 479–82b: FDA letters approving Abbott NDA, July 1, 1947, Squibb NDA, July 10, 1947, Lilly NDA, September 16, 1947, and McNeil NDA, October 13, 1948.

105. See FDA to Abbott, Def. App. (n29), p. 479b; also FDA to Lilly, May 29, 1947, reprinted in Abel App. (n91), Vol. II, pp. 232–3b.

106. FDA to McNeil, Def. App. (n29), p. 481b.
107. FDA to Lilly, Abel App. (n91), Vol. II, pp. 232–3b.
108. See for example, "Prostate Cancer Yields to a Drug"; "The Case of Benjamin Twaddle," *Time* (August 29, 1949), pp. 34, 37; "Mump Medicine for Men," *Science Digest 26* (September 1949), 48–9; Alan F. Guttmacher, "The Truth About Miscarriage," *Parents Magazine* (October 1955), pp. 48–9, 127–8.
109. The first quote is from Smith and Smith (n89), "The Influence of Diethylstilbestrol," 1003. The second is from Smith (n85), "Diethylstilbestrol in the Prevention and Treatment of Complications of Pregnancy," 829.
110. Grant Chemical Company's advertisement for DES, *American Journal of Obstetrics and Gynecology 73* (June 1957), 14.
111. Quote is from Dr. Willard Allen, discussant of W. J. Dieckmann, M. E. Davis, L. M. Rynkiewicz, et al., "Does the Administration of Diethylstilbestrol During Pregnancy Have Therapeutic Value?" *American Journal of Obstetrics and Gynecology 66* (November 1953), 1080. A 1955 article in *Parents Magazine* referred to "the usual orthodox hormone treatment" (Guttmacher [n108], "The Truth About Miscarriage," p. 128). Although aggregate sales data for 25-mg DES over time are not available, estimates based on a single company's sales (Herbst et al.), and on a review of medical records (Noller et al.) suggest that the peak usage of prenatal DES occurred between 1947 and 1953. See Arthur L. Herbst, Philip Cole, Marija J. Norusis, et al., "Epidemiologic Aspects and Factors Related to Survival in 384 Registry Cases of Clear Cell Adenocarcinoma of the Vagina and Cervix," *American Journal of Obstetrics and Gynecology 135* (December 1, 1979), 882; and Kenneth L. Noller, Duane E. Townsend, and Raymond H. Kaufman, "Genital Findings, Colposcopic Evaluation, and Current Management of the Diethylstilbestrol-Exposed Female," in Arthur L. Herbst and Howard A. Bern, eds., *Developmental Effects of Diethylstilbestrol (DES) in Pregnancy* (New York: Thieme–Stratton, 1981), p. 82. For data on hormone sales in general, see "Hormone Sales Climbing," *Chemical and Engineering News* (August 20, 1956), 4078–81.
112. R. E. Crowder, E. S. Bills, and J. S. Broadbent, "The Management of Threatened Abortion: A Study of 100 Cases," *American Journal of Obstetrics and Gynecology 60* (October 1950), 899. For studies casting doubt on the Smiths' theories, see I. F. Sommerville, G. F. Marrian, and B. E. Clayton, "Effect of Diethylstilboestrol on Urinary Excretion of Pregnanediol and Endogenous Oestrogen During Pregnancy," *Lancet 1* (April 23, 1949), 680–2; M. Edward Davis and Nicholas W. Fugo, "Steroids in the Treatment of Early Pregnancy Complications," *Journal of the American Medical Association 142* (March 18, 1950), 778–85.
113. Davis and Fugo (n112), "Steroids in the Treatment of Early Preg-

nancy Complications"; Crowder et al. (n112), "The Management of Threatened Abortion"; David Robinson and Landrum B. Shettles, "The Use of Diethylstilbestrol in Threatened Abortion," *American Journal of Obstetrics and Gynecology 63* (June 1952), 1330–3; James H. Ferguson, "Effect of Stilbestrol on Pregnancy Compared to the Effect of a Placebo," *American Journal of Obstetrics and Gynecology 65* (March 1953), 592–601. Ferguson's study was presented at the Annual Meeting of the Central Association of Obstetricians and Gynecologists, Memphis, Tenn., October 30–November 1, 1952.

114. Ferguson (n113), "Effect of Stilbestrol on Pregnancy," 600.

115. Dr. R. R. Greene, commenting on Ferguson (n113), "Effect of Stilbestrol on Pregnancy," 601. Greene had coauthored the series of articles on experimental intersexuality a decade earlier, which showed that DES caused structural abnormalities in animals born of mothers injected with it.

116. Dieckmann et al. (n111), "Does the Administration of Diethylstilbestrol During Pregnancy Have Therapeutic Value?" Quote is by Dr. M. E. Davis in discussion following the paper, p. 1081.

117. Yvonne Brackbill and Heinz W. Berendes, "Dangers of Diethylstilboestrol: Review of a 1953 Paper" (letter), *Lancet 2* (September 2, 1978), 520.

118. Dr. C. Frederick Irving commenting on Dieckmann et al. (n111), "Does the Administration of Diethylstilbestrol During Pregnancy Have Therapeutic Value?" 1080. For a classical study of the role of collegial networks in the diffusion of medical innovations, see James Coleman, Elihu Katz, and Herbert Menzel, *Medical Innovation: A Diffusion Study* (Indianapolis: Bobbs-Merrill, 1966).

119. Clyde L. Randall, Richard W. Baetz, Donald W. Hall, et al., "Pregnancies Observed in the Likely-to-Abort Patient With or Without Hormone Therapy Before or After Conception," *American Journal of Obstetrics and Gynecology 69* (March 1955), 643–56; Report of the Conference on Diabetes and Pregnancy, "The Use of Hormones in the Management of Pregnancy in Diabetics," *Lancet 2* (October 22, 1955), 833–6; Joseph W. Goldzieher and Benedict B. Benigno, "The Treatment of Threatened and Recurrent Abortion: A Critical Review," *American Journal of Obstetrics and Gynecology 75* (June 1958), 1202–14.

120. Estimated prescription rates are from Olli P. Heinonen, "Diethylstilbestrol in Pregnancy: Frequency of Exposure and Usage Patterns," *Cancer 31* (March 1973), 576. Sales of DES are discussed in Fenichell and Charfoos (n5), *Daughters at Risk*, p. 66.

121. For a ringing defense of the value of prenatal DES, see George V. Smith and Olive W. Smith, "Prophylactic Hormone Therapy: Relation to Complications of Pregnancy," *Obstetrics and Gynecology 4* (August 1954), 129–41.

122. Council on Pharmacy and Chemistry, "Too Many Drugs?" *Journal of the American Medical Association 139* (February 5, 1949), 378.

123. The *PDR* has been published annually since 1947 by Medical Economics, Inc. in cooperation with the drug industry. Until 1962, the product descriptions included a detailed dosage schedule (corresponding to the Smiths' originally recommended regime) for use in "accidents of pregnancy," which ended with a daily dose of 125 mg in the thirty-fifth week. In 1962 that schedule was dropped and thereafter dosages were specified only as "25 to 100 mg daily." In 1969, the product descriptions dropped "accidents of pregnancy" as an indicated use and added a strong warning concerning risks.

124. Cited in Russell R. Miller, "Prescribing Habits of Physicians," *Drug Intelligence and Clinical Pharmacy 8* (February 1974), 87. The *PDR* remains a critical source of information for physicians; see Temin (n15), *Taking Your Medicine*, pp. 88–9.

125. Lilly pamphlet, "Diethylstilbestrol: A Crystalline Synthetic Estrogen," (1953), in Pla. App. (n40) 43.

126. For a review of some of these studies, see Russell R. Miller, "Prescribing Habits of Physicians, Parts I–VIII," *Drug Intelligence and Clinical Pharmacy 7* (November 1973), 492–500; 7 (December 1973), 557–64; *8* (February 1974), 81–91.

127. Quotes from Walter S. Measday, "The Pharmaceutical Industry," in Walter Adams, ed., *The Structure of American Industry* (New York: Macmillan, 1977), pp. 270–1. See also Morton Mintz, "Squibb Doctor Faced Built-In Conflict," *Washington Post*, June 8, 1969, p. A9.

128. Dr. Dale Console, quoted in Mintz (n127), "Squibb Doctor Faced Built-In Conflict."

129. Irving Sider to W. R. McHargue, November 22, 1950, reprinted in Def. App. (n29), p. 510b.

130. Evan M. Wylie, "Why You Won't Lose Your Baby," *Good Housekeeping* (March 1960), pp. 82–3, 118–19.

131. Quote is from Robinson and Shettles (n113), "The Use of Diethylstilbestrol in Threatened Abortion," 1330. See Apfel and Fisher (n5), *To Do No Harm*, for an insightful discussion of the nature of the doctor–patient relationship and some of the reasons why women might have wanted to be given DES.

132. The use of "Dear Doctor" letters began in the 1960s; see Milton Silverman and Philip R. Lee, *Pills, Profits, and Politics* (Berkeley: University of California Press, 1974), p. 64.

133. For evidence of Larrick's friendly ties with the drug industry, see Morton Mintz, *The Therapeutic Nightmare* (Boston: Houghton Mifflin, 1965), pp. 95–182.

134. National Academy of Sciences, National Research Council [here-

after NAS], *Drug Efficacy Study* (Washington, D.C.: Government Printing Office, 1969), p. 1.

135. The original 1942 approval of DES in veterinary practice is cited in Bell (n5), *The Synthetic Compound DES*, p. 300.

136. "Case of the Barren Mink," *Time* (February 19, 1951), 65. See also, Lawless (n5), *Technology and Social Shock*, p. 72; Harrison Wellford, *Sowing the Wind* (New York: Grossman, 1972), p. 159; Seaman and Seaman (n5), *Women and the Crisis in Sex Hormones*, pp. 50–2; and Abel App. (n91), Vol. III, pp. 523–5b.

137. Wise Burroughs, C. C. Culbertson, Joseph Kastelic, et al., "The Effects of Trace Amounts of Diethylstilbestrol in Rations of Fattening Steers," *Science 120* (July 9, 1954), 66–7; John A. Rohlf, "Stilbestrol Ok'd for Fattening Beef!" *Farm Journal* (December 1954), pp. 34–5.

138. "Pellets Ok'd for Beef," *Farm Journal* (January 1956), 14; "Latest Feeding Tests with Stilbestrol," *Farm Journal* (June 1955), 124–5; "Stilbestrol for Milkers?" *Farm Journal* (August 1956), 55; "Stilbestrol-Treated Meat is Declared Safe," *American Druggist* (October 22, 1956), 71.

139. John A. Rohlf, "Two Million Head on Stilbestrol!" *Farm Journal* (March 1955), 38.

140. See, for example, the testimony of Robert K. Enders, Professor of Zoology, Swarthmore College, in *Chemicals in Food Products*, hearings before the House Select Committee to Investigate the Use of Chemicals in Food Products, U.S. House, 82nd Congress, 1st Session, April, May, and June 1951, pp. 429–43.

141. W. J. Hadlow and Edward F. Grimes, "Stilbestrol-Contaminated Feed and Reproductive Disturbances in Mice" (letter), *Science 122* (October 7, 1955), 643–4.

142. Granville F. Knight, W. Coda Martin, Rigoberta Iglesias, et al., "Possible Cancer Hazard Presented by Feeding Diethylstilbestrol to Cattle," reprinted in *Regulation of Food Additives and Medicated Animal Feeds*, hearings before a Subcommittee of the Committee on Government Operations, U.S. House, 92nd Congress, 1st Session, March 16–18, 29–30, 1971, pp. 568–76. These findings drew national attention at the time; see Waldemar Kaempffert, "Possible Danger in Feeds Now Given to Meat Animals," *New York Times*, January 29, 1956, p. E9. Additional evidence of the greater carcinogenic effect of minute traces of DES over a long time period had also been reported in Congressional hearings in 1951; see the testimony of Enders, in hearings on *Chemicals in Food Products* (n140), U.S. House, April, May, June 1951, p. 430.

143. *Congressional Record: Appendix*, February 21, 1957, p. A1352–4.

144. "Now It's Beef and Mutton," *Newsweek* (November 8, 1971), 85.

145. This trend continued: By 1966, the FDA's budget was ten times

larger than it had been a decade earlier and the agency employed five times as many people; Temin (n15), *Taking Your Medicine*, p. 121.

146. Food Additives Amendment of 1958, PL 85–929. For a summary of the history of the Delaney Amendment, see Vincent A. Klein-feld, "The Delaney Proviso – Its History and Prospects," *Food Drug Cosmetic Law Journal 28* (September 1973), 556–66. More generally, see Edward J. Allera, "An Overview of How the FDA Regulates Carcinogens Under the Federal Food, Drug, and Cosmetic Act," *Food Drug Cosmetic Law Journal 33* (February 1978), 59–77.

147. Distribution to "welfare agencies" is reported in "Fleming Says 1% of Chickens Have Drug That Causes Cancer; Sale of Fowl Halted," *Wall Street Journal*, December 11, 1959, p. 3. See also "Hormones and Chickens," *Time* (December 21, 1959), 32; "U.S. Sets Payment for Berry Losses," *New York Times*, May 5, 1960, p. 23. The FDA's ban of DES in fowl was appealed by the product manufacturers and in essence upheld by the U.S. District Court. The final ruling, containing the quoted excerpt, was published in the *Federal Register 30*, No. 35 (February 20, 1965), 2315.

148. This clause was Section 512(d)(1)(H) of the Food, Drug and Cosmetic Act. See Allera (n146), "An Overview of How the FDA Regulates," 63; and Wade (n5), "DES: A Case Study of Regulatory Abdication," 335.

149. George H. Gass, Don Coats, and Nora Graham, "Carcinogenic Dose-Response Curve to Oral Diethylstilbestrol," *Journal of the National Cancer Institute 33* (December 1964), 971–7. The number of cattle tested was increased to 6,000 under congressional pressure, although still no residues were detected. See Hadden (n5), "DES and the Assessment of Risk."

150. G. David Wallace, "FDA Eases Curb on Beef Fattener," *Washington Post*, September 20, 1970, p. E2. G. David Wallace, "Monitoring of Meat is Reduced Sharply," *Washington Post*, November 22, 1970, p. A2.

151. David G. Hawkins to Dr. Delphis C. Goldberg, March 5, 1971, reprinted in hearings on *Regulation of Food Additives and Medicated Animal Feeds* (n142), U.S. House, March 1971, pp. 513–15.

152. Charles C. Edwards, M.D., FDA Commissioner, in hearings on *Regulation of Food Additives and Medicated Animal Feeds* (n142), U.S. House, March 1971, p. 176.

153. For a detailed account of the passage of the Kefauver–Harris Amendments, see Silverman and Lee (n132), *Pills, Profits, and Politics*, pp. 107–37.

154. Temin (n15), *Taking Your Medicine*, p. 127.

155. Mintz (n133), *The Therapeutic Nightmare*, pp. 107–8.

156. Jonathan Spivak, "FDA Plans New Steps to Avoid Harm from

Drugs in Pregnancies," *Wall Street Journal*, September 20, 1963, p. 1. A month later, the potential dangers of estrogens again made headlines when Dr. Roy Hertz, chief endocrinologist for the National Cancer Institute, was quoted as saying that some cases of cancer in women "are circumstantially related to the use of estrogens for a long period of time," a charge that makers of birth control pills promptly challenged. See "Drug Concern Rejects Claim That Estrogen May Produce Cancer," *Wall Street Journal*, October 24, 1963, p. 13.

157. NAS (n134), *Drug Efficacy Study*. The full NAS review was not completed until 1969, but the NAS gave the FDA evaluations for individual drugs (including DES) as they were completed. For useful discussions of the implementation of the NAS study and its consequences, see "Drug Efficacy and the 1962 Drug Amendments," *Georgetown Law Journal 60* (1971), 185–224; and Silverman and Lee (n132), *Pills, Profits, and Politics*, pp. 121–37.

158. "Conditions for Marketing New Drugs Evaluated in Drug Efficacy Study: Drugs for Human Use; Drug Efficacy Study Implementation," *Federal Register 35*, No. 135 (July 14, 1970), 11273–4.

159. The FDA's receipt of the NAS evaluation of DES in 1967 is reported in *Regulation of Diethylstilbestrol, Part 1: Its Use as a Drug for Humans and in Animal Feeds*, hearing before a Subcommittee of the Committee on Government Operations, U.S. House, 92nd Congress, 1st Session, November 11, 1971, p. 79. Consumer criticisms of the FDA's overly close ties with the drug industry are described in Thomas P. Southwick, "FDA: Efficiency Drive Stumbles over the Issue of Drug Efficacy," *Science 169* (September 18, 1970), 1188–9; also, Anita Johnson, "The Bureau of Drugs: Professional Working Conditions and Their Effect on Substantive Decisions," *Journal of Drug Issues 5* (Spring 1975), 115–19. The major groups suing the FDA were the American Public Health Association and the National Council of Senior Citizens; see Thomas P. Southwick, "Faster FDA Action Asked in Lawsuit," *Science 168* (June 26, 1970), 1560. The FDA responded to this escalation in public criticism by publishing a list of 369 drugs that *had* been withdrawn because of inefficacy or lack of safety, and continued slowly to release the ratings (Harold M. Schmeck, Jr., "F.D.A. Lists 369 Drugs as Ineffective or Perilous," *New York Times*, November 28, 1970, pp. 1, 35). The lawsuit dragged on until 1980, when the FDA agreed to complete the removal of the remaining ineffective drugs from the market by 1984; "APHA Settles Drug Case," *Nation's Health* (November 1980), 1, 5.

160. Report of the Panel on Drugs Used in Disturbances of the Reproductive System, NAS (n134), *Drug Efficacy Study*. One panel member, Dr. Gilbert Gordan, did recall that the panel thought DES

extremely unlikely to be effective against accidents of pregnancy but considered the definition of "ineffective" too absolute to be suitable (personal communication to Diana Dutton, August 24, 1987). The definition specifies that "there is no acceptable evidence . . . to support a claim of effectiveness [and little chance that additional supporting evidence might be developed]. If there is clear evidence of ineffectiveness, the Panel should cite it. The recommendations to the FDA could be that no useful purpose is served by continuing to make this product available for the indication in question, and that immediate administrative action would appear to be justified" (NAS, *Drug Efficacy Study*, p. 43). The panel cited six references as documentation for its rating, including the 1948 paper by Smith (n85), "Diethylstilbestrol in the Prevention and Treatment of Complications of Pregnancy." Of the various studies published since the 1950s demonstrating that DES was ineffective in preventing miscarriages, only the paper by Dieckmann et al. (n111), "Does the Administration of Diethylstilbestrol During Pregnancy Have Therapeutic Value?" was included. The panel made no mention of any animal studies showing the hazards of DES.

161. *Physicians' Desk Reference*, 1969 (Rutherford, N.J.: Medical Economics, Inc.), pp. 819–20.

162. J. McLean Morris and Gertrude van Wagenen, "Compounds Interfering with Ovum Implantation and Development," *American Journal of Obstetrics and Gynecology* 96 (November 15, 1966), 804–15; J. McLean Morris and Gertrude van Wagenen, "Post-Coital Oral Contraception," *Proceedings of the Eighth International Conference of the International Planned Parenthood Federation,* Santiago, Chile, April 9–15, 1967, pp. 256–9. Quotes are from 1966 paper, pp. 806–7.

163. First quote is from Dr. Somers H. Sturgis, commenting on Morris and van Wagenen (n162), "Compounds Interfering with Ovum Implantation," 815. Second quote from Edward Edelson, "That Morning-After Pill," *Pageant* (July 1967), 55.

164. Dr. Charles E. McLennan, commenting on Morris and van Wagenen (n162), "Compounds Interfering with Ovum Implantation," 814.

165. Howard Ulfelder, "The Stilbestrol Disorders in Historical Perspective," *Cancer* 45 (June 15, 1980), 3009.

166. Arthur L. Herbst and Robert E. Scully, "Adenocarcinoma of the Vagina in Adolescence," *Cancer* 25 (April 1970), 745–57.

167. Testimony of Arthur L. Herbst, M.D., Department of Gynecology, Massachusetts General Hospital, in hearing on *Regulation of Diethylstilbestrol, Part 1* (n159), U.S. House, November 1971, pp. 7–9.

168. Alexander D. Langmuir, "New Environmental Factor in Congenital Disease," *New England Journal of Medicine 284* (April 22,

1971), 912–13. For Herbst's article, see Arthur L. Herbst, Howard Ulfelder, and David C. Poskanzer, "Adenocarcinoma of the Vagina," *New England Journal of Medicine 284* (April 22, 1971), 878–81.

169. Peter Greenwald, Joseph J. Barlow, Philip C. Nasca, et al., "Vaginal Cancer after Maternal Treatment with Synthetic Estrogens," *New England Journal of Medicine 285* (August 12, 1971), 390–2.

170. Hollis S. Ingraham, M.D. to all N.Y. physicians, June 22, 1971, reprinted in hearing on *Regulation of Diethylstilbestrol, Part 1* (n159), U.S. House, November 1971, pp. 14–15.

171. Ingraham to Edwards, June 15, 1971, reprinted in ibid., p. 15.

172. Testimony of Peter Greenwald, M.D., Director, New York Cancer Control Bureau, in hearing on *Regulation of Diethylstilbestrol, Part 1* (n159), U.S. House, November 1971, pp. 16–17.

173. Robert W. Miller, "Transplacental Chemical Carcinogenesis in Man" (editorial), *Journal of the National Cancer Institute 47* (December 1971), 1169–71; Kathryn S. Huss, "Maternal Diethylstilbestrol: A Time Bomb for Child?" (editorial), *Journal of the American Medical Association 218* (December 6, 1971), 1564–5; Judah Folkman, "Transplacental Carcinogenesis by Stilbestrol" (letter), *New England Journal of Medicine 285* (August 12, 1971), 404–5.

174. E. M. Ortiz, M.D., "Comments on Unpublished Article," April 12, 1971, reprinted in *Regulation of Diethylstilbestrol, Part 2: Its Use as a Drug for Humans and in Animal Feeds*, hearing before a Subcommittee of the Committee on Government Operations, U.S. House, 92nd Congress, 1st Session, December 13, 1971, pp. 192–3.

175. The account of this episode is based on Wade (n5), "DES: A Case Study of Regulatory Abdication;" and Wellford (n136), *Sowing the Wind*. For additional evidence of suppression of data by the USDA, see John N. White, "The Stilbestrol Conspiracy," *National Health Federation Bulletin* (January 1970), 10–14.

176. Quoted in hearing on *Regulation of Diethylstilbestrol, Part 1* (n159), U.S. House, November 1971, p. 1. On the NRDC's lawsuit, see "Court Asked to Ban Disputed Hormone," *New York Times*, October 29, 1971, p. 25.

177. Lucile K. Kuchera, "Postcoital Contraception with Diethylstilbestrol," *Journal of the American Medical Association 218* (October 25, 1971), 562–3.

178. "Certain Estrogens for Oral or Parenteral Use," *Federal Register 36*, No. 217 (November 10, 1971), 21537–8.

179. This figure is derived from Heinonen (n120), "Diethylstilbestrol in Pregnancy," by multiplying 8/12 (based on the eight-month delay between March, when the FDA received Herbst's galleys, and November, when the FDA finally acted) times the 100,000 prescriptions for DES that Heinonen reports were written annually for pregnant women during the 1960s.

180. Goldberg, in hearing on *Regulation of Diethylstilbestrol, Part 1* (n159), U.S. House, November 1971, p. 80. On the FDA's delayed response to Ingraham, see pp. 109–10.

181. Edwards, in hearing on *Regulation of Diethylstilbestrol, Part 1* (n159), U.S. House, November 1971, pp. 80, 110.

182. Gerald F. Meyer, Director, Office of Legislative Services, to Fountain, March 29, 1972, reprinted in hearing on *Regulation of Diethylstilbestrol, Part 2* (n174), U.S. House, December 1971, p. 194.

183. This point is made in Gillam (n5), "Prenatal DES: The Vagaries of Public Policy," p. 31.

184. In Congressional testimony, FDA employees later confirmed the agency's bias toward approving rather than disapproving drugs and blamed it on drug industry pressure. They reported that every time they decided a drug should not be marketed, they were called on the carpet, had their decisions overridden and their objections removed from the files, and in many cases were transferred to other positions. See Judith A. Turner, "Consumer Report/FDA Pursues Historic Role Amid Public, Industry Pressures," *National Journal Reports* 7 (February 15, 1975), 256. Also, Anita Johnson (n159), "The Bureau of Drugs."

185. "Health Research Group Report on the Morning-After Pill," December 8, 1972, reprinted in *Quality of Health Care: Human Experimentation, 1973*, hearings before the Subcommittee on Health of the Committee on Labor and Public Welfare, U.S. Senate, 93rd Congress, lst Session, Part 1, February 21 and 22, 1973, p. 202. Total sales of DES presumably represent all uses of DES, including livestock and postcoital uses as well as prenatal and other medical uses.

186. Roy Hertz, M.D., Rockefeller University, in hearing on *Regulation of Diethylstilbestrol, Part 1* (n159), U.S. House, November 1971, p. 67.

187. David Hawkins, quoted in Lane Palmer, "Court Suit Asks Ban on Stilbestrol," *Farm Journal* (December 1971), 25.

188. Hertz, in hearing on *Regulation of Diethylstilbestrol, Part 1* (n159), U.S. House, November 1971, p. 68.

189. Edwards, in hearing on *Regulation of Diethylstilbestrol, Part 1* (n159), U.S. House, November 1971, p. 52.

190. See hearing on *Regulation of Diethylstilbestrol, Part 1* (n159), U.S. House, November 1971, p. 104.

191. "The Dangers of DES" (editorial), *Washington Post*, November 12, 1971, p. A18.

192. Barry Kramer, "Stalking a Killer: How Cancer in Women Was Linked to a Drug Their Mothers Took," *Wall Street Journal*, December 23, 1975, p. 1.

193. Testimony of Robert H. Pantell, M.D., Stanford University School of Medicine, in *Regulation of Diethylstilbestrol, 1975*, joint hearing before the Subcommittee on Health of the Committee on

Labor and Public Welfare and the Subcommittee on Administrative Practice and Procedure of the Committee on the Judiciary, U.S. Senate, 94th Congress, 1st Session, February 27, 1975, pp. 19–20.

194. "Warning On Use of Sex Hormones in Pregnancy," *FDA Drug Bulletin 5* (January–March 1975), 4.

195. Edwards, in hearing on *Regulation of Diethylstilbestrol, Part 1* (n159), U.S. House, November 1971, p. 100.

196. One such brochure was entitled "Were You or Your Daughter Born after 1940?" (later updated to "Were You or Your Daughter or Son Born after 1940?"). There was also a more detailed public information pamphlet, "Questions and Answers About DES Exposure During Pregnancy and Before Birth." While these publications were better than nothing, they were, in the eyes of many consumer groups, incomplete. DES Action repeatedly expressed concern to the National Cancer Institute that they glossed over or did not address some important issues, such as whether DES daughters should take oral contraceptives (which would increase their estrogen intake), whether DES sons were at risk, and the specifics of what a "DES examination" for DES daughters should include. Although NCI officials welcomed these suggestions and met with DES Action to discuss them, none of the proposed changes were incorporated into later publications.

197. Huss (n173), "Maternal Diethylstilbestrol: A Time Bomb for Child?" 1564–5.

198. Don H. Mills, "Prenatal Diethylstilbestrol and Vaginal Cancer in Offspring" (editorial), *Journal of the American Medical Association 229* (July 22, 1974), 472. On the medical profession's historical opposition to patient information, see Bell (n5), *The Synthetic Compound DES*, pp. 173, 210–11. On reasons physicians are reluctant to inform patients, see Apfel and Fisher (n5), *To Do No Harm*, pp. 107–25.

199. Joseph W. Scott, "The Management of DES-Exposed Women: One Physician's Approach," *Journal of Reproductive Medicine 16* (June 1976), 285–8. For a description of a large, government-funded study's difficulties identifying and locating DES-exposed patients, see Sally Nash, Barbara C. Tilley, Leonard T. Kurland, et al., "Identifying and Tracing a Population at Risk: The DESAD Project Experience," *American Journal of Public Health 73* (March 1983), 253–9.

200. Fenichell and Charfoos (n5), *Daughters at Risk*, p. 164. In private communications to DES Action, a number of women described their difficulties learning of or confirming DES exposure, ranging from outright denial to suspicious gaps in files at doctors' offices (they found the charts in front of and behind theirs but not theirs).

201. "HEW Recommends Followup on DES Patients," *FDA Drug*

Bulletin 8 (October–November 1978), 31; Julius B. Richmond, M.D., Surgeon General, Department of Health, Education and Welfare, "Physician Advisory," October 4, 1978.

202. Hans Neumann, "Patients Have the Right Not to Know" (editorial), *Medical World News* (June 23, 1980), 63. Neumann argues that given the small chance of cancer developing, patients are better off *not* knowing they are at risk so that they can continue "enjoying themselves to the limit of their capacity for a good life." He assumes that if problems develop, they will come to light during routine examinations or because of symptoms. This is simply untrue. Vaginal cancer in its early stages – when it is most amenable to successful treatment – is symptomless and is unlikely to be diagnosed unless the physician performs a special examination to look for it. Even the most common symptom of vaginal cancer, abnormal vaginal bleeding, can easily be mistaken for any number of conditions. Moreover, other DES-related problems, such as abnormal pregnancy, require the patient's advance knowledge so that she can receive appropriate care to prevent preterm labor.

203. E. R. Greenberg, A. B. Barnes, L. Resseguie, et al., "Breast Cancer in Mothers Given Diethylstilbestrol in Pregnancy," *New England Journal of Medicine 311* (November 29, 1984), 1393–8. Stanley J. Robboy, Kenneth L. Noller, Peter O'Brien, et al., "Increased Incidence of Cervical and Vaginal Dysplasia in 3,980 Diethylstilbestrol-Exposed Young Women," *Journal of the American Medical Association 252* (December 7, 1984), 2979–83.

204. Sidney M. Wolfe, M.D. to Margaret Heckler, Secretary of Health and Human Services, July 9, 1985 (photocopy from Public Citizen's Health Research Group, Washington, D.C.).

205. Department of Health and Human Services, National Cancer Institute, *Report of the 1985 DES Task Force* (Bethesda, Md.: National Institutes of Health, 1985), pp. 19, 31. Staff at the National Cancer Institute blamed the delay in producing revised documents for professionals and the public on inadequate staffing levels in the relevant NCI offices (personal communication, Alice Hamm, Office of Cancer Communications, National Cancer Institute, NIH, to Diana Dutton, August 24, 1987).

206. Edwards, quoted in Stephen M. Johnson, "Diethylstilbestrol Issue Exploited," *Des Moines Register*, October 23, 1972, reprinted in *Agriculture – Environmental, and Consumer Protection Appropriations for 1975*, hearing before a Subcommittee of the Committee on Appropriations, U.S. House, 93rd Congress, 2nd Session, May 6, 1974, pp. 72–4.

207. "Diethylstilbestrol; Notice of Opportunity for Hearing on Proposal to Withdraw Approval of New Animal Drug Applications," *Federal Register 37*, No. 120 (June 21, 1972), 12251–3. Edwards

quote is from Wade (n5), "DES: A Case Study of Regulatory Abdication," 337.

208. The full text of Fountain's address is reprinted in *Regulation of Diethylstilbestrol, Part 3: Its Use as a Drug for Humans and In Animal Feeds*, hearing before a Subcommittee of the Committee on Government Operations, U.S. House, 92nd Congress, 2nd Session, August 15, 1972, pp. 400–2.

209. *Regulation of Diethylstilbestrol, 1972*, hearing before the Subcommittee on Health of the Committee on Labor and Public Welfare, U.S. Senate, 92nd Congress, 2nd Session, July 20, 1972.

210. Frank J. Rauscher, quoted in Morton Mintz, "Cancer Institute Head Urges FDA Ban DES," *Washington Post*, July 10, 1972, p. A2.

211. "Diethylstilbestrol, Order Denying Hearing and Withdrawing Approval of New Animal Drug Applications for Liquid and Dry Premixes, and Deferring Ruling on Implants," *Federal Register 37*, No. 151 (August 4, 1972), 15749. The new results discrediting the seven-day withdrawal period are reported on p. 15749. The use of ear implants was banned in April 1973 when a second study revealed liver residues 120 days after implantation.

212. Court decision quoted in Hadden (n5), "DES and the Assessment of Risk," p. 122. See also "DES Ban of the FDA Attacked by 3 Firms," *Chemical Marketing Reporter 202* (September 25, 1972), 4.

213. The full recommendations appear in *Agriculture – Environmental, and Consumer Protection Appropriations for 1975* (n206), U.S. House, May 1974, pp. 26–31.

214. See, for example, "Group says DES Safe, Hits Delaney Clause," *Chemical and Engineering News* (January 31, 1977), p. 5; Peter B. Hutt, "Public Policy Issues in Regulating Carcinogens in Food," *Food Drug Cosmetic Law Journal 33* (October 1978), 541–57; National Research Council [hereafter NRC], Board on Agriculture, *Regulating Pesticides in Food: The Delaney Paradox* (Washington, D.C.: National Academy Press, 1987).

215. *Regulation of New Drug R&D by the Food and Drug Administration, 1974*, joint hearings before the Subcommittee on Health of the Committee on Labor and Public Welfare and the Subcommittee on Administrative Practice and Procedure of the Committee on the Judiciary, U.S. Senate, 93rd Congress, 2nd Session, September 25 and 27, 1974, p. 440.

216. "Diethylstilbestrol; Withdrawal of Approval of New Animal Drug Applications; Commissioner's Decision," *Federal Register 44*, No. 185 (September 21, 1979), 54852–900.

217. "U.S. Proposes Jail Term for Illegal Use of DES Hormone," *Los Angeles Times*, April 16, 1980, Part 1, p. 10. See also Steve Frazier and Steve Weiner, "Many Cattlemen Ignored the Federal Ban on

Use of DES to Speed Animals' Growth," *Wall Street Journal*, July 15, 1980, p. 46.

218. Several Puerto Rican physicians reportedly found that most of the children recovered when beef, chicken, and milk were eliminated from their diets, leading them to attribute the problem to contamination of these products with DES and other hormones. Government scientists dispute this conclusion, however, and the issue remains unsettled. See "Maturing Early, a Puzzle in Puerto Rico," *Time* (October 25, 1982), p. 86; Manuel Suarez, "Inquiry on Early Cases of Sexual Development," *New York Times*, February 8, 1985, p. 7.

219. Synovex, made by Syntex, is the most commonly used growth promotant; other brands include Ralgro and Compudose (interview with David Liggett, Senior Product Manager, Agribusiness Division, Syntex Corporation, Palo Alto, California, August 4, 1987). When the 1972 ban on DES went into effect, *Business Week* reported that FDA and USDA scientists were "hurriedly testing these potential substitutes for DES" ("The Hormone Ban Spreads Confusion," *Business Week* [August 12, 1972], 29).

220. Keith Schneider, "Congress Looks to the American Table Amid Questions on Food Safety," *New York Times*, June 22, 1987, p. 10. See also Orville Schell, *Modern Meat* (New York: Vintage Books, 1985); NRC (n214), *Regulating Pesticides in Food*; and General Accounting Office, *Pesticides: Need to Enhance FDA's Ability to Protect the Public From Illegal Residues*, Publ. No. RCED–87–7 (Washington, D.C.: Government Printing Office, October 1986).

221. Minutes of the Ob-Gyn Advisory Committee Meeting, October 20, 1972, quoted in *Use of Advisory Committees by the Food and Drug Administration*, hearings before a Subcommittee of the Committee on Government Operations, U.S. House, 93rd Congress, 2nd Session, March 6–8, 12, 13; April 30; and May 21, 1974, p. 393.

222. William C. Drury to Director, Office of Scientific Evaluation, FDA, July 18, 1972.

223. See, for example, the testimony of Sidney Wolfe, M.D., and Anita Johnson of the Health Research Group in hearings on *Quality of Health Care: Human Experimentation, 1973* (n185), U.S. Senate, February 1973, pp. 194–212.

224. J. Richard Crout, M.D., Director, FDA Bureau of Drugs, quoted in hearings on *Use of Advisory Committees by the Food and Drug Administration* (n221), U.S. House, April 1974, p. 436.

225. Quote is from transcript of an interview with Richard Blye, M.D. by Becky O'Malley, February, 1980, p. 4. Representative Fountain expressed some surprise and skepticism about the legality of such a procedure, but Peter Hutt, Assistant General Counsel for the FDA, assured him that it was both legal and well precedented.

See hearings on *Use of Advisory Committees by the Food and Drug Administration* (n221), U.S. House, April 1974, pp. 390–2. For doubts on the legality, see *Use of Advisory Committees by the FDA*, Eleventh Report by the Committee on Government Operations, January 26, 1976, U.S. House Report No. 94–787 (Washington, D.C.: Government Printing Office, 1976), pp. 37, 57–8.

226. See Edwards's testimony in hearings on *Quality of Health Care: Human Experimentation, 1973* (n185), U.S. Senate, February 1973, p. 23.

227. Mortimer Lipsett, M.D., quoted in hearings on *Use of Advisory Committees by the Food and Drug Administration* (n221), U.S. House, April 1974, pp. 394–5.

228. Crout, quoted in hearings on *Use of Advisory Committees by the Food and Drug Administration* (n221), U.S. House, April 1974, p. 398.

229. Quote is Edwards's summary of the Ob-Gyn Advisory Committee's conclusions, in hearings on *Use of Advisory Committees by the Food and Drug Administration* (n221), U.S. House, April 1974, p. 410.

230. Interview with J. Richard Crout by Larry Molton, February 3, 1981.

231. Lucile K. Kuchera, "The Morning-After Pill" (letter to the editor), *Journal of the American Medical Association 224* (May 14, 1973), 1038. Also see Joe B. Massey, Celso-Ramon Garcia, and John P. Emich, Jr., "Management of Sexually Assaulted Females," *Obstetrics and Gynecology 38* (July 1971), 29–36; Barbara B. Coe, "The Use of Diethylstilbestrol as a Post-Coital Contraceptive," *Journal of the American College Health Association 20* (April 1972), 286; and J. McLean Morris and Gertrude van Wagenen, "Interception: The Use of Postovulatory Estrogens to Prevent Implantation," *American Journal of Obstetrics and Gynecology 115* (January 1, 1973), 101–6.

232. Based on the New York State Cancer Registry, the range of exposure among mothers and daughters who developed vaginal cancer was 135 mg to 15,750 mg of DES. The normal morning-after pill dosage was 250 mg. See the testimony of Greenwald, in hearing on *Regulation of Diethylstilbestrol, 1975* (n193), U.S. Senate, February 1975, p. 12.

233. See hearings on *Quality of Health Care: Human Experimentation, 1973* (n185), U.S. Senate, February 1973; hearings on *Use of Advisory Committees by the Food and Drug Administration* (n221), U.S. House, March, April, May 1974; hearing on *Regulation of Diethylstilbestrol, 1975* (n193), U.S. Senate, February 1975; *Diethylstilbestrol*, hearing before the Subcommittee on Health and the Environment of the Committee on Interstate and Foreign Commerce, U.S. House, 94th Congress, 1st Session, December 16,

1975; and *Food and Drug Administration Practice and Procedure, 1976*, joint hearing before the Subcommittee on Health of the Committee on Labor and Public Welfare and the Subcommittee on Administrative Practice and Procedure of the Committee on the Judiciary, U.S. Senate, 94th Congress, 2nd Session, July 20, 1976.

234. "Postcoital Diethylstilbestrol," *FDA Drug Bulletin 3* (May 1973).

235. Proposed patient labeling requirements were first published in "Diethylstilbestrol; Use as Postcoital Contraceptive; Patient Labeling," *Federal Register 38*, No. 186 (September 26, 1973), 26809–11. Final approval and marketing conditions appeared in "Diethylstilbestrol as Postcoital Oral Contraceptive; Patient Labeling," *Federal Register 40*, No. 25 (February 5, 1975), 5351–5, and "Estrogens for Oral or Parenteral Use; Drugs for Human Use; Drug Efficacy Study Implementation; Amended Notice," *Federal Register 40*, No. 39 (February 26, 1975), 8242.

236. Hearing on *Regulation of Diethylstilbestrol, 1975* (n193), U.S. Senate, February 1975.

237. *Federal Register*, February 5, 1975, 5352, 5354–5.

238. See testimony of Robert H. Furman, M.D., Vice President, Eli Lilly, Inc., in hearing on *Regulation of Diethylstilbestrol, 1975* (n193), U.S. Senate, February 1975, p. 33.

239. See testimony of Robert L. Pillarella, President, Tablicaps, Inc., in hearing on *Regulation of Diethylstilbestrol, 1975* (n193), U.S. Senate, February 1975, pp. 36–9.

240. Marvin Seife, M.D., Director, FDA Generic Drug Division, in hearing on *Regulation of Diethylstilbestrol, 1975* (n193), U.S. Senate, February 1975, p. 27.

241. See testimony of Seife and Vincent Karusaitis, M.D., Medical Officer, FDA Generic Drug Division, in hearing on *Regulation of Diethylstilbestrol, 1975* (n193), U.S. Senate, February 1975, pp. 26–31; and in hearing on *Food and Drug Administration Practice and Procedure, 1976* (n233), U.S. Senate, July 1976, pp. 42–60.

242. Interview by Becky O'Malley with Vincent Karusaitis and Marvin Seife, February 1980, p. 20.

243. "DES and Breast Cancer," *FDA Drug Bulletin 8* (March–April 1978), 10.

244. In 1984, a medical advice column in the *San Francisco Chronicle* wrongly informed readers: "To date, only DES (diethylstilbestrol) has been approved for [postcoital contraception] in the United States." (Dr. June Reinisch, "The Kinsey Report: A Pill for the Morning After?" *San Francisco Chronicle*, March 26, 1984, p. 14.) The continuing use of DES as a morning-after pill is documented in the EVIST survey cited in note 246. Also see Dianne Glover, Meghan Gerety, Shirley Bromberg, et al., "Diethylstilbestrol in the Treatment of Rape Victims," *Western Journal of Medicine 125* (October 1976), 331–4.

245. Christopher Norwood, *At Highest Risk* (New York: McGraw-Hill, 1980), p. 143. Morning-after pill failure could occur because the treatment itself failed, because it was discontinued before completion, or because the woman was already pregnant from a prior sexual contact.

246. This figure is based on a survey of sixty-two U.S. college and university health services in the United States conducted in 1985–6 by the Stanford EVIST Project. Of the fifty-six that responded, nine (16.1%) indicated that they used DES as a morning-after pill or referred patients to other providers that used it.

247. Personal communication, Dr. Philip Corfman, Medical Officer, Division of Metabolism and Endocrine Drug Products, FDA, August 25, 1987.

248. Through June 30, 1985, the international DES registry included a total of 519 women with confirmed clear-cell adenocarcinoma, but DES exposure was verified in only 311 of these cases. See Sandra Melnick, Philip Cole, Diane Anderson, et al., "Rates and Risks of Diethylstilbestrol-Related Clear-Cell Adenocarcinoma of the Vagina and Cervix," *New England Journal of Medicine 316* (February 26, 1987), 514–16.

249. For doubts on the causal role of DES, see Michael J. McFarlane, Alvan R. Feinstein, and Ralph I. Horwitz, "Diethylstilbestrol and Clear Cell Vaginal Carcinoma: Reappraisal of the Epidemiologic Evidence," *American Journal of Medicine 81* (November 1986), 855–63. These researchers criticized the original Herbst and Greenwald case control studies for not matching cases and controls according to the mother's history of problem pregnancies. Since the case patients' mothers turned out to have had more problem pregnancies than those of the controls as well as a far higher use of DES, McFarlane and colleagues suggest that this poor maternal history, rather than DES itself, may account for the case patients' clear-cell cancer. Even researchers who believe that the evidence supports a causal interpretation of the association between DES and cancer call it an "incomplete carcinogen," meaning that other factors besides DES may also be part of the cancer-causing process.

250. Most important, the ability of DES to cause cancer in many different species of animals is well established. The findings for humans also strongly suggest causation. First, clear-cell adenocarcinoma was essentially unknown in young women prior to DES, whereas mothers have presumably always been having problem pregnancies. Second, the prevalence of clear-cell cancer cases peaked in the mid-1970s and declined thereafter, just as the volume of DES sales peaked in the early to mid-1950s and then declined; this twenty-or-so-year lag corresponds roughly to the latency period necessary for the cancers to develop. Third, 60 percent of the

clear-cell cancer cases reported to the international DES cancer registry reveal in utero DES exposure, confirming the original Herbst and Greenwald findings. (The other 40 percent where DES exposure cannot be documented might be explained by inadequate records, or by unwitting exposure through the consumption of DES-treated meat.) Fourth, the prevalence of cases reported has generally been highest in areas of both the United States and other countries where DES was most commonly prescribed during pregnancy. Finally, the likelihood of cancer appears to be greatest when exposure occurred early in pregnancy; see Arthur L. Herbst, Sharon Anderson, Marian M. Hubby, et al., "Risk Factors for the Development of Diethylstilbestrol-Associated Clear Cell Adenocarcinoma: A Case-Control Study," *American Journal of Obstetrics and Gynecology 154* (April 1986), 814–22. For additional arguments and evidence in support of a causal interpretation of the relationship between DES and cancer, see Fenichell and Charfoos (n5), *Daughters at Risk*, pp. 160–1; Herbst, "The Epidemiology of Vaginal and Cervical Clear Cell Adenocarcinoma," in Herbst and Bern (n111), *Developmental Effects of Diethylstilbestrol*, pp. 63–70.

251. See Herbst and Bern (n111), *Developmental Effects of Diethylstilbestrol*, p. 194.

252. John A. Jefferies, Stanley J. Robboy, Peter C. O'Brien, et al., "Structural Anomalies of the Cervix and Vagina in Women Enrolled in the Diethylstilbestrol Adenosis (DESAD) Project," *American Journal of Obstetrics and Gynecology 148* (January 1, 1984), 59–66; Robboy et al. (n203), "Increased Incidence of Cervical and Vaginal Dysplasia." Clear-cell adenocarcinoma can develop in women with nonmalignant DES-related abnormalities, even under medical surveillance. In 1981, Dr. Norma Veridiano reported on a case of vaginal clear-cell adenocarcinoma occurring in a DES daughter after four years of continuous observation for adenosis. Pap smears had been negative, including the one taken at diagnosis. See Norma P. Veridiano, M. Leon Tancer, Isaac Delke, et al., "Delayed Onset of Clear Cell Adenocarcinoma of the Vagina in DES-Exposed Progeny," *Obstetrics and Gynecology 57* (March 1981), 395–8. Mitchell S. Kramer, Vicki Seltzer, Burton Krumholz, et al., "Diethylstilbestrol-Related Clear-Cell Adenocarcinoma in Women with Initial Examinations Demonstrating No Malignant Disease," *Obstetrics and Gynecology 69* (June 1987), 868–70, describe similar cases at screening clinics, and conclude: "These cases emphasize the need for frequent follow-up for these patients using careful inspection and palpation, vaginal and cervical cytology, colposcopy, and colposcopically directed biopsy of suspicious areas" (p. 868). Also see B. Anderson, W. G. Watring,

D. D. Edinger, Jr., et al., "Development of DES-Associated Clear-Cell Carcinoma: The Importance of Regular Screening," *Obstetrics and Gynecology 53* (March 1979), 293–9.

253. See, for example, Raymond H. Kaufman, Gary L. Binder, Paul M. Gray, Jr., et al., "Upper Genital Tract Changes Associated With Exposure in Utero to Diethylstilbestrol," *American Journal of Obstetrics and Gynecology 128* (May 1, 1977), 51–9; Alan H. DeCherney, Ina Cholst, and Frederick Naftolin, "Structure and Function of the Fallopian Tubes Following Exposure to Diethylstilbestrol (DES) During Gestation," *Fertility and Sterility 36* (December 1981), 741–5; Raymond H. Kaufman, Kenneth Noller, Ervin Adam, et al., "Upper Genital Tract Abnormalities and Pregnancy Outcome in Diethylstilbestrol-Exposed Progeny," *American Journal of Obstetrics and Gynecology 148* (April 1, 1984), 973–84.

254. Of three large controlled studies comparing fertility rates, two found significantly lower fertility among DES daughters than among comparable nonexposed women: See Arthur L. Herbst, Marian M. Hubby, Freidoon Azizi, et al., "Reproductive and Gynecologic Surgical Experience in Diethylstilbestrol-Exposed Daughters," *American Journal of Obstetrics and Gynecology 141* (December 15, 1981), 1019–28; and Marluce Bibbo, William B. Gill, Freidoon Azizi, et al., "Follow-Up Study of Male and Female Offspring of DES-Exposed Mothers," *Journal of Obstetrics and Gynecology 49* (January 1977), 1–8. In the third, which was based on a subgroup of the DESAD Project, the mothers had received lower doses of DES during pregnancy (Ann B. Barnes, Theodore Colton, Jerome Gundersen, et al., "Fertility and Outcome of Pregnancy in Women Exposed in Utero to Diethylstilbestrol," *New England Journal of Medicine 302* [March 13, 1980], 609–13). All studies show poorer pregnancy outcomes among DES daughters, especially those with documented anatomical abnormalities. For a comprehensive review of the literature on pregnancy outcomes in DES-exposed women, see Joan Emery, *Reproductive Outcomes in Women Exposed in Utero to Diethylstilbestrol* (1986), published and distributed by DES Action National, New Hyde Park, N.Y. 11040.

255. The 1985 DES Task Force reviewed all of the available data and concluded that "the weight of the evidence in January 1985 now indicates that women who used DES during their pregnancies may subsequently experience an increased risk of breast cancer" (1985 *DES Task Force Report* [n205], p. 12). The estimate of 25,000 excess breast cancer cases is from a letter Sidney Wolfe wrote to Margaret Heckler, December 3, 1984 (photocopy from Public Citizen's Health Research Group, Washington, D.C.). Wolfe assumed that the rate of breast cancer was 41 percent higher among DES moth-

ers than other women and applied this to the estimated 2 million or more American women who were given DES to obtain the number of excess cases among DES mothers. DES Action estimates exposure at 4.8 million women, which yields 62,000 excess breast cancer cases (Pat Cody, letter to Sidney Wolfe, June 24, 1987).

256. There has been much less research on DES sons than daughters. The findings are mixed, although most studies in which the mothers had received higher doses of DES have reported elevated rates of structural abnormalities of the reproductive tract in DES sons compared with nonexposed men (these are reviewed in William B. Gill, Gebhard F. Schumacher, Marian M. Hubby, et al., "Male Genital Tract Changes in Humans Following Intrauterine Exposure to Diethylstilbestrol," in Herbst and Bern (n111), *Developmental Effects of Diethylstilbestrol*, pp. 103–19). See also Robert J. Stillman, "In Utero Exposure to Diethylstilbestrol: Adverse Effects on the Reproductive Tract and Reproductive Performance in Male and Female Offspring," *American Journal of Obstetrics and Gynecology 142* (April 1, 1982), 905–21. A more recent study conducted at the Mayo Clinic in Minnesota reported conflicting findings, but the authors noted that this may have been due to the lower total dosage of DES the exposed mothers had received (1,429 mg) compared with the total dosage in another study (12,046 mg) which found an excess of reproductive anomalies in DES-exposed sons. See Frank J. Leary, Laurence J. Resseguie, Leonard T. Kurland, et al., "Males Exposed in Utero to Diethylstilbestrol," *Journal of the American Medical Association 252* (December 7, 1984), 2988.

257. Some surveillance of DESAD participants by mailed questionnaire continues, but this will provide only limited clinical information and no laboratory or examination data (Dr. Leonard Kurland, Mayo Clinic, to Diana Dutton, June 23, 1987).

4. The artificial heart

1. J. J. C. Legallois, *Experiments on the Principle of Life*, trans. N. C. and J. C. Nancrede (Philadelphia: 1813).
2. H. H. Dale and E. H. J. Schuster, "A Double Perfusion Pump," *Journal of Physiology (London) 64* (February 1928), 356–64.
3. Reviewed in William C. DeVries, "The Total Artificial Heart," in David C. Sabiston and Frank C. Spencer, eds., *Gibbon's Surgery of the Chest*, 4th ed. (Philadelphia: W. B. Saunders, 1983), p. 1629.
4. John H. Gibbon, Jr., "Application of a Mechanical Heart and Lung Apparatus to Cardiac Surgery," *Minnesota Medicine 37* (March 1954), 171–80.
5. See, for example, Tetsuze Akutsu, Charles S. Houston, and Wil-

lem J. Kolff, "Artificial Hearts Inside the Chest, Using Small Electro-motors," *Transactions of the American Society for Artificial Internal Organs 6* (1960), 299; Domingo Liotta, Tomas Taliani, Antonio H. Giffoniello, et al., "Artificial Heart in the Chest: Preliminary Report," *Transactions of the American Society for Artificial Organs 7* (1961), 318; Frank W. Hastings, W. H. Potter, and J. W. Holter, "Artificial Intracorporeal Heart," *Transactions of the American Society for Artificial Internal Organs 7* (1961), 323–6; Willem J. Kolff, "An Intrathoracic Pump to Replace the Human Heart: Current Developments at the Cleveland Clinic," *Cleveland Clinic Quarterly 29* (1962), 107; Yukihiko Nosé, Martin Schamaun, and Adrian Kantrowitz, "Experimental Use of an Electronically Controlled Prosthesis as an Auxiliary Left Ventricle," *Transactions of the American Society for Artificial Internal Organs 9* (1963), 269–73.

6. Tetsuze Akutsu and Willem J. Kolff, "Permanent Substitutes for Valves and Hearts," *Transactions of the American Society for Artificial Internal Organs 4* (1958), 230–5; William C. DeVries and Lyle D. Joyce, "The Artificial Heart," *Ciba Foundation Symposia 35*, No. 6 (1983), 4–5.

7. DeVries and Joyce (see n6), "The Artificial Heart."

8. See Deborah P. Lubeck and John P. Bunker, "Case Study #9: The Artificial Heart: Costs, Risks, and Benefits," in *The Implications of Cost-Effectiveness Analysis of Medical Technology*, U.S. Congress, Office of Technology Assessment (May 1982), p. 4; Bert K. Kusserow, "Artificial Heart Research – Survey and Prospectus," *Transactions of the New York Academy of Sciences 27* (1965), 309–23; Fabio Guion, William C. Birtwell, Wendell B. Thrower, et al., "A Prosthetic Left Ventricle Energized by an External Replaceable Actuator, and Capable of Intermittent or Continuous Pressure Assist for the Left Ventricle," *Surgical Forum 16* (1965), 146–8; Adrian Kantrowitz, Tetsuze Akutsu, P. A. Chaptal, et al., "A Clinical Experience with an Implanted Mechanical Auxiliary Ventricle," *Journal of the American Medical Association 197* (1966), 525; C. William Hall, Domingo Liotta, and Michael E. DeBakey, "Artificial Heart – Present and Future," in *Research in the Service of Man: Biomedical Knowledge, Development, and Use* (Washington, D.C.: Government Printing Office, 1967), pp. 201–16; idem, "Bioengineering Efforts in Developing Artificial Hearts and Assistors," *American Journal of Surgery 114* (1967), 24–30; Willem J. Kolff, "Artificial Organs – Forty Years and Beyond," *Transactions of the American Society for Artificial Internal Organs 29* (1983), 6–7.

9. Robert K. Jarvik, "The Total Artificial Heart," *Scientific American 244*, No. 1 (January 1981), 77; National Heart and Lung Institute, Artificial Heart Assessment Panel Report, *The Totally Implantable Artificial Heart: Legal, Social, Ethical, Medical, Economic, and Psy-*

chological Implications (Bethesda, Md.: NHLBI, DHEW Publ. No. [NIH] 74–191, June 1973), pp. 37–40 [hereafter cited as Assessment Panel, *The Totally Implantable Artificial Heart*].

10. National Heart, Lung, and Blood Institute, Working Group on Mechanical Circulatory Support, *Artificial Heart and Assist Devices: Directions, Needs, Costs, and Societal and Ethical Issues* (Bethesda, Md.: NHLBI, May 1985), pp. 9–10 [hereafter cited as NHLBI Working Group, *Artificial Heart and Assist Devices*]; John T. Watson, "Past, Present, and Future of Mechanical Circulatory Support," presented at the Third International Symposium on Heart Substitution, Rome, Italy, May 17, 1982.

11. Lubeck and Bunker (n8), "The Artificial Heart: Costs, Risks, and Benefits."

12. U.S. Congress, Senate, Subcommittee of the Committee on Appropriations for 1964, *Hearings on Department of Health, Education, and Welfare Appropriations* (Washington, D.C.: Government Printing Office, 1963), p. 1401 [hereafter cited as *Senate Hearings on DHEW Appropriations* (1963)].

13. Ibid., p. 1402.

14. U.S. Congress, House, Subcommittee on Appropriations, *Hearings on Department of Health, Education, and Welfare Appropriations for 1966* (Washington, D.C.: Government Printing Office, 1965), pp. 504–6 [hereafter cited as *House Hearings on DHEW Appropriations* (1965)].

15. Lubeck and Bunker (n8), "The Artificial Heart: Costs, Risks, and Benefits," p. 4, note 4.

16. *House Hearings on DHEW Appropriations* (n14) (1965).

17. Reviewed in Assessment Panel (n9), *The Totally Implantable Artificial Heart*, p. 205.

18. Michael Strauss, "The Political History of the Artificial Heart," *New England Journal of Medicine 310* (1984), 332–6.

19. See Stephen P. Strickland, *Politics, Science, and Dread Diseases: A Short History of the United States Medical Research Policy* (Cambridge, Mass.: Harvard University Press, 1972); Richard A. Rettig, *Cancer Crusade* (Princeton, N.J.: Princeton University Press, 1977); N. D. Spingarn, *Heartbeat: The Politics of Health Research* (Washington, D.C.: Robert B. Luce, 1976).

20. See Strauss (n18), "The Political History of the Artificial Heart."

21. *House Hearings on DHEW Appropriations* (n14) (1965), p. 504.

22. Discussed in Strauss (n18), "The Political History of the Artificial Heart."

23. Quoted in Barton J. Bernstein, "The Artificial Heart Program," *Center Magazine* (May–June 1981), 22–33.

24. Documented in Strauss (n18), "The Political History of the Artificial Heart."

25. See Bernstein (n23), "The Artificial Heart Program."

26. Harvey M. Sapolsky, "Here Comes the Artificial Heart," *The Sciences 13* (1978), 26.
27. *House Hearings on DHEW Appropriations* (n14) (1965); Bernstein (n23), "The Artificial Heart Program."
28. *House Hearings on DHEW Appropriations* (n14) (1965).
29. From N. Lindgren, "The Artificial Heart – Example of Mechanical Engineering Enterprise," *IEEE Spectrum* (September 1967).
30. Ibid.
31. *House Hearings on DHEW Appropriations* (n14) (1965).
32. Frank Hastings, in *Hearings on HEW Appropriations for 1968*, Senate Subcommittee on Appropriations, 90th Congress, 1st Session (Washington, D.C.: Government Printing Office, 1967).
33. *House Hearings on DHEW Appropriations* (n14) (1965).
34. Ibid.
35. Bernstein (n23), "The Artificial Heart Program."
36. Hittman Associates, *Final Summary Report on Six Studies Basic to the Consideration of the Artificial Heart Program* (Baltimore: Hittman, 1966).
37. Ibid.
38. Strauss (n18), "The Political History of the Artificial Heart," 334–5.
39. Given in Bernstein (n23), "The Artificial Heart Program."
40. Quoted in Strauss (n18), "The Political History of the Artificial Heart."
41. See Willem J. Kolff, "Comments on a draft entitled 'Artificial Heart: Costs, Risks, and Benefits,' " March 12, 1980; and Assessment Panel (n9), *The Totally Implantable Artificial Heart*, p. 208.
42. Kolff (n41), "Comments."
43. Discussed in Lubeck and Bunker (n8), "The Artificial Heart: Costs, Risks, and Benefits," p. 5; and Frank D. Altieri, "Status of Implantable Energy Systems to Actuate and Control Ventricular Assist Devices," *Artificial Organs 7* (1983), 5–20.
44. Assessment Panel (n9), *The Totally Implantable Artificial Heart*, pp. 103–4; Iurig M. Kiselev, Gregory P. Dubrovskii, Tengiz G. Mosidze, et al., "Prospects for Using Implanted Systems of Assisted Circulation and Artificial Heart with a Radioisotope Power Source: Biomedical, Thermal, and Radiation Aspects," *Artificial Organs 7* (1983), 117–21; National Heart, Lung, and Blood Institute, *Quarterly Report* (Division of Heart and Vascular Diseases, Contract No. NO1-HV–92908, JCGS- 8209–232, March 1982) [hereafter cited as NHLBI, *Quarterly Report*].
45. Assessment Panel (n9), *The Totally Implantable Artificial Heart*, pp. 209–10.
46. Ibid.
47. Bernstein (n23), "The Artificial Heart Program."
48. Assessment Panel (n9), *The Totally Implantable Artificial Heart*, p. 207.

49. Lubeck and Bunker (n8), "The Artificial Heart: Costs, Risks, and Benefits," p. 5.

50. For commentary on this event, see Christiaan Barnard and C. Pepper, *Christiaan Barnard: One Life* (New York: Macmillan, 1969); Philip Blaiberg, *Looking at My Heart* (New York: Stein and Day, 1968); Renée C. Fox and Judith P. Swazey, *The Courage to Fail* (Chicago: University of Chicago Press, 1974), pp. 109–21; and Jay Katz, *The Silent World of Doctor and Patient* (New York: Free Press, 1984), pp. 130–42.

51. Fox and Swazey (n50), *The Courage to Fail*, pp. 122–48.

52. Denton A. Cooley, Domingo Liotta, G. L. Hallman, et al., "Orthotopic cardiac prosthesis for two-staged cardiac replacement," *American Journal of Cardiology 24* (1969), 723–30.

53. Michael DeBakey, C. W. Hall, et al., "Orthotopic cardiac prosthesis: Preliminary Experiments in Animals with Biventricular Artificial Heart," *Cardiovascular Research Center Bulletin 7* (April–June 1969), 127–42.

54. Reviewed in Fox and Swazey (n50), *The Courage to Fail*, pp. 149–211; T. Thompson, "The Texas Tornado Vs. Dr. Wonderful," *Life* (April 10, 1970); J. Quinn, "I'm Qualified to Judge What Is Right," *New York Daily News*, September 10, 1969, p. 2.

55. Fox and Swazey (n50), *The Courage to Fail*, p. 167; Bernstein (n23), "The Artificial Heart Program," pp. 27–8.

56. Fox and Swazey (n50), *The Courage to Fail*, p. 179.

57. Ibid., p. 190.

58. Willem Kolff, personal communication; Michael DeBakey, personal communication.

59. National Heart and Lung Institute, *Cardiac Replacement: Medical, Ethical, Psychological, and Economic Implications* (Bethesda, Md.: DHEW Publ. No. [NIH] 77–1240, October 1969).

60. See Barton J. Bernstein, "The Artificial Heart: Is It a Boon or a High-Tech Fix?" *Nation* (January 22, 1983), 71–2.

61. NHLBI Working Group (n10), *Artificial Heart and Assist Devices*, p. 9.

62. Jarvik (n9), "The Total Artificial Heart," 78.

63. Reviewed in Bernstein (n23), "The Artificial Heart Program," 28.

64. Clifford Kwan-Gett, H. H. J. Zwart, A. C. Kralios, et al., "A Prosthetic Heart with Hemispherical Ventricles Designed for Low Hemolytic Action," *Transactions of the American Society for Artificial Internal Organs 16* (1970), 409; Jarvik (n9), "The Total Artificial Heart," 79.

65. Ibid.

66. Assessment Panel (n9), *The Totally Implantable Artificial Heart*, p. iv.

67. Harold P. Green, "An NIH Panel's Early Warnings," *Hastings Center Report* (October 1984), 13.
68. Ibid.; Assessment Panel (n9), *The Totally Implantable Artificial Heart*, p. 14.
69. Assessment Panel (n9), *The Totally Implantable Artificial Heart*, p. 16.
70. Ibid.; Green (n67), "An NIH Panel's Early Warnings," p. 14.
71. Assessment Panel (n9), *The Totally Implantable Artificial Heart*, p. 15.
72. Ibid., p. 2.
73. Ibid., pp. 58, 89.
74. Ibid., pp. 71–2.
75. Ibid., pp. 98–101, 215–21.
76. Reviewed in Green (n67), "An NIH Panel's Early Warnings," p. 15.
77. Ibid., pp. 13–14; Assessment Panel (n9), *The Totally Implantable Artificial Heart*, pp. 122, 199–200.
78. Assessment Panel (n9), *The Totally Implantable Artificial Heart*, p. 127.
79. Ibid., pp. 111–23.
80. NHLBI (n44), *Quarterly Report*, p. 60.
81. Assessment Panel (n9), *The Totally Implantable Artificial Heart*, pp. 182, 197.
82. NHLBI Working Group (n10), *Artificial Heart and Assist Devices*, p. 10.
83. National Heart and Lung Institute, *Report on Left Ventricular Assist Device* (DHEW Publ. No. [NIH] 75–626, January 1974).
84. See Watson (n10), "Past, Present, and Future of Mechanical Circulatory Support"; and Lubeck and Bunker (n8), "The Artificial Heart: Costs, Risks, and Benefits," p. 7.
85. NHLBI Working Group (n10), *Artificial Heart and Assist Devices*, p. 10.
86. Bernstein (n23), "The Artificial Heart Program," 31.
87. National Heart, Lung, and Blood Institute, Advisory Council, *Report of the Artificial Heart Working Group* (May 21, 1981).
88. Described in Willem J. Kolff and J. Lawson, "Perspectives for the Total Artificial Heart," *Transplantation Proceedings 11*, No. 1 (March 1979), 317; and H. Sanders, "Artificial Organs," *Chemical and Engineering News 49* (1971), 68.
89. National Heart, Lung, and Blood Institute, Advisory Council, Report of the Working Group on Circulatory Assist and the Artificial Heart, *Mechanically Assisted Circulation* (August 1980).
90. Discussed in Watson (n10), "Past, Present, and Future of Mechanical Circulatory Support"; and William S. Pierce, "The Implantable Ventricular Assist Pump," *Journal of Thoracic and Cardiovascular Surgery 87* (1984), 811–13.

91. William S. Pierce, "Artificial Hearts and Blood Pumps in the Treatment of Profound Heart Failure," *Circulation 68* (1983), 883–8.

92. See "An Incredible Affair of the Heart," *Newsweek* (December 13, 1982), 70–9; "Living on Borrowed Time," *Time* (December 13, 1982), 72–3; and Thomas A. Preston, "The Case Against the Artificial Heart," *Seattle Weekly* (March 30–April 5, 1983), 25.

93. See Kolff (n8), "Forty Years and Beyond," 7.

94. Kolff has spent a great deal of his career in developing artificial kidneys and loves to tell interviewers success stories about patients who have lived useful lives dependent on the artificial kidney. His chief criterion for clinical success of technology is the happiness of the patient. He told me, as he has told many, "If the patient is not happy, we shouldn't do it. I have always said that."

95. See Strickland (n19), *Politics, Science, and Dread Diseases*; Rettig (n19), *Cancer Crusade*; and Spingarn (n19), *Heartbeat*.

96. Betty Gilson, personal communication; Robert G. Wilson, personal communication.

97. See Malcolm N. Carter, "The Business Behind Barney Clark's Heart," *Money* (April 1983), 130.

98. DeVries and Joyce (n6), "The Artificial Heart," 5.

99. Willem J. Kolff, personal communication.

100. Kolff (n5), "An Intrathoracic Pump to Replace the Human Heart," 7.

101. See Carter (n97), "The Business Behind Barney Clark's Heart."

102. Ibid., 132.

103. Ibid., 134.

104. Kolff, personal communication.

105. Jarvik (n9), "The Total Artificial Heart," 79.

106. See Carter (n97), "The Business Behind Barney Clark's Heart," 132. When I asked Kolff why he decided on a human implant then, he answered, "Because we wanted to." To the same question, Jarvik replied, "There was nothing more to do."

107. "High Spirits on a Plastic Pulse," *Time* (December 10, 1984), p. 83.

108. See F. Ross Woolley, "Ethical Issues in the Implantation of the Total Artificial Heart," *New England Journal of Medicine 310* (1984), 292–6.

109. William C. DeVries, "The Total Artificial Heart Replacement in Man: A Proposal to the Review Committee for Research with Human Subjects," University of Utah, February 27, 1981.

110. The interactions between members of the NIH and FDA set forth in this section are documented in transcriptions of telephone conversations and letters made public through the Freedom of Information Act.

111. Extracted from the telephone log of Mr. Glenn Rahmoeller, di-

rector, Division of Cardiovascular Devices, Bureau of Medical Devices, FDA, Silver Springs, Maryland, January 23, 1981.

112. Kolff, personal communication.

113. Peter L. Frommer, deputy director, National Heart, Lung, and Blood Institute, letter to Glenn Rahmoeller, FDA, February 13, 1981.

114. Peter L. Frommer, letter to Thomas A. Preston, September 9, 1983.

115. In a private discussion, Kolff confirmed the element of independence from the NIH as a reason for forming his private company.

116. Glenn A. Rahmoeller, director, Division of Cardiovascular Devices, FDA, letter to Peter Frommer, deputy director, National Heart, Lung, and Blood Institute, March 5, 1981; see also Arthur L. Caplan, "The Artificial Heart," *Hastings Center Report* (February 1982), p. 24.

117. Denton A. Cooley, Tetsuze Akutsu, John C. Norman, et al., "Total Artificial Heart in Two-Staged Cardiac Transplantation," *Cardiovascular Diseases Bulletin of Texas Heart Institute 8* (1981), 305.

118. For example, see "The Artificial Heart Is Here," *Life* (September 1981), 28–31.

119. Caplan (n116), "The Artificial Heart," 22–4.

120. The FDA responded with a letter from Victor Zafra, acting director, Bureau of Medical Devices, to Newell France, Executive Director, St. Luke's Episcopal Hospital. In the letter, Mr. Zafra said, "We believe that the clinical use of a total artificial heart requires an investigational device exemption (IDE) approved by the FDA." He went on to say," "If you anticipate implanting another artificial heart in a patient in the future, please submit an IDE to FDA."

121. Glenn A. Rahmoeller, director, Division of Cardiovascular Devices, FDA, personal communication.

122. Ibid. In his reply to Zafra's letter, Newell France wrote: "It is our position that the unique circumstances presented by the instant case warranted the employment of all extraordinary, life-sustaining measures available to us."

123. For example, on August 29, 1985, Dr. Jack G. Copeland implanted a Jarvik–7 artificial heart into a patient without prior FDA approval. Dr. Copeland used the artificial heart as a "bridge" to a heart transplantation, and justified his act as necessary therapy in an emergency. The FDA investigated the incident, but took no formal action.

124. See Carter (n97), "The Business Behind Barney Clark's Heart," 130–44.

125. Ibid.

126. Kolff, personal communication.

127. *New York Times*, March 20, 1983, p. 14.

128. Prospectus, Kolff Medical, Inc., July 15, 1983, p. 27.

129. Daniel Kehrer, "High Goals of Artificial-Heart Firm Bring Low Stock-Market Payoff," *Medical Month* (December 1983), 83–4.

130. "An Incredible Affair of the Heart (n92); "Living on Borrowed Time" (n92); Preston (n92), "The Case Against the Artificial Heart," 25; William C. DeVries, Jeffrey L. Anderson, Lyle D. Joyce, et al., "Clinical Use of the Total Artificial Heart," *New England Journal of Medicine 310* (1984), 273–8; Pierre M. Galletti, "Replacement of the Heart with a Mechanical Device," *New England Journal of Medicine 310* (1984), 312–14.

131. Claudia K. Berenson and Bernard I. Grosser, "Total Artificial Heart Implantation," *Archives of General Psychiatry 41* (1984), 911–16.

132. Ibid., 910; Woolley (n108), "Ethical Issues in the Implantation of the Total Artificial Heart," 292–6.

133. Barton J. Bernstein, "The Misguided Quest for the Artificial Heart," *Technology Review* (November/December 1984), 19.

134. "Barney Clark's MD-Son: 'I'm Ambivalent,' " *American Medical News* (April 22/29, 1983), 1.

135. Berenson and Grosser (n131), "Total Artificial Heart Implantation." Barney Clark's son Stephen said of his father: "His interest in going ahead, he told this to me, was to make this contribution, whereas the only other way was to die of the disease" (*New York Times*, March 25, 1983, p. 11).

136. "Barney Clark's MD-Son: 'I'm Ambivalent' " (n134).

137. DeVries, et al. (n130), "Clinical Use of the Total Artificial Heart."

138. "Barney Clark's MD-Son: 'I'm Ambivalent' " (n134).

139. Lawrence K. Altman, "Heart Team See Lessons in Dr. Clark's Experience," *New York Times*, April 17, 1983, p. 20.

140. See Renée C. Fox, "It's the Same, But Different: A Sociological Perspective on the Case of the Utah Artificial Heart," in Margery W. Shaw, ed., *After Barney Clark* (Austin: University of Texas Press, 1984), p. 79.

141. Preston (n92), "The Case Against the Artificial Heart."

142. Thomas A. Preston, "Who Benefits from the Artificial Heart?" *Hastings Center Report* (February 1985), 5–6.

143. Albert R. Jonsen, "The Selection of Patients," in Shaw (n140), *After Barney Clark*, pp. 7–10. In the subsequent case of Baby Fae, who received a baboon heart, the surgeon based his act on an "intent" to benefit the patient, although the majority of professional opinion held there was no scientific basis for expecting success. The point is important in establishing therapies and setting public policy. On the one hand, health care workers ranging from qualified physicians to quacks have long insisted on their personal

authority in establishing beneficial therapies. This is in contrast, on the other hand, to more recent government-induced requirements for scientific evidence of benefit.

144. Preston (n92), "The Case Against the Artificial Heart."
145. See Bernstein (n133), "The Misguided Quest for the Artificial Heart," 62.
146. Lawrence K. Altman, "After Barney Clark: Reflections of a Reporter on Unresolved Issues," in Shaw (n140), *After Barney Clark*, pp. 113–28.
147. Personal and confidential communication.
148. Altman (n146), "After Barney Clark: Reflections of a Reporter"; David Zinman, "Improving the Artificial Heart," *Newsday* (August 1, 1983).
149. See Chase N. Peterson, "A Spontaneous Reply to Dr. Lawrence Altman," in Shaw (n140), *After Barney Clark*, pp. 129–38.
150. See Berenson and Grosser (n131), "Total Artificial Heart Implantation."
151. Ibid.
152. "Barney Clark's MD-Son: 'I'm Ambivalent' " (n134).
153. Zinman (n148), "Improving the Artificial Heart"; "Insurers Debating Costs of Transplanting Organs," *New York Times*, November 21, 1983, p. 9.
154. Albert R. Jonsen, "The Artificial Heart's Threat to Others," *Hastings Center Report* (February 1986), 9–11.
155. Denise Grady, "Artificial Heart Skips a Beat," *Medical Month* (December 1983), p. 44; George J. Annas, "Consent to the Artificial Heart: The Lion and the Crocodiles," *Hastings Center Report* (April 1983), 20–2.
156. Altman (n146), "After Barney Clark: Reflections of a Reporter."
157. John A. Bosso, "Deliberations of the Utah Institutional Review Board Concerning the Artificial Heart," in Shaw (n140), *After Barney Clark*, pp. 139–45.
158. Lawrence K. Altman, "The Artificial Heart Mired in Delay and Uncertainty," *New York Times*, October 25, 1983, p. C2.
159. See "High Spirits on a Plastic Pulse" (n107); "Heart Implants May Move with DeVries," *Salt Lake City Tribune*, August 1, 1984, p. B1.
160. "Earning Profits, Saving Lives," *Time* (December 10, 1984), 85.
161. See "Artificial Heart Program Prompts Debate," *Cardiovascular News 3* (November 1984), 1–6; "The Critics Turn Thumbs Down, Humana Keeping Its Thumbs Up on Adequacy of Bionic Bill's Heart," *Medical Tribune 26* (January 23, 1985), 9; Peter Tarr, "Did DeVries Get a Rubber Stamp?" *Times Union (Albany)*, March 17, 1985, p. A1.

162. Comment made at Symbion Symposium on the Artificial Heart, Park City, Utah, April 2, 1985.
163. Robert Bazell, "Hearts of Gold: Winners and Losers in the Celebrity Surgery Sweepstakes," *New Republic* (February 18, 1985), 19.
164. "Humana: Making the Most of Its Place in the Spotlight," *Business Week* (May 6, 1985), 68–9.
165. Lawrence Altman, "Health Care as Business," *New York Times*, November 27, 1984, p. 1.
166. F. Ross Woolley, personal communication.
167. See, for example, "Artificial Heart Program Prompts Debate" (n161); Philip M. Boffey, "Artificial Heart: Should It Be Scaled Back?" *New York Times*, December 3, 1985, p. C1.
168. Grady (n155), "Artificial Heart Skips a Beat."

5. The swine flu immunization program

1. Quoted in Arthur J. Viseltear, "Immunization and Public Policy: A Short Political History of the 1976 Swine Influenza Legislation," in June E. Osborn, ed., *History, Science, and Politics: Influenza in America, 1918–1976* (New York: Prodist, 1977), p. 29.
2. Interview with Martin Goldfield, M.D., by Ralph Silber, November 7, 1979.
3. Lawrence Wright, "Sweating Out the Swine Flu Scare," *New Times*, June 11, 1976, p. 30. CDC is the federal government's "preventive medicine" organization, with responsibility for tracking the course of diseases through the population, offering recommendations as to whether a new vaccine is needed and if so what kind, and dealing with the state and local health departments.
4. For a detailed description of these events, see Wright (n3), "Sweating Out the Swine Flu Scare."
5. Quoted in Wright (n3), "Sweating Out the Swine Flu Scare," 33.
6. The original work was reported in Richard E. Shope, "Swine Influenza: III. Filtration Experiments and Etiology," *Journal of Experimental Medicine 54* (September 1931), 373–85; Thomas Francis, Jr. and Richard Shope, "Neutralization Tests with Sera of Convalescent or Immunized Animals and the Viruses of Swine and Human Influenza," *Journal of Experimental Medicine 63* (May 1936), 645–53; and Richard E. Shope, "The Incidence of Neutralizing Antibodies for Swine Influenza Virus in the Sera of Human Beings of Different Ages," *Journal of Experimental Medicine 63* (May 1936), 669–84. Some virologists took the fact that the virus isolated at Fort Dix did not spread to the surrounding community and turned out not to grow well in eggs to indicate that it was a relatively weak pathogen and hence only distantly related, if at all, to the

lethal virus of 1918–19. For research that challenged the conclusion that the 1918–19 pandemic was caused by the swine flu virus, see Paul Brown, D. Carleton Gajdusek, and J. Anthony Morris, "Virus of the 1918 Influenza Pandemic Era: New Evidence About Its Antigenic Character," *Science 166* (October 3, 1969), 117–19. Additional doubts are discussed in Albert B. Sabin, "Swine Flu: What Happened?" *The Sciences 17* (March/April 1977), 14–15, 26–7, 30.

7. Influenza viruses are identified by their surface proteins or "antigens." When a virus appears with antigens differing in composition from those of the virus previously circulating in the population, an "antigenic shift" is said to have occurred. Although it is still widely believed that antigenic shifts produce influenza pandemics, some experts, such as Albert Sabin, cite contrary evidence. See Sabin (n6), "Swine Flu: What Happened?"

8. See, for example, Nic Masurel and William M. Marine, "Recycling of Asian and Hong Kong Influenza A Virus Hemagglutinins in Man," *American Journal of Epidemiology 97* (1973), 44–9. For a popular summary of the eleven-year cycle thought to characterize influenza pandemics and the prediction of one in the late 1970s, see Edwin D. Kilbourne, "Flu to the Starboard! Man the Harpoons! Fill 'em With Vaccine! Get the Captain! Hurry!" *New York Times*, February 13, 1976, p. 33.

9. Harold M. Schmeck, Jr., "U.S. Calls Flu Alert on Possible Return of Epidemic's Virus," *New York Times*, February 20, 1976, p. 1.

10. Harry M. Meyer, Jr., M.D., quoted in Center for Disease Control, Bureau of Biologics/National Institute of Allergy and Infectious Disease, *Influenza Workshop* (Bethesda, Md.: February 20, 1976), transcript, p. 9.

11. Maurice R. Hilleman, quoted in ibid., p. 103. Several Public Health Service officials picked up on the idea of "heroic" policymaking some six weeks later in their congressional testimony on appropriations for the swine flu immunization program.

12. The most definite way to identify a virus is by isolating the virus from a throat culture. A second method that is nearly as certain is to show an increase in antibody levels in successive blood samples, which presumably reflects the body's response to the virus. The third and least satisfactory method is simply to measure the absolute antibody concentration in a single blood sample, with high concentrations being interpreted as evidence of previous infection. However, there may be disagreement about how "high" antibody levels have to be to indicate infection; furthermore, since antibodies can rise in response to more than one virus, the interpretation of high concentrations is not always straightforward. People with Victoria or other flu, for example, could show a rise in swine flu antibodies. Hence serological evidence alone provides

an ambiguous basis for inferring the presence of swine flu (Sabin [n6], "Swine Flu: What Happened?").

13. Interview with Walter R. Dowdle, M.D., by Ralph Silber, November 1, 1979. See also Richard E. Neustadt and Harvey V. Fineberg, *The Swine Flu Affair: Decision-Making on a Slippery Disease* (Washington, D.C.: Government Printing Office, 1978), pp. 10, 22; and Arthur M. Silverstein, *Pure Politics and Impure Science: The Swine Flu Affair* (Baltimore: Johns Hopkins University Press, 1981), pp. 34–5, 41.

14. Silverstein (n13), *Pure Politics and Impure Science*, p. 30.

15. Neustadt and Fineberg (n13), *The Swine Flu Affair*, p. 11.

16. Center for Disease Control, Immunization Practices Advisory Committee, Summary Minutes of Meeting, Atlanta, Ga., March 10, 1976, pp. 4–5.

17. Department of Health, Education and Welfare, General Accounting Office, *The Swine Flu Program: An Unprecedented Venture in Preventive Medicine*, Publ. No. HRD–77–115 (Washington, D.C.: Government Printing Office, June 27, 1977), pp. 38–41. See also Silverstein (n13), *Pure Politics and Impure Science*, pp. 18–23.

18. Interview with Dowdle (n13), November 1, 1979.

19. Reuel A. Stallones, M.D., quoted in Neustadt and Fineberg (n13), *The Swine Flu Affair*, p. 12.

20. E. Russell Alexander, M.D., quoted in Neustadt and Fineberg (n13), *The Swine Flu Affair*, p. 13.

21. Interview with Dowdle (n13), November 1, 1979. In June 1976, Dr. Gary Noble, chief of the Respiratory Virology Branch at CDC, stated that only two people in the history of influenza vaccines in the United States had died as a result of flu shots, and this occurred in the 1940s when vaccines were comparatively unrefined. However, he did admit that although "everyone is well agreed that the risk of modern vaccine is quite minimal, . . . a mass vaccination program would generate thousands, perhaps millions, of reactions ranging from sore arms to disabling headaches, fevers, convulsions, malaise or nausea" (quoted in Wright [n3], "Sweating Out the Swine Flu Scare," p. 34).

22. Alexander, quoted in Neustadt and Fineberg (n13), *The Swine Flu Affair*, p. 13.

23. Interview with Goldfield (n2), November 7, 1979.

24. John R. Seal, M.D. or Harry M. Meyer, Jr., M.D. to David J. Sencer, M.D., quoted in Neustadt and Fineberg (n13), *The Swine Flu Affair*, p. 14.

25. Also, by some accounts, the ACIP was practically a house organ and generally satisfied Sencer's wishes. Quote is from interview with David J. Sencer, M.D., Director of CDC, by Ralph Silber, November 6, 1979.

26. Barton J. Bernstein, "The Swine Flu Immunization Program", *Medical Heritage 1* (July–August 1985), 241–2. It is possible that one of the two who reported that he had spoken in favor of watchful waiting was less clear in mid-March than he later recalled, since in late April he seemed quite comfortable with mass immunization (William Elsea to Sencer, April 21, 1976, CDC Papers, quoted in Bernstein, 262, note 40).

27. Memorandum signed by Dr. James F. Dickson on behalf of Dr. Theodore Cooper to David Mathews, Secretary of HEW, March 18, 1976, p. 1, reprinted in Neustadt and Fineberg (n13), *The Swine Flu Affair*, pp. 147–55.

28. Ibid., pp. 2, 9.

29. Ibid., p. 4.

30. Interview with Sencer (n25), November 6, 1979. On Mathews's reputation, see Neustadt and Fineberg (n13), *The Swine Flu Affair*, p. 17. See also Silverstein (n13), *Pure Politics and Impure Science*, pp. 40–1.

31. Research notes lent to the Kennedy School Case Program by Professor Richard Neustadt.

32. Mathews, quoted in Neustadt and Fineberg (n13), *The Swine Flu Affair*, p. 18.

33. Dickson, quoted in J. Bradley O'Connell, "Swine Flu," case study prepared for the Kennedy School of Government Case Program under the supervision of Laurence E. Lynn, Jr., Professor of Public Policy at the John Fitzgerald Kennedy School of Government, Harvard University, Cambridge, Mass., 1979, p. 9.

34. Research notes lent to the Kennedy School Case Program by Professor Richard Neustadt.

35. Memorandum from David Mathews, Secretary of HEW, to James T. Lynn, Director of OMB, March 15, 1976, reprinted in Neustadt and Fineberg (n13), *The Swine Flu Affair*, p. 156.

36. Walter R. Dowdle, "The Swine Flu Vaccine Program," *American Society for Microbiology News 43* (May 1977), 244.

37. Victor Zafra quotes are from O'Connell (n33), "Swine Flu," p. 11.

38. In fact, it later came to light that after the recruit who died of swine flu had collapsed, his sergeant had given him mouth-to-mouth resuscitation, but had not subsequently contracted the disease.

39. Zafra, quoted in Bernstein (n26), "The Swine Flu Immunization Program," 244.

40. Zafra, quoted in O'Connell (n33), "Swine Flu," p. 11. The internal OMB memo cited is from Nancy E. Bateman, Budget Examiner, to Paul O'Neill, Deputy Director of OMB, March 15, 1976 (photocopy of OMB memorandum).

41. Zafra, quoted in O'Connell (n33), "Swine Flu," p. 16.

42. Interview with Sencer (n25), November 6, 1979.
43. Memorandum from John D. Young, Comptroller of HEW, to Secretary Mathews, March 18, 1976, quoted in O'Connell (n33), "Swine Flu," p. 13.
44. Theodore Cooper, M.D., quoted in O'Connell (n33), "Swine Flu," p. 6.
45. Silverstein (n13), *Pure Politics and Impure Science*, p. 43. Quotes by William H. Taft and HEW Assistant Secretary William Morrill are from O'Connell (n33), "Swine Flu," p. 14.
46. Paul O'Neill, quoted in O'Connell (n33), "Swine Flu," p. 16.
47. James Cavanaugh quotes are from O'Connell (n33), "Swine Flu," pp. 16–17.
48. Harold M. Schmeck, Jr., "Flu Experts Soon to Rule on Need of New Vaccine," *New York Times*, March 21, 1976, pp. 1, 39.
49. O'Connell (n33), "Swine Flu," p. 18.
50. James Cannon, Director of Domestic Council, quoted in O'Connell (n33), "Swine Flu," p. 17.
51. Shortly thereafter, Morris gave a series of public lectures at NIH on problems with the swine flu program (interview with J. Anthony Morris, Ph.D., by Ralph Silber, November 2, 1979). Morris was fired from his position at the FDA's Bureau of Biologics in July, 1976. He believes this action resulted from his opposition to the swine flu program, although government health officials contend there were other reasons. See Philip M. Boffey, "Vaccine Imbroglio: The Rise and Fall of a Scientist-Critic," *Science 194* (December 3, 1976), 1021–5.
52. Neustadt and Fineberg (n13), *The Swine Flu Affair*, p. 27.
53. Wright (n3), "Sweating Out the Swine Flu Scare," p. 35.
54. About a month later one of the federal officials, Dr. John Seal, deputy director of NIAID, finally agreed to offer his own ballpark estimate of the probability of swine flu epidemic as a favor for someone who was writing a journal article on the subject. His estimate was 2% (research notes lent to the Kennedy School Case Program by Professor Richard Neustadt).
55. Interview with Harry M. Meyer, M.D., Director, FDA Bureau of Biologics, by Ralph Silber, November 2, 1979.
56. President Gerald R. Ford, quoted in O'Connell (n33), "Swine Flu," p. 21.
57. Quoted in Neustadt and Fineberg (n13), *The Swine Flu Affair*, p. 28.
58. Interview with Sencer (n25), November 6, 1979.
59. Memorandum from James Cannon to Max Friedersdorf, March 24, 1976, quoted in O'Connell (n33), "Swine Flu," p. 20.
60. Office of the White House Press Secretary, "Fact Sheet, Swine Influenza Immunization Program," March 24, 1976.
61. Since no one could numerically estimate the probability, "a very

real possibility" became a common description of the seriousness of the threat.

62.   "Swine Flu Inoculations, The President's Remarks Announcing Actions to Combat the Influenza, March 24, 1976," *Weekly Compilation of Presidential Documents 12* (March 29, 1976), 484.
63.   See Silverstein (n13), *Pure Politics and Impure Science*, pp. 64–5.
64.   Robert C. Pierpoint, quoted in Neustadt and Fineberg (n13), *The Swine Flu Affair*, p. 30.
65.   Neustadt and Fineberg (n13), *The Swine Flu Affair*, p. 29.
66.   "Excerpts from Cronkite Show of 3/25/76," quoted in Bernstein (n26), "The Swine Flu Immunization Program," 247.
67.   Quoted in Neustadt and Fineberg (n13), *The Swine Flu Affair*, p. 30.
68.   Bernstein (n26), "The Swine Flu Immunization Program," 247.
69.   *Swine Flu Immunization Program, 1976*, hearings before the Subcommittee on Health of the Committee on Labor and Public Welfare, U.S. Senate, 94th Congress, 2nd Session, April 1 and August 5, 1976, pp. 1–5. For a detailed description of the Congressional deliberations on swine flu, see Viseltear (n1), "Immunization and Public Policy," pp. 29–58.
70.   Testimony of C. Joseph Stetler in *Proposed National Swine Flu Vaccination Program*, hearing before the Subcommittee on Health and the Environment of the Committee on Interstate and Foreign Commerce, U.S. House, 94th Congress, 2nd Session, March 31, 1976, p. 34.
71.   Testimony of Theodore Cooper, M.D., in ibid., pp. 13–22.
72.   C. Schoenbaum, S. R. Mostow, W. R. Dowdle, et al., "Studies with Inactivated Influenza Vaccines Purified by Zonal Centrifugation," *Bulletin of the World Health Organization 41* (1969), 531. For the GAO report, see General Accounting Office (n17), *The Swine Flu Program*, pp. 38–9. Clinical studies question the ability of vaccines to prevent disease, especially compared with the immunity resulting from natural infection. See Timothy F. Nolan, Jr., "Influenza Vaccine Efficacy" (editorial), *Journal of the American Medical Association 245* (May 1, 1981), 1762.
73.   There is even substantial disagreement about how to define and measure vaccine efficacy; different methods and definitions yield different results. For a summary of the problems in producing effective flu vaccines and in measuring efficacy, see Neustadt and Fineberg (n13), *The Swine Flu Affair*, pp. 104–14. The difficulties for swine flu vaccine in particular are discussed in General Accounting Office (n17), *The Swine Flu Program*, pp. 41–2. Also, see Philip M. Boffey, "Swine Flu Vaccination Campaign: The Scientific Controversy Mounts," *Science 193* (August 13, 1976), 559–63.

74. Testimony of Merritt Low, M.D., in hearing on *Swine Flu Immunization Program, 1976* (n69), U.S. Senate, April 1976, p. 57. See also statement of Anthony Robbins, M.D., in hearing on *Proposed National Swine Flu Vaccination Program* (n70), U.S. House, March 1976, p. 49.

75. Testimony of Bruce Brennan, in hearings on *Swine Flu Immunization Program, 1976* (n69), U.S. Senate, April 1976, p. 73. Stetler quote is from p. 72.

76. Wright (n3), "Sweating Out the Swine Flu Scare," p. 35.

77. Jonathan E. Fielding, M.D., quoted in Neustadt and Fineberg (n13), *The Swine Flu Affair*, p. 38.

78. Associated Press (Jack Stillman), April 2, 1976, cited in Bernstein (n26), "The Swine Flu Immunization Program," 247.

79. Richard Friedman, HEW Regional Director, to Cooper, May 5, 1976, cited in Bernstein (n26), "The Swine Flu Immunization Program," 247.

80. Goldfield, quoted in Wright (n3), "Sweating Out the Swine Flu Scare," p. 36.

81. Goldfield, transcript of CBS Evening News, April 2, 1976, quoted in Neustadt and Fineberg (n13), *The Swine Flu Affair*, p. 40.

82. Interview with Goldfield (n2), November 7, 1979.

83. Quoted in Neustadt and Fineberg (n13), *The Swine Flu Affair*, p. 40.

84. Alexander to Sencer, April 7, 1976, quoted in Neustadt and Fineberg (n13), *The Swine Flu Affair*, p. 39.

85. "Flu Vaccine" (editorial), *New York Times*, April 6, 1976, p. 34.

86. HEW press releases, cited in Silverstein (n13), *Pure Politics and Impure Science*, pp. 84–5. These responses, as Silverstein notes, were undoubtedly influenced by a letter that Assistant Secretary for Health Cooper sent to many newspapers describing the program and urging a favorable response to it.

87. "Influenza," *Weekly Epidemiological Record of the World Health Organization*, No. 16 (April 15, 1976), p. 123.

88. "Experts in Europe Question U.S. Plan for Mass Flu Shots," *New York Times*, June 9, 1976, pp. 1, 8. On Canada's response, see A. B. Morrison, A. J. Liston, and John D. Abbott, "The Canadian Influenza Decision, 1976," *Canadian Medical Association Journal 115* (November 6, 1976), A–D. Some provinces elected to hold the vaccines until further evidence of an epidemic, and therefore never began a swine flu immunization program for the general population (J. R. Waters, M.D., Director, Communicable Disease Control and Epidemiology, Alberta, Canada, to Barton J. Bernstein, June 16, 1982, copy provided by Bernstein). On the British response, see David Perlman, "The Flu Shot Program Poses a Dilemma," *San Francisco Chronicle*, April 5, 1976, p. 2.

89. Nancy Bateman to Mr. McGurk, "Rethinking of Swine Influenza Immunization Program," June 10, 1976 (photocopy of OMB memorandum).
90. Bernstein (n26), "The Swine Flu Immunization Program," 250.
91. Interview with Sencer (n25), November 6, 1979.
92. Dr. John Seal, Scientific Director, National Institute for Allergy and Infectious Diseases, quoted in Neustadt and Fineberg (n13), *The Swine Flu Affair*, p. 42. Actually, two-dose tests had been suggested by several outside scientists at the March 25 Bureau of Biologics meeting, but the suggestion was not picked up by the field test planners.
93. Silverstein (n13), *Pure Politics and Impure Science*, p. 79.
94. Interview with Morris (n51), November 2, 1979.
95. Silverstein (n13), *Pure Politics and Impure Science*, pp. 79–80.
96. "Split" vaccines are generally less toxic, but may also be less effective. For a technical discussion of these methods, see Philip Selby, ed., *Influenza: Virus, Vaccines, and Strategy* (New York: Academic Press, 1976), pp. 137–77.
97. Predictions of pandemic influenza spread were highly uncertain, since data existed for only two previous pandemics, 1957 and 1968, on which they could be based. The period from the first virus isolation to documented outbreak in one-third or more states was ten weeks in 1957 and twelve weeks in 1968 (CDC, "Analysis of Vaccine Stockpile Option," reprinted in Neustadt and Fineberg [n13], *The Swine Flu Affair*, p. 162). These figures suggested that even using CDC's more pessimistic estimates of the time it would take to move from stockpiling to immunization if evidence of a pandemic appeared, there might be enough time if spread occurred at the same rate as in 1968, but not if it occurred as quickly as in 1957. If the program could be mobilized more rapidly, as Sabin was arguing, the likelihood of timely intervention was even greater.
98. Interview with Dowdle (n13), November 1, 1979. See also Neustadt and Fineberg (n13), *The Swine Flu Affair*, p. 46, citing an interview with ACIP member Dr. Reuel Stallones.
99. Interview with Dowdle (n13), November 1, 1979.
100. Alexander, transcript of CBS Evening News, June 22, 1976, quoted in Neustadt and Fineberg (n13), *The Swine Flu Affair*, p. 46.
101. "Swine Flu Scare" (editorial), *New York Times*, July 3, 1976, p. 20. Although swine flu officials and some medical experts bitterly resented what they felt was unfair and overly critical coverage of the program by the media, experts in journalism judged it to be accurate and rarely sensational. See David M. Rubin and Val Hendy, "Swine Influenza and the News Media," *Annals of Internal Medicine 87* (December 1977), 769–74.

102. The first paper referred to is A. S. Beare and J. W. Craig, "Virulence for Man of a Human Influenza-A Virus Antigenically Similar to 'Classical' Swine Viruses," *Lancet 2* (July 3, 1976), 4–5. Some American experts discounted these results on the grounds that because the experiment involved putting the swine flu virus in chick embryos before it was administered to the volunteers, it had lost virulence. The quote is from Charles Stuart-Harris, "Swine Influenza Virus in Man: Zoonosis or Human Pandemic?" *Lancet 2* (July 3, 1976), 31–2. See also "Planning for Pandemics" (editorial), *Lancet 2* (July 3, 1976), 25–6.

103. H. Bruce Dull, M.D., quoted in Boyce Rensberger, "Flu Vaccine Drive Meets Snags That Could Limit It," *New York Times*, July 23, 1976, p. A22.

104. Testimony of Sidney Wolfe, M.D., in *Swine Flu Immunization Program*, supplemental hearings before the Subcommittee on Health and the Environment of the Committee on Interstate and Foreign Commerce, U.S. House, 94th Congress, 2nd Session, June 28, July 20, 23, and September 13, 1976, p. 141.

105. See Philip M. Boffey, "Swine Flu Campaign: Should We Vaccinate the Pigs?" *Science 192* (May 28, 1976), 870–1.

106. The two cases were *Davis* v. *Wyeth* (1968) and *Reyes* v. *Wyeth* (1974). For further information about these cases and the legal trend of which they were part, see Chapter 8.

107. Testimony of David J. Sencer, M.D., in hearings on *Swine Flu Immunization Program, 1976* (n69), U.S. Senate, August 1976, p. 131.

108. Representative Paul G. Rogers, in hearings on *Swine Flu Immunization Program* (n104), U.S. House, July 1976, p. 259.

109. Leslie Cheek, quoted in Neustadt and Fineberg (n13), *The Swine Flu Affair*, p. 57.

110. Representative John Moss, Chairman of the Oversight and Investigations Subcommittee, to Representative Paul Rogers, July 1, 1976, cited in Bernstein (n26), "The Swine Flu Immunization Program," 252.

111. Cooper, in his deposition by the Justice Department, was asked whether, prior to October 1976, he had had discussions with Sencer regarding the probability of a swine flu epidemic that fall. He responded: "It came up often, but they never indicated that the possibility had diminished" (p. 2, "Deposition of Theodore R. Cooper," February 1, 1979, Department of Justice).

112. President Gerald Ford, letter to Representative Paul Rogers, July 23, 1976, cited in Neustadt and Fineberg (n13), *The Swine Flu Affair*, p. 59 and reprinted in hearings on *Swine Flu Immunization Program* (n104), U.S. House, July 1976, pp. 266–7.

113. See Neustadt and Fineberg (n13), *The Swine Flu Affair*, p. 60; also, Silverstein (n13), *Pure Politics and Impure Science*, pp. 98–106.

114. "Swine Flu Immunization Program: The President's Remarks Urging Congressional Enactment of the Program, August 6, 1976," *Weekly Compilation of Presidential Documents 12* (August 9, 1976), 1249

115. Senator Ted Kennedy, quoted in "Shots in the Arm," *Time* (August 23, 1976), 41. For other objections, see Elizabeth Bowman, "Swine Flu: New Efforts to Break Impasse," *Congressional Quarterly Weekly Report 34* (August 7, 1976), 2125 and Prudence Crewdson, "Ford Signs Swine Flu Bill Into Law," *Congressional Quarterly Weekly Report 34* (August 14, 1976), 2235.

116. General Accounting Office (n17), *The Swine Flu Program*, pp. 20–1. As of 1987, the government still had recovered only a minimal amount from insurers for negligence (Jeffrey Axelrad, Director, Torts Branch, Civil Division, U.S. Department of Justice, letter to Diana Dutton, June 22, 1987).

117. "National Swine Flu Immunization Program of 1976: The President's Remarks on Signing S.3735 Into Law, August 12, 1976," *Weekly Compilation of Presidential Documents 12* (August 16, 1976), 1256–7.

118. Interview with Dowdle (n13), November 1, 1979.

119. Neustadt and Fineberg (n13), *The Swine Flu Affair*, p. 62.

120. Philip M. Boffey, "Swine Flu Vaccine: A Component Is Missing," *Science 193* (September 24, 1976), 1224–5.

121. Harold M. Schmeck, Jr., "Production Lags on Flu Vaccines," *New York Times*, September 2, 1976, p. 27. For an analysis of the various errors and obstacles that impeded production, see General Accounting Office (n17), *The Swine Flu Program*, pp. 48–54.

122. Cited in Silverstein (n13), *Pure Politics and Impure Science*, p. 107.

123. Silverstein (n13), *Pure Politics and Impure Science*, p. 108.

124. Actually, two nearly identical forms were developed, one for monovalent vaccine (swine flu only) and one for bivalent vaccine (swine flu and A/Victoria). They are reprinted in Neustadt and Fineberg (n13), *The Swine Flu Affair*, Appendix D, pp. 164–6; the bivalent form is also reprinted in hearings on *Swine Flu Immunization Program* (n104), U.S. House, July 1976, p. 221.

125. EVIST interview with Professor Jay Katz, Yale Law School, January 15, 1980. For a more complete description of problems in the informed consent form, see General Accounting Office (n17), *The Swine Flu Program*, pp. 22–9. Also "Informed Consent" (editorial), *New York Times*, August 22, 1976, Section 4, p. 16. The criticisms of consumer groups are summarized in testimony of Marcia Greenberger, in hearings on *Swine Flu Immunization Program* (n104), U.S. House, September 1976, pp. 340–50.

126. Silverstein (n13), *Pure Politics and Impure Science*, p. 109; also Greenberger testimony, ibid.

127. Interview with Wendell Bradford, Associate Director, Bureau of State Services, CDC, by Malcolm Goggin, February 11, 1980.

128. Bernstein (n26), "The Swine Flu Immunization Program," 253.

129. Public Health Service Advisory Committee on Immunization Practice, "Influenza Vaccine – Supplemental Statement," *Morbidity and Mortality Weekly Report 25* (July 23, 1976), 227. In the largest published review of Guillain–Barré syndrome based on 1,100 cases, only 32 were found to be associated with vaccine inoculations, and only one of those with swine flu vaccine. See Felix Leneman, "The Guillain-Barré Syndrome," *Archives of Internal Medicine 118* (August 1966), 139–44.

130. The Hattwick quote is from the transcript of CBS Television, *60 Minutes: Swine Flu*, Vol. 12, No. 8 (November 4, 1979), p. 5. Although Hattwick claims to have been aware of the possibility of neurological complications, according to Dr. Lawrence Schonberger of CDC, Hattwick was, like most experts at the time, extremely dubious that Guillain–Barré syndrome – the paralyzing neurological disorder that would eventually halt the program – was etiologically linked to flu shots (Lawrence B. Schonberger, M.D., letter to Diana Dutton, June 16, 1987). For CDC's rationale for not including neurological disorders on the informed consent form, see General Accounting Office (n17), *The Swine Flu Program*, pp. 25–6.

131. "Phantom Flu" (editorial), *New York Times*, October 14, 1976, p. 36.

132. Philip H. Dougherty, "Flu-Shot Ad Campaign Ready To Go," *New York Times*, November 17, 1976, p. D12.

133. Harold M. Schmeck, Jr., "Most States Resume Flu Shots or Plan To," *New York Times*, October 15, 1976, pp. A1, 18.

134. Dougherty (n132), "Flu-Shot Ad Campaign."

135. See Neustadt and Fineberg (n13), *The Swine Flu Affair*, p. 67.

136. One new case of swine flu not directly traceable to pigs was reported in Concordia, Missouri (Neustadt and Fineberg [n13], *The Swine Flu Affair*, p. 68). But one case did not an epidemic make, or even a full-fledged "outbreak." Meanwhile, millions were coming down with other respiratory ailments that fall, including other forms of infectious influenza.

137. Harold M. Schmeck, Jr., "U.S. Discloses Shortage of Swine Flu Vaccine for Children 3 to 17," *New York Times*, November 16, 1976, p. 18. Figures on the proportion of children receiving shots are from Department of Health, Education and Welfare, "Administration of the National Swine Flu Immunization Program of 1976: Final Report to Congress," 1978, p. 19.

138. Department of Health, Education and Welfare (n137), "Administration of the National Swine Flu Immunization Program of 1976," p. 17.

139. For high- and low-income adults aged 45–64, the incidence of influenza is 36 vs. 64/100 persons per year, respectively. Department of Health, Education and Welfare, Public Health Service,

Health, United States: 1978, Publ. No. (PHS)78–1232 (Hyattsville, Md.: December 1978), p. 241. See also *Health Status of Minorities and Low-Income Groups*, prepared by Melvin H. Rudov and Nancy Santangelo, Department of Health, Education and Welfare, Publ. No. (HRA) 79–627 (Washington, D.C.: Government Printing Office, 1979), p. 91. For a description of special appeals to blacks, see Bernstein (n26), "The Swine Flu Immunization Program," 255. On reasons for not getting shots, see Ronald Sullivan, "New York City Residents Resist Flu Shots and a Poll Learns Why," *New York Times*, November 21, 1976, Section 1, p. 59.

140. Neustadt and Fineberg (n13), *The Swine Flu Affair*, p. 68. Percent of the population vaccinated is from Department of Health, Education and Welfare (n137), "Administration of the National Swine Flu Immunization Program of 1976," p. 19.

141. J. Davenport, "Flu Staff Meeting," November 10, 1976, quoted in Bernstein (n26), "The Swine Flu Immunization Program," 256.

142. J. Donald Millar to state health officials November 12, 1976, quoted in Bernstein (n26), "The Swine Flu Immunization Program," 256.

143. J. Davenport, "Flu Staff Meeting," November 10, 1976, quoted in Bernstein (n26), "The Swine Flu Immunization Program," 256.

144. Schonberger to Dutton (n130), June 16, 1987.

145. Ibid.

146. Quoted in Philip M. Boffey, "Guillain–Barré: Rare Disease Paralyzes Swine Flu Campaign," *Science 195* (January 14, 1977), 156.

147. The main reason for the change was that the estimated incidence of GBS in the normal population had been revised sharply downward (from 14,000 cases in the initial estimate to only 4,000 cases in the revised estimate). This revision resulted from Schonberger's sudden recognition – at 2:00 AM, December 15, 1976 – that the risk of GBS was most accurately expressed in terms of cases per person-weeks of risk rather than cases per size of population at risk (Schonberger to Dutton [n130], June 16, 1987). Compared to this new, much lower estimate, the rates of GBS found among swine flu vaccinees looked far more ominous. See Boffey (n146), "Guillain–Barré: Rare Disease Paralyzes Swine Flu Campaign," 155–9.

148. Cooper, quoted in Lawrence K. Altman, "Swine Flu Program Suspended in Nation; Disease Link Feared," *New York Times*, December 17, 1976, p. A1.

149. Interview with Lawrence B. Schonberger, M.D., Viral Diseases Division, Bureau of Epidemiology, CDC, by Malcolm Goggin, February 11, 1980.

150. First Califano quote is from Neustadt and Fineberg (n13), *The Swine Flu Affair*, p. 74. Second quote is from Joseph A. Califano, Jr., *Governing America: An Insider's Report from the White House and the Cabinet* (New York: Simon and Schuster, 1981), p. 176.

151. "Swine Flu: Letting the Sunshine In" (editorial), *Washington Post*, February 13, 1977, p. C6.
152. Silverstein (n13), *Pure Politics and Impure Science*, p. 126.
153. "The Califano Prescription for Flu" (editorial), *New York Times*, February 10, 1977, p. 38.
154. "Swine Flu: Letting the Sunshine In" (editorial) (n151).
155. The results of the original CDC investigation were published in "Guillain–Barré Syndrome – United States," *Morbidity and Mortality Weekly Report 25* (December 24, 1976), 401–2, and later in Lawrence B. Schonberger, Dennis J. Bregman, John Z. Sullivan-Bolyai, et al., "Guillain–Barré Syndrome Following Vaccination in the National Influenza Immunization Program, United States, 1976–1977," *American Journal of Epidemiology 110* (1979), 105–23. For additional evidence implicating swine flu vaccine in GBS, see Joel G. Breman and Norman S. Hayner, "Guillain–Barré Syndrome and Its Relationship to Swine Influenza Vaccination in Michigan, 1976–1977," *American Journal of Epidemiology 119* (June 1984), 880–9; and Richard L. Greenstreet, "Estimation of the Probability That Guillain–Barré Syndrome Was Caused by the Swine Flu Vaccine: U.S. Experience (1976–77)," *Medicine, Science and the Law 24* (1984), 61–7. GBS does *not* appear to be associated with other types of influenza vaccines; see Eugene S. Hurwitz, Lawrence B. Schonberger, David B. Nelson, et al., "Guillain–Barré Syndrome and the 1978–1979 Influenza Vaccine," *New England Journal of Medicine 304* (June 25, 1981), 1557–61.
156. For criticisms of CDC's methods and findings, see, for example, Leonard T. Kurland, Wigbert C. Wiederholt, James W. Kirkpatrick, et al., "Swine Influenza Vaccine and Guillain–Barré Syndrome: Epidemic or Artifact?" *Archives of Neurology 42* (November 1985), 1089–90. The reevaluation occurred in two steps: first, a reexamination of the national data previously analyzed by CDC, and second, an intensive reevaluation of data from two states (forthcoming). Both supported CDC's original conclusions (interview with Leonard Kurland, M.D., Mayo Clinic, by Diana Dutton, June 23, 1987). The reanalysis of national data concluded that the relative risk of "extensive" disease from GBS was estimated to be between 3.96 and 7.75 times higher for swine flu vaccine recipients than for the unvaccinated; see Alexander D. Langmuir, Dennis J. Bregman, Leonard T. Kurland, et al., "An Epidemiologic and Clinical Evaluation of Guillain–Barré Syndrome Reported in Association with the Administration of Swine Influenza Vaccines," *American Journal of Epidemiology 119* (June 1984), 841–79.
157. Silverstein (n13), *Pure Politics and Impure Science*, p. 127. Also, see Neustadt and Fineberg (n13), *The Swine Flu Affair*, p. 142.
158. Secretary Joseph A. Califano of Health, Education and Welfare, *HEW News*, press release, June 20, 1978, MDL Doc. 534, pp. 2,4.

159. "Swine Flu Statistics," Torts Branch, Civil Division, U.S. Department of Justice, June 4, 1987. On denial of claims, see Chapter 8, note 58.

160. For a detailed analysis of one state's expenditures, see Allen N. Koplin, Byron J. Francis, Russell J. Martin, et al., "Administrative Costs of the Influenza Control Program of 1976–1977 in Illinois," *Medical Care 27* (February 1979), 201–9. Somewhat lower estimates of aggregate state and local costs are reported in Department of Health, Education and Welfare (n137), "Administration of the National Swine Flu Immunization Program of 1976."

161. CDC Director Sencer and Assistant Secretary Cooper both lost their jobs shortly after Carter took office. Goldfield, formerly chief epidemiologist for the New Jersey State Health Department, was forced out of this post in 1977 and attributes the acute hypertension he suffered during that time to this experience. Another critic, Anthony Morris, a Bureau of Biologics bacteriologist, was also fired from his job following his opposition to the swine flu program and pursued a long and bitter legal battle with the government.

162. In this survey, conducted in late 1977 in southeastern Pennsylvania, slightly more people intended to get flu shots in the future than had gotten swine flu shots in 1976 (about 60% vs. 52%). See William A. Pearman, "Participation in Flu Immunization Projects: What Can We Expect in the Future?" *American Journal of Public Health 68* (July 1978), 674–5.

163. Department of Health and Human Services, Public Health Service, *Health, United States: 1980*, Publ. No. (PHS)81-1232 (Hyattsville, Md.: December 1980), p. 302.

164. Department of Health, Education and Welfare, *Healthy People: The Surgeon General's Report on Health Promotion and Disease Prevention*, Publ. No. (PHS)79-55071 (Washington, D.C.: Government Printing Office, 1979), p. Ch8-16. On diversion of state resources, see Jonathan E. Fielding, "Managing Public Health Risks: The Swine Flu Immunization Program Revisited," *American Journal of Law & Medicine 4* (Spring 1978), 35–43.

165. Childhood immunization levels in 1979 are from Department of Health and Human Services (n163), *Health, United States: 1980*, p. 302. For Reagan administration cuts, see Richard E. Neustadt and Harvey V. Fineberg, *The Epidemic That Never Was* (New York: Vintage Books, 1983), p. 226.

166. See, for example, Harry Schwartz, "Swine Flu Fiasco," *New York Times*, December 21, 1976, 33; George A. Silver, "Lessons of the Swine Flu Debacle," *Nation* (February 12, 1977), p. 166–9; Cyril H. Wecht, "The Swine Flu Immunization Program: Scientific Venture or Political Folly?" *American Journal of Law & Medicine 3* (Winter 1977–78), 425–45.

167. Sencer and Califano quotes are from Nicholas Wade, "1976 Swine Flu Campaign Faulted Yet Principals Would Do It Again," *Science* *202* (November 24, 1978), 851–2.
168. Quoted in Neustadt and Fineberg (n13), *The Swine Flu Affair*, p. 93.
169. This suggestion is based on Neustadt and Fineberg's discussion of "thinking about doing" and "thinking of the media" (n13, *The Swine Flu Affair*, pp. 91–7), although they place more emphasis on the focus of the thinking (implementation) and somewhat less on the composition of the group doing it. Silverstein (n13, *Pure Politics and Impure Science*, pp. 138–40) offers a similar suggestion.
170. Interview with Dowdle (n13), November 1, 1979.
171. See, for example, Kenneth D. Rosenberg, "Swine Flu: Play It Again, Uncle Sam," *Health/PAC Bulletin* (November/December 1976), pp. 1–6, 10–20; Silver (n166), "Lessons of the Swine Flu Debacle;" and Schwartz (n166), "Swine Flu Fiasco." This pattern is also discussed in Philip M. Boffey, "Anatomy of a Decision: How the Nation Declared War on Swine Flu," *Science* *192* (May 14, 1976), 636–41.

6. Genetic engineering

1. Quoted in Sharon McAuliffe and Kathleen McAuliffe, *Life for Sale* (New York: Coward, McCann & Geoghegan, 1981), p. 11.
2. Lewis Thomas, "Oswald Avery and the Cascade of Surprises," *Technology in Society 6* (1984), 37.
3. Jeremy Rifkin, *Algeny* (New York: Penguin Books, 1984), p. 6.
4. James D. Watson and John Tooze, *The DNA Story: A Documentary History of Gene Cloning* (San Francisco: W. H. Freeman, 1981), p. viii.
5. Watson and Tooze (see n4), *The DNA Story*, p. xii.
6. Ibid., p. vii.
7. Ibid., p. viii.
8. Robert Pollack, transcript of an interview by Mary Terrall, March 26, 1976, MIT Oral History Program, Project on the Development of Recombinant DNA Research Guidelines, Institute Archives, MIT, Cambridge, Mass. [hereafter cited as MIT Oral History Program], p. 34.
9. See Sheldon Krimsky, *Genetic Alchemy: The Social History of the Recombinant DNA Controversy* (Cambridge, Mass.: MIT Press, 1982), pp. 13–69, for a detailed description of incidents that gave rise to these concerns.
10. Krimsky (n9), *Genetic Alchemy*, p. 67.
11. Robert Pollack, quoted in Nicholas Wade, "Microbiology: Hazardous Profession Faces New Uncertainties," *Science 182* (November 9, 1973), 566.

12. John Lear, *Recombinant DNA – The Untold Story* (New York: Crown, 1978), p. 70.

13. Maxine Singer and Dieter Soll, "Guidelines for DNA Hybrid Molecules" (letter), *Science 181* (September 21, 1973), 1114.

14. Paul Berg, quoted in Michael Rogers, *Biohazard* (New York: Alfred A. Knopf, 1977), p. 44.

15. Paul Berg, David Baltimore, Herbert W. Boyer, et al., "Potential Biohazards of Recombinant DNA Molecules" (letter), *Science 185* (July 26, 1974), 303.

16. *Washington Post, New York Times*, July 18, 1974.

17. Paul Berg, quoted in Rogers (n14), *Biohazard*, p. 45.

18. The parliamentary group, chaired by Lord Ashby, submitted its report in January 1975 (Ashby Working Party, "Report of the Working Party on the Experimental Manipulation of the Genetic Composition of Micro-Organisms," [HMSO Cmnd. 5880], January 1975). The Ashby report did not support continuation of the moratorium and was considered by many to be too lax (Bernard Dixon, "Not Good Enough" [comment], *New Scientist 65* [January 23, 1975], 86).

19. The organizing committee's desire to end the moratorium quickly was evident in a statement prepared by David Baltimore, a member of the organizing committee and cosigner of the moratorium letter, which (with the approval of other committee members) he released to the press when the news about the proposed moratorium broke. The statement outlined the elements of the controversy, but noted that: "The committee hopes that it will be possible to proceed with the experimentation in the near future when the potential for hazard has been overcome" (quoted in Lear [n12], *Recombinant DNA*, p. 93).

20. David Baltimore, transcript of an interview by Charles Weiner and Rae Goodell, May 13, 1975, MIT Oral History Program (n8), p. 53.

21. Paul Berg, David Baltimore, Sydney Brenner, et al., "Asilomar Conference on Recombinant DNA Molecules," *Science 188* (June 6, 1975), 991–4.

22. Alex Capron, quoted in Lear (n12), *Recombinant DNA*, p. 139.

23. Andrew Lewis, quoted in Lear (n12), *Recombinant DNA*, p. 142.

24. Krimsky (n9), *Genetic Alchemy*, p. 106.

25. Ibid., pp. 156–7. Notable in his absence from the RAC was Paul Berg, who wanted to continue his recombinant DNA experiments and viewed this as a conflict of interest for him on the committee. Three other members involved in recombinant DNA work felt no such conflict.

26. Nicholas Wade, "Recombinant DNA: NIH Group Stirs Storm by Drafting Laxer Rules," *Science 190* (November 21, 1975), 767–9;

idem, "Recombinant DNA: NIH Sets Strict Rules to Launch New Technology," *Science 190* (December 19, 1975), 1175–7.

27. Regulatory approaches in different countries varied greatly with respect to scope, mechanisms of enforcement, and structure of decisionmaking. They are described in U.S. Congress, Office of Technology Assessment [hereafter cited as OTA], *Impacts of Applied Genetics*, Publ. No. OTA-HR-132 (Washington, D.C.: Government Printing Office, April 1981), pp. 322–8; also, Watson and Tooze (n4), *The DNA Story*, pp. 305–35.

28. Lear (n12), *Recombinant DNA*, p. 162. Clifford Grobstein, a respected scientist and commentator on the recombinant DNA policy process, describes the Asilomar meeting as a "nuts and bolts issue" for the people involved. The effort to construct a taxonomy of risks, says Grobstein, unavoidably reflected the personal concerns and biases of the meeting's participants, who "gave entirely different levels of importance to the various scenarios. In one sense they were formally considering... the kinds of problems that might occur, and in another sense they were concerned about the scenarios in terms of what their particular research was and how it would be affected... " (Clifford Grobstein, personal communication, June 3, 1985).

29. Genetic Engineering Group of Science for the People, "Open Letter to the Asilomar Conference on Hazards of Recombinant DNA," February 1975, reprinted in Watson and Tooze (n4), *The DNA Story*, p. 49.

30. Jonathan King, transcript of an interview by Rae Goodell, August 26, 1975, MIT Oral History Program (n8), p. 46.

31. Robert Sinsheimer, "Troubled Dawn for Genetic Engineering," *New Scientist 68* (October 16, 1975), 151.

32. Erwin Chargaff, "On the Dangers of Genetic Meddling" (letter to the editor), *Science 192* (June 4, 1976), 938, 940.

33. Boston Area Recombinant DNA Group, letter to DeWitt Stetten, Deputy Director for Science, NIH, November 24, 1975, reprinted in *Recombinant DNA Research*, Vol. 1: *Documents Relating to "NIH Guidelines for Research Involving Recombinant DNA Molecules"* (Washington, D.C.: Department of Health, Education and Welfare, August 1976) [hereafter cited as USDHEW–NIH, *Documents Relating to NIH Guidelines*], p. 356.

34. Richard Goldstein, letter to Donald Fredrickson, February 13, 1976, reprinted in USDHEW–NIH, *Documents Relating to NIH Guidelines* (n33), Vol. 1, p. 453.

35. Nicholas Wade, "Gene-Splicing: Critics of Research Get More Brickbats than Bouquets," *Science 195* (February 4, 1977), 466–9.

36. James Watson, quoted in Lear (n12), *Recombinant DNA*, p. 159.

37. *Genetic Engineering, 1975*, hearing before the Subcommittee on

Health of the Committee on Labor and Public Welfare, U.S. Senate, 94th Congress, 1st Session, April 22, 1975.

38. Senator Adlai Stevenson, "The Status of Recombinant DNA Research," *Congressional Record: Senate*, October 14, 1978, p. 37766.

39. See, for example, the testimony of Halsted Holman, in hearing on *Genetic Engineering, 1975* (n37), U.S. Senate, April 1975, pp. 13–18.

40. The coalition's 1977 Position Statement is reprinted in Watson and Tooze (n4), *The DNA Story*, pp. 207–8.

41. Testimony of Pamela Lippe, in *Recombinant DNA Research Act of 1977*, hearings before the Subcommittee on Health and the Environment of the Committee on Interstate and Foreign Commerce, U.S. House, 95th Congress, 1st Session, March 15, 16, and 17, 1977, p. 304.

42. Nicholas Wade, "Recombinant DNA at White House," *Science 193* (August 6, 1976), 468.

43. Testimony of Tony Mazzocchi, in hearings on *Recombinant DNA Research Act of 1977* (n41), U.S. House, March 1977, p. 467.

44. Janet L. Hopson, "Recombinant Lab for DNA and My 95 Days in It," *Smithsonian 8* (June 1977), 62.

45. Nicholas Wade, "Recombinant DNA: NIH Rules Broken in Insulin Gene Project," *Science 197* (September 30, 1977), 1342–4.

46. Nicholas Wade, "Harvard Gene Splicer Told to Halt," *Science 199* (January 6, 1978), 31; idem, "UCSD Gene Splicing Incident Ends Unresolved," *Science 209* (September 26, 1980), 1494–5.

47. "Statement on Recombinant DNA Reserach," Bishops' Committee for Human Values, National Conference of Catholic Bishops, May 2, 1977 (U.S. Catholic Conference, 1312 Massachusetts Avenue, N.W., Washington, D.C. 20005).

48. See Jeremy Rifkin, "Who Should Play God?" in *Recombinant DNA Regulation Act, 1977*, hearing before the Subcommittee on Health and Scientific Research of the Committee on Human Resources, U.S. Senate, 95th Congress, 1st Session, April 6, 1977, pp. 301–9.

49. Testimony of Jeremy Rifkin, in hearings on *Recombinant DNA Research Act of 1977* (n41), U.S. House, March 1977, p. 285.

50. Kennedy's bill was S. 1217 (Calendar No. 334, Report No. 95–359), 95th Congress, 1st Session, April 1, 1977. Rogers introduced a series of bills in the same session: H.R. 4759 (March 9, 1977), H.R. 6158 (April 6, 1977), H.R. 7418 (May 24, 1977), and H.R. 7897 (June 20, 1977). H.R. 7897, the most restrictive in terms of the freedom accorded local communities to set their own standards, ultimately received the most legislative support in the House. (These bills are all reprinted in USDHEW–NIH, *Documents Relating to NIH Guidelines* [n33], Vol. 2 [March 1978], pp. 543–740).

51. For contrasting perspectives on the issues involved in federal legislation, see Krimsky (n9), *Genetic Alchemy*, pp. 312–37, and Watson and Tooze (n4), *The DNA Story*, pp. 137–201.

52. "Philip Handler on Recombinant DNA Research" (editor's page), *Chemical and Engineering News* (May 9, 1977), p. 3.

53. Cities and towns besides Cambridge that passed laws regulating recombinant DNA research in 1977–8 were Emeryville and Berkeley (California), Princeton (New Jersey), and Amherst (Massachusetts).

54. The following discussion of events in Cambridge is based primarily on Krimsky (n9), *Genetic Alchemy*, pp. 298–307; Lear (n12), *Recombinant DNA*, pp. 152–8; and Watson and Tooze (n4), *The DNA Story*, pp. 91–135.

55. Alfred Vellucci, quoted in Dorothy Nelkin, "Threats and Promises: Negotiating the Control of Research," *Daedalus 107* (Spring 1978), 199.

56. Barbara Culliton, "Recombinant DNA Bills Derailed: Congress Still Trying to Pass a Law," *Science 199* (January 20, 1978), 274. Rae Goodell, in a detailed analysis of the Cambridge process, argues that the public's role in the controversy was significantly weakened by lack of technical expertise (Rae S. Goodell, "Public Involvement in the DNA Controversy: The Case of Cambridge, Massachusetts," *Harvard Newsletter on Science, Technology, and Human Values 27* [Spring 1979], 36–43).

57. Quoted in Krimsky (n9), *Genetic Alchemy*, p. 307.

58. Alfred E. Vellucci, transcript of an interview by Rae Goodell, May 9, 1977, MIT Oral History Program (n8), p. 4.

59. Alfred Vellucci, letter to Philip Handler, May 16, 1977; Philip Handler, letter to Alfred Vellucci, July 7, 1977, reprinted in Watson and Tooze (n4), *The DNA Story*, p. 206.

60. See, for example, James D. Watson, "An Imaginary Monster," *Bulletin of the Atomic Scientists 33* (May 1977), 12–13.

61. For an analysis and comparison of the various state and local laws that were enacted, see Sheldon Krimsky, "A Comparative View of State and Municipal Laws Regulating the Use of Recombinant DNA Molecules Technology," *Recombinant DNA Technical Bulletin 2* (November 1979), 121–5; also Nicholas Wade, "Gene-Splicing: At Grass-Roots Level a Hundred Flowers Bloom," *Science 195* (February 11, 1977), 558–60.

62. For a detailed description of these laws, see Sheldon Krimsky, Anne Baeck, and John Bolduc, *Municipal and State Recombinant DNA Laws: History and Assessment* (Medford, Mass.: Boston Neighborhood Network, June 1982). State legislation was passed by New York and Maryland.

63. Sherwood Gorbach, letter to Donald Fredrickson, July 14, 1977, reprinted in *Regulation of Recombinant DNA Research*, hearings before

the Subcommittee on Science, Technology and Space of the Committee on Commerce, Science and Transportation, U.S. Senate, 95th Congress, 1st Session, November, 2, 8, and 10, 1977, p. 133.

64.   James D. Watson, "In Defense of DNA," *New Republic 176* (June 25, 1977), 14.

65.   Watson (n60), "An Imaginary Monster," 12.

66.   Richard Goldstein, letter to Donald Fredrickson, August 30, 1977, reprinted in hearings on *Regulation of Recombinant DNA Research* (n63), U.S. Senate, November 1977, pp. 135–6, citing August 22, 1977 letter from Jonathan King and Richard Goldstein to Phage Workers, reprinted in ibid., pp. 48–9.

67.   See Bruce R. Levin, Associate Professor of Zoology, University of Massachusetts, letter to Donald Fredrickson, July 29, 1977, reprinted in hearings on *Regulation of Recombinant DNA Research* (n63), U.S. Senate, November 1977, pp. 131–2. Among the reasons for continuing concern was evidence reported at Falmouth that although *E. coli* K12 generally fails to colonize the intestines of laboratory animals because the normal intestinal flora is antagonistic to invaders, it did colonize the intestines of germ-free mice (mice raised in a sterile environment); moreover, once the K12 strain had adapted itself to the germ-free mice, it was then capable of colonizing normal mice. Another paper reported that the addition of virulence-associated plasmids to *E. coli* K12 greatly increased its lethal impact on baby chickens. And, of course, no amount of empirical investigation could reveal whether new genetically engineered strains of K12 might behave in unpredictable ways. See Krimsky (n9), *Genetic Alchemy*, pp. 215–32, for a detailed analysis of these and other points omitted from the Gorbach letter.

68.   The "Gorbach report" was published in an NIH quarterly newsletter and picked up by the *Washington Post* and *Science* (Lear [n12], *Recombinant DNA*, pp. 176–7).

69.   For a description of the various lobbying tactics employed, see Watson and Tooze (n4), *The DNA Story*, pp. 137–261; also Lear (n12), *Recombinant DNA*, pp. 167–214.

70.   Watson and Tooze (n4), *The DNA Story*, pp. 139–42.

71.   Nicholas Wade, "Gene Splicing: Senate Bill Draws Charges of Lysenkoism," *Science 197* (July 22, 1977), 348–9. Lysenko, an agricultural biologist in the Soviet Union, falsified data and argued for the inheritance of acquired characteristics. His views came to dominate aspects of Soviet biology.

72.   Memorandum from Robert Rosenzweig, Stanford University, to "DNA Fans," June 7, 1977.

73.   Norton Zinder, letter to "Bergetal," September 6, 1977, reprinted in Watson and Tooze (n4), *The DNA Story*, p. 259.

74. Pamela Lippe, transcript of an interview by Aaron Seidman, January 13, 1978, MIT Oral History Program (n8), p. 42.

75. Stanley Cohen, letter to Donald Fredrickson, September 6, 1977, reprinted in hearings on *Regulation of Recombinant DNA Research* (n63), U.S. Senate, November 1977, pp. 136–7.

76. Richard Novick, quoted in Krimsky (n9), *Genetic Alchemy*, p. 273.

77. Testimony of Jonathan King, in hearings on *Regulation of Recombinant DNA Research* (n63), U.S. Senate, November 1977, p. 44.

78. "Scientists and Environmentalists Accuse Pro-DNA Forces of Deception and Urge Strong Regulatory Legislation," press release from Friends of the Earth (Pamela Lippe), Sierra Club (Nancy Pfund), Environmental Defense Fund (Leslie Dach), photocopy, September 23, 1977.

79. James Watson, quoted in "Doomsday: Tinkering with Life," *Time* (April 18, 1977), p. 32.

80. Edward M. Kennedy, Remarks to American Medical Writers Association, New York City, September 27, 1977, reprinted in Watson and Tooze (n4), *The DNA Story*, pp. 173–4.

81. Norton Zinder, letter to "Bergetal," September 6, 1977, reprinted in Watson and Tooze (n4), *The DNA Story*, p. 259.

82. USDHEW–NIH, *Documents Relating to NIH Guidelines* (n33), Vol. 4 (December 1978), pp. 30–53.

83. See USDHEW–NIH, *Documents Relating to NIH Guidelines* (n33), Vol. 3 (September 1978), Vol. 3–Appendices (September 1978), and Vol. 4, for correspondence to NIH and minutes of meetings during this period.

84. Stanley Falkow to Donald Fredrickson, letter of April 19, 1978, reprinted in USDHEW–NIH, *Documents Relating to NIH Guidelines* (n33), Vol. 3: Appendix A, pp. 284–6.

85. The proceedings of the Falmouth Workshop, for example, still had not been published. See letter from Richard Hartzman to Donald Fredrickson, September 30, 1977, reprinted in USDHEW–NIH, *Documents Relating to NIH Guidelines* (n33), Vol. 3: Appendix A, pp. 4–5.

86. Quote is from James D. Watson, "DNA Folly Continues," *New Republic* (January 13, 1979), p. 15.

87. Krimsky (n9), *Genetic Alchemy*, p. 234.

88. Diana B. Dutton and John L. Hochheimer, "Institutional Biosafety Committees and Public Participation: Assessing an Experiment," *Nature 297* (May 6, 1982), 11–15.

89. Department of Health, Education and Welfare, National Institutes of Health, "U.S.–EMBO Workshop to Assess Risks for Recombinant DNA Experiments Involving the Genomes of Animal, Plant, and Insect Viruses," *Federal Register 43*, No. 63 (March 31, 1978), 13749.

90. See, for example, Falkow's letter to Fredrickson, April 19, 1978 (n84).

91. Eleanor Lawrence, "Guidelines Should Go, DNA Meeting Concludes," *Nature 278* (April 12, 1979), 590–1.

92. The proposal, authored by Wallace Rowe of National Institutes of Health and Allen Campbell of Stanford University, was first published in Department of Health, Education and Welfare, National Institutes of Health, "Recombinant DNA Research: Proposed Actions Under Guidelines," *Federal Register 44*, No. 73 (April 13, 1979), 22314–16. The 80–85 percent figure is from Richard Goldstein, letter to Donald Fredrickson, September 14, 1979, reprinted in USDHEW–NIH, *Documents Relating to NIH Guidelines* (n33), Vol. 5 (March 1980), p. 324.

93. Wallace Rowe, quoted in "Scientists Debate Safety of Research on *E. coli* Strain," *Nature 279* (May 31, 1979), 360.

94. Hardy W. Chan, Mark A. Israel, Claude F. Garon, et al., "Molecular Cloning of Polyoma Virus DNA in *Escherichia coli*: Lambda Phage Vector System," *Science 203* (March 2, 1979), 892. See also Mark A. Israel, Hardy W. Chan, Wallace P. Rowe, et al., "Molecular Cloning of Polyma Virus DNA in *Escherichia coli*: Plasmic Vector System," *Science 203* (March 2, 1979), 883–7; and Mark A. Israel, Hardy W. Chan, Malcolm A. Martin, et al., "Molecular Cloning of Polyoma Virus DNA in *Escherichia coli*: Oncogenicity Testing in Hamsters," *Science 205* (September 14, 1979), 1140–2.

95. This basic argument was anticipated in an earlier article published by King entitled "New Diseases in New Niches," *Nature 276* (November 2, 1978), 4–7. King and his colleague Ethan Signer summarized their views in correspondence to the *New York Times* and *Science*, although *Science* refused to publish their letter (see Krimsky [n9], *Genetic Alchemy*, p. 256). Sloan-Kettering scientist Barbara Rosenberg and Lee Simon, a Rutgers microbiologist, agreed with King's interpretation. The "main import" of the Rowe–Martin results, they wrote in *Nature*, "is to give new importance to the question of new routes of genetic access that could be provided by *E. coli* . . . Until this has been fully studied there is no basis for reassurance on the risks of recombinant DNA . . . The new data actually confirm several aspects of hazard mechanisms that have been postulated, but they do not address other aspects that must be investigated." Barbara Rosenberg and Lee Simon, "Recombinant DNA: Have Recent Experiments Assessed All the Risks?" *Nature 282* (December 20–7, 1979), 773–4. See also Stuart A. Newman, "Tumour Virus DNA: Hazards No Longer Speculative" (letter), *Nature 281* (September 20, 1979), 176.

96. B. P. Sagik and C. A. Sorber, "The Survival of Host–Vector Systems in Domestic Sewage Treatment Plants," *Recombinant*

DNA Technical Bulletin 2 (July 1979), 55–61. See Krimsky (n9), *Genetic Alchemy*, pp. 233–63, for an insightful analysis of the shifting perceptions of risks and the role of different risk-assessment studies.

97. Mark A. Chatigny, Melvin T. Hatch, H. Wolochow, et al., "Studies on Release and Survival of Biological Substances Used in Recombinant DNA Laboratory Procedures," *Recombinant DNA Technical Bulletin 2* (July 1979), 62–7.

98. Stuart B. Levy and Bonnie Marshall, "Survival of *E. coli* Host–Vector Systems in the Human Intestinal Tract," *Recombinant DNA Technical Bulletin 2* (July 1979), 77–80.

99. Paul S. Cohen, Robert W. Pilsucki, M. Lynn Myhal, et al., "Colonization Potentials of Male and Female *E. coli* K12 Strains, *E. coli* B and Human Fecal *E. coli* Strains in the Mouse GI Tract," *Recombinant DNA Technical Bulletin 2* (November 1979), 106–13.

100. One letter endorsing the exemption had 183 signatories. See letter to Donald Fredrickson from 183 signatories, June 14, 1979, reprinted in USDHEW–NIH, *Documents Relating to NIH Guidelines* (n33), Vol. 5, pp. 279–83.

101. Roy Curtiss, III, letter to Donald Fredrickson, October 4, 1979, reprinted in USDHEW–NIH, *Documents Relating to NIH Guidelines* (n33), Vol. 5, pp. 339–40.

102. National Institutes of Health, "Recombinant DNA Research: Proposed Actions Under Guidelines," *Federal Register 44*, No. 232 (November 30, 1979), 69234–51.

103. Marjorie Sun, "DNA Rules Kept to Head Off New Laws," *Science 215* (February 26, 1982), 1079–80.

104. James D. Watson, "Trying to Bury Asilomar" (editorial), *Clinical Research 26* (April 1978), 113–15.

105. See, for example, the articles by Frances E. Sharples, "Regulation of Products from Biotechnology," and Bernard D. Davis, "Bacterial Domestication: Underlying Assumptions," *Science 235* (March 13, 1987), 1329–35. In addition, some scientists are concerned about the safety of research involving oncogenes and other highly toxic agents. See Ditta Bartels, "Occupational Hazards in Oncogene Research," *Genewatch 1* (September–December 1984), 6–8; Marcel Blanc, "Has Genetic Manipulation Caused Two Deaths at the Pasteur Institute?" *Genewatch 4* (March–April 1987), 1, 8. More generally, see OTA (n27), *Impacts of Applied Genetics*.

106. Statement of Philip Handler in hearings on *Regulation of Recombinant DNA Research* (n63), U.S. Senate, November 1977, pp. 13–14. See also "A Commercial Debut for DNA Technology," *Business Week* (December 12, 1977), pp. 128, 132.

107. David Perlman, "Scientific Announcements" (letter to the editor), *Science 198* (November 25, 1977), 782.

108. Cetus Corporation, "Background Material: Special Report," October 1975, pp. 6–7, photocopy.

109. Ronald Cape, Cetus Corporation, interview by Nancy Pfund and Joel Gurin, April 17, 1980. See, for example, Hal Lancaster, "Most Brokers Shun Research on Gene Firms," *Wall Street Journal*, December 29, 1980, pp. 1, 9, 13.

110. Nancy Pfund and Laura Hofstadter, "Biomedical Innovation and the Press," *Journal of Communication 31* (Spring 1981), 138–54.

111. Nicholas Wade, "Gene Splicing Company Wows Wall Street," *Science 210* (October 31, 1980), 506–7.

112. Sheldon Krimsky, "Corporate Academic Ties in Biotechnology," *Genewatch 1* (September–December 1984), 3–5.

113. "The Potential of Gene Splicing," *Business Week* (November 10, 1980), p. 89. For a description of major industries' investments in smaller biotechnology firms, see "Big Firms Gain Profitable Foothold in New Gene-Splicing Technologies," *Wall Street Journal*, November 5, 1980, p. 31.

114. *Diamond v. Chakrabarty*, Supreme Court of the United States, 100 S. Ct. 2204 (1980).

115. Stephen Feldman, quoted in Mitchel Zoler, "Chakrabarty Case Still Debatable," *Genetic Engineering News 1* (May/June 1981), 1.

116. See, for example, Panel of Bioethical Concerns, National Council of the Churches of Christ/USA, *Genetic Engineering: Social and Ethical Consequences* (New York: Pilgrim Press, 1984). Also Charles Austin, "Ethics of Gene Splicing Troubling Theologians," *New York Times*, July 5, 1981, pp. 1, 12. In 1985, the U.S. Patent and Trademark Office ruled that genetically engineered plants, seeds, and tissue cultures could also be patented. See Marjorie Sun, "Plants Can be Patented Now," *Science 230* (October 18, 1985), 303.

117. Dr. Claire Randall, Rabbi Bernard Mandelbaum, and Bishop Thomas Kelly, "Message from Three General Secretaries," National Council of the Churches of Christ (110 Maryland Avenue, N.E., Washington, D.C. 20002), July 1980.

118. President's Commission for the Study of Ethical Problems in Medicine and Biomedical and Behavioral Research, *Splicing Life: A Report on the Social and Ethical Issues of Genetic Engineering with Human Beings* (Washington, D.C.: Government Printing Office, November 1982), p. 4.

119. See, for example, Brief on Behalf of the Peoples' Business Commission as Amicus Curiae (*Diamond v. Chakrabarty*, 100 S. Ct. 2204 [1980]), pp. 6–13. For a more recent discussion, see Jean-Pierre Berland and Richard Lewontin, "Breeders' Rights and Patting Life Forms," *Nature 322* (August 28, 1986), 785–8. On fears about the inhibition of academic communication, see "Fresh Debate over the Life-Form Ruling," *Chemical Week* (August 6, 1980),

pp. 47–8. See also OTA (n27), *Impacts of Applied Genetics*, pp. 237–54; and Robert F. Acker and Moselio Schaechter, eds., *Patentability of Microorganisms: Issues and Questions* (Washington, D.C.: American Society for Microbiology, 1981).

120. OTA (n27), *Impacts of Applied Genetics*, p. 154.

121. John C. Fletcher, chief of the bioethics program at the NIH, quoted in Keith Schneider, "Patenting Life," *New York Times*, April 18, 1987, p. 6. See also Keith Schneider, "Bill Seeks a Delay on Patents for Animals," *New York Times*, August 6, 1987, p. 13.

122. Paul Berg, quoted in Sharon Begley and Pamela Abramson, "The DNA Industry," *Newsweek* (August 20, 1979), p. 53.

123. See Leslie Roberts, "Who Owns the Human Genome?" *Science 237* (July 24, 1987), 358–61.

124. The preceding account is based on Nicholas Wade, "University and Drug Firm Battle over Billion-Dollar Gene," *Science 209* (September 26, 1980), 1492–4.

125. Barbara J. Culliton, "Drug Firm and UC Settle Interferon Suit," *Science 219* (January 28, 1983), 372. No claims were made in this case for the rights of the leukemia victim from whom the cells had been taken, although in 1984 another leukemia patient sued Golde for a share in the profits from a patented cell line derived from his spleen. See Barbara J. Culliton, "Patient Sues UCLA over Patent on Cell Line," *Science 225* (September 28, 1984), 1458. The issue of patients' rights to share in the royalties from patents on their own cells has continued to generate discussion. See Marjorie Sun, "Who Should Have Rights to a Patient's Cells?" *Science 231* (February 7, 1986), 543–4.

126. Quoted in Barbara J. Culliton, "Biomedical Research Enters the Marketplace," *New England Journal of Medicine 304* (May 14, 1981), 1199.

127. Woodland Hastings, quoted in Nicholas Wade, "Gene Goldrush Splits Harvard, Worries Brokers," *Science 210* (November 21, 1980), 878.

128. Otto Solbrig, quoted in ibid., 878.

129. Philip J. Hilts, "Ivy-Covered Capitalism," *Washington Post*, November 10, 1980, pp. A1, 30, 32.

130. "Harvard, Inc.?" *Washington Star*, November 11, 1980.

131. Alfred Vellucci, quoted in Joel Gurin and Nancy E. Pfund, "Bonanza in the Bio Lab," *Nation* (November 22, 1980), 529.

132. For a description of the various types of industry–university collaboration, see OTA, *Commercial Biotechnology: An International Analysis*, Publ. No. OTA-BA–218 (Washington, D.C.: Government Printing Office, January 1984), pp. 411–31. See also the seven-part series of articles entitled, "The Academic–Industrial Complex," *Science 216* (May 1982–January 1983); Sandra Blakeslee, "Another Joint Venture," *Nature 313* (January 24, 1985), 261;

and John Walsh, "New R&D Centers Will Test University Ties," *Science 227* (January 11, 1985), 150–2.

133. David Blumenthal, Michael Gluck, Karen Seashore Louis, et al., "Industrial Support of University Research in Biotechnology," *Science 231* (January 17, 1986), 242–6.

134. Jonathan King, quoted in Colin Norman, "MIT Agonizes over Links with Research Unit," *Science 214* (October 23, 1981), 416.

135. David F. Noble, "The Selling of the University," *Nation* (February 6, 1982), 144.

136. Colin Norman, "Whitehead Link Approved," *Science 214* (December 4, 1981), 1104.

137. The faculty member was Dr. Ray Valentine of the University of California at Davis, whose ties to Calgene, a biotechnology company he started while at Davis, led to considerable controversy. See William Boly, "The Gene Merchants," *California* (September 1982), pp. 76–9, 170–9; also the testimony of Albert Meyerhoff, Natural Resources Defense Council, and Charles Hess, Dean of College of Agricultural and Environmental Sciences, University of California at Davis, in *University/Industry Cooperation in Biotechnology*, hearings before the Subcommittee on Investigations and Oversight and the Subcommittee on Science, Research and Technology of the Committee on Science and Technology, U.S. House, 97th Congress, 2nd Session, June 16 and 17, 1982, pp. 48–89.

138. Reply Brief for the Petitioner (*Diamond v. Chakrabarty*, 100 S. Ct. 2204 [1980]), p. 7.

139. Testimony of Donald Kennedy, in *Commercialization of Academic Biomedical Research*, hearings before the Subcommittee on Investigations and Oversight and the Subcommittee on Science, Research and Technology of the Committee on Science and Technology, U.S. House, 97th Congress, 1st Session, June 8 and 9, 1981, p. 8. See also the incidents described by Jonathan King in the same hearings, pp. 61–76.

140. Donald Kennedy, quoted in Barbara J. Culliton, "Pajaro Dunes: The Search for Consensus," *Science 216* (April 9, 1982), 155.

141. Leon Wofsy, "Biology and the University on the Market Place: What's for Sale?" transcript of a lecture given at the University of California, Berkeley, March 16, 1982, p. 2. See also David Dickson, *The New Politics of Science* (New York: Pantheon, 1984).

142. Robert Luciano, quoted in Mariann Hansen, "A Special Union: Science, Business Join Forces at DNAX Center," *Palo Alto Times Tribune*, June 21, 1985, pp. E–6, E–8.

143. Barbara J. Culliton, "Academe and Industry Debate Partnership," *Science 219* (January 14, 1983), 150–1.

144. Critics' arguments are described in Lee Randolph Bean, "Entre-

preneurial Science and the University," *Hastings Center Report 12* (October 1982), 5–9.

145. Robert K. Merton, *The Sociology of Science* (Chicago: University of Chicago Press, 1973), pp. 273–4.

146. Stanley Cohen, quoted in Gurin and Pfund (n131), "Bonanza in the Bio Lab," p. 544.

147. Marjorie Sun, "Mixed Signals on Stanford Biotech Patent," *Science 228* (April 26, 1985), 478.

148. For a fuller discussion of these arguments, see Nicholas Wade, Background Paper for the Report of the Twentieth Century Fund Task Force on the Commercialization of Scientific Research, *The Science Business* (New York: Priority Press, 1984), pp. 17–84.

149. "Biotech Comes of Age," *Business Week* (January 23, 1984), pp. 84–94.

150. Krimsky (n112), "Corporate Academic Ties in Biotechnology," p. 4.

151. Mark Crawford, "Biotech Market Changing Rapidly," *Science 231* (January 3, 1986), 12–14. See also Marjorie Sun, "Hot Market for Biotech Stocks in 1986," *Science 233* (August 1, 1986), 516–17.

152. Don Clark, "First Genetic Vaccine OKd by FDA – For Hepatitis B," *San Francisco Chronicle*, July 24, 1986, pp. 1, 20; "U.S. to Approve Interferon for a Rare Cancer," *New York Times*, June 5, 1986, p. 19; Harold M. Schmeck, Jr., "Gene-Spliced Hormone for Growth Is Cleared," *New York Times*, October 19, 1985, p. 8.

153. For the "value-added" explanation, see Roger E. Shamel, "Biotechnology: The New Growth Industry," *USA Today* (March 1985), 38–41.

154. Emily A. Arakaki, "A Study of the U.S. Competitive Position in Biotechnology," in Department of Commerce, International Trade Administration, *High Technology Industries: Profiles and Outlooks–Biotechnology* (Washington, D.C.: Government Printing Office, July 1984), p. 47.

155. Crawford (n151), "Biotech Market Changing Rapidly."

156. OTA (n132), *Commercial Biotechnology*.

157. Ibid., Summary, p. 19.

158. Ibid., pp. 307–28.

159. Arakaki (n154), "A Study of the U.S. Competitive Position in Biotechnology," p. 152.

160. Colin Norman, "Business to Boost R&D," *Science 222* (December 9, 1983), 1103.

161. George A. Keyworth, II, "Four Years of Reagan Science Policy: Notable Shifts in Priorities," *Science 224* (April 6, 1984), 9.

162. Dr. Bernadine Healy, director of the Cabinet Work Group on Biotechnology, quoted in Martin Kenney, "Biotech Conference

Targets Regulatory Questions," *Genetic Engineering News* 5 (June 1985), 3, 12.

163. The initial proposal was issued in 1984 and a final plan adopted in 1986. See Office of Science and Technology Policy, "Proposal for a Coordinated Framework for Regulation of Biotechnology; Notice," *Federal Register 49*, No. 252 (December 31, 1984), 50856–907; and Office of Science and Technology Policy, "Coordinated Framework for Regulation of Biotechnology; Announcement of Policy and Notice for Public Comment," *Federal Register 51*, No. 123 (June 26, 1986), 23302–93.

164. Marjorie Sun, "Biotech Guidelines Challenged by Rifkin," *Science 233* (August 1, 1986), 516; Mark Crawford, "Court Rejects Rifkin in Biotech Cases," *Science 235* (January 9, 1987), 159.

165. Representative John Dingell, quoted in Roger S. Johnson, "Biotech Regulations Debated by Congressional Committee," *Genetic Engineering News* 5 (January 1985), 1, 48. A House Committee on Science and Technology subcommittee also held a series of hearings in 1985 and 1986 focusing on weakness in the regulation of biotechnology. See *Issues in the Federal Regulation of Biotechnology: From Research to Release*, Report of the Subcommittee on Investigations and Oversight, Committee on Science and Technology, U.S. House, 99th Congress, 2nd Session (Washington, D.C.: Government Printing Office, December 1986).

166. "Gene-Spliced Drug for Heart Attacks," *San Francisco Chronicle*, November 14, 1984, p. 4; Harold M. Schmeck, Jr., "The New Age of Vaccines," *New York Times Magazine*, April 29, 1984, pp. 58–9, 81–7; David Perlman, "Scientists Isolate a Gene That Controls Blood Pressure," *San Francisco Chronicle*, June 22, 1984, p. 4. For a comprehensive summary of the many possible applications of biotechnology in the pharmaceutical, agricultural, chemical, energy, and electronics industries, see OTA (n132), *Commercial Biotechnology*, pp. 119–257.

167. Nina Kramer, "Companies Producing Interferon: Who's Doing What," *Genetic Engineering News 2* (January/February 1982), 17–19.

168. For a detailed description of applications of biotechnology to animal breeding, see OTA (n27), *Impacts of Applied Genetics*, pp. 167–92. The less stringent regulation of veterinary products has allowed relatively rapid commercialization; the very first product of biotechnology to reach the market in 1982 was a vaccine to prevent diarrhea in pigs (Casper Schuuring, "New Era Vaccine," *Nature 296* [April 29, 1982], 792).

169. OTA, "Technology, Public Policy, and the Changing Structure of American Agriculture," Publ. No. OTA-F-285 (Washington, D.C.: Government Printing Office, March 1986). See also Edward

Yoxen, *The Gene Business* (London: Pan Books Limited, 1983), pp. 146–7.

170. For a critical analysis of alternative uses of biotechnology in the chemical industry, see Yoxen (n169), *The Gene Business*, pp. 183–210. Also, Dick Schneider, "Biomass Fuels Versus Soil Fertility: Is There a Middle Ground?" Bay Area Sierra Club *Yodeler* (February 1982), p. 5; Constance Holden, "Is Bioenergy Stalled?" *Science 227* (March 1, 1985), 1018.

171. David Baltimore, quoted in "The Ties That Bind or Benefit," *Nature 283* (January 10, 1980), 130–1.

172. Walter Gilbert, quoted in Marc Lappé, *Broken Code: The Exploitation of DNA* (San Francisco: Sierra Club Books, 1984), p. 273.

173. Testimony of Jonathan King, in hearings on *Commercialization of Academic Biomedical Research* (n139), U.S. House, June 1981, p. 64.

174. Testimony of Paul Gray, in hearings on *Commercialization of Academic Biomedical Research* (n139), U.S. House, June 1981, p. 30.

175. See National Science Foundation, Productivity Improvement Research Section, Division of Industrial Science and Technological Innovation, *Cooperative Science: A National Study of University and Industry Researchers*, Vols. I and II, Publ. No. NSF–84–39A,B (Washington, D.C.: Government Printing Office, November 1984). Even scientists who expressed discomfort with commercial ties in earlier years, such as Paul Berg (see, for example, Charles Petit, "The Bold Entrepreneurs of Gene Engineering," *San Francisco Chronicle*, December 2, 1977, p. 2), now endorse such arrangements.

176. World Health Organization (Geneva, 1982), cited in Lappé (n172), *Broken Code*, pp. 80–1.

177. Representative John D. Dingell, "Biotechnology: What Are the Problems Beyond Regulation?" transcript of statement before the Brookings Institution Conference, January 15, 1985, p. 7.

178. Quote is by David Martin, in Lappé (n172), *Broken Code*, p. 250. See also Eliot Marshall, "Genentech Bows Out of NYU's Malaria Project," *Science 220* (April 29, 1983), 485; Eliot Marshall, "NYU's Malaria Vaccine: Orphan at Birth?" *Science 219* (February 4, 1983), 466–7. On Hoffman–La Roche's role, see John Walsh, "Human Trials Begin for Malaria Vaccine," *Science 235* (March 13, 1987), 1319–20.

179. For a critical analysis of this and related problems, see Jack Doyle, *Altered Harvest: Agriculture, Genetics, and the Fate of the World's Food Supply* (New York: Viking, 1985). Also, Kathleen Selvaggio, "Cashing in on DNA," *Multinational Monitor 5* (March 1984), 13–15, 20.

180. James Brooke, "Scientists Seek to Foment Green Revolution in Africa," *New York Times*, January 18, 1987, p. E30.

181. Testimony of Ethan Signer, in hearings on *Recombinant DNA Research Act of 1977* (n41), U.S. House, March 1977, p. 79.

182. This attempt is discussed more fully in Chapter 7 ("Risks and Rights"). Also, see Nicholas Wade, "Gene Therapy Caught in More Entanglements," *Science 212* (April 3, 1981), 24–5; Marjorie Sun, "Cline Loses Two NIH Grants," *Science 214* (December 11, 1981), 1220.

183. Jean L. Marx, "Gene Therapy – So Near and Yet So Far Away," *Science 232* (May 16, 1986), 824–5.

184. Working Group on Human Gene Therapy, National Institutes of Health, Department of Health and Human Services, "Points to Consider in the Design and Submission of Human Somatic-Cell Gene Therapy Protocols," in *Federal Register 50* (January 22, 1985), 2940–5. NIH's receipt of grant applications is reported in OTA, *Human Gene Therapy, Background Paper*, Publ. No. OTA-BP-BA-32 (Washington, D.C.: Government Printing Office, December 1984), p. 2.

185. National Science Board, National Science Foundation, *Science Indicators 1980* (Washington, D.C.: Government Printing Office, 1981), p. 166; Tabitha M. Powledge, "Public Says Genetic Engineers Should Proceed Cautiously," *Bio/Technology* (October 1983), 645–6.

186. See OTA, *New Developments in Biotechnology – Background Paper: Public Perceptions of Biotechnology*, Publ. No. OTA-BP-BA-45 (Washington, D.C.: Government Printing Office, May 1987), p. 71. People from all walks of life react more negatively to human genetic engineering than to any other foreseeable application. See Jon D. Miller, *The Attitudes of Religious, Environmental and Science Policy Leaders Toward Biotechnology*, Public Opinion Laboratory, Northern Illinois University (April, 1985).

187. Panel of Bioethical Concerns (n116), *Genetic Engineering*, p. 66. At least one religious group, however, has taken a stand affirming genetic research on humans, particularly "those who by reason of their genetic inheritance face early death, prolonged pain and suffering or great physical or intellectual disadvantages." So stated a resolution passed in 1984 by the Connecticut Conference of the United Church of Christ (the Congregational church). The resolution also declared, "we believe that God is working God's purpose out in enabling people to treat and prevent disease." See Barbara J. Culliton, "Connecticut Church Passes Genetics Resolution," *Science 226* (November 9, 1984), 674.

188. Representative Albert Gore, in *Human Genetic Engineering*, hearings before the Subcommittee on Investigations and Oversight of the Committee on Science and Technology, U.S. House, 97th Congress, 2nd Session, November 16, 17, and 18, 1982, p. 182.

189. "Congress' 1985 Agenda," *Genewatch 1* (September–December 1984), 10–13.
190. Colin Norman, "Clerics Urge Ban on Altering Germline Cells," *Science 220* (June 24, 1983), 1360–1. The resolution was reprinted in the *Congressional Record: Senate*, June 10, 1983, pp. S8203–5.
191. OTA (n184), *Human Gene Therapy*, p. iii. See pp. 6–7 concerning the possibility of germline therapy at some point in the future. The rationale for germline therapy is illustrated in the testimony of Dr. Theodore Friedmann, in hearings on *Human Genetic Engineering* (n188), U.S. House, November 1982, p. 277.
192. Jeffrey G. Williams, "Mouse and Supermouse," *Nature 300* (December 16, 1982), 575; Richard D. Palmiter, Ralph L. Brinster, Robert E. Hammer, et al., "Dramatic Growth of Mice That Develop from Eggs Microinjected with Metallothionein: Growth Hormone Fusion Genes," *Nature 300* (December 16, 1982), 611–15. Researchers have also inserted functioning human growth genes into rabbits, pigs, and sheep, although, as of mid-1985, these experiments had not produced unusual growth in the animals ("Livestock Given Human Gene for First Time," *New York Times*, June 27, 1985, p. 15).
193. Jeremy Rifkin, quoted in Michael Bowker, "The Hawkers of Heredity," *Sierra Club Bulletin* (January/February 1985), p. 27. See also Jeffrey L. Fox, "Rifkin Takes Aim at USDA Animal Research," *Science 226* (October 19, 1984), 321; Harold M. Schmeck, Jr., "Bid to Ban Transfers of Genes Is Rejected by Panel," *New York Times*, October 30, 1984, pp. 19–20.
194. These and other examples of ecological disruption are discussed in Subcommittee on Investigations and Oversight, Committee on Science and Technology, U.S. House, Staff Report on *The Environmental Implications of Genetic Engineering*, 98th Congress, 2nd Session (Washington, D.C.: Government Printing Office, February 1984). See also *Environmental Implications of Genetic Engineering*, hearings before the Subcommittee on Investigations and Oversight and the Subcommittee on Science, Research and Technology of the Committee on Science and Technology, U.S. House, 98th Congress, 1st Session, June 22, 1983 (especially Appendix A).
195. Jeremy Rifkin, quoted in David Perlman, "An Attempt to Block Gene-Splicing Tests," *San Francisco Chronicle*, April 13, 1984, p. 6.
196. Judge J. Skelly Wright, quoted in Marjorie Sun, "Rifkin and NIH Win in Court Ruling," *Science 227* (March 15, 1985), 1321.
197. Marjorie Sun, "NIH Bows to Part of Rifkin Suit," *Science 226* (November 30, 1984), 1508.
198. Marjorie Sun, "Field Test of Altered Microbe Still in Limbo,"

Science 232 (June 13, 1986), 1340; "UC Tries to Salvage 'Ice-Minus' Test," *San Francisco Chronicle*, May 28, 1987, p. 4.

199. Marjorie Sun, "Local Opposition Halts Biotechnology Test," *Science 231* (February 14, 1986), 667–8.

200. Anonymous, quoted in Philip J. Hilts, "Bacteria Tested Outdoors Without U.S. Approval," *Washington Post*, February 26, 1986, p. A3.

201. Steven Schatzow, Director of the Office of Pesticide Programs, quoted in Keith Schneider, "E.P.A. to Move in Tests of a Genetic Chemical," *New York Times*, March 22, 1986, p. 10.

202. Jeremy Rifkin, quoted in Keith Schneider, "E.P.A. Faults Company for Outdoor Use of Gene-Altered Agent," *New York Times*, February 27, 1986, p. 10.

203. Jeremy Rifkin, quoted in Philip Shabecoff, "Fine Cut in Test of Gene Altering," *New York Times*, June 7, 1986, p. 7.

204. Anonymous, quoted in Keith Schneider, "U.S. Quietly Approved the Sale of Genetically Altered Vaccine," *New York Times*, April 4, 1986, pp. 1, 9.

205. Keith Schneider, "Release of a Gene-Altered Virus Is Halted by U.S. After Challenge," *New York Times*, April 9, 1986, pp. 1, 11; idem, "U.S. Ends Curb on a Vaccine Using Altered Virus," *New York Times*, April 23, 1986, pp. 1, 7.

206. "A Novel Strain of Recklessness" (editorial), *New York Times*, April 6, 1986, p. 22.

207. Mark Crawford, "Regulatory Tangle Snarls Agricultural Research in the Biotechnology Arena," *Science 234* (October 17, 1986), 275–7. Also, *Issues in the Federal Regulation of Biotechnology* (n165), U.S. House, December 1986.

208. Mark Crawford, "USDA Research Rules Killed; NIH Panel To Rewrite Standards," *Science 234* (November 7, 1986), 668.

209. See R. Jeffrey Smith, "The Dark Side of Biotechnology," *Science 224* (June 15, 1984), 1215–16 for a description of a 1984 symposium on this topic. Also, Susan Wright and Robert L. Sinsheimer, "Recombinant DNA and Biological Warfare," *Bulletin of the Atomic Scientists 39* (November 1983), 20–6.

210. Marjorie Sun, "NIH Sees No Need for DNA Weapons Ban," *Science 217* (July 9, 1982), 135; Elizabeth Milewski, "RAC Discussion on the Construction of Biological Weapons," *Recombinant DNA Technical Bulletin 5* (December 1982), 188–91. Letters in support of the proposal are reprinted in USDHEW–NIH, *Documents Relating to NIH Guidelines*, Vol. 7 (December 1982), pp. 775–826.

211. "Rifkin Takes on RAC, Wins Some, Loses Some; Toxin-Cloning Gets Okay," *McGraw-Hill's Biotechnology Newswatch 4* (February 20, 1984), 2–3; R. Jeffrey Smith, "Army Agrees to New Study of Biowarfare Laboratory," *Science 227* (February 8, 1985), 614.

212. For a comparison of the U.S. semiconductor industry and bio-technology, see OTA (n132), *Commercial Biotechnology*, pp. 531–41.

213. See, for example, Sheldon Krimsky, "Social Responsibility in an Age of Synthetic Biology," *Environment* 24 (July/August 1982), 2–11. At the international level, the United Nations has been trying for several years to create an international center for research and training in biotechnology directed specifically at the needs of the Third World, but progress has been hampered by lukewarm support from industrial nations and intense controversy among developing countries about the location of the proposed center. See David Dickson, "UNIDO Hopes for Biotechnology Center," *Science* 221 (September 30, 1983), 1351–3; "International Biotechnology Center," *Genewatch* 1 (May–August 1984), 6–7.

214. Dingell (n177), "Biotechnology: What Are the Problems Beyond Regulation?", p. 11.

7. Risks and rights

1. Amos Tversky and Daniel Kahneman, "The Framing of Decisions and the Psychology of Choice," *Science* 211 (January 30, 1981), 453–8.

2. See generally, Paul Slovic, Baruch Fischhoff, and Sarah Lichtenstein, "Rating the Risks," *Environment* 21 (April 1979), 14–39; Paul Slovic, Baruch Fischhoff, and Sarah Lichtenstein, "Perceived Risk: Psychological Factors and Social Implications," in F. Warner and D. H. Slater, eds., *The Assessment and Perception of Risk* (London: The Royal Society, 1981); Paul Slovic, Baruch Fischhoff, and Sarah Lichtenstein, "Facts and Fears: Understanding Perceived Risk," in R. C. Schwing and W. A. Albers, Jr., eds., *Societal Risk Assessment: How Safe Is Safe Enough?* (New York: Plenum Press, 1980).

3. The phenomenon of professional optimism has been convincingly demonstrated in analyses of published reports of clinical trials. The publications were classified according to the quality of the experiment design: well designed, controlled studies on the one hand; poorly designed, uncontrolled studies on the other. The analyses showed that the investigator's enthusiasm for the procedure being evaluated was inversely related to the quality of the experimental design. Poorly designed studies almost always were interpreted as more favorable to the treatment than were well-controlled studies (John P. Gilbert, Bucknam McPeek, and Frederick Mosteller, "Statistics and Ethics in Surgery and Anesthesia," *Science* 198 [November 18, 1977], 684–9; John P. Gilbert, Bucknam McPeek, and Frederick Mosteller, "Progress in Surgery and Anesthesia: Benefits and Risks of Innovative Therapy," in John P. Bunker, Benjamin

A. Barnes, and Frederick Mosteller, eds., *Costs, Risks, and Benefits of Surgery* [New York: Oxford University Press, 1977], pp. 124–69).

4. David M. Eddy, "Clinical Policies and the Quality of Clinical Practice," *New England Journal of Medicine 307* (August 5, 1982), 343–7.

5. Louis Harris Poll, 1980. For the first survey, see Martin Baron, "$100,000 Survey Says U.S. Public Will Take Risks for the Good Life," *Los Angeles Times*, June 28, 1980, p. 1.

6. "Surgical Spectacular" (editorial), *Nation* (May 9, 1966), 540–1.

7. Milton Silverman and Philip R. Lee, *Pills, Profits and Politics* (Berkeley: University of California Press, 1974), pp. 262–4.

8. Knight Steel, Paul M. Gertman, Caroline Crescenzi, et al., "Iatrogenic Illness on a General Medical Service at a University Hospital," *New England Journal of Medicine 304* (March 12, 1981), 638–42.

9. Charlotte Muller, "The Overmedicated Society: Forces in the Marketplace for Medical Care," *Science 176* (May 5, 1972), 488–92; John H. Lavin, "What's Your Role in 'The Overmedicated Society'?" *Medical Economics* (October 13, 1980), 191–203; Silverman and Lee (see n7), *Pills, Profits, and Politics*. Ivan Illich presents a wide-ranging critique of Western medicine in *Medical Nemesis: The Expropriation of Health* (New York: Pantheon, 1976).

10. Theodore Cooper, "Untoward Outcomes of Health Care: Who Is Liable?" *Public Health Reports 91* (November–December 1976), 492.

11. V. W. J. Gunn and N. H. Golding, "The Swine Flu Immunization Program: National Opinion Surveys and Media Coverage of Events," unpublished manuscript, no date. On the failure of the informed consent forms to shield the federal government from liability, see Peter A. Pavarini, "Registration Is Not Informed Consent: The Swine Flu Program Revisited," *Public Health Reports 96* (May–June 1981), 287–8.

12. See, for example, Elizabeth F. Loftus and James F. Fries, "Informed Consent May Be Hazardous to Health," *Science 204* (April 6, 1979), 11. Also Jay Katz, "Why Doctors Don't Disclose Uncertainty," *Hastings Center Report 14* (February 1984), 35–44.

13. Renée C. Fox, "Training for Uncertainty," in Robert K. Merton, George G. Reader, and Patricia L. Kendall, eds., *The Student-Physician: Introductory Studies in the Sociology of Medical Education* (Cambridge, Mass.: Harvard University Press, 1969), pp. 207–41; President's Commission for the Study of Ethical Problems in Medicine and Biomedical and Behavioral Research [hereafter cited as President's Commission], *Making Health Care Decisions: The Ethical and Legal Implications of Informed Consent in the Patient–Practitioner Relationship*, Vol. 1: *Report* (Washington, D.C.: Gov-

ernment Printing Office, October 1982), pp. 85–9; also, Fred Davis, "Uncertainty in Medical Prognosis, Clinical and Functional," in Eliot Freidson and Judith Lorber, eds., *Medical Men and Their Work* (Chicago: Aldine Atherton, 1972), pp. 239–48.

14. Irving L. Janis, *Psychological Stress: Psychoanalytic and Behavioral Studies of Surgical Patients* (New York: John Wiley, 1958).

15. Ruth R. Faden, Catherine Becker, Carol Lewis, et al., "Disclosure of Information to Patients in Medical Care," *Medical Care 19* (July 1981), 718–33; President's Commission (n13), *Making Health Care Decisions*, p. 75.

16. Quoted in Jay Katz, "Disclosure and Consent: In Search of Their Roots," in Aubrey Milunsky and George Annas, eds., *Genetics and the Law II* (New York: Plenum Press, 1980), p. 124.

17. Charles W. Lidz, Alan Meisel, Marian Osterweis, et al., "Barriers to Informed Consent," *Annals of Internal Medicine 99* (October 1983), 539–43.

18. Alexander M. Capron, "Informed Consent in Catastrophic Disease Research and Treatment," *University of Pennsylvania Law Review 123* (1974), 340–438.

19. Jay Katz, "Informed Consent: A Fairy Tale? Law's Vision," *University of Pittsburgh Law Review 39* (Winter 1977), 137–74.

20. *Salgo v. Leland Stanford Jr. University Board of Trustees*, 154 Cal. App. 2d 560,578, 317 P.2d 170,181 (1957); *Natanson v. Kline*, 187 Kan. 186, 354 P.2d 670 (1960); *Canterbury v. Spence*, 464 F.2d 772 (D.C. Cir. 1972). For a discussion of these cases, see Sidney A. Shapiro, "Limiting Physician Freedom to Prescribe a Drug for Any Purpose: The Need for FDA Regulation," *Northwestern University Law Review 73* (December 1978), 801–72.

21. *Wilkinson v. Vesey*, 110 R.I. 606,627, 295 A.2d 676,689 (1972). For a list of courts following the reasonable-man standard, see Leonard L. Riskin, "Informed Consent: Looking for the Action," *Law Forum 1975*, No. 4 (1975), 580–611.

22. *Cobbs v. Grant*, 8 Cal. 3d 229, 502 P.2d 1, 104 Cal. Rptr. 505 (1972).

23. President's Commission (n13), *Making Health Care Decisions*, p. 23.

24. Dennis H. Novack, Robin Plumer, Raymond L. Smith, et al., "Changes in Physicians' Attitudes Toward Telling the Cancer Patient," *Journal of the American Medical Association 241* (March 2, 1979), 897–900. See also F. J. Ingelfinger, "Informed (But Uneducated) Consent," *New England Journal of Medicine 287* (August 31, 1972), 465–6.

25. Katz (n16), "Disclosure and Consent," p. 122.

26. Glen D. Mellinger, "Progress Report: Public Judgments About Ethical Issues in Research," Grant No. MH 27337–03, National Institute of Mental Health, National Institute on Alcohol Abuse

and Alcoholism, and National Institute on Drug Abuse, October 1978.

27. Interview with Phyllis Wetherill by Diana Dutton, November 7, 1981. See also, Marlene Cimons, "DES Suit Highlights Victims' Plight," *Los Angeles Times*, May 5, 1977, Part IV, pp. 1, 8–9.

28. See generally Jay Katz, *Experimentation with Human Beings* (New York: Russell Sage Foundation, 1972), pp. 9–65. Also Bernard Barber, "The Ethics of Experimentation with Human Subjects," *Scientific American 234* (February 1976), 25–31; and Henry K. Beecher, "Ethics and Clinical Research," *New England Journal of Medicine 274* (June 16, 1966), 1354–60.

29. "Requirements for Review to Insure the Rights and Welfare of Individuals: Public Health Service Policy and Procedure No. 129, Revised July 1, 1966," in Henry K. Beecher, *Research and the Individual: Human Studies* (Boston: Little, Brown, 1970), Appendix A, pp. 293–6.

30. Kathleen Weiss, "Fact Sheet," in *Quality of Health Care: Human Experimentation, 1973*, hearings before the Subcommittee on Health of the Committee on Labor and Public Welfare, U.S. Senate, 93rd Congress, 1st Session, Part 1, February 21 and 22, 1973, pp. 300–13; testimony of Belita Cowan in *Regulation of Diethylstilbestrol (DES), 1975*, joint hearing before the Subcommittee on Health of the Committee on Labor and Public Welfare and the Subcommittee on Administrative Practice and Procedure of the Committee on the Judiciary, U.S. Senate, 94th Congress, 1st Session, February 27, 1975, pp. 21–6. See also Belita Cowan, "Ethical Problems in Government-Funded Contraceptive Research," in Helen B. Holmes, Betty B. Hoskins, and Michael Gross, eds., *Birth Control and Controlling Birth* (Clifton, N.J.: Humana Press, 1979), pp. 37–46.

31. Bradford H. Gray, "Complexities of Informed Consent," *Annals of the American Academy of Political and Social Sciences 437* (May 1978), 37–48. For additional examples of lack of informed consent, see Barber (n28), "The Ethics of Experimentation."

32. Gray (n31), "Complexities of Informed Consent."

33. Quoted in Renée C. Fox and Judith P. Swazey, "The Case of the Artificial Heart," *The Courage to Fail: A Social View of Organ Transplants and Dialysis* (Chicago: University of Chicago Press, 1974), p. 204.

34. *Karp v. Cooley*, 493 F.2d 408 (5th Cir. 1974). See also Capron (n18), "Informed Consent."

35. Gina Bari Kolata and Nicholas Wade, "Human Gene Treatment Stirs New Debate," *Science 210* (October 24, 1980), 407; Nicholas Wade, "UCLA Gene Therapy Racked by Friendly Fire," *Science 210* (October 31, 1980), 509–11; W. French Anderson and John C. Fletcher, "Gene Therapy in Human Beings: When Is It Ethical

to Begin?" *New England Journal of Medicine 303* (November 27, 1980), 1293–7.

36. Bob Williamson, "Gene Therapy," *Nature 298* (July 29, 1982), 416–18.

37. Marjorie Sun, "Cline Loses Two NIH Grants," *Science 214* (December 11, 1981), 1220.

38. "U.S. Stalls Plan to Require Drug Information," *San Francisco Chronicle*, April 9, 1981, p. 7; David Perlman, "A New Law for Breast Cancer Victims," *San Francisco Chronicle*, March 23, 1981, p. 2.

39. Lynn C. Epstein and Louis Lasagna, "Obtaining Informed Consent: Form or Substance," *Archives of Internal Medicine 123* (1969), 682–8. The difficulty of many informed consent forms is documented in Gary R. Morrow, "How Readable Are Subject Consent Forms?" *Journal of the American Medical Association 244* (July 4, 1980), 56–8. See also Barrie R. Cassileth, Robert V. Zupkis, Katherine Sutton-Smith, et al., "Informed Consent: Why Are Its Goals Imperfectly Realized?" *New England Journal of Medicine 302* (April 17, 1980), 896–900; and T. M. Grundner, "On the Readability of Surgical Consent Forms," *New England Journal of Medicine 302* (April 17, 1980), 900–2.

40. Joel Brinkley, "Disciplinary Cases Rise for Doctors," *New York Times*, January 20, 1986, pp. 1, 14; also, Ronald Sullivan, "Cuomo's Plan for Testing Doctors Is Part of Growing National Effort," *New York Times*, June 9, 1986, pp. 1, 11.

41. Stephen Budiansky, "Regulation Issue Is Resurrected by EPA," *Nature 304* (August 18, 1983), 572.

42. See Marjorie Sun, "EPA Approves Field Test of Altered Microbes," *Science 230* (November 29, 1985), 1015–16. A National Academy of Sciences panel concluded that the risks of genetically altered organisms depended on the nature of the organism itself rather than the artificial method by which it had been created. See Harold M. Schmeck, Jr., "Report Discounts Gene Splicing Peril," *New York Times*, August 15, 1987, pp. 1, 7.

43. Keith Schneider, "Biology's Unknown Risks," *New York Times*, April 5, 1986, pp. 1,6.

44. Ann Crittenden, "The Gene Machine Hits the Farm," *New York Times*, June 28, 1981, Section 3, pp. 1, 20.

45. A. M. Chakrabarty, quoted in "Scientists Make a Bug That Eats Toxic Spills," *San Francisco Chronicle*, December 1, 1981, p. 2.

46. Paul Berg, quoted in Paul Jacobs, "Genetic Engineers Urge Patience, Not Patients," *San Jose Mercury News*, October 19, 1980, p. G-2.

47. Martin Cline, quoted in Eden Graber and Janice Burg, "NIH Panel Advocates Harsh Price for Cline," *Genetic Engineering News 1* (May/June 1981), 1, 10–11.

48. Quoted in Charles Austin, "Ethics of Gene Splicing Troubling Theologians," *New York Times*, July 5, 1981, pp. 1, 12. For an excellent review of the social and moral concerns raised by genetic engineering, see *Splicing Life: A Report on the Social and Ethical Issues of Genetic Engineering with Human Beings*, Report of the President's Commission for the Study of Ethical Problems in Medicine and Biomedical and Behavioral Research (Washington, D.C.: Government Printing Office, November 1982).

49. Interview with Philip Handler, President of the National Academy of Sciences, "In Science, 'No Advances Without Risks,' " *U.S. News and World Report* (September 15, 1980), 60–1.

50. Reginald Jones, quoted in Amitai Etzioni, "How Much Is a Life Worth?" *Social Policy 9* (March/April 1979), 4.

51. Aaron Wildavsky, "No Risk Is the Highest Risk of All," *American Scientist 67* (January–February 1979), 32–7.

52. Judith Randal, "A Jackpot for Insurers of Flu Shots," *San Francisco Sunday Examiner and Chronicle*, July 10, 1977, p. 3.

53. Rhonda Rundle, "Genetic Engineering Spawns a New Risk," *Business Insurance* (November 17, 1980), 2, 47.

54. David Dickson and David Noble, "The New Corporate Technocrats," *Nation* (September 12, 1981), 209.

55. Quoted in David F. Noble, "Cost–Benefit Analysis," *Health/Pac Bulletin 11* (July/August 1980), 2.

56. Maria Shao, "Concerns Seeking Credibility Put Scientists in Spotlight," *Wall Street Journal*, September 22, 1981, p. 29.

57. Noble (n55), "Cost–Benefit Analysis," 33.

58. Bruce M. Owen and Ronald Braeutigam, "The Regulation Game: Strategic Use of the Administrative Process (Cambridge, Mass.: Ballinger, 1978), p. 7.

59. Quoted in Gregory C. Staple, "Free-Market Cram Course for Judges," *Nation* (January 26, 1980), 78–81. Other information is from the LEC's 1982–3 Annual Report, Law and Economic Center, Emory University, Atlanta, Georgia 30322.

60. Quoted in LEC 1981–2 Annual Report, University of Miami, P.O. Box 248000, Coral Gables, Florida 33124.

61. National Council on Health Care Technology, Department of Health and Human Services, Office of the Assistant Secretary for Health, Office of Health Research, Statistics, and Technology, "Minutes of Meeting, August 13, 1981," Hubert H. Humphrey Building, Washington, D.C., pp. 1–2.

62. Some of the Center's technology assessment functions were assigned to the National Center for Health Services Research, but with little increase in funding or staff. For useful insights into the factors contributing to the Center's demise, see David Blumenthal, "Federal Policy Toward Health Care Technology: The Case of

the National Center," *Milbank Memorial Fund Quarterly/Health and Society 61* (Fall 1983), 584–613.

63. Representative Harold Volkmer quoted in "Worry About Biotechnology," *Palo Alto Times Tribune*, April 30, 1986, p. A-10.

64. U.S. General Accounting Office, *Biotechnology: Agriculture's Regulatory System Needs Clarification*, Report to the Chairman, Committee on Science and Technology, House of Representatives, GAO/RCED-86-59, March 1986, pp. 3–4.

65. Jeremy Rifkin, quoted in "Genetic Swine Project Assailed," *San Francisco Chronicle*, April 4, 1986, p. 25.

66. "FDA to Fast-Track Recombinant Drugs," *American Pharmacy NS21* (February 1981), 19.

67. Charles Perrow, "Not Risk but Power," *Contemporary Sociology 11* (May 1982), 298–300.

68. For a superb discussion of this and other problems with quantitative methods, see Laurence H. Tribe, "Technology Assessment and the Fourth Discontinuity: The Limits of Instrumental Rationality," *Southern California Law Review 46* (1973), 617–60.

69. Harold P. Green, "The Risk–Benefit Calculus in Safety Determinations," *George Washington Law Review 43* (March 1975), 791–807.

70. Mark Green, quoted in Noble (n55), "Cost–Benefit Analysis," 11.

71. Steven C. Schoenbaum, Barbara J. McNeill, and Joel Kavet, "The Swine-Influenza Decision," *New England Journal of Medicine 295* (September 30, 1976), 759–65.

72. Office of Technology Assessment, *The Implications of Cost-Effectiveness Analysis of Medical Technology*, Publ. No. OTA-H–176 (Washington, D.C.: Government Printing Office, August 1980), p. 5.

73. Executive Order 12291, "Federal Regulation," February 17, 1981 (Washington, D.C.: White House).

74. Burke K. Zimmerman, "Cost–Benefit Analysis: The Cop-Out of Governmental Regulation," *Trial Magazine* (February 1978), 43–7.

75. Charles Perrow, *Normal Accidents: Living with High Risk Technologies* (New York: Basic Books, 1984).

76. See Norman Kaplan, "Separate Views of Dr. Kaplan," *The Totally Implantable Artificial Heart: Legal, Social, Ethical, Medical, Economic, Psychological Implications*, A Report of the Artificial Heart Assessment Panel, National Heart and Lung Institute (Bethesda, Md.: Department of Health, Education and Welfare Publ. No. [NIH] 74–191, June 1973), Appendix C, pp. 223–9. For a discussion of the notion of societal informed consent, see Laurence R. Tancredi and Arthur J. Barsky, "Technology and Health Care Decision Making: Conceptualizing the Process for Societal Informed Consent," *Medical Care 12* (October 1974), 845–59.

77. Tversky and Kahneman (n1), "The Framing of Decisions."
78. Office of Technology Assessment, *Strategies for Medical Technology Assessment*, Publ. No. OTA-H–181 (Washington, D.C.: Government Printing Office, September 1982), p. 5.
79. Green (n69), "The Risk–Benefit Calculus in Safety Determinations," 807.

8. Compensating the injuries of medical innovation

1. "Swine Flu Statistics," Torts Branch, Civil Division, U.S. Department of Justice, June 4, 1987.
2. Lilly's legal expenses promise to dwarf its profits from DES sales. One estimate puts the total sales of DES from 1947 to 1971 at $2.5 million (Michael Kinsley, "Fate and Lawsuits," *New Republic* [June 14, 1980], p. 21). Legal expenses are hard to measure, but $5–$10 million is a reasonable guess. Figures on DES lawsuits are from Harriet Chiang, "Women Await Word on DES Suit," *San Francisco Chronicle*, April 9, 1987, p. 34. It may also not always be easy for DES victims to find lawyers willing to invest the estimated $50,000 or more it can cost to bring a case to trial – especially when there is no assurance of winning. In San Francisco, one DES daughter with cancer was turned down by eight or nine law firms before finally finding one willing to represent her (Wesley Wagnon, Attorney, Hersh and Hersh, San Francisco, personal communication, September 21, 1984).
3. Insurance Services Office, *Special Malpractice Review: 1974 Closed Claim Survey, Report of the All-Industry Committee* (New York: Insurance Services Office, 1976). See also Jeffrey O'Connell, "Contracting for No-Fault Liability Insurance Covering Doctors and Hospitals," *Maryland Law Review 36* (1977), note 7, 556.
4. "Uninsured Increase, Study Shows," *Washington Report on Medicine and Health/Perspectives* (Washington, D.C.: McGraw-Hill, April 30, 1984), p. 3.
5. Insurance Services Office (see n3), *Special Malpractice Review*, p. 17, cited in Eli P. Bernzweig, *By Accident Not Design* (New York: Praeger, 1980), p. 107.
6. "DES User's Daughter Gets $1.75 Million Award," *National Law Journal 4* (April 12, 1982), 11. This verdict was appealed, and subsequently settled out of court for what is believed to be $2.1 million (Wesley Wagnon [n2], personal communication, September 21, 1984). The $150,000 figure is from Kinsley (n2), "Fate and Lawsuits."
7. Lawrence Charfoos, quoted in Kinsley (n2), "Fate and Lawsuits."
8. Paul D. Carrington and Barbara A. Babcock, *Civil Procedure*, 2nd ed. (Boston: Little, Brown, 1977), p. 700. For an example of the high costs of complex litigation, see Tamar Lewin, "Business and

the Law: Manville's Big Legal Fees," *New York Times*, June 12, 1984, p. D2.

9. Department of Commerce, *Interagency Task Force on Product Liability: Final Report*, Publ. No. PB–273–220 (Washington, D.C.: Government Printing Office, 1978), p. V-67.

10. Strict liability, developed in the 1960s, allows someone who has been injured by a defective product to recover from the manufacturer without having to prove negligence. In practice, however, proof that a product was in fact "defective" can involve many of the same arguments and evidence that would be necessary for proof of negligence. See Jeffrey O'Connell, *Ending Insult to Injury: No-Fault Insurance for Products and Services* (Urbana: University of Illinois Press, 1975), pp. 56–69.

11. Quoted in Douglas E. Rosenthal, *Lawyer and Client: Who's in Charge?* (New York: Russell Sage Foundation, 1974), p. 82.

12. Liability insurance might not weaken deterrence if premium costs reflected the risks of particular products. Pharmaceutical companies, however, usually insure all product lines as a part of general liability coverage; thus the deterrent effect of financial liability for a particularly risky product is diluted by the lower risks of the company's other products. In fact, because of insurance rating practices, premium costs may not even reflect the risk experiences of particular companies very accurately (see Department of Health, Education and Welfare, *Liability Arising out of Immunization Programs: A Final Report to Congress* [Washington, D.C., May 1978] [hereafter cited as HEW, *Final Report*], pp. 16–24).

13. See Jeffrey O'Connell, *The Lawsuit Lottery: Only the Lawyers Win* (New York: Free Press, 1979); also Marc A. Franklin, "Replacing the Negligence Lottery: Compensation and Selective Reimbursement," *Virginia Law Review 53* (1967) 774–814.

14. Richard Smith, "Compensation for Drug Injury: Two Solutions to an Insoluble Problem," *British Medical Journal 282* (May 16, 1981), 1610–12.

15. Arthur H. Downey and Kenneth G. Gulley, "Theories of Recovery for DES Damage," *Journal of Legal Medicine 4* (1983), 167–200.

16. The need for expert testimony has long been a major stumbling block for the injured. See W. Page Keeton, gen. ed., *Prosser and Keeton on the Law of Torts*, 5th ed. (St. Paul, Minn.: West, 1984), p. 188. Also see *Salgo* v. *Leland Stanford Jr. University Board of Trustees*, 154 Cal. App. 2d 560, 317 P.2d 170 (1957).

17. "Strict liability" in tort is a hybrid of contract and tort theories. It was first adopted in California in *Greenman* v. *Yuba Power Products, Inc.*, 59 Cal. 2d 57, 377 P.2d 897, 27 Cal. Rptr. 697 (1963), and has been adopted in almost all jurisdictions. Automobile and tool cases account for many of the key cases that have followed.

See *Vandermark* v. *Ford Motor Company*, 61 Cal. 2d 256, 391 P.2d 168, 37 Cal. Rptr. 896 (1964) and *Elmore* v. *American Motors Corporation*, 70 Cal. 2d 578, 451 P.2d 84, 75 Cal. Rptr. 652 (1969).

18. *Restatement of the Law of Torts, Second*, Appendix Vol. 3, Section 402A, Comment k, 1963–64.

19. *Basko* v. *Sterling Drug, Inc.*, 416 F.2d 417 (2d Cir. 1969). A few jurisdictions have held manufacturers liable even though they could not necessarily have known that the product was defective. See, for example, *Green* v. *American Tobacco Company*, 154 So. 2d 169 (Fla. 1963) [question certified to Florida Supreme Court by the Fifth Circuit in *Green* v. *American Tobacco Company*, 304 F.2d 70 (5th Cir. 1962), *reversed and remanded*, 325 F.2d 673 (5th Cir. 1963), *cert. denied*, 377 U.S. 943 (1964)].

20. See *Brody* v. *Overlook Hospital*, 121 N.J. Super. 299, 296 A.2d 668 (1972).

21. *Bichler* v. *Eli Lilly and Company*, 436 N.Y.S.2d 625 (N.Y. App. Div. 1981). The jury was asked to address seven questions, all of which were answered in plaintiff's favor: "(1) Was DES reasonably safe in the treatment of accidents of pregnancy when it was ingested by plaintiff's mother in 1953? (2) Was DES a proximate cause of plaintiff's cancer? (3) In 1953, when plaintiff's mother ingested DES, should the defendant, as a reasonably prudent drug manufacturer, have foreseen that DES might cause cancer in the offspring of pregnant women who took it? (4) Foreseeing that DES might cause cancer in the offspring of pregnant women who took it, would a reasonably prudent drug manufacturer test it on pregnant mice before marketing it? (5) If DES had been tested on pregnant mice, would the tests have shown that DES causes cancer in their offspring? (6) Would a reasonably prudent drug manufacturer have marketed DES for use in treating accidents of pregnancy at the time it was ingested by the plaintiff's mother if it had known that DES causes cancer in the offspring of pregnant mice? (7) Did defendant and the other drug manufacturers act in concert with each other in the testing and marketing of DES for use in treating accidents of pregnancy?" [See *Bichler* v. *Eli Lilly and Company*, 450 N.Y.S.2d 776 (N.Y. 1982), pp. 783–4].

22. *Bichler* v. *Eli Lilly and Company*, 450 N.Y.S.2d 776 (N.Y. 1982), p. 776. In explaining its reasons for rejecting Lilly's defense, the Court of Appeals pointed out that in 1947 a researcher had conducted an experiment in which pregnant mice were injected with urethane, an anesthetic administered to women during delivery, in order to determine if the offspring developed cancer from the drug (C. D. Larsen, Lucille L. Weed, and Paul B. Rhoads, Jr., "Pulmonary-Tumor Induction by Transplacental Exposure to Urethane," *Journal of the National Cancer Institute 8* [October 1947], 63–70). "Obviously, this researcher," the Court noted, "was con-

cerned about a human transplacental carcinogen" (p. 783). In 1987, a Federal Court in Wisconsin ruled otherwise, finding Eli Lilly not negligent in marketing DES in 1960 and the drug itself not defective. See Ken Wysocky, "Pharmaceutical Firm Found Not Negligent on DES," *Milwaukee Sentinel*, March 14, 1987, p. 6.

23. A recent example involved Eli Lilly and Company, the maker of Oraflex, an antiarthritis drug. Government investigators found that Lilly had been informed of twenty-seven "serious" adverse reactions in Great Britain, including five deaths, linked to Oraflex, but had failed to report them to the FDA before the drug went on sale in the United States in April 1982. It was pulled off the world market four months later, but by then, twenty-six American patients had died. Eliot Marshall, "Guilty Plea Puts Oraflex Case to Rest," *Science 229* (September 13, 1985), 1071.

24. *Bichler v. Eli Lilly and Company*, 436 N.Y.S.2d 625 (N.Y. App. Div. 1981), p. 629.

25. A glaring example is the antibiotic chloramphenicol, the preferred treatment for typhoid fever and certain serious infections, even though it causes a potentially fatal blood disorder in a very small number of users. Explicit warnings on the label and cautionary letters to physicians did little to curb the use of chloramphenicol; drug companies aggressively marketed it, and millions of doses were prescribed during the 1950s and 1960s for a wide range of problems, including some as trivial as sore throats and acne. Publicity from 1967 Senate hearings finally led to a decline in dangerous and indiscriminate prescribing. See Milton Silverman and Philip R. Lee, *Pills, Profits and Politics* (Berkeley: University of California Press, 1974), pp. 283–8.

26. Market value from Pamela G. Hollie, "The Shift to Generic Drugs," *New York Times*, July 23, 1984, pp. 19, 24. Profit levels from Leonard G. Schifrin, "Economics and Epidemiology of Drug Use," in Kenneth Melmon and Howard Morrelli, eds., *Clinical Pharmacology: Basic Principles in Therapeutics*, 2nd ed. (New York: Macmillan, 1978), p. 1100.

27. Donald Kennedy, "Creative Tension: FDA and Medicine," *New England Journal of Medicine 298* (April 13, 1978), 846. Drug industry estimates of sales efforts run around $400 million, whereas those from more objective sources put total promotional expenditures at roughly $1 billion (Silverman and Lee [n25], *Pills, Profits and Politics*, pp. 54–5). For a summary of studies of physicians' prescribing behavior, see Silverman and Lee (n25), *Pills, Profits and Politics*, pp. 75–6. Also, Ingrid Waldron, "Increased Prescribing of Valium, Librium and Other Drugs – An Example of the Influence of Economic and Social Factors on the Practice of Medicine," *International Journal of Health Services 7* (1977), 37–62; and Jerry Avorn, Milton Chen, and Robert Hartley, "Scientific Versus

Commercial Sources of Influence on the Prescribing Behavior of Physicians," *American Journal of Medicine 73* (July 1982), 4–8.

28. See, for example, Joe Collier and Linda New, "Illegibility of Drug Advertisements," *Lancet 1* (February 11, 1984), 341–2.

29. For a summary of these studies and their lack of impact on prescription practices, see Thomas C. Chalmers, "The Impact of Controlled Trials on the Practice of Medicine," *The Mount Sinai Journal of Medicine 41* (November–December 1974), 754–5.

30. Forty-three states and two territories passed laws modifying at least one significant substantive or procedural aspect of tort law dealing with medical malpractice claims (Department of Health, Education and Welfare, Public Health Service, *Legal Topics Relating to Medical Malpractice* [Washington, D.C.: Government Printing Office, January 1977], p. 1). *Florida Statutes Annotated*, section 95.031, p. 15: twelve-year maximum (1978 Laws: 78–418); *Annotated Indiana Code*, section 33-1-1.5-5, pp. 24–5: ten-year maximum (1978 Acts: P.L.141); *Code of Alabama*, section 6-5-502, p. 101: ten-year maximum (1979 Acts: 79–468).

31. Sybil Shainwald, Attorney, Fuchsberg and Fuchsberg, New York, New York, personal communication, August 5, 1985.

32. Cynthia A. Feigin, "Statutes of Limitations: The Special Problem of DES Suits," *American Journal of Law and Medicine 7* (Spring 1981), 91–106; Sidney Draggan, *Compensation for Victims of Toxic Pollution: Assessing the Scientific Knowledge Base*, PRA Research Report 83–6, prepared for the National Science Foundation, Directorate for Scientific, Technological, and International Affairs (Washington, D.C., March 1983), p. 124. As of 1984, the large majority of states set the statute of limitations from the time of discovery of the injury rather than from the time of exposure. Mary Anne Galante, "Statute of Limitations Expanded in DES Cases," *National Law Journal 7* (September 9, 1985), 7. On the New York law, see William R. Greer, "New Law Opens Courts to DES Suits," *New York Times*, July 30, 1986, pp. 7, 12.

33. Alfred S. Julien and Sybil Shainwald, "Special Feature: Litigation in DES Cases," *DES Action Voice 2* (Winter 1980), 4–5.

34. *Sindell v. Abbott Laboratories*, 26 Cal. 3d 588, 607 P.2d 924, 163 Cal. Rptr. 132 (1980), *cert. denied*, 449 U.S. 912 (1980). See also "Market Share Liability: An Answer to the DES Causation Problem," *Harvard Law Review 94* (1981), 668–80.

35. Besides market-share liability, other theories courts have accepted in allowing recovery of damages, even when the plaintiff cannot identify the specific DES manufacturer alleged to have caused the injury, include alternative liability, concert of action, enterprise liability, and risk contribution. The New York decision, based on a theory of concerted action, is *Bichler v. Eli Lilly and Company*, 436 N.Y.S.2d 625 (N.Y. App. Div. 1981), *aff'd*, 450 N.Y.S.2d

776 (N.Y. 1982). The Michigan decision, based on the theory of alternative liability, is *Abel* v. *Eli Lilly and Company*, 418 Mich. 311, 343 N.W.2d 164 (1984). The Washington decision is *Martin* v. *Abbott Laboratories*, 102 Wash. 2d 581, 689 P.2d 368 (1984). The Wisconsin decision, based on the theory of "risk contribution," is *Collins* v. *Eli Lilly and Company*, 116 Wis. 2d 166, 342 N.W.2d 37 (1984). For an analysis of these decisions, see Sheila L. Birnbaum and Barbara Wrubel, "Agent Orange Class Certification and Industrywide Liability for DES," *National Law Journal 6* (February 27, 1984), 38–41.

36. *Collins* v. *Eli Lilly and Company*, 116 Wis. 2d 166, 342 N.W.2d 37 (1984), p. 49.

37. Lawrence Charfoos, quoted in Lynne Reaves, "DES Dynamite," *American Bar Association Journal 70* (July 1984), 27.

38. Lawrence Rout, "Product-Liability Law Is in Flux as Attorneys Test a Radical Doctrine," *Wall Street Journal*, December 30, 1980, pp. 1, 8. One objection to market-share liability and similar theories is that implementation will be difficult because evidence concerning companies' market shares may not be available. In 1978, however, a Superior Court Judge established a formula for determining each manufacturer's share of liability for the 25 years that DES was prescribed (Harriet Chiang, "S.F. Judge Opens Door for Drug Liability Suits," *San Francisco Chronicle*, July 3, 1987, p. 3). For additional criticisms, see Draggan (n32), *Compensation for Victims of Toxic Pollution*, p. 119.

39. "A Legislative Victory for DES Victims," *San Francisco Chronicle*, January 21, 1982, p. 24.

40. *Payton* v. *Abbott Laboratories*, 83 F.R.D. 382 (D. Mass. 1979), p. 390.

41. *Payton* v. *Abbott Laboratories*, 100 F.R.D. 336, 38 Fed. R. Serv. 2d (Callaghan) 315 (D. Mass. 1983). The Federal Rules of Civil Procedure specify requirements for class certification. Classes have been rejected for various reasons, the most common being that potential members of the class were deemed too dissimilar to allow joint resolution (see Larry Tell, "DES Class: The Reviews Are Mixed," *National Law Journal 4* [July 26, 1982], 3, 13. Also, Susan E. Silbersweig, "*Payton* v. *Abbott Laboratories*: An Analysis of the Massachusetts DES Class Action Suit," *American Journal of Law and Medicine 6* [Summer 1980], 243–82). Besides Massachusetts, there have been filings to get class action status in Illinois, California, New Jersey, Connecticut, Michigan, and Maryland, but these have been denied (Pat Cody, Treasurer of DES Action, San Francisco, California, letter to Diana Dutton, November 29, 1984).

42. *Bichler* v. *Eli Lilly and Company*, 436 N.Y.S.2d 625 (N.Y. App. Div. 1981), *aff'd*, 450 N.Y.S.2d 776 (N.Y. 1982). The legal term

for this ruling is *collateral estoppal*. The only question left open by New York's highest court was whether DES producers acted "in concert."

43. "Lawsuit Won on Pregnancy Failure," *DES Action Voice*, No. 23 (Winter 1985), p. 2.

44. Checkups are generally recommended every six months; see, for example, B. Anderson, W. G. Watring, D. D. Edinger, et al., "Development of DES-Associated Clear-Cell Carcinoma: The Importance of Regular Screening," *Obstetrics and Gynecology 53* (March 1979), 293–9. Cost estimates are based on a survey reported in "The Medical Costs of DES-Exposed Daughters," *DES National Quarterly 2* (Spring 1980), 6. The treasurer of DES Action estimated that medical costs for all DES victims run somewhere between $300 and $600 million per year (Pat Cody, letter to Diana Dutton, November 29, 1984).

45. *Davis* v. *Wyeth Laboratories*, 399 F.2d 121 (9th Cir. 1968); *Reyes* v. *Wyeth Laboratories*, 498 F.2d 1264 (5th Cir. 1974).

46. Leslie C. Ohta, "Immunization Injuries: Proposed Compensatory Mechanisms – An Analysis," *Connecticut Law Review 11* (Fall 1978), 152–7.

47. Although the term *absolute liability* has often been used interchangeably with strict liability, it is used here to indicate a standard that, especially for unavoidably unsafe products, goes a step beyond strict liability in eliminating such theories as assumption of risk and contributory negligence as appropriate defenses. With absolute liability, the fact of injury alone is the basis for liability. See Thomas E. Baynes, Jr., "Liability for Vaccine Related Injuries: Public Health Considerations and Some Reflections on the Swine Flu Experience," *St. Louis University Law Journal 21* (1977), 70. Frank M. McClellan, "Strict Liability for Drug Induced Injuries: An Excursion Through the Maze of Products Liability, Negligence and Absolute Liability," *Wayne Law Review 25* (November 1978), 1–36. Nor would warnings constitute an appropriate defense under absolute liability, as they generally do under strict liability for unavoidably unsafe products (see Draggan [n32], *Compensation for Victims of Toxic Pollution*, pp. 8–11).

48. Richard E. Neustadt and Harvey V. Fineberg, *The Swine Flu Affair: Decision-Making on a Slippery Disease* (Washington, D.C.: Government Printing Office, 1978), p. 54.

49. Richard Gaskins, "Equity in Compensation: The Case of Swine Flu," *Hastings Center Report 10* (February 1980), 5–8. This was also the conclusion of a report by the Office of Technology Assessment [OTA], *Compensation for Vaccine-Related Injuries*, Publ. No. OTA-TM-H–6 (Washington, D.C.: Government Printing Office, November 1980).

50. HEW (n12), *Final Report*, pp. 16–18.

51. Ibid., p. 21.

52. Department of Health, Education and Welfare, General Accounting Office, *The Swine Flu Program: An Unprecedented Venture in Preventive Medicine*, Publ. No. HRD–77–115 (Washington, D.C.: Government Printing Office, June 27, 1977), p. 21.

53. See, for example, *Congressional Record: House*, August 10, 1976, pp. 26795–6, 267800–5.

54. For an excellent discussion of this issue, see Bernzweig (n5), *By Accident Not Design*, pp. 102–45.

55. Arthur M. Silverstein, *Pure Politics and Impure Science: The Swine Flu Affair* (Baltimore: Johns Hopkins University Press, 1981). See also, Gaskins (n49), "Equity in Compensation," 6.

56. Gaskins (n49), "Equity in Compensation," 7.

57. Jeffrey Axelrad, U.S. Department of Justice, letter to Diana Dutton, March 16, 1983. Judges have frequently ruled against people with an undisputed diagnosis of Guillain–Barré syndrome on the grounds that their injury was not proven to be *causally* related to swine flu immunization. Typically in such cases, Guillain–Barré did not appear within ten weeks after immunization, the time period often used by the government in assessing causation, emerging instead after other adverse reactions to the shot. See, for example, *Varga* v. *United States*, 566 F. Supp. 987 (N.D. Ohio 1983) and *Gates* v. *United States*, 707 F.2d 1141 (10th Cir. 1983).

58. "Swine Flu Statistics," Torts Branch, Civil Division, U.S. Department of Justice, June 4, 1987.

59. Jeffrey Axelrad, quoted in Philip Taubman, "Claims of Flu Shot Victims Unpaid Despite U.S. Pledge," *New York Times*, June 10, 1979, p. 30.

60. Mark P. Friedlander, Jr., Attorney, Friedlander, Friedlander and Brooks, Arlington, Virginia, letter to Barton Bernstein, July 25, 1983.

61. "$285,000 for Swine Flu Death," *San Francisco Chronicle*, August 30, 1979, p. 5. In the *Cardillo* (1984) case, a wrongful death suit, damages of $5 million were awarded; that judgment was also appealed (Jeffrey Axelrad, personal communication, July 10, 1986).

62. "Big Award in Swine Flu Vaccine Case," *San Francisco Chronicle*, February 2, 1983, p. 2.

63. OTA (n49), *Compensation for Vaccine-Related Injuries*.

64. The six nations are Great Britain, Japan, France, West Germany, Switzerland, and Denmark (ibid., pp. 61–8).

65. California set up a state program for compensation of vaccine injuries in 1977, and the Rhode Island legislature considered doing so in 1979. Interestingly, the major problem so far with Califor-

nia's compensation program appears to be underuse, not abuse. As of 1980, only a single claim had been filed! (OTA [n49], *Compensation for Vaccine-Related Injuries*, p. 61).

66. Gina Kolata, "Litigation Causes Huge Price Increases in Childhood Vaccines," *Science 232* (June 13, 1986), 1339; Barbara J. Culliton, "Omnibus Health Bill: Vaccines, Drug Exports, Physician Peer Review," *Science 234* (December 12, 1986), 1313–14; "Reagan Seeks Rewrite of Vaccine Law," *Nation's Health* (May–June 1987), 4.

67. Institute of Medicine, Division of Health Promotion and Disease Prevention, *Vaccine Supply and Innovation* (Washington, D.C.: National Academy Press, 1985); Health and Public Policy Committee, American College of Physicians, "Compensation for Vaccine-Related Injuries," *Annals of Internal Medicine 101* (October 1984), 559–61; "Hearings Look at Threats to Vaccine Caused by Liability Problems," *Nation's Health* (October 1984), 1, 4.

68. The foregoing arguments are discussed in President's Commission for the Study of Ethical Problems in Medicine and Biomedical and Behavioral Research [cited hereafter as President's Commission], *Compensating for Research Injuries*, Vol. 1: *Report* (Washington, D.C.: Government Printing Office, June 1982), Chapter 3. Although patient-subjects in "therapeutic research" hope to benefit personally from their participation, it has been found that they are also more likely than healthy subjects in "nontherapeutic research" to suffer injury (President's Commission, *Compensating for Research Injuries*, Vol. 1, p. 65).

69. National survey data are from the report *HEW Secretary's Task Force on the Compensation of Injured Research Subjects*, Department of Health, Education and Welfare, Publ. No. OS-77-003 (Bethesda, Md.: National Institutes of Health, January 1977) [cited hereafter as *Task Force Report*], p. IV-1-15. More recent studies are discussed in President's Commission (n68), *Compensating for Research Injuries*, Vol. 1, pp. 65–80.

70. Information on research institutions is from Bradford H. Gray, Robert A. Cooke, and Arnold S. Tannenbaum, "Research Involving Human Subjects," *Science 201* (September 22, 1978), 1101. Information on private companies is from Pharmaceutical Manufacturers Association, "Report on Survey Concerning Availability of Compensation for Injuries Resulting from Participation in Biomedical Research," in President's Commission (n68), *Compensating for Research Injuries*, Vol. 2: *Appendices*, pp. 302–7.

71. William J. Curran, "Legal Liability in Clinical Investigations," *New England Journal of Medicine 298* (April 6, 1978), 778–9.

72. *Mink v. University of Chicago*, 460 F. Supp. 713 (N.D. Ill. 1978), pp. 715–16. What study participants were actually told is difficult to determine, since no formal or written informed consent pro-

cedures were used, and in view of studies that have documented poor patient recall of even recent informed consent procedures (see, for example, George Robinson and Avraham Merav, "Informed Consent: Recall by Patients Tested Postoperatively," *Annals of Thoracic Surgery 22* [September 1976], 209–12; Dianne Leeb, David G. Bowers, and John B. Lynch, "Observations on the Myth of 'Informed Consent,' " *Plastic and Reconstructive Surgery 58* [September 1976], 280–2).

73. "A Bittersweet Victory for DES Victims," *San Francisco Chronicle*, February 28, 1982, p. A13.

74. For a history of these five groups, see President's Commission (n68), *Compensating Research Injuries*, Vol. 1, pp. 37–43.

75. *Task Force Report* (n69), p. II-2.

76. Department of Health, Education and Welfare, "Protection of Human Subjects: Informed Consent: Definition Amended to Include Advice on Compensation," *Federal Register 43*, No. 214 (November 3, 1978), 51559.

77. Department of Health and Human Services, "Final Regulations Amending Basic HHS Policy for the Protection of Human Research Subjects," *Federal Register 46*, No. 16 (January 26, 1981), 8366–91.

78. Charles McCarthy, Office of Research Risks, National Institutes of Health, letter to John Bunker, October 11, 1984.

79. At Stanford, for example, the informed consent form states that reimbursement for necessary medical expenses is available "in the event of physical injury that arises solely out of the negligence of the Stanford University Medical Center or its staff . . . "

80. Quotes are from President's Commission (n68), *Compensating for Research Injuries*, Vol. 1, pp. 1, 3, 132–5. The commission recommended considering as a partial exception a limited class of injuries arising from certain "non-beneficial" procedures employed in therapeutic research. Regarding costs and feasibility, the commission's report stated that "it would be inordinately expensive to 'compensate' all patient-subjects for all illness or injury they experience after participating in research." The commission rejected the 1977 task force's approach to identifying research-related injuries, the so-called on-balance test, arguing that it would "pose enormous burdens of administration which could not be justified in light of the relatively weaker moral claim for compensating patients who participate in research with the hope of deriving personal health benefits." To obtain more information on the need for and feasibility of no-fault compensation for injured research subjects, the commission advised the government to fund a small-scale experiment in several institutions.

81. John P. Gilbert, Richard J. Light, and Frederick Mosteller, "Assessing Social Innovations: An Empirical Base for Policy," in Carl

A. Bennett and Arthur A. Lumsdaine, eds., *Evaluation and Experiment* (New York: Academic Press, 1975), p. 182.

82. See, for example, Guido Calabresi, "Reflections on Medical Experimentation in Humans," in Paul A. Freund, ed., *Experimentation with Human Subjects* (New York: George Braziller, 1969), pp. 178–96; Clark C. Havighurst, "Compensating Persons Injured in Human Experimentation," *Science 169* (July 10, 1970), 153–7.

83. See, for example, Guido Calabresi, "Fault, Accidents and the Wonderful World of Blum and Kalven," *Yale Law Journal 75* (1965), 216–38.

84. The Federal Minister for Youth, Family Affairs and Health, *Law on the Reform of Drug Legislation of the Federal Republic of Germany*, August 24, 1976 (translation), Federal Law Gazette I, p. 2445, Bonn, West Germany.

85. Association of the British Pharmaceutical Association, "Compensation and Drug Trials," *British Medical Journal 287* (September 3, 1983), 675.

86. President's Commission (n68), *Compensating for Research Injuries*, Vol. 1, p. 65. See also Vol. 2: *Appendices*, pp. 239–333.

87. S. Leslie Misrock and Gideon D. Stern, "Responsibility, Liability for Gene Products, Part One and Part Two," *Genetic Engineering News 1* (July/August 1981), 1, 12 and (September/October 1981), 1, 24–5. Faced with exorbitant premiums, or unable to obtain coverage, biotechnology companies are insuring themselves or going uninsured. Mark Crawford, "Insurance Drought Fosters Self-Help Plan for Biotechnology Firms," *Science 232* (April 11, 1986), 154.

88. Dee Ziegler, "Long-Cause Cases: Shaw: Logjam May Close Civil Courts," *Recorder*, San Francisco, California, September 13, 1984, p. 1.

89. Streamlining can occur either through "multidistrict legislation," or "complex litigation" (Wesley Wagnon [n2], personal communication, September 21, 1984). Class action suits have been attempted, unsuccessfully, in cases involving DES, Bendectin, a drug for morning sickness (Gina Kolata, "Jury Clears Bendectin," *Science 227* [March 29, 1985], 1559), and the Dalkon Shield, a contraceptive intrauterine device (Susan Sward, "Unusual Class Action over IUDs," *San Francisco Chronicle*, June 20, 1983, p. 4). As of 1984, the only successful federal class action in products liability litigation involved suits over injuries caused by Agent Orange, the defoliant used in Vietnam (Gail Appleson, "Two 'Agent Orange' Classes Unique," *National Law Journal 6* [January 2, 1984], 3, 8).

90. These include a $615 million fund from the A. H. Robins Company to settle Dalkon Shield claims, and a $180 million fund from chemical companies for Agent Orange claims. For information on

these various settlements, see Peter H. Schuck, *Agent Orange on Trial: Mass Toxic Disasters in the Courts* (Cambridge, Mass: Harvard University Press, 1986); Steven E. Prokesch, "Manville Plan Seeks Cost Cuts – Officers Expect Weakening in Reorganization," *New York Times*, August 12, 1985, pp. 19, 22; and Stuart Diamond, "Drug Company Asks Protection from Creditors – Suits on Dalkon Shield Are Cited by Robins," *New York Times*, August 22, 1985, pp. 1, 36.

91. Because of the dramatic escalation in the number of claims, damage payments made from the fund, which with interest had grown to $200 million in the first year, were far lower than expected. Maximum payments for death are $3,400, and for total disability $12,800. No direct payments will be made for lesser injuries, but a foundation will be set up to finance services for people who claim they were injured by Agent Orange. Joseph P. Fried, "Judge Orders Allocation of Agent Orange Fund," *New York Times*, May 29, 1985, p. 17. Another mass settlement fund, offered by the maker of Bendectin, a drug for morning sickness suspected of causing birth defects, experienced a similar mushrooming in the number of claims after the settlement had occurred. See Kolata (n89), "Jury Clears Bendectin."

92. Amy Tarr, "Michigan DES Suits Settle for Secret Sum," *National Law Journal 8* (November 11, 1985), 3, 10.

93. See, for example, William L. Prosser, *Handbook of the Law of Torts*, 4th ed. (St. Paul, Minn.: West, 1971), pp. 236–90.

94. " 'Toxic Torts' Seen as Leading Way Toward New Injury Compensation System," *Product Safety & Liability Report 7* (June 22, 1979), 505. See also Jeffrey Trauberman, "Statutory Reform of 'Toxic Torts': Relieving Legal, Scientific, and Economic Burdens on the Chemical Victim," *Harvard Environmental Law Review 7* (1983), 177–296. For a critical analysis of these developments, see Dennis R. Honabach, "Toxic Torts – Is Strict Liability Really the 'Fair and Just' Way to Compensate the Victims?" *University of Richmond Law Review 16* (1982), 305–21; also, Alfred R. Light, *Compensation for Victims of Toxic Pollution – Assessing the Scientific Knowledge Base*, Addendum to Draggan (n32), *Compensation for Victims of Toxic Pollution*, PRA Research Report 83–8, prepared for the National Science Foundation, Directorate for Scientific, Technological, and International Affairs (Washington, D.C., April 1983).

95. Public Law 96-510, December 11, 1980, in *U.S. Statutes at Large 1980*, Vol. 94, Part 3, pp. 2767–811.

96. The probability of various cancers being caused by different doses of ionizing radiation is estimated from epidemiological data (Janet Raloff, "Compensating Radiation Victims," *Science News 124* [November 19, 1983], 330–1).

97. R. Jeffrey Smith, "Judge Says Atom Tests Caused Cancer," *Science* 224 (May 25, 1984), 853, 856. See also Tom Christoffel and Daniel Swartzman, "Nuclear Weapons Testing Fallout: Proving Causation for Exposure Injury," *American Journal of Public Health* 76 (March 1986), 290–2. An appeals court overturned this ruling in 1987 on the grounds that the Government could not be sued because of its discretionary function in making decisions ("Ruling of Negligence in U.S. Atomic Tests Overturned by Court," *New York Times*, April 22, 1987, pp. 1, 14). In the Agent Orange case, a federal judge ruled that the U.S. Government did not have to contribute to the chemical companies' mass settlement fund because the plaintiffs had not proved the "causal connection between Agent Orange" and their injuries (Joseph P. Fried, "Judge Says U.S. Need Not Aid Agent Orange Fund," *New York Times*, May 10, 1985, pp. 1, 16).

98. Quoted in Stephen Tarnoff, "Keene Ruling Applied to DES Coverage," *Business Insurance* (April 30, 1984), p. 30. Indeed, Lilly has flourished financially. In 1978, it increased sales by 19 percent to $1.8 billion, and net income by 24 percent, to $277.5 million. Its return on equity and margin on sales were well above the average for major drug companies, which in turn far surpass the averages for U.S. industry ("Eli Lilly: New Life in the Drug Industry," *Business Week* [October 29, 1979], pp. 134–45).

99. E. Claiborne Robins, Jr., quoted in Diamond (n90), "Drug Company Asks Protection."

100. See, for example, the proposal by Richard Merrill, a former FDA general counsel, for further expansion of drug manufacturers liability as a means of facilitating compensation as well as promoting deterrence of future drug injuries (Richard A. Merrill, "Compensation for Prescription Drug Injuries," *Virginia Law Review* 59 [January 1973], 1–120).

101. "Setting Limits on Lawsuits and Lawyers," *New York Times*, May 25, 1986, p. 18. James Barron, "40 Legislatures Act to Readjust Liability Rules," *New York Times*, July 14, 1986, pp. 1, 9. William Glaberson, "Liability Rates Flattening Out As Crisis Eases," *New York Times*, February 9, 1987, pp. 1, 25.

102. New Zealand's system is described in Bernzweig (n5), *By Accident Not Design*, pp. 191–209. For a description of Sweden's "patient-injury system," see James K. Cooper, "Sweden's No-Fault Patient-Injury Insurance," *New England Journal of Medicine* 294 (June 3, 1976), 1268–70.

103. Quoted in Richard Gaskins, "The New Zealand Accident Compensation Act: An Innovative Response to Technology-Related Risk," paper delivered at the Third Annual Meeting of the Society for Social Studies of Science (Bloomington, Indiana, November 4, 1978).

104. Richard Smith, "The World's Best System of Compensating Injury?" *British Medical Journal 284* (April 24, 1982), 1243–5; Richard Smith, "Compensation for Medical Misadventure and Drug Injury in the New Zealand No-Fault System: Feeling the Way," *British Medical Journal 284* (May 15, 1982), 1457–9. See also Bernzweig (n5), *By Accident Not Design*, pp. 205–6.

105. Although no-fault in theory, workers' compensation is very much an adversary system in practice, with delayed benefits and high legal fees as a result. A 1977 Federal Interdepartmental Policy Group study found that 39 percent of all permanent partial disability and death cases are contested by employers, that 52 percent of all permanent total disability cases end up in litigation, and that the average delay between injury and payment in contested cases is 134 days (Bernzweig [n5], *By Accident Not Design*, p. 18). A union-funded legal foundation has recently been established to provide legal and technical assistance with delayed-onset injuries and those due to toxic substances, since most workers' compensation laws are as ill-prepared as the courts to deal with such cases (Constance Holden, "Union Sets Up Legal Fund for Occupational Health," *Science 226* [October 19, 1984], 324).

106. U.S. Department of Transportation, *State No-Fault Automobile Insurance Experience, 1971–1977* (June 1977), p. 80, quoted in O'Connell (n13), *The Lawsuit Lottery*, p. 175.

107. These no-fault plans and their rationale are discussed in O'Connell (n10), *Ending Insult to Injury*; idem (n13), *The Lawsuit Lottery*; and Clark C. Havighurst and Laurence R. Tancredi, " 'Medical Adversity Insurance' – A No-Fault Approach to Medical Malpractice and Quality Assurance," *Milbank Memorial Fund Quarterly/Health and Society 51* (Spring 1973), 125–67.

108. Marshall S. Shapo, *A Nation of Guinea Pigs* (New York: Free Press, 1979), p. 56.

109. O'Connell (n13), *The Lawsuit Lottery*; idem (n10), *Ending Insult to Injury*. Marc Franklin has also argued for uncoupling financial liability and compensation (see Franklin [n13], "Replacing the Negligence Lottery").

110. See Julian Gresser, Koichiro Fujikura, and Akio Morishima, *Environmental Law in Japan* (Cambridge, Mass.: MIT Press, 1981), pp. 285–323. Toxic substance emissions have in fact fallen since the compensation system was introduced (Takefumi Kondo, M.D., Director of Environmental Statistics Section, Government of Japan, Tokyo, personal communication, October 15, 1984).

111. Similar approaches have been proposed for toxic substance compensation in the United States, although without much chance of success for the foreseeable future. A bill introduced by Congressman William Brodhead, H.R. 9616, was modeled closely on the Japanese system. See Stephen M. Soble, "A Proposal for the Ad-

ministrative Compensation of Victims of Toxic Substance Pollution: A Model Act," *Harvard Journal on Legislation 14* (June 1977), 683–824. Even Japan's system, enacted in an atmosphere of crisis following the discovery of severe health problems caused by pollution, has run into difficulties. No new diseases have been declared eligible for compensation, largely because of the system's steadily escalating costs. Cancer has never been included, for instance, despite known linkages with certain chemicals (Takefumi Kondo [n110], personal communication, October 15, 1984).

112. Insurance Bureau, Michigan Department of Commerce, *No-Fault Insurance in Michigan: Consumer Attitudes and Performance* (1978), pp. 15–18, quoted in O'Connell (n13), *The Lawsuit Lottery*, p. 173.

113. Franklin (n13), "Replacing the Negligence Lottery"; Trauberman (n94), "Statutory Reform of 'Toxic Torts' "; Melissa A. Turner, "Bearing the Burden of DES Exposure," *Oregon Law Review 60* (1981), 309–24.

114. Turner (n113), "Bearing the Burden of DES Exposure," 324.

9. What is fair?

1. "Patient Just What the Doctors Ordered," *San Francisco Chronicle*, December 3, 1982, p. 5.

2. "Barney Clark Called a Poor Choice," *San Francisco Chronicle*, September 17, 1983, p. 10.

3. Interview with Dr. Ross Woolley, Department of Community Medicine, University of Utah, by Diana Dutton, March 6, 1984.

4. Lawrence K. Altman, "Artificial Heart: Test of Technology," *New York Times*, April 8, 1985, p. B5.

5. Egalitarianism, strictly speaking, is a "mixed" theory of justice in that it seeks to establish equality as the paramount value to guide behavior, but permits some consideration of utilitarian consequences as well. See, for example, William K. Frankena, *Ethics*, 2nd ed. (Englewood Cliffs, N.J.: Prentice-Hall, 1973); Bernard A. O. Williams, "The Idea of Equality," in Hugo A. Bedau, ed., *Justice and Equality* (Englewood Cliffs, N.J.: Prentice-Hall, 1971), pp. 116–37. Another widely acclaimed mixed theory with an egalitarian bent is the "justice as fairness" framework developed by John Rawls, a political philosopher, in *A Theory of Justice* (Cambridge, Mass.: Harvard University Press, 1971). This framework is based on the principles which people would, according to Rawls, rationally choose if they had to establish the moral rules governing society while ignorant of their own social position. The resulting principles would permit inequality, Rawls concludes, but only if it were to the benefit of the least advantaged members of society. Although Rawls did not include health within the scope of his theory (he treated health as a given rather than as a good to be

distributed), others have applied the Rawlsian principles to health and have concluded that they would yield an equal distribution of access to health services. See Ronald M. Green, "Health Care and Justice in Contract Theory Perspective," in Robert M. Veatch and Roy Branson, eds., *Ethics and Health Policy* (Cambridge, Mass.: Ballinger, 1976), pp. 111–26.

6. Robert M. Veatch, "What Is a 'Just' Health Care Delivery?" in Veatch and Branson (see n5), *Ethics and Health Policy*, p. 134. Other philosophers have taken a different view of what "equality" requires. Frankena, for example, contends that it means "making the same relative contribution to the goodness of [people's] lives" (Frankena [n5], *Ethics*, p. 51).

7. Normal Daniels, "Health-Care Needs and Distributive Justice," *Philosophy and Public Affairs 10* (Spring 1981), 146–79.

8. See, for example, Alastair V. Campbell, *Medicine, Health and Justice* (Edinburgh, Scotland: Churchill Livingstone, 1978).

9. Public Law 89-749. Comprehensive Health Planning Act of 1966: Findings and Declaration of Purpose.

10. "Constitution of the World Health Organization," in *The First Ten Years of the World Health Organization* (Geneva, Switzerland: World Health Organization, 1958), p. 459.

11. Kenneth Stuart, "Health for All: Its Challenge for Medical Schools," *Lancet 1* (February 25, 1984), 441–2.

12. Department of Health, Education and Welfare, *Papers on the National Health Guidelines: Conditions for Change in the Health Care System*, Publ. No. (HRA) 78-642 (Washington, D.C.: Government Printing Office, September 1977), pp. 48–50.

13. For an excellent discussion of this point, see Dan E. Beauchamp, "Public Health as Social Justice," *Inquiry 13* (March 1976), 3–14.

14. See, for example, Victor R. Fuchs, *Who Shall Live?* (New York: Basic Books, 1974).

15. National Commission for the Protection of Human Subjects of Biomedical and Behavioral Research, *Report and Recommendations: Ethical Guidelines for the Delivery of Health Services by DHEW*, Department of Health, Education and Welfare, Publ. No. (OS) 78-0010 (Washington, D.C.: Government Printing Office, September 1978), p. 88.

16. President's Commission for the Study of Ethical Problems in Medicine and Biomedical and Behavioral Research, *Securing Access to Health Care: The Ethical Implications of Differences in the Availability of Health Services*, Vol. 1: *Report* (Washington, D.C.: Government Printing Office, March 1983), p. 4.

17. See for example Janet D. Perloff, Susan A. LeBailly, Phillip R. Kletke, et al., "Premature Death in the United States: Years of Life Lost and Health Priorities," *Journal of Public Health Policy* (June 1984), pp. 167–84; Theodore D. Woolsey, *Toward an Index of*

Preventable Mortality, Department of Health and Human Services, Publ. No. (PHS) 81-1359 (Washington, D.C.: Government Printing Office, May 1981).

18. *Health Status of Minorities and Low-Income Groups*, prepared by Melvin H. Rudov and Nancy Santangelo, Department of Health, Education and Welfare, Publ. No. (HRA) 79-627 (Washington, D.C.: Government Printing Office, 1979), p. 110.

19. Health Insurance Institute, "Health and Health Insurance: The Public's View," (December 1979), p. 13; "Public Is Found Not Willing to Take Cuts in Own Care," *Nation's Health* (April 1984), 1, 5.

20. "Health Spending Projected at 15% GNP at Turn of Century," *Nation's Health* (July 1987), 5.

21. Prevention is not without its own costs and risks, of course; see Louise B. Russell, *Is Prevention Better than Cure?* (Washington, D.C.: Brookings Institution, 1986).

22. Benjamin Freedman, "The Case for Medical Care, Inefficient or Not," *Hastings Center Report 7* (April 1977), 31–9.

23. Garrett Hardin, "The Tragedy of the Commons," *Science 162* (December 13, 1968), 1243–8. See also Howard H. Hiatt, "Protecting the Medical Commons: Who Is Responsible?" *New England Journal of Medicine 293* (July 31, 1975), 235–41.

24. Interview with Dr. Ross Woolley, Department of Community Medicine, University of Utah, by Diana Dutton, June 6, 1983.

25. Lawrence K. Altman, "Heart Team Sees Lessons in Dr. Clark's Experience," *New York Times*, April 17, 1983, pp. 1, 20.

26. Dr. Robert Jarvik, quoted in Denise Grady, "Summary of Discussion on Ethical Perspectives," in Margery W. Shaw, ed., *After Barney Clark* (Austin, Tex.: University of Texas Press, 1984), p. 46.

27. The Working Group on Mechanical Circulatory Support, National Heart, Lung, and Blood Institute, *Artificial Heart and Assist Devices: Directions, Needs, Costs, and Societal and Ethical Issues* (Bethesda, Md.: National Heart, Lung, and Blood Institute, May 1985) [hereafter cited as "1985 Working Group"], p. 26.

28. Quoted in Robert J. Levine, "Total Artificial Heart Implantation – Eligibility Criteria" (editorial), *Journal of the American Medical Association 252* (September 21, 1984), 1459.

29. Kidney dialysis allows blood to be circulated outside the body through a machine that removes toxic waste products before returning the blood to the person's body. It works like an artificial kidney, and is generally very effective. In one hospital, it was reported that during a three-year period, "120 suitable patients have been considered for chronic hemodialysis, but only 21 patients have been treated because of limitations of space and equipment. The 21 patients are all alive, whereas of the remaining 99,

only one is surviving and he was rejected only a month ago."
Stanley Shaldon, Christina M. Comty, and Rosemarie A. Baillod,
"Profit and Loss in Intermittent Haemodialysis" (letter to the ed-
itor), *Lancet 2* (December 4, 1965), 1183.

30. Shana Alexander, "They Decide Who Lives, Who Dies," *Life*
(November 9, 1962), 102–25.

31. People who had heart disease, diabetes or other chronic illnesses
were considered unsuitable for dialysis. See Jhan Robbins and June
Robbins, "The Rest Are Simply Left to Die," *Redbook 130* (November 1967), 81, 132–4.

32. Ibid., 133.

33. David Sanders and Jesse Dukeminier, Jr., "Medical Advance and
Legal Lag: Hemodialysis and Kidney Transplantation," *UCLA
Law Review 15* (February 1968), 378.

34. Nicholas Rescher, "The Allocation of Exotic Medical Lifesaving
Therapy," in *Unpopular Essays on Technological Progress* (Pittsburgh: University of Pittsburgh Press, 1980), p. 39.

35. Sanders and Dukeminier, Jr. (n33), "Medical Advance and Legal
Lag," 378.

36. Paul A. Freund, "Ethical Problems in Human Experimentation,"
New England Journal of Medicine 273 (September 23, 1965), 688.

37. Amendment 14, Section 1 of the U.S. Constitution states: "No
state shall make or enforce any law which shall abridge the priv-
ileges or immunities of citizens of the United States; nor shall any
state deprive any person of life, liberty, or property without due
process of law; nor deny to any person within its jurisdication the
equal protection of the law." See Sanders and Dukeminier, Jr.
(n33), "Medical Advance and Legal Lag," 373–6.

38. Richard A. Rettig, "The Policy Debate on Patient Care Financing
for Victims of End-Stage Renal Disease," *Law and Contemporary
Problems 40* (Autumn 1976), 220.

39. Guido Calabresi and Philip Bobbitt, *Tragic Choices* (New York:
Norton, 1978), p. 189.

40. Report of the Artificial Heart Assessment Panel, National Heart
and Lung Institute, *The Totally Implantable Artificial Heart: Legal,
Social, Ethical, Medical, Economic, Psychological Implications* (Bethesda, Md.: Department of Health, Education and Welfare Publ.
No. [NIH] 74-191, June 1973), p. 147.

41. Ibid., p. 198.

42. *Federal Register 44*, No. 76 (April 18, 1979), 23194.

43. "Report of the Artificial Heart Working Group," May 21, 1981,
photocopy obtained from the National Heart, Lung, and Blood
Institute, Washington, D.C., p. 20. It has been pointed out that
the Defense Department, much maligned for cost overruns and
inefficiency, has been considerably more candid about cost projections than the National Heart Institute (Harvey M. Sapolsky,

"Here Comes the Artificial Heart," *The Sciences* [December 1978], 25–7).

44. "U.S. Plans Cut in Dialysis Payments," *San Francisco Chronicle*, March 26, 1983, p. 7. By 1985 the costs of the kidney dialysis program were approaching $3 billion a year (Irvin Molotsky, "Panel Will Study Medicare Heart Fund," *New York Times*, May 3, 1985, p. 11).

45. 1985 Working Group (n27), *Artificial Heart and Assist Devices*, p. 19. These figures are in 1985 dollars; the Working Group reported figures in 1983 dollars.

46. "The ESRD Program: Health Policy Microcosm," *Washington Report on Medicine & Health/Perspectives* (Washington, D.C.: McGraw-Hill, March 8, 1982).

47. 1985 Working Group (n27), *Artificial Heart and Assist Devices*, p. 32.

48. Ibid., p. 3.

49. Barbara Greenberg and Robert A. Derzon, "Determining Health Insurance Coverage of Technology: Problems and Options," *Medical Care 19* (October 1981), 967–78.

50. 1985 Working Group (n27), *Artificial Heart and Assist Devices*, p. 32.

51. "The Heart of the Future," *San Francisco Chronicle*, December 6, 1982, p. 4.

52. Ibid.

53. See Roger W. Evans, "Health Care Technology and the Inevitability of Resource Allocation and Rationing Decisions, Part II" *Journal of the American Medical Association 249* (April 22/29, 1983), 2211. Furthermore, a significant number – some 35,000 – of patients that receive coronary artery bypass grafts derive little if any medical benefit. See CASS Principal Investigators and Their Associates, "Myocardial Infarction and Mortality in the Coronary Artery Surgery Study (CASS) Randomized Trial," *New England Journal of Medicine 310* (March 22, 1984), 750–8.

54. Nelda McCall, "Utilization and Costs of Medicare Services by Beneficiaries in Their Last Year of Life," *Medical Care 22* (April 1984), 329–42. For aggregate figures on Medicare expenditures for kidney disease, see Evans (n53), "Health Care Technology," 2208.

55. Mark R. Chassin, Robert H. Brook, R. E. Park, et al., "Variations in the Use of Medical and Surgical Services by the Medicare Population," *New England Journal of Medicine 314* (January 30, 1986), 285–90; John Wennberg and Alan Gittelsohn, "Variations in Medical Care Among Small Areas," *Scientific American 246* (April 1982), 120–9; *Background Report on Surgery in State Medicaid Programs*, Subcommittee on Oversight and Investigations of the Committee on Interstate and Foreign Commerce, U.S. House, 95th

Congress, 1st Session (Washington, D.C.: Government Printing Office, July 1977).

56. These data are reviewed in Diana B. Dutton, "Social Class, Health and Illness," in Linda H. Aiken and David Mechanic, eds., *Applications of Social Science to Clinical Medicine and Health Policy* (New Brunswick, N.J.: Rutgers University Press, 1986), pp. 31–62. Also, see Department of Health and Human Services, Public Health Service, *Health Status of the Disadvantaged: Chartbook 1986*, DHHS Publ. No. (HRSA) HRS-P-DV86-2 (Washington, D.C.: Government Printing Office, 1986).

57. Dutton (n56), "Social Class, Health, and Illness." For data on vitamins, see Jeffrey P. Koplan, Joseph L. Annest, Peter M. Layde, et al., "Nutrient Intake and Supplementation in the United States (NHANES II)," *American Journal of Public Health 76* (March 1986), 287–9.

58. Julian Tudor Hart, "The Inverse Care Law," *Lancet 1* (February 27, 1971), 405–12.

59. The influenza incidence rates for persons aged 45–64 are 36% versus 64% per year for the highest and lowest income groups, respectively; for those aged 17–44 years, the corresponding rates are 53% versus 72%. National Center for Health Statistics, *Health United States 1978*, Department of Health, Education and Welfare, Publ. No. (PHS) 78-1232 (Washington, D.C.: Government Printing Office, December 1978), pp. 240–1; for death rates see Rudov and Santangelo (n18), *Health Status of Minorities and Low Income Groups*, p. 91.

60. David Perlman, "The Flu Shot Program Poses a Dilemma," *San Francisco Chronicle*, April 5, 1976, p. 2.

61. Kenneth D. Rosenberg, "Swine Flu: Play It Again, Uncle Sam," *Health/PAC Bulletin* (November/December 1976), p. 17.

62. The 1985 Working Group (n27) projected a maximum of 35,000 artificial heart candidates. Figures on the prevalence of heart disease are from Alain Colvez and Madeleine Blanchet, "Disability Trends in the United States Population 1966–76: Analysis of Reported Causes," *American Journal of Public Health 71* (May 1981), 468. Proponents of partial assist devices ("VADs") claim they could aid as many as 80,000 victims of fatal heart disease each year, but this is still only about 0.5 percent of all people with heart disease.

63. Artificial heart program funding from Michael J. Strauss, "The Political History of the Artificial Heart," *New England Journal of Medicine 310* (February 2, 1984), 335; NHLBI funding for heart disease from *NIH Data Book 1983*, Department of Health and Human Services, NIH Publ. No. 83-1261 (June 1983), p. 16. The comparison of federal funding for the artificial heart versus heart transplants is from Norman Shumway, in John Langone, "The

Artificial Heart Is Really Very Dangerous," *Discover* (June 1986), p. 40.

64. Hittman Associates, Inc., *Final Summary Report on Six Studies Basic to Consideration of the Artificial Heart Program*, prepared for the National Institutes of Health under Contract PH 43-66-90 (Baltimore: Hittman Associates, October 24, 1966), p. III-7.

65. 1985 Working Group (n27), *Artificial Heart and Assist Devices*.

66. Dr. Allan Lansing, quoted in "Artificial Heart Program Shortcomings Admitted," *San Francisco Chronicle*, May 13, 1985, p. 19.

67. This argument is elaborated in George J. Annas, "No Cheers for Temporary Artificial Hearts," *Hastings Center Report 15* (October 1985), 27–8; and idem, "The Phoenix Heart: What We Have to Lose," *Hastings Center Report 15* (June 1985), 15–16.

68. 1985 Working Group (n27), *Artificial Heart and Assist Devices*, p. 56.

69. Ibid., pp. 19–20. The estimates for heart and liver transplants are from the *Report of the Massachusetts Task Force on Organ Transplantation*, George J. Annas, Chairman (Boston: Department of Public Health, October 1984), p. 58, and are roughly double those of the 1985 Working Group (n27).

70. Richard F. Harris, "Debate over Grim Results of Medi-Cal Liver Transplants," *San Francisco Sunday Examiner and Chronicle*, March 4, 1984, p. B1.

71. Ibid., p. B2.

72. These figures are based on the cost estimates provided in Annas (n69), *Report of the Massachusetts Task Force* and reported rates of transplantation and waiting lists. See Alan L. Otten, "Federal Support for Organ Transplant Programs Grows Amid Mounting Pressure from Congress," *Wall Street Journal*, June 24, 1986, p. 62. Also, Emanuel Thorne and Gilah Langner, "The Body's Value Has Gone Up," *New York Times*, September 8, 1986, p. 23; Shirley Kraus, "Human Transplants Pose Ethical Issues for Doctors, Patients," *Campus Report*, Stanford University, April 9, 1986, p. 21.

73. "HHS Questions Paying for Costly Procedures," *Nation's Health* (August 1980), p. 11.

74. Alexander Leaf, "The MGH Trustees Say No to Heart Transplants," *New England Journal of Medicine 302* (May 8, 1980), 1088. The MGH decision came shortly before Medicare's plans were announced, and MGH officials said financing was not an important part of its decision; according to one trustee, they "did not even explore the funding in any great depth" (Richard A. Knox, "A Reluctant 'No' at MGH: Hospital's Heart Transplant Decision Not an Easy One," *Boston Globe*, February 17, 1980, p. A2).

75. Mark Rust, "Blue Plans to Cover Transplants," *American Medical News* (February 17, 1984), pp. 1, 15; Charles Petit, "Blue Shield OKs Stanford Heart Transplant Insurance," *San Francisco Chron-*

icle, February 2, 1984, p. 1; Nancy G. Kutner, "Issues in the Application of High Cost Medical Technology: The Case of Organ Transplantation," *Journal of Health and Social Behavior 28* (March 1987), 23–36.

76. John K. Iglehart, "The Politics of Transplantation," *New England Journal of Medicine 310* (March 29, 1984), 864–8.

77. Thorne Gray, "Duke Orders State Aid for Child's Transplant," *Sacramento Bee,* March 17, 1984, p. A6; Assembly Bill 3266, introduced by Assemblyman Papan, California Legislature, 1983–4.

78. John K. Iglehart, "Transplantation: The Problem of Limited Resources," *New England Journal of Medicine 309* (July 14, 1983), 123–8.

79. "City, State Health Officers Testify: Effective Health Efforts Have Been Cut," *Nation's Health* (July 1983), p. 7; Milton Kotelchuck, Janet B. Schwartz, Marlene T. Anderka, et al., "WIC Participation and Pregnancy Outcomes: Massachusetts Statewide Evaluation Project," *American Journal of Public Health 74* (October 1984), 1086–92.

80. Data from the Texas Early Periodic Screening, Detection, and Treatment Program, cited in "The Costs of Not Providing Health Care: People Under Pressure: Cutback Fact Sheet," typescript from the National Health Law Program, Los Angeles, California, 1982.

81. National Health Law Program (n80), "The Costs of Not Providing Health Care."

82. Neil A. Holtzman, "Prevention: Rhetoric and Reality," *International Journal of Health Services 9* (1979), 25–39.

83. National Health Law Program (n80), "The Costs of Not Providing Health Care."

84. H. Jack Geiger, "Community Health Centers: Health Care as an Instrument of Social Change," in Victor W. Sidel and Ruth Sidel, eds., *Reforming Medicine: Lessons of the Last Quarter Century* (New York: Pantheon, 1984), pp. 11–32.

85. Thomas E. Kottke, Pekka Puska, Roger Feldman, et al., "A Decline in Earning Losses Associated with a Community-Based Cardiovascular Disease Prevention Project," *Medical Care 20* (July 1982), 663–75.

86. Geiger (n84), "Community Health Centers." For an eloquent plea against such funding cuts, see "For Children: A Fair Chance – Stop Wasting Lives, and Money" (editorial), *New York Times,* September 6, 1987, Section E, p. 14.

87. Austin Scott, "Country's Poor Seen Delaying Visits to Doctor," *Los Angeles Times,* December 19, 1982, Part II, pp. 1, 3; "11 Die After Losing Aid: U.S. Found Them Fit to Work," *San Francisco Chronicle,* September 18, 1982, p. 4.

88. "Federal Policy and Infant Health: A Debate," *Washington Report on Medicine & Health/Perspectives* (Washington, D.C.: McGraw-Hill, January 23, 1984).

89. In 1981, neonatal intensive care for a single premature infant averaged $60,000–$100,000, and the costs of institutionalizing a mentally retarded person run about $25,000 per year for as long as the person lives. "Factual Memorandum and Arguments in Support of Petition for Rulemaking: For Rules and Regulations to Declare Prenatal Care a Public Health Service and to Establish Standards for Access to Such Care by Low-Income Women," photocopy obtained from the National Health Law Program, Los Angeles, 1982.

90. Ronald M. Green, "Intergenerational Distributive Justice and Environmental Responsibility," *BioScience 27* (April 1977), 260–5.

91. June Axinn and Mark J. Stern, "Age and Dependency: Children and the Aged in American Social Policy," *Milbank Memorial Fund Quarterly/Health and Society 63* (Fall 1985), 665. According to the Congressional Budget Office, 39.2 percent of all poor people in 1983 were children (Robert Pear, "Increase Found in Child Poverty in Study by U.S.," *New York Times*, May 23, 1985, p. 1).

92. Samuel H. Preston, "Children and the Elderly in the U.S." *Scientific American 251* (December 1984), 45.

93. McCall (n54), "Utilization and Costs of Medicare Services," 329.

94. Fern Schumer Chapman, "Deciding Who Pays to Save Lives," *Fortune* (May 27, 1985), 60.

95. Jonathan A. Showstack, Mary Hughes Stone, and Steven A. Schroeder, "The Role of Changing Clinical Practices in the Rising Costs of Hospital Care," *New England Journal of Medicine 313* (November 7, 1985), 1201–7.

96. Edward B. Fiske, "Executives Urge a Rise in Aid for Poor Children," *New York Times*, September 6, 1987, p. 14.

97. Harry Schwartz, "We Need to Ration Medicine," *Newsweek* (February 8, 1982), 13.

98. Dr. Robert Jarvik, quoted in Marilyn Chase, "Firm That Developed Artificial Heart Seeks to Build Bionic Market," *Wall Street Journal*, July 24, 1984, p. 1.

99. Malcolm N. Carter, "The Business Behind Barney Clark's Heart," *Money* (April 1983), 143.

100. Lee Smith, quoted in Carter (n99), "The Business Behind Barney Clark's Heart," 143.

101. Chase (n98), "Firm That Developed Artificial Heart."

102. Kathleen Selvaggio, "Cashing in on DNA," *Multinational Monitor 5* (March 1984), 13–15, 20.

103. "The Health Care System in the Mid-1990s," study conducted by Arthur D. Little, Inc. for the Health Insurance Association of America, 1850 K Street N.W., Washington, D.C. 20006, January 1985, pp. 1–2.

104. David Talbot, quoted in James Barron, "Humana Focus: Tech-
     nology," *New York Times*, August 14, 1984, Business Section, p.
     33.
105. David Jones, quoted in "Hospital Sees Operation as a 'Wise In-
     vestment,' " *New York Times*, November 26, 1984, p. 11.
106. George Atkins, quoted in "Hospital Sees Operation" (n105).
107. Kathleen Deveny and John P. Tarpey, "Humana: Making the
     Most of Its Place in the Spotlight," *Business Week* (May 6, 1985),
     68.
108. Henry J. Werronen, quoted in Deveny and Tarpey (n107), "Hu-
     mana: Making the Most of Its Place," 69.
109. George Atkins, quoted in "Hospital Sees Operation" (n105).
110. Dr. Arnold Relman, quoted in Lawrence K. Altman, "Health Care
     as Business," *New York Times*, November 27, 1984, p. 24.
111. Arthur L. Caplan, "Should Doctors Move at a Gallop?" *New York
     Times*, August 17, 1984, p. 25.
112. Dr. William DeVries, quoted in Lawrence K. Altman, "Utah
     Heart Implant Program Moving to Kentucky Institute," *New York
     Times*, August 1, 1984, p. 6.
113. Altman (n110), "Health Care as Business," p. 24.
114. Office of Technology Assessment, *Federal Policies and the Medical
     Devices Industry*, Publ. No. OTA-H-229 (Washington, D.C.: Gov-
     ernment Printing Office, October 1984), p. 14.
115. Institute of Medicine, *For-Profit Enterprise in Health Care* (Wash-
     ington, D.C.: National Academy Press, 1986).
116. Lester C. Thurow, "Learning to Say 'No,' " *New England Journal
     of Medicine 311* (December 13, 1984), 1569–72.
117. Dr. Chase Peterson, quoted in Thomas A. Preston, "The Case
     Against the Artificial Heart," *Utah Holiday* (June 1983), 39.
118. Ibid., p. 39.
119. *Smoking and Health: Report of the Advisory Committee to the Surgeon
     General of the Public Health Service*, Department of Health, Edu-
     cation and Welfare, Publ. No. PHS 1103 (Washington, D.C.:
     Government Printing Office, January 1964). See also Weldon J.
     Walker, "Changing United States Lifestyle and Declining Vascular
     Mortality: Cause or Coincidence?" *New England Journal of Medicine
     297* (July 21, 1977), 163–5; and Michael P. Stern, "The Recent
     Decline in Ischemic Heart Disease Mortality," *Annals of Internal
     Medicine 91* (October 1979), 630–40.
120. Department of Health and Human Services, National Center for
     Health Statistics, *Health United States 1985*, Publ. No. (PHS) 86-
     1232 (Washington, D.C.: Government Printing Office, December
     1985), p. 1. See also Jeremiah Stamler, "Lifestyles, Major Risk
     Factors, Proof and Public Policy," *Circulation 58* (July 1978), 15.
121. Guido Calabresi, "Reflections on Medical Experimentation in Hu-
     mans," in Paul A. Freund, ed., *Experimentation with Human Subjects*
     (New York: George Braziller, 1969), pp. 178–96.

122. See Jonathan Glover, *Causing Death and Saving Lives* (New York: Penguin Books, 1977), pp. 210–12.
123. Department of Health, Education and Welfare, *Healthy People: The Surgeon General's Report on Health Promotion and Disease Prevention*, Publ. No. (PHS) 79-55071 (Washington, D.C.: Government Printing Office, 1979), p. Ch10-10.
124. Stamler (n120), "Lifestyles," 14.
125. See Kenneth E. Warner, *Selling Smoke: Cigarette Advertising and Public Health* (Washington, D.C.: American Public Health Association, 1986). Also, *Developing a Blueprint for Action*, Proceedings of the National Conference on Smoking *or* Health, November 18–20, 1981 (New York: American Cancer Society, 1981), p. 249.
126. George Gallup, The Gallup Poll: "One of Every 5 Adults in U.S. Can't Always Afford Food," *San Francisco Chronicle*, March 19, 1984, p. 7.
127. See, for example, Thomas McKeown, *The Role of Medicine: Dream, Mirage, or Nemesis?* (Princeton, N.J.: Princeton University Press, 1979). For more recent evidence, see "U.S. Held Losing Fight with Cancer; Prevention Is Stressed," *New York Times*, May 8, 1986, p. 16; John C. Bailar, III, and Elaine M. Smith, "Progress Against Cancer?" *New England Journal of Medicine 314* (May 8, 1986), 1226–32.
128. Howard Berliner and J. Warren Salmon, "Swine Flu, the Phantom Threat," *Nation* (September 25, 1976), p. 272.
129. Alison Maclure and Gordon T. Stewart, "Admission of Children to Hospitals in Glasgow: Relation to Unemployment and Other Deprivation Variables," *Lancet 2* (September 22, 1984), 682–5.
130. Gerald Weissmann, "AIDS and Heat," *New York Times*, September 28, 1983, p. 25.
131. Mary Sue Henifin and Ruth Hubbard, "Genetic Screening in the Workplace," *Genewatch* (November/December 1983), 5.
132. See, for example, Lewis Thomas, "Biomedical Science and Human Health: The Long-Range Prospect," *Daedalus 106* (Summer 1977), 163–71. Also Richard M. Krause, "The Beginning of Health Is to Know the Disease," *Public Health Reports 98* (November–December 1983), 531–5.
133. Lewis Thomas, "Who Will Be Saved? Who Will Pay the Cost?" *Discover* (February 1983), 30.
134. Lewis Thomas, *The Lives of a Cell: Notes of a Biology Watcher* (New York: Bantam Books, 1974), p. 40.
135. Department of Health and Human Services, *Research in Prevention: Fiscal Years 1981–1983* (Bethesda, Md.: National Institutes of Health, June 1984), p. 2.
136. Department of Health, Education and Welfare (n123), *Healthy People*, p. Ch1-10.
137. See, for example, Stephen P. Strickland, *Research and the Health of*

Americans: Improving the Policy Process (Lexington, Mass.: D. C. Heath, 1978).

138. Hastings, quoted in Paul Sullivan, Ronald Randall, Charles Sethness, et al., *Artificial Internal Organs: Promise, Profits, and Problems* (Boston: Nimrod Press, 1966), p. 160.

139. Quote is from Erich Bloch, "NSF's Budget and Economic Competitiveness" (editorial), *Science 235* (February 6, 1987), 621. On the proposed expansion of the Commerce Department's role, see Mark Crawford, "Broader R&D Role Sought for Commerce," *Science 237* (July 3, 1987), 19.

140. Willard Gaylin, "Harvesting the Dead: The Potential for Recycling Human Bodies," *Harpers* (September 1974), 23–30.

141. "Brain-dead Man's Body Kept Alive by Artificial Heart," *San Jose Mercury*, May 14, 1982, p. 16A.

142. David Perlman, "Doctors Blast Human Organ Broker: 3 California Experts Are Outraged," *San Francisco Chronicle*, October 22, 1983, p. 2. See also Marjorie Sun, "Organs for Sale," *Science 222* (October 7, 1983), 34.

143. George J. Annas, "Life, Liberty, and the Pursuit of Organ Sales," *Hastings Center Report 14* (February 1984), 22–3.

144. See, for example, Lori B. Andrews, "My Body, My Property," *Hastings Center Report 16* (October 1986), 28–38. "Reagan Administration Fights Organ-Donor Bill," *San Francisco Chronicle*, October 18, 1983, p. 11; Phil Gunby, "Organ Transplant Improvements, Demands Draw Increasing Attention," *Journal of the American Medical Association 251* (March 23/30, 1984), 1521–7.

145. Dr. Claire Randall, Rabbi Bernard Mandelbaum, and Bishop Thomas Kelly, "Message from Three General Secretaries," National Council of the Churches of Christ (110 Maryland Avenue, N.E., Washington, D.C. 20002), July 1980.

146. Constance Holden, "Ethics Panel Looks at Human Gene Splicing," *Science 217* (August 6, 1982), 516–17.

147. Colin Norman, "Clerics Urge Ban on Altering Germline Cells," *Science 220* (June 24, 1983), 1360–1.

148. Gina Kolata, "Gene Therapy Method Shows Promise," *Science 223* (March 30, 1984), 1376–9; Harold M. Schmeck, "Gene Therapy for Humans Moves Nearer to Reality," *New York Times*, May 26, 1987, pp. 15, 19.

149. Kathleen McAuliffe and Sharon McAuliffe, "Keeping Up with the Genetic Revolution," *New York Times Magazine*, November 6, 1983, p. 96.

150. Ibid., pp. 95–6.

151. Government surveys have shown that more than a third of the public believes that scientific discoveries tend to undermine people's ideas of right and wrong. National Science Board, National

Science Foundation, *Science Indicators 1980* (Washington, D.C.: Government Printing Office, 1981), pp. 162–3.

152. Gaylin (n140), "Harvesting the Dead," 30.

153. Hanna Fenichel Pitkin and Sara M. Shumer, "On Participation," *Democracy 2* (Fall 1982), 44.

10. The role of the public

1. Material and quotes in this section are from Richard A. Knox, "Layman Is Center of Scientific Controversy," *Boston Globe*, January 9, 1977, pp. 29, 42; idem, "Cambridge Panel OK's Genetic Experiments at Harvard, MIT," *Boston Globe*, January 6, 1977, pp. 1, 9.

2. For a critique of the review board's performance and conclusions, see Rae S. Goodell, "Public Involvement in the DNA Controversy: The Case of Cambridge, Massachusetts," *Harvard Newsletter on Science, Technology and Human Values 27* (Spring 1979), 36–43.

3. "Can Gene-Splicers Make Good Neighbors?" *Business Week* (August 10, 1981), 32.

4. Todd R. La Porte and Daniel Metlay, "Technology Observed: Attitudes of a Wary Public," *Science 188* (April 11, 1975), 121–7.

5. See, for example, Carole Patemen, *Participation and Democratic Theory* (Cambridge, U.K.: Cambridge University Press, 1970); and, Martin Carnoy and Derek Shearer, *Economic Democracy: The Challenge of the 1980s* (New York: M. E. Sharpe, 1980).

6. Interviews with panel members Albert Jonsen (January 23, 1980) and Jay Katz (January 15, 1980). The panel's final report states, "We were not asked to consider specifically the question whether a totally implantable artificial heart should be developed or whether there is reasonable likelihood that it will be" (Report of the Artificial Heart Assessment Panel, National Heart and Lung Institute, *The Totally Implantable Artificial Heart: Legal, Social, Ethical, Medical, Economic, Psychological Implications* [Bethesda, Md.: Department of Health, Education and Welfare Publ. No. (NIH) 74-191, June 1973], p. 15).

7. Based on a tally of pages of testimony and discussion in House and Senate Subcommittees on Appropriations, 1964–79, and especially 1967–72 (Barton J. Bernstein, "The Artificial Heart Program," *Center Magazine* [May/June 1981], 28).

8. In 1962, the *Consumer Bulletin* stated: "DES *should not be permitted* in treatment of animals, in their feed or by implantation" ("Harmful Drug in Meats?" *Consumer Bulletin* [May 1962], 29). In 1971, the National Resources Defense Council sued both the U.S. Department of Agriculture and the Food and Drug Administration to try to force a ban on DES as a feed additive. DES was banned in 1972, but the ban was phased in over a six-month period and

was overturned two years later. A group called DOOM (Drugs Out Of Meat) opposed the six-month delay and collected 25,000 signatures in support of an immediate total ban. The group demanded that the "FDA *immediately ban all forms of diethylstilbestrol, liquid, powder, implant,* etc. from being used in livestock meant for our dinner tables," and concluded: "We are fed up with this complacency in the face of increased cancer deaths" (reprinted in *Regulation of Diethylstilbestrol, Part 3*, hearing before a Subcommittee of the Committee on Government Operations, U.S. House, 92nd Congress, 2nd Session, August 15, 1972, pp. 438–9).

9. See, for example, Barbara Seaman and Gideon Seaman, *Women and the Crisis in Sex Hormones* (New York: Rawson Associates, 1977).

10. Testimony of Anita Johnson and Sydney Wolfe in *Quality of Health Care: Human Experimentation, 1973*, hearings before the Subcommittee on Health of the Committee on Labor and Public Welfare, U.S. Senate, 93rd Congress, 1st Session, Part 1: February 21 and 22, 1973, p. 199.

11. "Fact Sheet," by Kathleen Weiss, reproduced in hearings on *Quality of Health Care: Human Experimentation, 1973* (see n10), U.S. Senate, February 1973, pp. 300–13.

12. Testimony of Belita Cowan, in *Regulation of Diethylstilbestrol, 1975*, joint hearing before the Subcommittee on Health of the Committee on Labor and Public Welfare and the Subcommittee on Administrative Practice and Procedure of the Committee on the Judiciary, U.S. Senate, 94th Congress, 1st Session, February 27, 1975, pp. 94–9. Some critics argued for a total ban on the use of DES as a morning-after pill, whereas others argued for proper experimental safeguards, including informed consent, with use restricted to physicians who had filed an Investigational New Drug Application with the FDA (Health Research Group on the Morning After Pill, December 8, 1972, reprinted in *Quality of Health Care: Human Experimentation, 1973* [n10], U.S. Senate, February 1973, pp. 201–8). Although the DES morning-after pill has been the most controversial, similar concerns have been raised about other estrogens. Summarizing evidence on the relation between estrogens and cancer, Sydney Wolfe told the Food and Drug Administration Ob-Gyn Advisory Committee: "The animal evidence certainly doesn't let one pick and choose amongst estrogens . . . we do not know of any estrogen, natural or synthetic, that is safe and that protects someone from the risk of cancer. They all need to be assumed to be both animal and human carcinogens" (transcript of the FDA Ob-Gyn Advisory Committee meeting, January 30–1, 1978, pp. 28–9).

13. The study in question was by Lucile K. Kuchera, "Postcoital Contraception with Diethylstilbestrol," *Journal of the American Medical*

Association 218 (October 25, 1971), 562–3. Cowan testified: "Dr. Kuchera's study of more than 1,000 college students at the University of Michigan was retrospective. There was no control group. I have interviewed many of these students. They were not told they were part of an experimental study. Further, many were not even followed up to see if they were pregnant. Dr. Kuchera's claim of 100 percent effectiveness is wishful thinking, and the FDA knows this" (*Diethylstilbestrol*, hearing before the Subcommittee on Health and the Environment of the Committee on Interstate and Foreign Commerce, U.S. House, 94th Congress, 1st Session, December 16, 1975, p. 51).

14. Lucile K. Kuchera, "The Morning-After Pill" (letter to the editor), *Journal of the American Medical Association 224* (May 14, 1973), 1038; Kuchera, "Postcoital Contraception with Diethylstilbestrol – Updated," *Contraception 10* (July 1974), 47–54.

15. Testimony of Vicki Jones, transcript of the Food and Drug Administration Ob-Gyn Advisory Committe meeting, January 30–1, 1978, pp. 72–8.

16. Arthur Herbst, quoted in "Disagree on DES Data," *Chemical Week* (January 18, 1978), p. 16. The report was later published by Marluce Bibbo, William M. Haenszel, George L. Wied, et al., "A Twenty-five-Year Follow-up Study of Women Exposed to Diethylstilbestrol During Pregnancy," *New England Journal of Medicine 298* (April 6, 1978), 763–7.

17. Sidney M. Wolfe, Public Citizen's Health Research Group, to Joseph Califano, Secretary of Health, Education and Welfare, letter of December 12, 1977 (photocopy from Health Research Group, 2000 P Street, N.W., Washington, D.C. 20036).

18. Adrian Gross to Sidney Wolfe, letter of December 21, 1978, quoted in Wolfe's statement before the Food and Drug Administration Ob-Gyn Advisory Committee, "Evidence of Breast Cancer from DES and Current Prescribing of DES and Other Estrogens," typescript, January 30, 1978. Technically, the disagreement was over whether a one- or two-tailed t-test was the appropriate measure of statistical significance. The general rule is that a one-tailed test is appropriate when differences are expected in only one direction (e.g., DES would be expected only to increase, not decrease, the risk of breast cancer), whereas the two-tailed test is appropriate when differences in both directions are equally likely. Wolfe argued that, given animal studies and other evidence, it was higher rates of breast cancer among the DES-exposed women that were expected and that lower rates were biologically implausible.

19. Nancy Adess, personal communication based on participation in the Task Force meeting.

20. Of the five additional studies of the association between DES and breast cancer that appeared between 1978 and 1985, four showed higher rates of breast cancer among women exposed to DES than

among the nonexposed; two (Hadjimichael et al., and Greenberg et al.) indicated an excess risk of 40–50 percent among the DES-exposed, a differential that was statistically significant. See Valerie Beral and Linda Colwell, "Randomised Trial of High Doses of Stilboestrol and Ethisterone in Pregnancy: Long-Term Follow-up of Mothers," *British Medical Journal 281* (October 25, 1980), 1098–101; O. C. Hadjimichael, J. W. Meigs, F. W. Falcier, et al., "Cancer Risk Among Women Exposed to Exogenous Estrogens During Pregnancy," *Journal of the National Cancer Institute 73* (October 1984), 831–4; E. R. Greenberg, A. B. Barnes, L. Resseguie, et al., "Breast Cancer in Mothers Given Diethylstilbestrol in Pregnancy," *New England Journal of Medicine 311* (November 29, 1984), 1393–8; Marian M. Hubby, William M. Haenszel, and Arthur L. Herbst, "Effects on the Mother Following Exposure to Diethylstilbestrol During Pregnancy," in Arthur L. Herbst and Howard A. Bern, eds., *Developmental Effects of Diethylstilbestrol (DES) in Pregnancy* (New York: Thieme-Stratton, 1981), pp. 120–8. The only study not showing a higher risk of breast cancer among DES-exposed women was that of M. P. Vessey, D. V. I. Fairweather, Beatrice Norman-Smith, et al., "A Randomized Double-Blind Controlled Trial of the Value of Stilboestrol Therapy in Pregnancy: Long-Term Follow-up of Mothers and Their Offspring," *British Journal of Obstetrics and Gynecology 90* (November 1983), 1007–17.

21. Department of Health and Human Services, National Cancer Institute, *Report of the 1985 DES Task Force* (Bethesda, Md.: National Institutes of Health, 1985), p. 4.

22. For further discussion of scientists' responses to the risks of recombinant DNA research, see Sheldon Krimsky, *Genetic Alchemy: The Social History of the Recombinant DNA Controversy* (Cambridge, Mass.: Massachusetts Institute of Technology Press, 1982).

23. Coalition for Responsible Genetic Research, "Position Statement on Recombinant DNA Technology," undated typescript.

24. Testimony of Anthony Mazzocchi, of the Oil, Chemical, and Atomic Workers Union, in *Recombinant DNA Research Act of 1977*, hearings before the Subcommittee on Health and the Environment of the Committee on Interstate and Foreign Commerce, U.S. House, 95th Congress, 1st Session, March 15, 16, and 17, 1977, p. 467.

25. See testimony of Margaret Mead, in *Regulation of Recombinant DNA Research*, hearings before the Subcommittee on Science, Technology, and Space of the Committee on Commerce, Science and Transportation, U.S. Senate, 95th Congress, 1st Session, November 2, 8, and 10, 1977, p. 166.

26. James Watson, quoted in Nicholas Wade, "Gene-Splicing Rules: Another Round of Debate," *Science 199* (January 6, 1978), 33.

27. Statement of Harlyn Halvorson, president of the American Society

for Microbiology and Director of the Basic Medical Science Research Institute, Brandeis University, in *Recombinant DNA Regulation Act, 1977*, hearing before the Subcommittee on Health and Scientific Research of the Committee on Human Resources, U.S. Senate, 95th Congress, 1st Session, April 6, 1977, pp. 139–42. The Watson quote is from "Recombinant DNA Research: A Debate on the Benefits and Risks," *Chemical and Engineering News* (May 30, 1977), p. 26. For description of the attacks on scientist critics, see Nicholas Wade, "Gene-Splicing: Critics of Research Get More Brickbats than Bouquets," *Science 195* (February 4, 1977), 466–9.

28. Statement of Oliver Smithies, professor of Medical Genetics, University of Wisconsin, in hearings on *Regulation of Recombinant DNA Research* (n25), U.S. Senate, November 1977, p. 248.

29. Marjorie Sun, "Local Opposition Halts Biotechnology Test," *Science 231* (February 14, 1986), 667–8.

30. Glenn Church, quoted in Seth Shulman, "Biotech Industry Still Working Out the Bugs," *In These Times*, April 16–22, 1986, pp. 3, 8.

31. Marjorie Sun, "Biotech Field Test Halted by State Court," *Science 233* (August 22, 1986), 838.

32. Testimony of Mazzocchi, in hearings on *Recombinant DNA Research Act of 1977* (n24), U.S. House, March 1977, p. 467.

33. Paul Slovic, "Perception of Risk," *Science 236* (April 17, 1987), 285.

34. See, for example, Amos Tversky and Daniel Kahneman, "The Framing of Decisions and the Psychology of Choice," *Science 211* (January 30, 1981), 453–8. Fear of unknown risks was dramatically illustrated in a 1986 opinion poll in which more respondents were willing to approve the release of genetically engineered organisms at a relatively high level of known risk to the environment such as 1 in 1,000 (55 percent) than at a risk described as "unknown but very remote" (45 percent). See U.S. Congress, Office of Technology Assessment, *New Developments in Biotechnology – Background Paper: Public Perceptions of Biotechnology*, Publ. No. OTA-BP-BA-45 (Washington, D.C.: Government Printing Office, May 1987) [hereafter cited as "OTA, *Public Perceptions of Biotechnology*"], p. 64.

35. Testimony of Ethan Signer, in hearings on *Recombinant DNA Research Act of 1977* (n24), U.S. House, March 1977, p. 79.

36. Quoted by Fran Fishbane in "The Role of DES Action in Federal Legislation," paper presented at the American Association for the Advancement of Science Annual Meeting, January 7, 1979, Houston, Texas.

37. Laurence H. Tribe, "Policy Science: Analysis or Ideology?" *Philosophy and Public Affairs 2* (Fall 1972), 66–110.

38. Harold P. Green, "Allocation of Resources: The Artificial Heart.

An NIH Panel's Early Warnings," *Hastings Center Report 14* (October 1984), 13–15.

39. Quote is from the "Statement on Recombinant DNA Research," Bishops' Committee for Human Values, National Conference of Catholic Bishops, May 2, 1977 (U.S. Catholic Conference, 1312 Massachusetts Avenue, N.W., Washington, D.C. 20005), p. 8. Looking ahead to the prospect of human genetic engineering at a Congressional hearing in 1976, Clifford Grobstein asked, "Is it safe, in the present state of our society, to provide means to intervene in the very essence of human individuality ... Can genetic destiny, whether of human or other species, wisely be governed by human decision?" (Clifford Grobstein, "Recombinant DNA Research: Beyond the NIH Guidelines," *Science 194* [December 10, 1976], 1134).

40. Dr. Claire Randall, Rabbi Bernard Mandelbaum, and Bishop Thomas Kelly, "Message from Three General Secretaries," National Council of the Churches of Christ (110 Maryland Avenue, N.E., Washington, D.C. 20002), July 1980.

41. Mark Crawford, "Religious Groups Join Animal Patent Battle," *Science 237* (July 31, 1987), 480–1.

42. OTA (n34), *Public Perceptions of Biotechnology*, pp. 53, 71. For attitudes on human applications, see Jon D. Miller, "Attitudes Toward Genetic Modification Research: An Analysis of the Views of the Sputnik Generation," *Science, Technology and Human Values 7* (Spring 1982), 37–43.

43. Ethan R. Signer, "Recombinant DNA: It's Not What We Need," *Research with Recombinant DNA: An Academy Forum*, March 7–9, 1977 (Washington, D.C.: National Academy of Sciences, 1977), p. 232.

44. "Statement on Recombinant DNA Research," National Conference of Catholic Bishops (n39), 1977.

45. See, for example, the statement of Ruth Hubbard, "Potential Risks," *Research with Recombinant DNA: An Academy Forum* (n43), pp. 165–9.

46. John D. Crawford, "Meat, Potatoes, and Growth Hormone" (editorial), *New England Journal of Medicine 305* (July 16, 1981), 163–4. The article was by Daniel Rudman, Michael H. Kutner, R. Dwain Blackston, et al., "Children with Normal-Variant Short Stature: Treatment with Human Growth Hormone for Six Months," *New England Journal of Medicine 305* (July 16, 1981), 123–31.

47. David L. Bazelon, "Risk and Responsibility," *Science 205* (July 20, 1979), 277–80.

48. Artificial Heart Assessment Panel (n6), *The Totally Implantable Artificial Heart*, pp. 2, 197.

49. Testimony of David Clem, member of the Cambridge City Coun-

cil, in hearing on *Recombinant DNA Regulation Act, 1977* (n27), U.S. Senate, April 1977, p. 173. For a discussion of local actions, see Sheldon Krimsky and David Ozonoff, "Recombinant DNA Research: The Scope and Limits of Regulation," *American Journal of Public Health* 69 (December 1979), 1252–9.

50. Testimony of Senator Edward Kennedy, in hearing on *Recombinant DNA Regulation Act, 1977* (n27), U.S. Senate, April 1977, pp. 2–3.

51. "Policing the Gene-Splicers" (editorial), *New York Times*, October 10, 1977, p. 28.

52. Testimony of Harlyn Halvorson, in hearing on *Recombinant DNA Regulation Act, 1977* (n27), U.S. Senate, April 1977, p. 141. For a full description of the lobbying effort, see John Lear, *Recombinant DNA: The Untold Story* (New York: Crown, 1978), pp. 172–203. Whereas some prominent scientists remained opposed to any form of federal legislation, Halvorson and his allies chose to support limited federal legislation consistent with certain agreed-on principles, rather than pursue what then seemed to be the hopeless goal of defeating all legislation. The seven recommended principles are summarized in Halvorson's testimony (p. 142).

53. Barbara J. Culliton, "Recombinant DNA Bills Derailed: Congress Still Trying to Pass a Law," *Science* 199 (January 20, 1978), 274–7.

54. Jonathan King, quoted in Nicholas Wade, "Recombinant DNA: NIH Group Stirs Storm by Drafting Laxer Rules," *Science* 190 (November 21, 1975), 769.

55. Jane Setlow, quoted in Nicholas Wade, *The Ultimate Experiment*, 2nd ed. (New York: Walker, 1979), p. 89.

56. David L. Porter, M.D., chairman, University of California at Los Angeles Institutional Biosafety Committee, letter to Diana Dutton, July 28, 1981. Also, Diana B. Dutton and John L. Hochheimer, "Institutional Biosafety Committees and Public Participation: Assessing an Experiment," *Nature* 297 (May 6, 1982), 11–15.

57. For example, Ruth Hubbard, a leading critic, testified before Congress, "Obviously one has to balance risks against benefits, but one has to also balance to whom the risks accrue and to whom the benefits accrue, and as long as these decisions are made by the people who are most likely to reap the immediate benefits and the people who are equally likely to reap the risks are not involved in the decision-making process, then the process itself is flawed" (testimony of Ruth Hubbard, professor of biology, Harvard University, in hearings on *Recombinant DNA Research Act of 1977* (n24), U.S. House, March 1977, p. 93).

58. Sidney M. Wolfe, Public Citizen's Health Research Group, to F. David Mathews, Secretary of Health, Education and Welfare, let-

ter of October 14, 1976 (photocopy from Health Research Group, 2000 P Street, N.W., Washington, D.C. 20036).

59. William Bennet and Joel Gurin, "Science That Frightens Scientists: The Great Debate over DNA," *Atlantic* (February 1977), p. 62.

60. Glenn Church, quoted in Shulman (n30), "Biotech Industry Still Working Out the Bugs," p. 3.

61. Douglas Sarojak, quoted in Sun (n29), "Local Opposition Halts Biotechnology Test," 668.

62. A 1986 Harris poll found that 43 percent of the respondents thought that "the degree of control that society has over science and technology" should be increased, whereas only 8 percent thought it should be decreased. In 1976, by contrast, 31 percent thought control should be increased, and in 1972 only 28 percent did. See OTA (n34), *Public Perceptions of Biotechnology*, p. 32.

63. Dorothy Bryant, "The DES Odyssey of Pat Cody," *San Francisco Sunday Examiner and Chronicle*, March 18, 1979, California Living, pp. 18, 21–2.

64. "DES Action Groups in Action," *DES Action Voice 1* (January 1979), 3.

65. *DES Action Voice* and *DES National Quarterly*.

66. "DES Legislation: Effective and Far-Reaching," *DES Action Voice 3* (Fall 1981), 1, 7. For an analysis of the New York legislation, see Donna M. Glebatis and Dwight T. Janerich, "A Statewide Approach to Diethylstilbestrol: The New York Program," *New England Journal of Medicine 304* (January 1, 1981), 47–50.

67. Kari Christianson, " 'National DES Awareness Week' Proclaimed by Congress and President," *DES Action Voice*, No. 25 (Summer 1985), pp. 1, 8.

68. "A Legislative Victory for DES Victims," *San Francisco Chronicle*, January 21, 1982, p. 24; "Victims of DES Win Victory in Assembly," *San Francisco Chronicle*, August 27, 1980, p. 10.

69. Susan P. Helmrich, "Victory for the Victims in New York State," *DES Action Voice*, No. 30 (Fall 1986), pp. 1, 6.

70. For example, Dr. Geraldine Oliva has worked closely with DES Action in San Francisco, collecting medical information for a fact sheet on DES that has been used widely. Nancy Adess, ex-president of DES Action National, notes that Oliva's expertise helped educate lay members, who now have a good command of the relevant technical issues themselves (personal communication). Adess also believes that Oliva's participation added to the group's legitimacy, especially in the early stages.

71. For further discussion of the role of activist-scientists in modern technical controversies, see Dorothy Nelkin, "Science, Technology, and Political Conflict: Analyzing the Issues," in Dorothy Nelkin, ed., *Controversy: Politics of Technical Decisions* (Beverly

Hills: Sage Publications, 1979); also Joel Primack and Frank von Hippel, *Advice and Dissent: Scientists in the Political Arena* (New York: Basic Books, 1974).

72. Despite the existence in 1975 of over 1200 public advisory committees, a Senate committee found that in more than half of all formal federal regulatory agency meetings, there was no public participation at all, nor was there virtually any public involvement in informal agency proceedings (U.S. Senate Committee on Governmental Affairs, "Public Participation in Regulatory Agency Proceedings," *Study on Federal Regulation*, Vol. 3 [Washington, D.C.: Government Printing Office, July 1977], p. vii). A Congressional investigation of FDA advisory committees concluded that their predominantly medical composition was "conducive to resolving issues in a manner more favorable to the practicing physician than to the public at large," and that the "FDA has increasingly used advisory committees to gain support in the medical community for regulatory decisions thought desirable by the agency's leadership" (*Use of Advisory Committees by the FDA*, Eleventh Report by the Committee on Government Operations, January 26, 1976, U.S. House Report No. 94-787 [Washington, D.C.: Government Printing Office, 1976], pp. 5–6).

73. One study of public participation in science and technology observed that expert "decision-making bodies traditionally accustomed to receiving technically-competent, legally reasoned briefs often find it difficult to cope with qualitatively different types of testimony. Intervenors expressing strong social, political, or emotional points of view are therefore often termed 'technically incompetent' or, simply, 'misinformed' " (K. Guild Nichols, *Technology on Trial: Public Participation in Decision-Making Related to Science and Technology* [Paris, France: Organisation for Economic Co-operation and Development, 1979], p. 26).

74. "Doomsday: Tinkering with Life," *Time* (April 18, 1977), p. 33.

75. *Study on Federal Regulation* (n72), 1977, p. vii. Although the Federal Advisory Committee Act requires that committees not be "inappropriately influenced by the appointing authority or by any special interest," there are no enforcement provisions. A 1976 survey of 16 federal energy agency advisory boards revealed that almost half of the board members represented industry, whereas consumer and environmental representatives each comprised less than 5 percent of the members (Dorothy Nelkin, "Science and Technology Policy and the Democratic Process," *Studies in Science Education 9* [1982], 55).

76. Harvey M. Sapolsky, personal communication, June 30, 1980; and interview with Jay Katz, January 15, 1980.

77. David Dickson and David Noble, "By Force of Reason: The Politics of Science and Technology Policy," in Thomas Ferguson and

Joel Rogers, eds., *The Hidden Election* (New York: Pantheon, 1982), p. 292.

78. Adlai E. Stevenson, quoted in Steven V. Roberts, "Liberals' Gloomy Good-by," *San Francisco Chronicle*, December 18, 1980, p. 19.

79. Sherry R. Arnstein, "A Ladder of Citizen Participation," *American Institute of Planning Journal 35* (July 1969), 216–24.

80. James A. Morone and Theodore R. Marmor, "Representing Consumer Interests: The Case of American Health Planning," *Ethics 91* (April 1981), 431–50. Interests, it should be noted, arise in the context of concrete policy outcomes. There would be no "interests," or at least no need to represent them, if all of the policy outcomes were to everyone's satisfaction. See Eugene Bardach, "On Representing the Public Interest," *Ethics 91* (April 1981), 486–90. However, because this is, unfortunately, clearly a hypothetical state, it does not vitiate the argument for representing interests. The dependence of interests on the policy context also does not imply that we must all agree on which policies are desirable in order to determine which interests are legitimate; representational claims exist for all groups who believe that their needs and interests, however they define them, are not being met.

81. Sheldon Krimsky, "Beyond Technocracy: New Routes for Citizen Involvement in Social Risk Assessment," in James C. Petersen, ed., *Citizen Participation in Science Policy* (Amherst, Mass.: University of Massachusetts Press, 1984), pp. 43–61.

82. John Rawls, *A Theory of Justice* (Cambridge, Mass: Harvard University Press, 1971).

83. At present, agencies do not always appoint individuals favored by consumer constituencies, especially those known to be forceful critics of government policies. Most advisory committees skirt the issue of representing opposing viewpoints, concentrating instead on the less controversial representation of specified demographic and geographical categories. (See, for example, Kit Gage and Samuel S. Epstein, "The Federal Advisory Committee System: An Assessment," *Environmental Law Reporter 7* [1977], 50001–12; also, *Study on Federal Regulation* [n72], 1977).

84. Bardach (n80), "On Representing the Public Interest."

85. Philip Handler, quoted in "In Science, 'No Advances Without Risks,' " *U.S. News and World Report* (September 15, 1980), 60.

86. These arguments are elaborated in Carole Pateman (n5), *Participation and Democratic Theory*.

87. Michel J. Crozier, Samuel P. Huntington, and Joji Watanuki, *The Crisis of Democracy: Report on the Governability of Democracies to the Trilateral Commission* (New York: New York University Press, 1975), p. 114; also, Joseph A. Schumpeter, *Capitalism, Socialism and Democracy*, 3rd ed. (New York: Harper, 1950); Harry Eckstein,

Division and Cohesion in Democracy (Princeton, N.J.: Princeton University Press, 1966).
88. Robert A. Dahl, *A Preface to Democratic Theory* (Chicago: University of Chicago Press, 1956), p. 89.
89. These arguments are discussed in Nelson M. Rosenbaum, "The Origins of Citizen Involvement in Federal Programs," in Clement Bezold, ed., *Anticipatory Democracy: People in the Politics of the Future* (New York: Random House, 1978), pp. 139–55.
90. See Nelkin (n75), "Science and Technology Policy and the Democratic Process," 47–64. See also OTA (n34), *Public Perceptions of Biotechnology.*
91. Thomas Jefferson, letter to William Charles Jarvis, September 28, 1820, Presidential Papers of Thomas Jefferson, Library of Congress, Washington, D.C. (1944), microfilm Series 1: 1820–2.

11. What is possible?

1. On foreign policy, see Alexander L. George and Associates, *Towards A More Soundly Based Foreign Policy: Making Better Use of Information*, Vol. 2, Appendices, Commission on the Organization of the Government for the Conduct of Foreign Policy (Washington, D.C.: Government Printing Office, Stock No. 022-000-00112-4, June 1975). On nuclear power, see Report of the President's Commission on the Accident at Three Mile Island, *The Need for Change: The Legacy of TMI*, John G. Kemeny, Chairman (Washington, D.C.: Government Printing Office, October 1979). Also, James C. Petersen, ed., *Citizen Participation in Science Policy* (Amherst: University of Massachusetts Press, 1984).
2. See, for example, George and Associates (see n1), *Towards A More Soundly Based Foreign Policy*, p. 10.
3. Marilyn T. Baker and Harvey A. Taub, "Readability of Informed Consent Forms for Research in a Veterans Administration Medical Center," *Journal of the American Medical Association 250* (November 18, 1983), 2646–8.
4. Louis Harris survey, cited in Morris B. Abram, "To Curb Medical Suits," *New York Times*, March 31, 1986, p. 17.
5. See William J. Curran, "Package Inserts for Patients: Informed Consent in the 1980s," *New England Journal of Medicine 305* (December 24, 1981), 1564–6. According to the Commerce Department, the patient package insert requirement was one of the top twenty regulations most odious to American industry (R. Jeffrey Smith, "Hayes Intends Modest Reforms at FDA," *Science 213* [August 28, 1981], 984–6). The evaluation of patient drug leaflets was conducted by the Rand Corporation. See David E. Kanouse, Sandra H. Berry, Barbara Hayes-Roth, et al., "Informing Patients About Drugs: Summary Report on Alternative Designs for Pre-

scription Drug Leaflets" (Santa Monica, Cal.: Rand Corporation, R-2800-FDA, August 1981). The Reagan administration formally abandoned the pilot program in 1982 (*Federal Register 47*, No. 32 [February 17, 1982], 7200–2).

6. Thomas G. Gutheil, Harold Bursztajn, and Archie Brodsky, "Malpractice Prevention Through the Sharing of Uncertainty," *New England Journal of Medicine 311* (July 5, 1984), 50.

7. Abram (n4), "To Curb Medical Suits."

8. Report of the Panel on the General Professional Education of the Physician and College Preparation for Medicine, *Physicians for the Twenty-first Century: The GPEP Report* (Washington, D.C.: Association of American Medical Colleges, 1984).

9. For a more detailed discussion of this suggestion, see President's Commission for the Study of Ethical Problems in Medicine and Biomedical and Behavioral Research, *Making Health Care Decisions: The Ethical and Legal Implications of Informed Consent in the Patient–Practitioner Relationship*, Vol. 1: *Report* (Washington, D.C.: Government Printing Office, October 1982), pp. 151–5.

10. John P. Bunker, "When Doctors Disagree," *New York Review of Books*, April 25, 1985, pp. 7ff. A similar point is made by Lester C. Thurow in "Medicine Versus Economics," *New England Journal of Medicine 313* (September 5, 1985), 611–14.

11. Louis Harris survey, cited in Don Colburn, "Weighing Risks in a Risky Time" *San Francisco Sunday Examiner and Chronicle*, June 1, 1986, Sunday Punch, p. 6.

12. Baruch Fischhoff, Paul Slovic, Sarah Lichtenstein, et al., "How Safe Is Safe Enough? A Psychometric Study of Attitudes Towards Technological Risks and Benefits," *Policy Sciences 9* (1978), 127–52. Also Paul Slovic, Baruch Fischhoff, and Sarah Lichtenstein, "Rating the Risks," *Environment 21* (April 1979), 14–39.

13. Louis Harris and Associates, Inc., *The Road After 1984*, Harris Study No. 832033, conducted for Southern New England Telephone, December 1983, p. 144.

14. A 1986 study by the Government Accounting Office, conducted at Congressional request, reported that the federal government supported only a small number of biotechnology research projects that could be classified as "direct risk assessment." See U.S. General Accounting Office, *Biotechnology: Analysis of Federally Funded Research*, Report to the Chairman, Subcommittee on Oversight and Investigations, Committee on Energy and Commerce, House of Representatives, Publ. No. RCED-86-187, August 1986, p. 5.

15. Office of Technology Assessment, *Federal Policies and the Medical Devices Industry*, Publ. No. OTA-H-229 (Washington, D.C.: Government Printing Office, October 1984), pp. 6, 14. On changes in FDA regulations, see "New Regulations to Speed Drug Approvals, Improve Safety Monitoring," *FDA Drug Bulletin* (April

1985), 2–3. Also, see Irvin Molotsky, "Critics Say F.D.A. Is Unsafe in Reagan Era," *New York Times*, January 4, 1987, p. 4.

16. Mark Crawford, "R&D Eroding at EPA," *Science 236* (May 22, 1987), 904–5. Support for occupational safety regulations is reported in Vicente Navarro, "Where Is the Popular Mandate?" *New England Journal of Medicine 307* (December 9, 1982), 1516–18. See also Perry D. Quick, "Businesses: Reagan's Industrial Policy," in John L. Palmer and Isabel V. Sawhill, eds., *The Reagan Record* (Cambridge, Mass.: Ballinger, 1984), pp. 287–316. The administration also took a number of other steps to benefit private industry, including a substantial reduction of corporate tax rates (while taxes on the working poor rose) and revision of antitrust legislation to permit monopolistic and anticompetitive practices. Such actions were justified on the grounds of promoting international competitiveness and domestic prosperity.

17. Postmarketing surveillance has been recommended by the Joint Commission on Prescription Drug Use (Final Report submitted to Senator Edward Kennedy, January 23, 1980). See Eliot Marshall, "A Prescription for Monitoring Drugs," *Science 207* (February 22, 1980), 853–5. A number of agencies currently perform, or contract for, technology assessments, including the Congressional Office of Technology Assessment (OTA), the National Center for Health Services Research and Health Care Technology Assessment (part of the Department of Health and Human Services), and the Institute of Medicine. However, none of these agencies has the purview or funding necessary to provide for comprehensive evaluation of emerging technologies, and all operate in an advisory capacity with no power of enforcement. For a review of the status of technology assessment in early 1986, see Seymour Perry, "Technology Assessment: Continuing Uncertainty," *New England Journal of Medicine 314* (January 23, 1986), 240–3. For a discussion of the need for systematic evaluation, see Stephen P. Strickland, *Research and the Health of Americans: Improving the Policy Process* (Lexington, Mass.: D.C. Heath, 1978). Also, John B. McKinlay, "From 'Promising Report' to 'Standard Procedure': Seven Stages in the Career of a Medical Innovation," *Milbank Memorial Fund Quarterly/Health and Society 59* (Summer 1981), 374–411.

18. Harold P. Green, "The Adversary Process in Technology Assessment" in Raphael G. Kasper, ed., *Technology Assessment: Understanding the Social Consequences of Technological Applications* (New York: Praeger, 1972), p. 51.

19. See Nancy E. Abrams and Joel R. Primack, "The Public and Technological Decisions," *Bulletin of the Atomic Scientists 36* (June 1980), 44–8. Consensus conferences, as the name implies, seek to establish a consensus among a group of appointed experts, who

may or may not represent the full spectrum of opinion on the issue in question.

20. Quoted in Robert Pear, "Reagan Aides Draft Bill to Pre-empt States' Product-Liability Laws," *New York Times*, April 21, 1986, p. 14.

21. See, for example, "The Manufactured Crisis," *Consumer Reports* (August 1986), pp. 544–9. Also, Irvin Molotsky, "Drive to Limit Product Liability Awards Grows as Consumer Groups Object," *New York Times*, March 2, 1986, p. 20.

22. Polly Ross Hughes, "Facts and Myths About the Insurance Crunch," *San Francisco Chronicle*, May 28, 1986, pp. 1, 4. In 1986, after-tax profits were reportedly $19 billion, far outstripping 1985, and even higher in 1987 (Eugene I. Pavalon, "High Damage Awards Don't Raise Rates" [letter to the editor], *New York Times*, September 9, 1987, p. 22).

23. Ernest J. Weinrib, "Liability Law Beyond Justice" (letter to the editor), *New York Times*, May 16, 1986, p. 26.

24. "Half a Response on Insurance" (editorial), *New York Times*, June 7, 1986, p. 14.

25. Robert Pear, "Federal Payment for Transplants for Poor Studied," *New York Times*, May 18, 1986, pp. 1, 20.

26. Mary O'Neil Mundinger, "Health Service Funding Cuts and the Declining Health of the Poor," *New England Journal of Medicine* *313* (July 4, 1985), 44–7.

27. Robert Pear, "Increase Found in Child Poverty in Study by U.S.," *New York Times*, May 23, 1985, pp. 1, 14.

28. Mundinger (n26), "Health Service Funding Cuts."

29. Quote is from "The Disabled 9, the Administration 0" (editorial), *New York Times*, June 4, 1986, p. 22. See also Robert Pear, "Disability Reviews Spur Legal 'Crisis,' " *New York Times*, September 9, 1984, pp. 1, 19.

30. Dorothy Rice, Special Session on Aging, Meeting of the American Public Health Association, Anaheim, California, November 12, 1984, cited in Mundinger (n26), "Health Service Funding Cuts."

31. "Health Care Found in 'Deterioration,' " *San Francisco Chronicle*, October 18, 1984, p. 14. Reported problems affording care are from "Money for Health Care," *Gallup Reports*, No. 220/221 (January 1984), p. 25.

32. David U. Himmelstein, Steffie Woolhandler, Martha Harnley, et al., "Patient Transfers: Medical Practice as Social Triage," *American Journal of Public Health 74* (May 1984), 494–7; George J. Annas, "Your Money or Your Life: 'Dumping' Uninsured Patients from Hospital Emergency Wards," *American Journal of Public Health 76* (January 1986), 74–7; Robert L. Schiff, David A. Ansell, James E. Schlosser, et al., "Transfers to a Public Hospital," *New England Journal of Medicine 314* (February 27, 1986), 552–7; Arnold S. Rel-

man, "Texas Eliminates Dumping," *New England Journal of Medicine 314* (February 27, 1986), 578–9.

33. Nicole Lurie, Nancy B. Ward, Martin F. Shapiro, et al., "Termination from Medi-Cal – Does It Affect Health?" *New England Journal of Medicine 311* (August 16, 1984), 480–4.

34. Kathleen Maurer Smith and William Spinrad, "The Popular Political Mood," *Social Policy 11* (March/April 1981), 37–45.

35. The 1984 ABC News/Washington Post poll is quoted in Alexander Cockburn, "State of the Union Is Enough to Make You Ill," *Wall Street Journal*, February 13, 1986, p. 27. Other survey results are from Robert J. Blendon and Drew E. Altman, "Public Attitudes About Health-Care Costs," *New England Journal of Medicine 311* (August 30, 1984), 613–16. Also, Navarro (n16), "Where Is the Popular Mandate?"

36. Robert Pear, "Reagan's Budget Asking Cutbacks in Health Plans," *New York Times*, February 4, 1986, pp. 1, 13.

37. Budget figures are from Colin Norman, "Science Escapes Brunt of Budget Ax," *Science 231* (February 21, 1986), 785–8; and Barbara J. Culliton, "Congress Boosts NIH Budget 17.3%," *Science 234* (November 14, 1986), 808–9. Increased military spending is a major cause of the huge federal deficit accumulated under Reagan, and the draconian steps being taken to reduce it, such as the Gramm–Rudman–Hollings budget-balancing legislation, will put enormous pressure on almost all federal domestic programs for years to come.

38. Seymour Martin Lipset, "Poll After Poll After Poll After Poll Warns President on Program," *New York Times*, January 13, 1982, p. A23. Also, George Gallup, Jr., "Cut Defense, Not Social Programs," *San Francisco Chronicle*, August 13, 1987, p. 13.

39. Blendon and Altman (n35), "Public Attitudes."

40. "Public Favors Major Restructuring of Health Care System: Harris Poll," *American Medical News 26* (October 21, 1983), 2, 8.

41. Office of Technology Assessment (n15), *Federal Policies and the Medical Devices Industry*.

42. Sources on trends in mortality are provided in David U. Himmelstein and Steffie Woolhandler, "Cost Without Benefit: Administrative Waste in U.S. Health Care," *New England Journal of Medicine 314* (February 13, 1986), 441–5. As a proportion of gross national product, Great Britain spends roughly half of what the United States does on health care: 6 percent versus 11 percent in 1984. However, because the per capita British GNP is about half that of the United States, the actual amount spent on health per person in Britain is roughly one-quarter of the amount spent in the United States.

43. Ibid.

44. "Bishops Issue Draft Letter on Economy," *San Francisco Chronicle*, June 3, 1986, p. 9.

45. 1981 *Washington Post*–"ABC News" Poll, cited in Lipset (n38), "Poll After Poll."

46. Income distribution figures for 1986 are from "U.S. Poverty Down, Family Income Up," *San Francisco Chronicle*, July 31, 1987, p. 6. Analyses of census data showed that government benefit programs lifted a smaller proportion of families with children out of poverty in 1986 than they did in 1979 ("The Poor Get Poorer, Researchers Say," *San Francisco Chronicle*, September 3, 1987, p. 24). More people were living below poverty in 1983 than at any time since 1965, although the number has declined slightly since 1983. Federal tax and benefit legislation enacted between 1980 and 1984 resulted in a net gain of roughly $17,000 for the most affluent and a net loss of roughly $1000 for the poor, taking inflation into account. See Thomas B. Edsall, *The New Politics of Inequality* (New York: W. W. Norton, 1984), pp. 21–2, 221.

47. Barry Checkoway, "Public Participation in Health Planning Agencies: Promise and Practice," *Journal of Health Politics, Policy and Law 7* (Fall 1982), 723–33. For a discussion of the problems faced by Health Systems Agencies, see Theodore R. Marmor and James A. Morone, "Representing Consumer Interests: Imbalanced Markets, Health Planning, and the HSAs," *Milbank Memorial Fund Quarterly/Health and Society 58* (Winter 1980), 125–65; and Bruce C. Vladeck, "Interest-Group Representation and the HSAs: Health Planning and Political Theory," *American Journal of Public Health 67* (January 1977), 23–9.

48. Institute of Medicine, *DHEW's Research Planning Principles: A Review*, Publ. No. IOM-79-02 (Washington, D.C.: National Academy of Sciences, March 1979).

49. See, for example, Strickland (n17), *Research and the Health of Americans*. Also, John Walsh, "NSF to Formulate 5-Year Plan, Budget," *Science 235* (February 27, 1987), 967; William D. Carey, "Dilemmas of Scale" (editorial), *Science 216* (June 18, 1982) 1277; and Linda E. Demkovich, "HEW Presses for Spending Priorities in Slicing NIH's Basic Research Pie," *National Journal 12* (February 16, 1980), 276–80.

50. The European experiences are described in Dorothy Nelkin, "Science and Technology Policy and the Democratic Process," *Studies in Science Education 9* (1982), 47–64. The Canadian approach is described in "Thousands Consulted on Future Technologies," *In Touch 2* (December 1984), Science Council of Canada, Ottawa, Ontario.

51. This suggestion was made by Clifford Grobstein, an authority on biology and public policy. For a description, see William Allen,

"Public Policy Expert Proposes New Mechanism for Funding Basic Research," *Genetic Engineering News* 7 (July/August 1987), 36, 49.

52. Archibald L. Cochrane, *Effectiveness and Efficiency* (London: The Nuffield Provincial Hospitals Trust, 1972).

53. Quoted in Henry J. Aaron and William B. Schwartz, *The Painful Prescription: Rationing Hospital Care* (Washington, D.C.: Brookings Institution, 1984), p. 49.

54. In 1987, a new law took effect requiring distribution of organs according to uniform criteria that set priorities for deciding who should get them. Criteria include such factors as medical need, length of time waited, and the proximity and compatibility of an organ donor (Robert Pear, "New Law May Spur Organ Donations," *New York Times*, September 6, 1987, p. 1). For a discussion of the guidelines recommended for Massachusetts, see George J. Annas, Chairman, *Report of the Massachusetts Task Force on Organ Transplantation* (Boston: Department of Public Health, October 1984).

55. Joseph Lelyveld, "Danes Seek to Help the Elderly," *New York Times*, May 1, 1986, pp. 21–2.

56. Deborah P. Lubeck, Byron W. Brown, and Halsted R. Holman, "Chronic Disease and Health System Performance," *Medical Care* 23 (March 1985), 266–77.

57. H. Jack Geiger, "Community Health Centers: Health Care as an Instrument of Social Change," in Victor W. Sidel and Ruth Sidel, eds., *Reforming Medicine: Lessons of the Last Quarter Century* (New York: Pantheon, 1984), pp. 11–32.

58. For an excellent discussion of the WHO recommendation and ethical problems of health and medicine worldwide, see M. H. King, "Medicine in an Unjust World," in *Oxford Textbook of Medicine* (Oxford: Oxford University Press, 1983), pp. 3.3–3.11.

59. Report of the New York–based Committee for Economic Development, quoted in Edward B. Fiske, "Executives Urge a Rise in Aid for Poor Children," *New York Times*, September 6, 1987, p. 14.

60. E. E. Schattschneider, *Politics, Pressures and the Tariff* (New York: Prentice-Hall, 1935).

61. Marmor and Morone (n47), "Representing Consumer Interests."

62. Howard Metzenbaum, quoted in Martin Tolchin, "The Outlook for Lobbying: Critical Capital Is Pondering the Deaver Effect," *New York Times*, May 3, 1986, p. 11. See Edsall (n46), *The New Politics of Inequality*, for a detailed description of the mobilization of the business community. For similar views from one long-time lobbyist, see Kenneth Schlossberg, "The Greening of Washington: Money Talks Louder than Ever," *New York Times*, May 14, 1986, p. 23. Figures on lobbyists are from Tolchin, "The Outlook for

Lobbying." Political action committee data are from Edsall (n46), *The New Politics of Inequality*, p. 131.

63. Data on voter turnout rates are from Edsall (n46), *The New Politics of Inequality*, pp. 181–4. An analysis by G. Bingham Powell, Jr. ("Voting Turnout in 30 Democracies: Partisan, Legal and Socioeconomic Influences," in Richard Rose, ed., *Electoral Participation* [Beverly Hills, Cal.: Sage Publications, 1980], cited in Edsall [n46], *The New Politics of Inequality*, p. 197) of voting patterns in thirty democracies on every continent ranks the United States twenty-seventh in voter turnout. Voter turnout in the United States today is far below what it was during the second half of the previous century, and considerably lower than the turnout in most contemporary democratic societies.

64. A seventy-two-page report published in 1983 by the Alliance for Justice, an organization representing twenty-three consumer, public interest, and environmental groups, described nearly 100 instances in which the Reagan administration had quietly violated provisions of the Administrative Procedures Act intended to ensure public accountability. Violations included postponing effective dates for regulations without public notice, suspending and redefining rules without announcement, issuing interim rules before soliciting public comment, and issuing final rules before the period for public comment had ended (Alliance for Justice, *Contempt for Law: Excluding the Public from the Rulemaking Process*, 600 New Jersey Avenue, N.W., Washington, D.C. 20001, 1983).

65. William Kruskal, quoted in Constance Holden, "Statistics Suffering Under Reagan," *Science 216* (May 21, 1982), 833. Federal data-gathering programs that have been cut back include the Library of Congress, the Department of Energy's Energy Information Administration, the reporting arm of the Environmental Protection Agency, the Office of Management and Budget's Statistical Policy Branch, and the Smithsonian Science Information Exchange. Also, see Eliot Marshall, "Library Cutbacks: An Information Deficit," *Science 232* (May 9, 1986), 700–1; Constance Holden, "Research Information Service Imperiled," *Science 213* (September 11, 1981), 1232; and "U.S. Restricts Information Act," *San Francisco Chronicle*, May 5, 1981, p. 12.

66. Daniel Boorstin, quoted in Marshall (n65), "Library Cutbacks."

67. David F. Salisbury, "Many Bills in Congress Worry Supporters of States' Rights," *San Francisco Sunday Examiner and Chronicle*, January 3, 1982, p. A15. Areas in which the White House has supported, or not opposed, preemptive federal legislation include the "Baby Doe" rules forbidding denial of life support measures for severely malformed infants (subsequently struck down by the Supreme Court), product liability, interstate banking, hazardous waste disposal, and product safety laws. For a description of the

effects of Reagan administration policies on state and local governments, see George E. Peterson, "Federalism and the States: An Experiment in Decentralization," in Palmer and Sawhill (n16), *The Reagan Record*, pp. 217–59.

68. Interested readers will find many excellent ideas and insights in Benjamin R. Barber, *Strong Democracy: Participatory Politics for a New Age* (Berkeley: University of California Press, 1984); Sara M. Evans and Harry C. Boyte, *Free Spaces: The Sources of Democratic Change in America* (New York: Harper and Row, 1986); Ronald Beiner, *Political Judgment* (Chicago: University of Chicago Press, 1983); Martin Carnoy, Derek Shearer, and Russell Rumberger, *A New Social Contract* (New York: Harper and Row, 1983); and Tom Hayden, *The American Future* (New York: Washington Square Press, 1980).

69. Ralph Crawshaw, Michael J. Garland, Brian Hines, et al., "Oregon Health Decisions: An Experiment with Informed Community Consent," *Journal of the American Medical Association 254* (December 13, 1985), 3213–16.

70. Brian Hines, "Health Policy on the Town Meeting Agenda," *Hastings Center Report 16* (April 1986), 5–7.

71. David A. Danielson and Arthur Mazer, "Results of the Massachusetts Referendum for a National Health Program," *Journal of Public Health Policy 8* (Spring 1987), 28–35.

72. Richard C. Atkinson, "Reinvigorating the Contract Between Science and Society," *Chronicle of Higher Education 18* (March 19, 1979), 80.

73. For arguments in favor of the science court idea, see Arthur Kantrowitz, "Controlling Technology Democratically," *American Scientist 63* (September–October 1975), 505–9; and idem, "The Science Court Experiment: Criticisms and Responses," *Bulletin of the Atomic Scientists 33* (April 1977), 43–50. For a critique and discussion of alternative, citizen-based approaches, see Sheldon Krimsky, "A Citizen Court in the Recombinant DNA Debate," *Bulletin of the Atomic Scientists 34* (October 1978), 37–43; Dorothy Nelkin, "Thoughts on the Proposed Science Court," *Newsletter on Science, Technology and Human Values 18* (January 1977), 20–41; and Sheldon Krimsky, "Public Participation in the Formation of Science and Technology Policy," unpublished paper prepared for the National Science Foundation, Washington, D.C., March 1979.

74. Dorothy Nelkin and Arie Rip, "Distributing Expertise: A Dutch Experiment in Public Interest Science," *Bulletin of the Atomic Scientists 35* (May 1, 1979), 21.

75. Diana B. Dutton and John L. Hochheimer, "Institutional Biosafety Committees and Public Participation: Assessing an Experiment," *Nature 297* (May 6, 1982), 11–15.

76. Henry Waxman, quoted in Barbara J. Culliton, "Reagan Vetoes NIH Bill; Override Is Likely," *Science 230* (November 29, 1985), 1021.

77. John Walsh, "Peer Review – oops – Merit Review in for Some Changes at NSF," *Science 235* (January 9, 1987), 153.

78. A proposal for a National Health Council to oversee health policy and planning is discussed by Henrik L. Blum in *Expanding Health Care Horizons: From a General Systems Concept of Health to a National Health Policy*, 2nd ed. (Oakland, Cal.: Third Party, 1983), pp. 153–207.

79. In recognition of the difficulties many citizen groups have paying for experts, attorneys, travel to Washington, and the like, the Carter administration, having affirmed its commitment to public participation in Federal agency proceedings (Executive Order 12044, March 23, 1978), further directed all agencies to establish public participation funding programs to provide financial assistance for persons otherwise unable to participate. See Memorandum from the President, May 16, 1979, "Public Participation in Federal Agency Proceedings" in *Weekly Compilation of Presidential Documents 15*, No. 20 (Washington, D.C.: White House, May 21, 1979), 867–8. Although these programs were limited in many ways, they illustrate the type of compensatory efforts that could help to broaden participation.

80. The majority of people today believe they are less happy than people were fifty years ago, that they have more to worry about, and have less "inner happiness" and "peace of mind." Furthermore, these impressions appear to be valid; the population's self-perceived "happiness" has been gradually declining over time. See Nicholas Rescher, "Technological Progress and Human Happiness," *Unpopular Essays on Technological Progress* (Pittsburgh: University of Pittsburgh Press, 1980), pp. 3–22.

81. Amy Gutmann, "Communitarian Critics of Liberalism," *Philosophy and Public Affairs* (Summer 1985), 308–22.

82. See, for example, Frank Riessman, "What's New in the New Populism?" *Social Policy 16* (Summer 1985), 2; and S. M. Miller, "Challenges for Populism," *Social Policy 16* (Summer 1985), 3–6.

Index

Index